Advances in Soil Science

# SOIL PROCESSES AND WATER QUALITY

*Edited by*

R. Lal
B. A. Stewart

**Library of Congress Cataloging-in-Publication Data**

Soil processes and water quality / edited by R. Lal, B. A. Stewart.
p. cm. — (Advances in soil science)
Includes bibliographical references and index.
ISBN 0-87371-980-8
1. Soil management—Environmental aspects. 2. Water—Pollution.
I. Lal, R. II. Stewart, B. A. (Bobby Alton), 1932– . III. Series: Advances in soil science (Boca Raton, Fla.).
TD428.S64S64 1994
628.1'684—dc20 94-13702
CIP



Direct all inquiries to CRC Press, Inc., 2000 Corporate Blvd., N.W., Boca Raton, Florida 33431.



Lewis Publishers is an imprint of CRC Press



International Standard Book Number 0-87371-980-8
Library of Congress Card Number 94-13702
Printed in the United States of America 1 2 3 4 5 6 7 8 9 0
Printed on acid-free paper

## Preface

Fresh water is a scarce commodity, and high quality of fresh water is crucial to the health of human, domestic livestock, and wildlife. High quality of water is also essential to minimizing risks of salinization or alkalization of irrigated land in semi-arid and arid regions. Several soil processes, as affected by a wide range of agricultural activities, play a major role in regulating quality of surface and ground waters. Quality of surface waters (comprising streams, rivers and lakes) is affected by use of agricultural chemicals e.g., fertilizers, pesticides, and manures. Nutrients applied as inorganic and organic amendments are easily carried in the overland flow either in solution or absorbed on sediments. Similarly, many pesticides are carried in the surface runoff to streams, rivers and lakes. Decline in soil structure, compaction, decrease in CEC, and accelerated soil erosion play an important role in determining quality of surface waters. Significant advances have been made in our understanding of the processes involved in soil degradation, and in our ability to predict movement of chemicals in surface waters. Fertilizers and pesticides are an essential component of the package of intensive agriculture, and their use is rapidly increasing especially in the developing countries. Important pollutants in fertilizers are nitrogen and phosphorus which are responsible for eutrophication of surface waters. Important pesticides that pose risks to water quality are atrazine, aldicarb, lindane, and 2,4-D among others. Soils of the tropics have less capacity to hold these chemicals and risk of pollution of surface waters are high.

Agricultural practices also affect the quality of ground water through leaching of fertilizers and pesticides applied to the soil surface. Chemicals can be readily transported into the ground water with water percolating from surface into the subsoil. A possible mechanism of pollutant transport to the ground water is the macropore flow. Macropores or biopores, created by root growth or the activity of soil fauna such as earthworms, enhance infiltration and may serve as preferential pathways for transport of agricultural chemicals from soil surface to the ground water. However, the mechanisms involved in transport of water and chemicals are not well understood.

This volume is a compilation of manuscripts addressing the role of soil processes in water quality. Ten chapters are selected to address important processes involved. The role of phosphorus and nitrogen applied as inorganic and organic fertilizers, on arable land and in pastures, on water quality is discussed in three chapters. Another chapter deals with the contamination of water by heavy metals. A case study from western Nigeria provides an example of the impact of tropical deforestation and use of intensive agricultural practices on water quality. A state-of-the-art review addresses the importance of macropore flow on water quality and highlights the mechanisms involved in transport of agricultural chemicals to the ground water. A separate chapter is devoted to water quality modeling in relation to soil management and agricultural practices.

This volume is an important compilation of the state-of-the-art review of principal soil processes that regulate quality of surface and ground waters. In addition, it outlines research and development priorities for furthering the knowledge on mechanisms of pollutant transport in natural waters, establishing critical limits of concentrations of pollutants with regard to human and animal health, and in standardizing methodology for field and laboratory techniques. The information contained is of interest to soil scientists, agronomists, environmentalists, hydrologists, ecologists, and the general public. We express our profound appreciation to all authors for their quality manuscripts, and for their excellent cooperation during the review and publication process.

| | |
|---|---|
| Rattan Lal | B.A. Stewart |
| Columbus, Ohio | Amarillo, Texas |

About the Editors:

Dr. R. Lal is a Professor of Soil Science in the Department of Agronomy at The Ohio State University, Columbus, Ohio. Prior to joining Ohio State in 1987, he served as a scientist for 18 years at the International Institute of Tropical Agriculture, Ibadan, Nigeria. Professor Lal is a fellow of the Soil Science Society of America, American Society of Agronomy, and the Third World Academy of Sciences. He is recipient of both the International Soil Science Award, and the Soil Science Applied Research Award of the Soil Science Society of America.

Dr. B.A. Stewart is a Distinguished Professor of Soil Science, and Director of the Dryland Agriculture Institute at West Texas A&M University, Canyon, Texas. Prior to joining West Texas A&M University in 1993, he was Director of the USDA Conservation and Production Research Laboratory, Bushland, Texas. Dr. Stewart is past president of the Soil Science Society of America, and was a member of the 1990-1993 Committee of Long Range Soil and Water Policy, National Research Council, National Academy of Sciences. He is a Fellow of the Soil Science Society of America, American Society of Agronomy, Soil and Water Conservation Society, a recipient of the USDA Superior Service Award.

# Contributors

*J.G. Arnold*, USDA Agricultural Research Service, 808 East Blackland Road, Temple, TX 76502, U.S.A.

*V.W. Benson*, USDA Soil Conservation Service, 808 East Blackland Road, Temple, TX 76502, U.S.A.

*C.W. Boast*, Department of Agronomy, University of Illinois, Urbana, IL 61801, U.S.A.

*R.H. Griggs*, Texas Agricultural Experiment Station, 808 East Blackland Road, Temple, TX 76502, U.S.A.

*Ardell Halvorson*, Central Great Plains Research Station, USDA Agricultural Research Service, Akron, CO 80720, U.S.A.

*C.A. Jones*, Texas Agricultural Experiment Station, 808 East Blackland Road, Temple, TX 76502, U.S.A.

*H. Kirchmann*, Department of Soil Sciences, Swedish University of Agricultural Sciences, 750 07 Uppsala, SWEDEN.

*E.J. Kladivko*, Agronomy Department, Purdue University, West Lafayette, IN 47907, U.S.A.

*R. Lal*, Department of Agronomy, The Ohio State University, Columbus, OH 43210-1086, U.S.A.

*E.L. McCoy*, Department of Agronomy, Ohio Agricultural Research and Development Center, The Ohio State University, Wooster, OH 44691, U.S.A.

*Rosa M.C. Muchovej*, Institute of Food and Agricultural Sciences, Agricultural Research and Education Center, University of Florida, Ona, FL 33865-9706, U.S.A.

*L.B. Owens*, North Appalachian Experimental Watershed, USDA Agricultural Research Service, Coshocton, OH 43812, U.S.A.

*Jack E. Rechcigl,* Institute of Food and Agricultural Sciences, Agricultural Research and Education Center, University of Florida, Ona, FL 33865-9706, U.S.A.

*Andrew Sharpley*, National Agricultural Water Quality Laboratory, USDA Agricultural Research Service, Durant, OK 74702-1430, U.S.A.

*Bal Ram Singh*, Department of Soil Sciences, Agricultural University of Norway, N-1432 Aas, NORWAY.

*R.C. Stehouwer*, Department of Agronomy, Ohio Agricultural Research and Development Center, The Ohio State University, Wooster, OH 44691, U.S.A.

*Eiliv Steinnes*, Department of Chemistry, University of Trondheim, AVH, N-7055 Dragvoll, NORWAY.

*B.A. Stewart*, Dryland Agriculture Institute, West Texas A&M University, Canyon, TX 79016, U.S.A.

*J.R. Williams*, USDA Agricultural Research Service, 808 East Blackland Road, Temple, TX 76502, U.S.A.

# Contents

**Soil Processes and Water Quality** . . . . . . . . . . . . . . . . . . . . . . . . . . . . . 1
*R. Lal and B.A. Stewart*

**The Management of Soil Phosphorus Availability and its Impact on Surface Water Quality** . . . . . . . . . . . . . . . . . . . . . . . . . . . . . . . . . . . 7
*A.N. Sharpley and A.D. Halvorson*

**Impact of Nitrogen Fertilization of Pastures and Turfgrasses on Water Quality** . . . . . . . . . . . . . . . . . . . . . . . . . . . . . . . . . . . . . . . 91
*Rosa M.C. Muchovej and Jack E. Rechcigl*

**Impacts of Soil N Management on the Quality of Surface and Subsurface Water** . . . . . . . . . . . . . . . . . . . . . . . . . . . . . . . . . . . . 137
*L.B. Owens*

**Animal and Municipal Organic Wastes and Water Quality** . . . . . . . 163
*H. Kirchmann*

**Soil and Water Contamination by Heavy Metals** . . . . . . . . . . . . . . . 233
*Bal Ram Singh and Eiliv Steinnes*

**Water Quality Effects of Tropical Deforestation and Farming System on Agricultural Watersheds in Western Nigeria** . . . . . . . . . . . . . . . . 273
*R. Lal*

**Macropore Hydraulics: Taking a Sledgehammer to Classical Theory** 303
*E.L. McCoy, C.W. Boast, R.C. Stehouwer, and E.J. Kladivko*

**Water Quality Models for Developing Soil Management Practices** . . 349
*J.R. Williams, J.G. Arnold, C.A. Jones, V.W. Benson, and R.H. Griggs*

**Research Priorities for Soil Processes and Water Quality in 21st Century** . . . . . . . . . . . . . . . . . . . . . . . . . . . . . . . . . . . . . . . . 383
*R. Lal and B.A. Stewart*

**Index** . . . . . . . . . . . . . . . . . . . . . . . . . . . . . . . . . . . . . . . . . . . . . 393

Advances in Soil Science

# SOIL PROCESSES AND WATER QUALITY

# Soil Processes and Water Quality

R. Lal and B.A. Stewart

I. Introduction . . . . . . . . . . . . 1
II. Principal Agricultural Activities . . . . . . . . . . . . 2
III. Soil Processes and Water Quality . . . . . . . . . . . . 3
IV. Water Quality Standards . . . . . . . . . . . . 4
References . . . . . . . . . . . . 5

## I. Introduction

Fresh water is a scarce resource. Estimates of the world water balance indicate that 94% (volume basis) of the world water is in oceans and seas (1370 m $km^3$). The fresh water accounts for merely 6% of the total volume comprising 4% in the aquifer as ground water (60 m $km^3$), and 2% in ice caps and glaciers (30 m $km^3$) (Nace, 1971; Edwards et al., 1983; van der Leeden et al., 1990). Other sources of fresh water include lakes and reservoirs (0.13 m $km^3$), soil moisture (0.07 m $km^3$), and atmospheric water (0.01 m $km^3$). Therefore, maintaining high quality of fresh water resources is important for health of human population, domestic livestock, and wildlife.

Agricultural activities can have adverse effects on quality on surface and ground waters. There are two principal types of farming: (a) extensive, and (b) intensive. Extensive agricultural practices (e.g. shifting cultivation) become detrimental only if length of the fallow phase is drastically reduced and soil degradation sets in. In addition, shifting cultivation, based on slash and burn techniques, can have severe adverse impact on water quality when it is practiced on marginal lands and in ecologically sensitive eco-regions. Presently, majority of resource poor farmers of the tropics use low rates of application of fertilizers and pesticides. The chemical based inputs, however, are likely to increase with rapidly increasing demographic pressure. With increase in chemical inputs, adverse impact on water quality is also likely to increase.

Intensive agricultural practices may involve bringing new land under agriculture or intensively farming on existing land. Both systems can have drastic adverse impact on consumptive use and quality of water resources. The magnitude of water discharged from agricultural lands is influenced by intensive

0-87371-980-8/94/$0.00+$.50

**Table 1.** Global fertilizer use and average grain yield

| | ---------Fertilizer use--------- | | ----Average grain yield---- | |
|---|---|---|---|---|
| Year | ($10^6$ Mg) | % Increase/Yr | (Mg/ha) | % Increase/Yr |
| 1950 | 15.1 | -- | 1.05 | -- |
| 1960 | 24.2 | 6.0 | 1.30 | 2.4 |
| 1970 | 59.2 | 14.5 | 1.35 | 1.9 |
| 1980 | 111.3 | 8.0 | 1.90 | 2.3 |
| 1990 | 142.9 | 2.8 | 2.20 | 1.6 |

(Modified from Brown, 1989.)

land use because of alterations in different components of the hydrologic cycle. The quality of water is affected because of disruptions in cycles of C and principal plant nutrients, and transport of sediments and dissolved chemicals to ground water, rivers and lakes.

## II. Principal Agricultural Activities

There are two principal agricultural activities which are globally important with regard to their impact on water quality. The first among these is conversion of tropical rainforest (TRF) and bring new land under cultivation. It is estimated that as much as 20 million ha of TRF is annually being converted to agricultural land uses. Conversion of TRF is often done by using heavy machinery that compacts the top soil, removes most biomass and leaf litter to the boundaries, and exposes structurally fragile soil to harsh tropical environments. Conversion of TRF in the Amazon and Congo Basins, West Africa, and Southeast Asia has caused ecological perturbations with severe adverse impact on water quality.

The second facet of intensive agriculture is heavy dependence on agricultural chemicals. About 19 billion kg of fertilizer and 450 million kg of pesticides are used annually on farmland in the USA (OTA, 1990). The global use of mineral fertilizers has broken the yield barriers and enhanced agricultural production (Table 1). In addition to mineral fertilizers, manures and other organic residues are also used to augment soil fertility. Fertilizer or nutrient use efficiency is generally low, ranging from 10 - 60% depending on the crop, fertilizer rate, tillage method, soil properties, etc. Depending on the nutrient chemistry a portion of the remainder may be released into natural waters.

The use of pesticides has also increased dramatically. For example, the total amount of pesticide active ingredients applied on farms in the United States increased 170 percent between 1964 and 1982, when total area under cultivation remained relatively constant. Herbicide use led the way, from 95 million kg in 1971 to 207 million kg in 1982 (NRC, 1989). The use of pesticides has also increased rapidly in developing countries. However, integrated pest management (IPM) has reduced sharply the amounts of insecticides applied to some crops, particularly cotton, peanuts, and grain sorghum (Adkisson, 1988). In the United States, farm insecticide use on these crops decreased between 1971 and 1982 by 77, 81, and 58 percent, respectively. Adkisson (1988) stated that the greatest impact of IPM may still be in the future and in the developing and lesser developing countries. In these countries, under the encouragement of the FAO Panel of Experts for Integrated Pest Management, the US/AID Consortium for International Crop Protection, the US/AID CRSP Programs, and the International Agricultural Research Centers, most plant protection specialists have accepted the IPM philosophy. They are developing and implementing simple IPM programs for major food crops and cotton utilizing pest resistant varieties, traditional cultural methods, physical methods of control, and selective use of insecticides. IPM already is being used on a wide scale by farmers in Asia and Latin America. It is not so widely used in Africa, but could be if there was an adequate infrastructure for producing and distributing seed of new pest resistant varieties and for providing technical assistance and credit to small farmers (Adkisson, 1988).

Impact of increased use of those chemicals on water resources are not known, especially in developing countries where infrastructure and resources needed for analyses are not available.

In addition, multiple cropping based on an intensive land use also leads to contamination of water resources. Intensive land use, with a little or no fallow period for soil restoration, can set in motion several soil degradative processes with adverse impact on water quality.

## III. Soil Processes and Water Quality

Principal soil processes affecting water quality include physical, chemical and biological, as outlined in Table 2. Important among physical processes are compaction, crusting, and accelerated soil erosion. Physical processes are set in motion by decline in soil structure with resultant decrease in water infiltration capacity and increase in surface runoff. Decline in structural stability also increases soil erodibility and risks of inter-rill and rill erosion. Surface runoff and erosion enhance transport of dissolved chemicals and sediment borne pollutants into natural waters. Leaching, transport of chemicals from surface into the sub-soil with percolating water, is another major process affecting water quality. Concentrations of soluble nutrients may be several orders of magnitude higher in seepage water than in surface runoff. Leaching is generally more severe during off-season when crops are not actively growing. Active crop growth removes readily soluble chemicals, e.g. $NO_3$, $NH_4$, P, K, Ca, Mg, etc. Leaching can be accentuated by macropore flow or bi-pass flow. This involves

**Table 2.** Soil processes affecting water quality

| Soil processes | Impact on water quality |
|---|---|
| 1. Soil erosion | * Transport of dissolved and suspended sediments in surface runoff. |
| 2. Leaching | * Movement of nutrients, agricultural chemicals and dissolved organic carbon in percolating water. |
| 3. Macropore flow | * Rapid transport of water and pollutants from surface to subsurface and into a drainage system. |
| 4. Mineralization of humus | * Release of readily soluble compounds that are easily washed away or leached out. |

rapid transport of water and chemicals from surface into the sub-soil through large pores made by biotic activity, e.g. worm holes, root channels, burrows by large animals. Macropores flow can also occur through cracks. Water flow through cracks is a dominant process of water movement in heavy textured soils, e.g. Vertisols. In general, the rate and magnitude of macropore flow are inversely related to those of surface runoff. In addition to nutrient application through fertilizers, plant nutrients also become available through mineralization of humus and soil organic matter. The process of humification releases plant nutrients immobilized in organic matter and makes them readily soluble and mobile. The biomass, active and dead, is a major buffer against nutrient loss (by erosion or leaching) out of the ecosystem. Therefore, decrease in total biomass, activity and species diversity of soil fauna can have serious adverse impact on water quality.

## IV. Water Quality Standards

With a few exceptions, water quality standards have not been established for principal pollutants and for different uses. Principal pollutants are sediments, nitrates, phosphates, dissolved organic carbon, and major pesticides, e.g. atrazine, aldicarb and ethylene dibromide. Major uses of water include consumption by human, livestock, wildlife, and industrial uses. In addition, water quality also has major ecological impact.

There is a need to establish and standardize critical or tolerable limits with regard to concentrations of major pollutants in natural waters. Presently, tolerable limits vary among regions and countries, and between intended uses,

**Table 3.** International water quality standards for human and livestock consumption

| Chemical element/compound | (Concentration mg/l) | |
|---|---|---|
| | Human | Livestock |
| Pb | < 0.1 | 0.05 |
| Mo | --- | 0.01 |
| As | < 0.05 | 0.05 |
| Se | < 0.01 | 0.01 |
| Zn | < 15 | < 20 |
| Cd | < 0.01 | 0.01 |
| Ba | < 1.0 | --- |
| Ca | < 200 | < 1000 |
| Hg | < 0.01 | 0.002 |
| $NO_3$ | < 45 | < 200 |
| $NH_4^{-2}$ | < 0.05 | --- |
| Cl | < 400 | < 1000 |
| | <200 | --- |

(From Edwards et al., 1983; van der Leeden et al., 1990.)

e.g. humans vs. livestock consumption. Some known values of water quality standards for human and livestock consumption are listed in Table 3. There are two problems with water quality standards. First, the impact of pollutants on human and animal health is not known. Secondly, it is difficult to impose stringent regulatory measures especially in developing countries.

While the demands for use of agricultural chemicals is rapidly increasing, development and implementation of environmental ethic and environmental laws have not been satisfactory. Implementation of environmental laws has not kept up the desired pace especially in developing countries. In these regions, drinking water supplies are often scarce, rarely treated, and seldom tested for pollutants and pathogens. There is an urgent need to educate rural and urban population about the dangers of using polluted water. Furthermore, agricultural industry should undertake the task of educating farmers about handling and disposal of farm chemicals.

## References

Adkisson, Perry L. 1988. The value of integrated pest management to crop production. p. 905-906. In: P.W. Unger, T.V. Sneed, W.R. Jordan, and Ric Jensen (eds.), Challenges in Dryland Agriculture: A Global Perspective. Proceedings International Conference on Dryland Farming, Amarillo-/Bushland, Texas, August 15-19, 1988. Texas Agricultural Experiment Station, College Station, Texas.

Brown, Lester R. 1989. Reexamining the World Food Prospect. p. 41-58. In: *State of the World*. W.W. Norton and Company, New York.

Edwards, K.A., G.A. Classen, and E.H.J. Schroeten. 1983. The water resource in tropical Africa and its exploitation. ILCA Res. Report No. 6, Addis Ababa, Ethiopia, 103 pp.

FAO/UNESCO. 1990. Water and sustainable agricultural development: A strategy for the implementation of the Mar del Plate Action Pan for the 1990's. FAO, Rome, Italy, 42 pp.

Lawrence, A.D. and K. Kuruppuarachchi. 1986. Impact of agriculture on ground water quality in Kalpitiya. Sri Lanka. Brit. Geological Survey, Open File Report, WD/OS/86.

Nace, N.R. (ed). 1971. Scientific framework of the world water balance. UNESCO Tech. Papers Hydrol. 7, UNESCO, Paris.

National Research Council. 1989. *Alternative Agriculture*. Board on Agriculture, National Research Council, National Academy Press, Washington, D.C. 448 pp.

Office for Technology Assessment. 1990. Beneath the bottom line: Agricultural approaches to reduce agricultural contamination of ground water. Summary. U.S. Govt. Printing Office, Washington, D.C.

Van der Leeden, F., F.L. Troise, and D.K. Tod. 1990. The water encyclopedia. Lewis Publishers, Chelsea, MI, 808 pp.

# The Management of Soil Phosphorus Availability and its Impact on Surface Water Quality

A.N. Sharpley and A.D. Halvorson

I. Introduction . . . . . 8
II. Dynamics of Soil P Availability . . . . . 12
A. Organic P Mineralization . . . . . 13
B. Analytical Limitations . . . . . 18
III. Management of Soil P Availability . . . . . 19
A. Fertilizer Management . . . . . 19
B. Manure Management . . . . . 25
C. Crop Yield Potential . . . . . 26
D. Nitrogen Required for Efficient Residual P Use . . . . . 28
E. Economics of Fertilizer Phosphorus Management . . . . . 30
IV. Soil Testing for Phosphorus Availability . . . . . 31
A. Positional Availability . . . . . 33
B. Chemical Availability . . . . . 33
C. Soil Test-Crop Yield Relationships . . . . . 34
D. Development of Nonconventional Procedures . . . . . 34
V. Transport of P in Runoff . . . . . 35
A. Forms . . . . . 35
B. Processes . . . . . 37
C. Amounts Transported . . . . . 40
VI. Predicting Phosphorus Availability and Transport . . . . . 45
A. Soil Availability . . . . . 46
B. Transport in Runoff . . . . . 48
C. Model Limitations . . . . . 53
D. Indexing Phosphorus Availability . . . . . 53
VII. Management Challenges . . . . . 59
A. Minimizing Phosphorus Transport . . . . . 59
B. Management Implications . . . . . 63
VIII. Research Needs . . . . . 69
Acknowledgement . . . . . 72
References . . . . . 72

# I. Introduction

An increased public perception of the role of agriculture in nonpoint source pollution, has prompted an urgency in obtaining information on the impact of current and proposed agricultural management practices on surface water quality. The transport of phosphorus (P) to surface waters can lead to accelerated eutrophication of these waters, which limits their use for fisheries, recreation, industry, or drinking. Although nitrogen (N) and carbon (C) are also associated with accelerated eutrophication, most attention has focused on P, because of the difficulty in controlling the exchange of N and C between the atmosphere and a water body, and fixation of atmospheric N by some blue green algae. Thus, P often limits eutrophication and its control is of prime importance in reducing the accelerated eutrophication of surface waters.

Inputs from point sources are easier to identify and have thus, seen more control than diffuse sources. Consequently, agricultural nonpoint sources now account for a larger share of all discharges than a decade ago (US EPA, 1984; Crowder and Young, 1988). In response to this, the Water Quality Act of 1987 (Section 319) increased attention on the need to control nonpoint sources of pollution to achieve the nation's water quality goals, with federal funds becoming available in 1990 for implementation of preventive and remedial control measures. In many states, the pollution control measures will focus on minimizing agricultural P losses through adoption of alternative or improved management practices to control soil erosion. This will require research on the long-term management of soil P availability in alternative or improved agricultural practices, in relation to soil productivity and water quality. In addition, it will be necessary to efficiently transfer to action agencies, existing information on the forms and amounts of P available to both crops and transport in runoff and predict short- and long-term management impacts on crop production and surface water quality.

Profitable crop production depends on many factors, including a sound P management program. Except for sunlight and water, soil fertility most frequently limits crop yields. Even with perfect weather and climatic conditions, a farmer that does everything right except meeting crop nutrient needs, will never reach maximum economic yield potential.

Water, N, and P are generally the dominant yield limiting factors for crops in the United States. Potassium, S, and micronutrients are usually not as limiting and like N and P, their needs can be assessed by soil testing (Halvorson, 1987; Halvorson et al., 1987a). However, soil P deficiency for cereal grains and other crops is common (Potash and Phosphate Institute, 1985, 1987a). Fertilizer P management varies with location and site specific conditions, such as initial soil test P level, soil type, soil pH, available application equipment, crop rotation, and tillage system. Soil testing is the best tool available to assess the need for P fertilization. Accurately assessing soil P availability status and the quantity of

P fertilizer required to alleviate P deficiency is necessary if maximum economic yields are to be obtained.

Fixation and immobilization of soil P in inorganic and organic forms unavailable for crop uptake, necessitates P amendments as fertilizer, animal manure, or crop residue material to achieve desired crop yield goals. Thus, P application has become an integral and essential part of crop production systems in order to provide adequate food and fiber for U.S. consumption and export demands. In addition, proper management of fertilizer P may reduce P enrichment of agricultural runoff via increased crop uptake and vegetative cover. On the other hand, a history of continual P applications via fertilizer and manure has increased soil test P levels in several states, to a point where the majority of tested soils were above sufficiency levels (Figure 1, T. Sims, Univ. Delaware, pers. commun.). A recent survey by Dr. T. Sims of soil test P levels in several mid Atlantic states indicates that the major portion of soils tested either high or excessive in soil test P (Table 1).

It must be remembered that many soils have low to medium soil test P levels which require annual applications of P to sustain profitable crop production (Potash and Phosphate Institute, 1987b). Overall, however, fertilizer P use in the U.S. has been declining since 1980 (Figure 1, from Berry and Hargett, 1989 and Potash and Phosphate Institute, 1987b). This trend reflects action agency efforts to reduce unnecessary applications and farmers' response to high soil test P levels, policy changes, regulating price support, P fertilizer cost, and production control measures.

Several environmental factors affect plant uptake of P from any source, soil or fertilizer (Munson and Murphy, 1986). These include temperature, soil compaction, soil moisture, soil aeration, soil pH, type and amount of clay content, P status of soil, and status of other nutrients in soil. When soil temperatures are low during early plant growth, P uptake is reduced. Soil compaction reduces pore space which reduces water and oxygen content which in turn reduces P uptake. Soil pH greatly affects plant available P, with P being fixed by Ca at high pH and by Fe and Al at low pH. Soils with high clay content tend to fix more P than low clay soils. Thus, more P needs to be added to raise the soil test level of clay soils than loam and sandy soils. The presence of ammonium-N ($NH_4$-N) enhances P uptake by creating an acid environment around the root when $NH_4^+$ ions are absorbed. High concentrations of $NH_4$-N in the soil with fertilizer P may interfere with and delay normal P fixation reactions, prolonging the availability of fertilizer P (Murphy, 1988). Thus, many factors can affect P availability to crops.

In spite of the recent decrease in fertilizer P use, there still exists a challenge to increase the utilization of both on- and off-farm sources of P, by identification and careful management of indigenous and amended forms of available soil P. The placement of P fertilizer or manure in the root zone will provide optimum available soil P levels for plant uptake during periods of maximum crop uptake and will maximize soil productivity and minimize potential P losses in runoff. However, climatic, edaphic, agronomic, and economic factors limit achievement

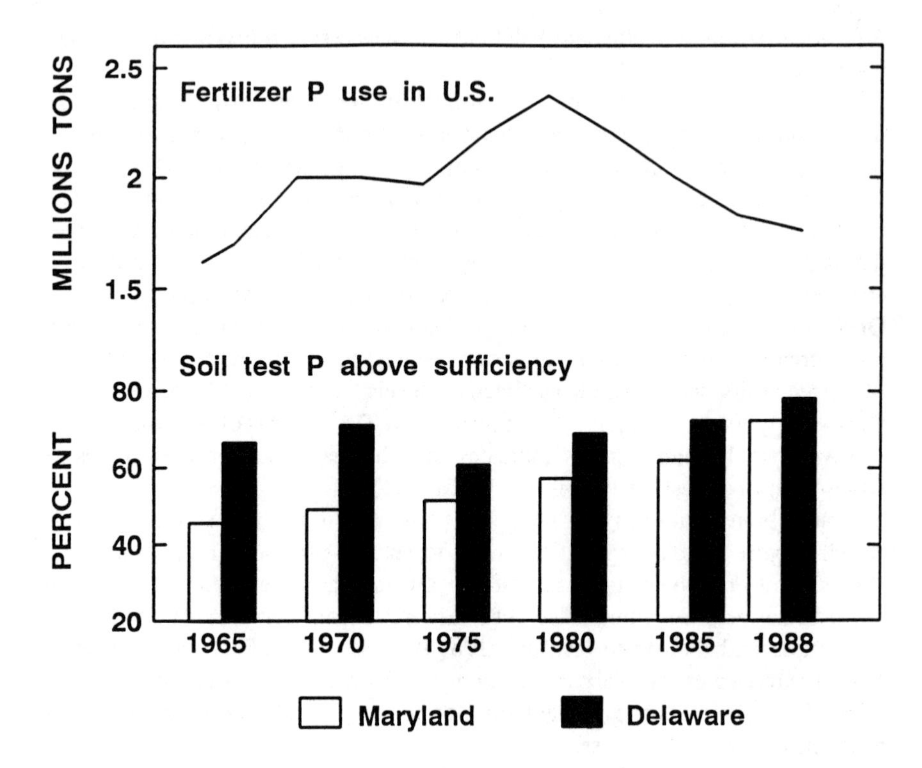


**Figure 1.** Fertilizer P use in the U.S. and percent soil samples testing above sufficiency level in Maryland and Delaware.

of this situation. In particular, the bulky nature and large amounts of animal manure that are produced throughout the year in localized areas, often limits subsurface placement of the manure. Careful consideration of these factors may lead to the development of more sustainable agricultural management systems that are environmentally sound.

This chapter presents the factors influencing P availability in soil and water systems, as shown in Figure 2. We discuss the role of agricultural management in maximizing soil P availability, while minimizing P losses in runoff. Considering this discussion, future challenges to the development of agronomically and environmentally sustainable P management systems and associated research needs are identified. With the current interest in the development and adoption of efficient and sustainable agricultural systems (Edwards et al., 1988; Francis et al., 1990; Potash and Phosphate Institute, 1989), there will be an increased reliance on efficiently utilizing indigenous soil P forms. Consequently, this chapter emphasizes the effect of soil management on the dynamics of organic and residual P availability.

**Table 1.** Phosphorus soil test survey for northeast, north central, and mid Atlantic states

| State | Soil test method | Summary year | Soil test P level for | | | | Percent samples testing [a] | | | |
|---|---|---|---|---|---|---|---|---|---|---|
| | | | Low | Medium | High | Excessive | Low | Medium | High | Excessive |
| | | | kg P ha$^{-1}$ | | | | % | | | |
| Northeast | | | | | | | | | | |
| CT | Modified Morgan | 1990 | 5.5 | 11 | 28 | >40 | 25 | 17 | 29 | 29 |
| ME | Modified Morgan | 1990 | 4 | 11 | 45 | >45 | 5 | 34 | 50 | 11 |
| NH | Mod. Morgan (pH 4.8) | 1990/91 | <7 | <17 | <27 | <27 | 51 | 21 | 10 | 18 |
| NY | Mod. Morgan (ph 4.8) | 1988 | <4.5 | 9 | 45 | >224 | 25 | 35 | 40 | 0 |
| PA | Mehlich-3 | 1989/90 | 34 | 68 | 112 | 170 | 33 | 23 | 18 | 26 |
| North Central | | | | | | | | | | |
| IA | Bray-I | 1983+ | 34 | 45 | 67 | | 31 | 13 | 19 | 0 |
| IN | Bray-I | 1988/89 | 22 | 34 | 56 | >56 | 11 | 11 | 21 | 50 |
| KS | Bray-I | 1990/91 | 28 | 56 | 112 | >112 | 25 | 27 | 24 | 21 |
| MI | Bray-I | 1990/91 | <34 | 67 | >90 | >224 | 10 | 27 | 40 | 23 |
| MN | Bray-I | 1968-76 | 11 | 17 | 45 | >45 | 12 | 21 | 18 | 49 |
| MO | Bray-I | 1990 | 25 | 50 | 78 | >150 | 19 | 30 | 23 | 25 |
| NE | Bray-I | 1990/91 | 5.5 | 17 | 27 | >27 | 35 | 23 | 22 | 20 |
| ND | Olsen | 1990/91 | 21 | 32 | 44 | >44 | 39 | 29 | 13 | 12 |
| SD | Bray-I and Olsen | 1990 | <17 | 28 | 45 | | 22 | 30 | 25 | 0 |
| Mid Atlantic | | | | | | | | | | |
| DE | Mehlich-1 | 1991 | 38 | 75 | 150 | >150 | 9 | 21 | 28 | 40 |
| MD | Mehlich-1 | 1990 | 29 | 50 | 100 | >100 | 8 | 18 | 74 | 0 |
| NJ | Mehlich-1 | 1989 | 30 | 50 | 99 | >99 | 12 | 8 | 15 | 65 |
| NC | Mehlich-3 | 1990/91 | 60 | 120 | 240 | >240 | 6 | 9 | 23 | 62 |
| SC | Mehlich-1 | 1988 | 22 | 44 | 88 | >270 | 17 | 25 | 34 | 5 |

[a] In some states, percentages may not total 100 due to incomplete data.
From unpublished data of T. Sims, University of Delaware.

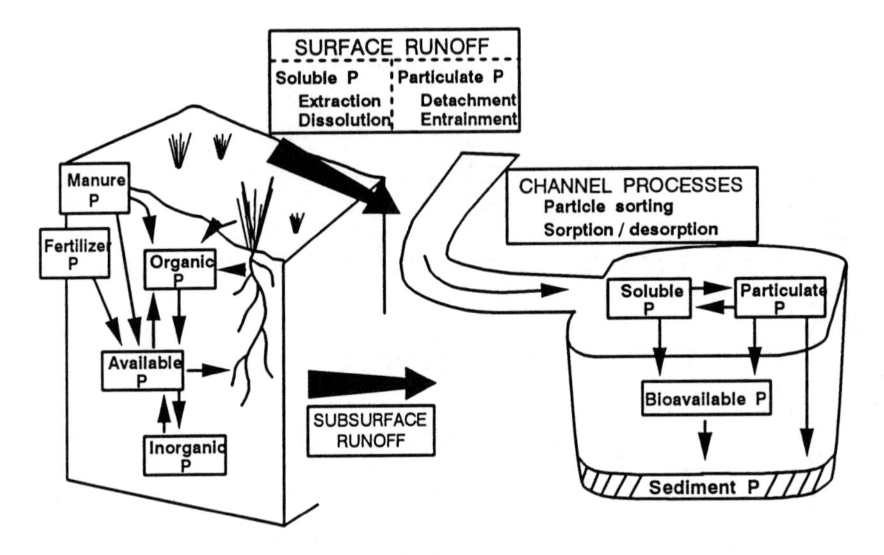


**Figure 2.** Factors influencing P availability in soils and water.

## II. Dynamics of Soil P Availability

Several studies on the effect of agricultural management on the dynamics of P cycling in soils, have found a differential behavior of inorganic and organic P forms (Agboola and Oko, 1976; Harrison, 1978; Sharpley, 1985a; Tiessen et al., 1983). With the application of P, available soil P content increases (Barber, 1979; Khasawneh et al., 1988; McCollum, 1991; Peterson and Kreuger, 1980). This increase is a function of certain physical and chemical soil properties (Barrow, 1980; Larsen et al., 1965; Lopez-Hernandez and Burnham, 1974). The portion of fertilizer P remaining as available P (resin P) 6 months after application, decreased as clay, organic C, Fe, Al, and $CaCO_3$ content increased for over 200 widely differing soils (Table 2: from Sharpley, 1991; Sharpley et al., 1984a; 1989). With an increase in degree of soil weathering, represented by soil taxonomic and other related properties, a general decrease in availability of applied fertilizer P was evident. Clearly, the dynamics of fertilizer P availability differs between soils, and influences the degree of soluble P enrichment of surface runoff.

Where no fertilizer P is added, a net loss of P from the system via removal in the harvested crop is often accounted for by a decrease in soil organic P, while inorganic P generally remains constant. For example, the growth of cotton on a Mississippi Delta soil, Dundee silt loam for 60 yr (1913-1973), with no reported fertilizer P applied, had little affect on inorganic P content (Sharpley

**Table 2.** Percent fertilizer P available (as resin P) 6 months after application

| Related properties | Number of soils | Availability | |
|---|---|---|---|
| | | Mean | Range |
| | | --------------%------------- | |
| *Calcareous* | | | |
| $CaCO_3$ | 56 | 45 | 11-72 |
| *Slightly weathered* | | | |
| Base saturation | 80 | 47 | 7-74 |
| Available P | | | |
| pH | | | |
| *Moderately weathered* | | | |
| Clay | 27 | 32 | 6-51 |
| Available P | | | |
| Organic C | | | |
| *Highly weathered* | | | |
| Clay | 40 | 27 | 14-54 |
| Extractable Al | | | |
| Extractable Fe | | | |

Data adapted from Sharpley, 1991 and Sharpley et al., 1984a, 1989.

and Smith, 1983). However, a decrease in the organic P content of the cultivated (93 mg $kg^{-1}$) compared to virgin analogue (223 mg $kg^{-1}$) surface soil (0-15 cm) was measured. Apparently, mineralization of organic P replenished the inorganic pool and provided adequate amounts of plant available P.

### A. Organic P Mineralization

Though inorganic P has generally been considered the major source of plant available P in soils, the incorporation of fertilizer P into soil organic P (McLaughlin et al., 1988) and lack of crop response to fertilizer P due to organic P mineralization (Doerge and Gardner, 1978), emphasize the need to consider organic P in the management of soil P availability, particularly with reduced tillage practices. For more detailed information on the dynamics of soil inorganic P transformations and availability as affected by management, the reader is referred to articles by Khasawneh et al. (1988) and Syers and Curtin (1988). The main variables controlling the dynamics of organic P mineralization

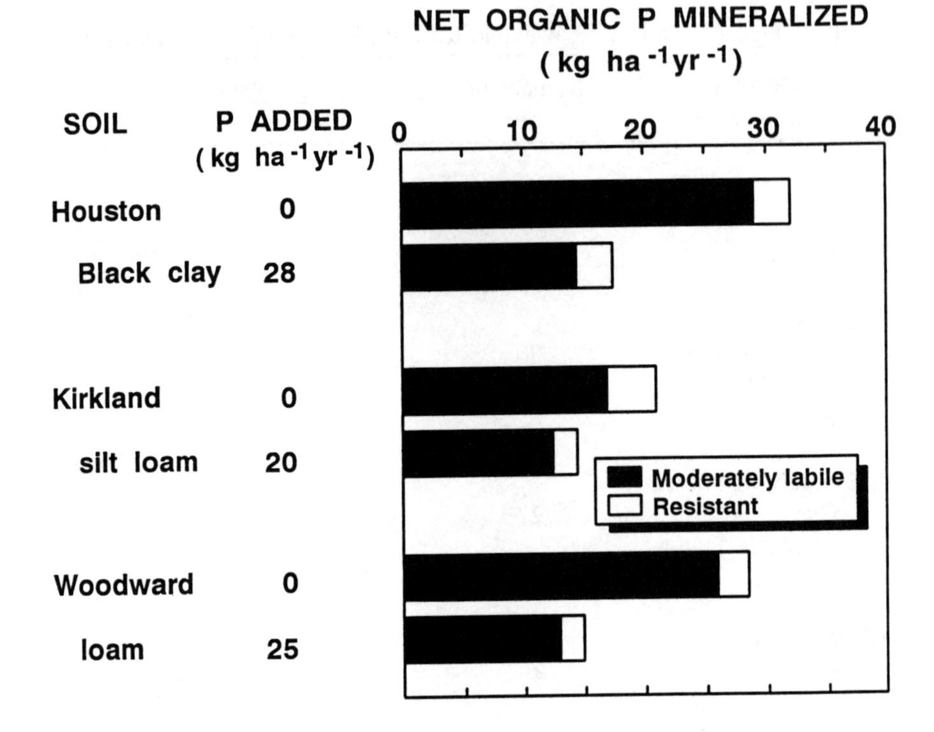


**Figure 3.** Annual net mineralization of moderately labile and resistant organic P in unfertilized and fertilized soils. (Data adapted from Sharpley, 1985a.)

can be divided into those related to climatic and soil factors and to crop residue factors.

1. Climatic and Soil Factors

Organic P mineralization in several unfertilized and P fertilized soils in the Southern Plains was quantified by Sharpley (1985a) as the decrease in soil organic P content during the period of maximum crop growth (spring and early summer). However, as soil organic P may be formed by plant residue incorporation, the value of net organic P mineralization will underestimate the actual value. Organic P was fractionated into labile, moderately labile, moderately resistant, and resistant pools by the sequential extraction procedure of Bowman and Cole (1978).

Averaged for each soil type, organic P mineralization ranged from 15 to 33 kg P $ha^{-1}$, with mineralization greater in unfertilized than P fertilized soils (Figure 3). Of this, moderately labile organic P contributed 83 to 93% of that mineralized. As labile and resistant organic P pools remained fairly constant

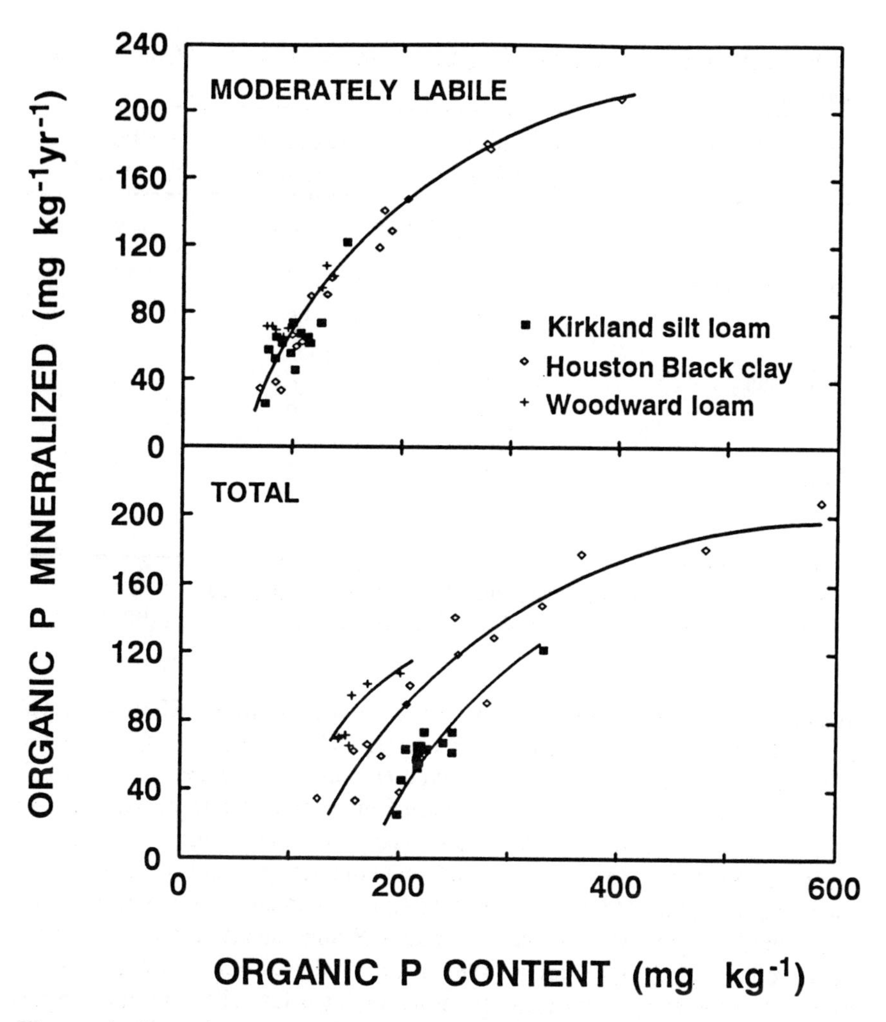


**Figure 4.** Annual net mineralization of organic P as a function of total and moderately labile organic P for three soils.

during the 2-yr study (Sharpley, 1985a), mineralization of moderately labile organic P replenished the available P pool, when it fell below a critical but as yet undefined level. Tate et al. (1991) also found labile organic P mineralization was an important source of P to pasture in both low- and high-P fertility soils in New Zealand. Both studies (Sharpley, 1985a; Tate et al., 1991), suggest that management practices maximizing the build-up of organic matter during autumn and winter, may reduce external P requirements for plant growth during the following spring and early summer.

Net organic P mineralization was related to total organic P content for both unfertilized and fertilized soil (Figure 4, from Sharpley, 1985a). For a given

**Table 3.** Average net amount of organic P mineralized for several climatic regions

| Region | Fertilizer P applied | Organic P mineralized | Percent mineralized [a] |
|---|---|---|---|
| | ------------kg P $ha^{-1}$ $yr^{-1}$ | ----------- | % $yr^{-1}$ |
| Southern Plains | 0 | 23 | 11 |
| | 25 | 17 | 8 |
| Temperate | 0 | 11 | 2 |
| | 34 | 6 | 1 |
| Tropics | 0 | 157 | 15 |
| | 40 | 67 | 18 |

[a] Percent of total soil organic P which is mineralized annually.
data adapted from Sharpley (1985a) and Stewart and Sharpley (1987).

organic P content, mineralization was greater for Woodward than Houston Black and Kirkland soils. As a function of moderately labile organic P, however, no difference between locations was observed (Figure 4). Apparently, organic P mineralization dynamics were a function of moderately labile organic P, the level of which is determined by climatic and soil factors. Further, organic P mineralization (15 to 33 kg $ha^{-1}$ $yr^{-1}$) was not completely inhibited by fertilizer P application (20 to 28 kg $ha^{-1}$ $yr^{-1}$), with similar amounts of P contributed by both sources (Figure 3). Amounts of organic P mineralized in the 3 Southern Plains soils studied by Sharpley (1985a) are similar to other temperate soils (Table 3). They are generally lower, however, than for soils from the tropics (67 to 157 kg $ha^{-1}$ $yr^{-1}$), where distinct wet and dry seasons and higher soil temperatures can increase the amounts of organic P mineralized.

## 2. Crop Residue Factors

Residue management can affect P cycling and availability as a function of residue amount, type, and degree of incorporation with tillage. A greater amount of residue will increase the amount of P being cycled and, particularly if left on the surface of the soil, will reduce evaporation losses and keep surface soil moist for more days during the growing season, thereby enhancing microbial activity and mineralization (Figure 5, from Sharpley and Smith, 1989a). Little difference in net organic P mineralization and mobility was observed between residue types during the 84-day incubation (Figure 5), even though residue C:P ratio ranged

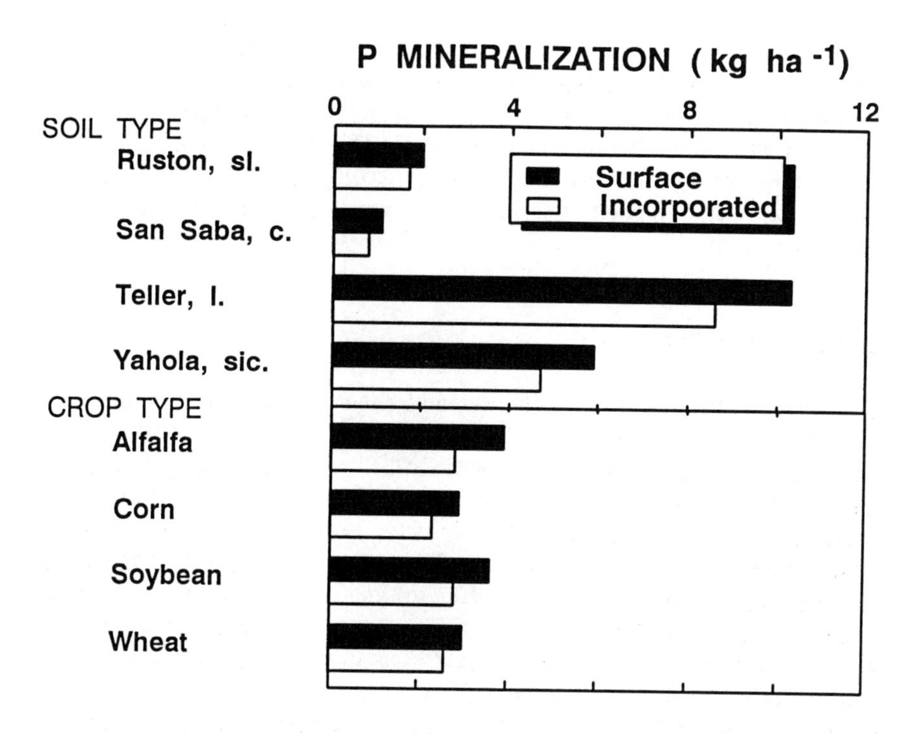


**Figure 5.** Mineralization of crop residue P in 84 days as a function of residue placement and soil and crop type.

from 260:1 (alfalfa) to 600:1 (corn). This suggests that the effect of C:P ratio may be more evident over a longer period of time as shown for crop residue N mineralization by Power et al. (1986). Net organic P mineralization and mobility was affected, however, by soil type (Figure 5), as a function of available P content and soil C:P ratio. This is consistent with the wide C:P ratio of the residues and resultant immobilization of soil P, prior to mineralization. Although amounts of residual P mineralized and leached are small (Figure 5) compared to normal fertilizer P applications to these soils and crops (30 kgP $ha^{-1}$ $yr^{-1}$), they may provide an important source of available P during initial crop growth when the residue will be either fresh (i.e., fall planting of winter wheat) or coming out of a relatively inactive microbial period (i.e. spring planting).

Continuously cultivated soils generally decline in organic matter content, such that the major source of P removed by a crop will be supplied by fertilizer. Under reduced tillage, enhanced microbial, faunal, and phosphatase activity may increase immobilization of fertilizer P as microbial organic P and subsequently, rates of organic P mineralization. These processes will also be important under crop rotations, which include a leguminous cover crop such as alfalfa, summer-fallow, or pasture phases, where fertilizer P may be rapidly incorporated into soil organic P, providing a potential source of available P for subsequent crops.

Further research is needed on the dynamics and magnitude of fertilizer P immobilization and mineralization as a function of climate (soil temperature and moisture), soil, and crop management. In addition, preliminary research (Sharpley and Smith, 1989a; Tate et al., 1991; Thibaud et al., 1988), indicates further information is needed on the relative decomposition and mineralization of residual P as a function of soil and crop type under different management practices.

## B. Analytical Limitations

One of the main limitations in evaluating the contribution of organic P mineralization to the dynamics of soil P cycling and soil availability under different management practices, has been the difficulty in accurately quantifying and identifying soil organic P. Total organic P is determined indirectly by difference (total - inorganic P) following either extraction or ignition methods (Olsen and Sommers, 1982). However, solubilization or complexation of mineral P during ignition and hydrolysis and/or incomplete removal of adsorbed or occluded organic P during extraction, may introduce errors in organic P estimation between and within soil types. Recent modifications of extraction and ignition methods (Bowman, 1989; Soltanpour et al., 1987) may simplify and increase the accuracy of organic P quantification. Although complete characterization of soil organic P has not been accomplished, application of solid state neutron magnetic resonance (NMR) spectroscopy may improve its identification. Alternatively, sequential extraction procedures may fractionate organic P according to its chemical stability and these fractions then related to bioavailability (Bowman and Cole, 1978; Hedley et al., 1982; Potter et al., 1991).

With the development of fumigation-extraction techniques to measure soil microbial biomass P (Brookes et al., 1982; Hedley and Stewart, 1982), its importance in P cycling has been recognized (McLaughlin et al., 1988; Stewart and Tiessen, 1987). For example, McLaughlin and Alston (1986) showed that microbial biomass can assimilate a similar portion of fertilizer P as taken up by wheat. In a study of P cycling through soil microbial biomass in England, Brookes et al. (1984) measured annual P fluxes of 5 and 23 kg P $ha^{-1}$ $yr^{-1}$ in soils under continuous wheat and permanent grass, respectively. Although biomass P flux under continuous wheat was less than P uptake by the crop (20 kg P $ha^{-1}$ $yr^{-1}$), annual P flux in the grassland soils were much greater than P uptake by the grass (12 kg P $ha^{-1}$ $yr^{-1}$).

Not all this biomass P is available for plant uptake each year; some will be directly transferred to subsequent microbes and some released to the soil solution, which can then be fixed by soil or taken up by microbes or plants (Brookes et al., 1984). However, the rate and extent of microbial P flux emphasizes the importance of microbial P in controlling the short-term dynamics of organic P transformations and thereby, management of soil P availability.

Because of the inaccuracies in organic P measurement, which are compounded in estimating organic P mineralization, attention has been given to direct measurement of native and residual organic P mineralization by isotopic dilution (Walbridge and Vitousek, 1987) or addition of $^{32}P$ labelled organic matter, respectively (Dalal, 1979; Harrison, 1982; McLaughlin et al., 1988). Further research is needed, however, on the validity and relative importance of assumptions concerning achievement of isotopic equilibrium and bidirectional and periodic movement of P between available and microbial pools in estimating organic P mineralization. Although the length of isotopic studies is limited due to the short half-life of $^{32}P$ and $^{33}P$ (14.3 and 24.4 days, respectively), they have indicated that the actual rate of organic P mineralization has been underestimated by nonisotopic difference methods. Improvement and application of these methods, will aid future research to elucidate the dynamics of organic P mineralization and role of microbial biomass as a function of soil, residue, and agricultural management.

## III. Management of Soil Availability

### A. Fertilizer Management

Fertilizer P management strategies that maximize soil P availability while minimizing surface soil accumulations which may increase P loss in surface runoff, must consider fertilizer application rate, timing, type, placement, and residual availability. Fertilizer rate is primarily determined by soil test P levels and desired crop yield goals, discussed in the next section. Because of the immobility of P in most soils, the timing of fertilizer P application is not as critical as its placement. In efforts to efficiently utilize P inputs in sustainable management systems, there has been renewed interest in the estimation and utilization of residual P availability from fertilizer or manure amendments (McCollum, 1991; Pierzynski et al., 1990; Yerokum and Christenson, 1990).

#### 1. Type

Except for the increased acidulation of P fertilizers, producing triple from single super P and partially acidulated rock P (RP), there has been less development of P sources than N. The treatment of fertilizer P to increase its solubility and thereby, crop-use efficiency, increases its cost. Consequently, farm management decisions regarding the type of fertilizer P to be used, have been based more on economic and agronomic considerations than on chemical availability and environmental aspects

Research has evaluated ways to broaden the use of slow release P fertilizers in the eastern U.S., such as RP, beyond soils which have low pH, Ca, and P content. For example, in soils of neutral pH, it may be possible to apply heavy initial dressing of finely ground RP and include a rotation of fine rooted legumes to generate a low pH rhizosphere with low Ca concentrations and thus increase RP dissolution. Other methods designed to increase acidity in the immediate RP-soil environment, and thereby its dissolution, include addition of elemental S (Muchovej et al., 1989), $NH_4^+$ fertilizers, or organic matter such as animal manure and crop residues (Hedley et al., 1989). More research is needed, however, to evaluate the effectiveness of these amendments to enhance P availability by root extraction in different soils and cropping practices. An increasing adoption of more efficient management systems and inclusion of forage legumes in crop rotations, along with the development of reactive RP sources, may increase the agronomic and economic value of RP. Thompson (1990) suggests that North Carolina reactive RP may have value for direct application, especially on moderately acid, medium to high organic matter soils low in available P and under forage legumes in low-input systems. Thompson (1990) also suggested that RP may be the best P amendment available for organic farmers and others using only naturally occurring soil amendments.

Less information is available on the effect of fertilizer type on the loss of P in runoff. For example, Sharpley et al. (1978) observed a slightly greater soluble P (SP) loss in runoff following the application of monocalcium P (MCP - main P component of super phosphate) to a permanent pasture in New Zealand (2.80 kg $ha^{-1}$), compared to that with dicalcium P (DCP) (2.17 kg $ha^{-1}$), a slow-release fertilizer. This difference was attributed to more rapid dissolution of MCP than DCP at the soil surface. However, an appreciably greater loss of sediment-bound P with DCP (4.92 kg $ha^{-1}$) than MCP (2.63 kg $ha^{-1}$), resulted from an increased loss of P by transport of the less soluble DCP particles in runoff. It is expected that RP will affect the enrichment of P in runoff in a similar manner as DCP.

## 2. Placement

Due to the general immobility of P in the soil profile, fertilizer placement is generally more critical for P than N. Not everyone agrees on the best method of P application. Fixen and Leikam (1989) stated that "contradictory recommendations for method and placement of phosphorus (P) often are due to the fact that conditions influencing P fertilizer response vary among studies." They discussed many factors affecting the effectiveness of P placement methods and addressed the following questions: a) Which is better, band or broadcast P applications?; b) Are all band P applications methods equal in effectiveness?; and c) How much can P recommendations be reduced if P is banded instead of broadcast? They list the following soil and crop factors as influencing fertilizer P response: a) soil test levels; b) P concentration of the fertilized soil solution; and c) root contact with the fertilized soil. Root contact with the fertilized soil is influenced by total root length, volume of soil fertilized (varies with placement method), and location of the fertilized soil in relation to plant roots.

Depending on soil and environmental factors, band applications of P may or may not be better than broadcast incorporated applications of P. In general, if there is a difference in crop response due to P application method, yield response to band applications will be equal to, or better than, broadcast applications. Long-term studies in the northern Great Plains have shown that high rates of broadcast P (90 kg P $ha^{-1}$) can have long-term effects (17 years) on soil test P, wheat yields, (Bailey et al., 1977; Halvorson and Black, 1985b; Roberts and Stewart, 1987) and profitability (Jose, 1981; Halvorson et al., 1986). Several studies have shown a greater yield response to surface or subsurface band application of fertilizer P at low rates, compared to broadcast or mixing (Alston, 1980; Bailey and Grant, 1989; Lammond, 1987; Yost et al., 1981). In fact, Welch et al. (1966) observed greater P uptake and yield of corn with a combined banded (50%) and broadcast (50%) application (40 kg $ha^{-1}$).

In addition to agronomics, other factors are equally important in selecting the best P application method. Equipment availability, labor requirements, product availability, and availability of operating capital all affect this decision.

Deep placement of N with P (pre-plant banding) has grown in popularity recently in the Great Plains of the U.S. and in the Canadian Prairie Provinces (Murphy, 1988). The placement of N with P under both conventional and reduced tillage systems has frequently been more effective for wheat than application methods which placed most of the N and P at different positions in the soil (Dahnke et al., 1986; Harapiak and Flore, 1984; Leikam et al., 1983), particularly at low P soil test levels. Generally, yield differences between deep banding or P placement near the seed have been relatively small. Consequently, fertilizer recommendations frequently do not differentiate between seed placement and deep banding in terms of P efficiency. Alessi and Power (1980) reported wheat yield increases resulting from banding P with the seed even when soil test levels were high. Environmental conditions should be considered in addition to soil test results as a part of the management decisions which go into recommending fertilizer rates for higher wheat yields. Cold, wet soil conditions compounded by heavy surface residue may be conducive to P responses, particularly from starter applications, even when P soil tests are high (Murphy, 1988).

#### a. Soil Test P

Method and rate of application can affect the response of wheat to P fertilization. If low rates of fertilizer P are applied to soils testing "low" in plant-available P, then banding the fertilizer P below or with the seed is generally more efficient and results in greater yield increases than broadcast P applications (Murphy and Dibb, 1986; Peterson et al., 1981; Sleight et al., 1984). However, if sufficient fertilizer P was to be added to attain maximum wheat yields on a soil testing "low" in P, then method of placement may not be as critical. On soil testing medium to high in available P, the difference in effectiveness between broadcast and band applications of any type is lessened (Peterson et al., 1981). The work of Wagar et al. (1986) supports this theory. They found that a single, broadcast P application of 80 kg P $ha^{-1}$ had a greater cumulative yield after 5 years than 20 kg P $ha^{-1}$ applied each crop year with the seed. Thus, the broadcast treatment produced at or near optimum yields each year, whereas the seed placed P treatment produced at less than optimum yield potential during the first several years. They also found that a combination of a residual 40 kg P $ha^{-1}$ broadcast one time plus 10 kg P $ha^{-1}$ applied each crop year with the seed produced near maximum wheat yields. The latter treatment would be desirable from the standpoint of spreading the P fertilizer costs out over a longer time frame and still being able to maintain near maximum yield potential. However, the recommendation of one-time, high P application rates at a particular site, must consider the potential vulnerability for P loss in runoff from the site. Site variables that should be considered include runoff and erosion potential and proximity to P-limited surface waters. The role of these factors in determining vulnerability to P loss is discussed in more detail in Section VI-D.

**b. Soil Type**

The effect of P application, however, varies with soil type. When P fertilizer is placed in a specific soil volume, root extraction of P depends on the rate of application, which affects soil P adsorption/desorption characteristics and diffusion, and on the stimulation of root growth in the fertilized soil volume. For six soils having a hundred-fold variation in P sorptivity, Holford (1989) found that fertilizer P effectiveness, as measured by yield response in the first crop (wheat), residual effect in the second crop (clover), or cumulative recovery of applied P, was consistently greater for shallow banding at 5 cm depth compared to banding at 15 cm and broadcast applications. The almost equal effect obtained by mixing P throughout the soil, regardless of P sorptivity, suggested that the important factor in maximizing fertilizer effectiveness is its positional availability in the root zone rather than reduction of chemical immobilization by concentration in bands (Holford, 1989).

**c. Crop Factors**

Positional availability will also be influenced by crop type. For banding or restricted fertilizer placement to increase potential root extraction of P, the rate of P absorption and growth of roots in fertilized soil must increase to compensate for roots in unfertilized soil. Increased root growth and P uptake in the P-fertilized volume of soil compared to unfertilized soil has been observed for corn (Anghioni and Barber, 1980), soybean (Borkert and Barber, 1985), and wheat

(Yao and Barber, 1986). In contrast, several studies have shown that flax does not respond to banded fertilizer due to an inability of its root system to expand and proliferate into and efficiently absorb P from high concentrations in the fertilized zone (Soper and Kalra, 1969; Strong and Soper, 1974). In the case of flax, increased P uptake and yield response was obtained when fertilizer P was placed 2 to 5 cm directly below the seed, ensuring adequate P levels during early growth (Bailey and Grant, 1989).

In the final analysis, P placement may enhance it's availability and increase yields and must be considered in formulating a management plan to maximize both crop yields and associated water quality. Growers attempting to improve crop yields and profitability should maintain recommended rates of P even if more efficient methods of application are used. Cutting back on P rates to cut production costs may result in lost profits (Murphy, 1988).

### 3. Residual Availability

A need for higher P application rates to optimize crop yield potentials, necessitates that the short- and long-term economic and environmental impacts of P fertilizer management be evaluated. Most research on soil P fertility in the Great Plains has been limited to evaluating wheat response to P fertilizer application from one crop harvest (Dahnke et al., 1986; Fiedler et al., 1987; Follett et al., 1987; Leikam et al., 1983; Peterson et al., 1981; Westfall et al., 1986). Effects of residual fertilizer P in the northern Great Plains have been positive in increasing small grain yields (Bailey et al., 1977; Black, 1982; Halvorson and Black, 1985a; Read et al., 1977; Roberts and Stewart, 1987; Wagar et al., 1986) as well as increasing farm profit potential (Halvorson et al., 1986; Jose, 1981; Roberts and Stewart, 1987). Many of these studies were conducted with conventional dryland tillage systems and a crop-fallow cropping sequence. On a long-term (4 crop years) basis, a single broadcast application of P fertilizer (80 kg P $ha^{-1}$) may be equally as effective in increasing wheat yields as annual band applications (20 kg P $ha^{-1}$) (Roberts and Stewart, 1987; Sleight et al., 1984). Long-term P studies conducted by Alessi and Power (1980), Bailey et al. (1977), Black (1982), Halvorson and Black (1985a and 1985b), and Read et al. (1977) in the northern Great Plains indicate that benefits from a single P fertilizer application at rates of 45 kg P $ha^{-1}$ or more may last as long as 16 years, depending on initial rate of application and cropping history. Halvorson (1989) reported that irrigated no-till winter wheat, grown annually on the same land, responded positively to residual broadcast fertilizer P (34 and 67 kg $ha^{-1}$).

Multiple year responses of alfalfa and grain sorghum to single applications of P fertilizer have been investigated in the central Great Plains (Havlin et al., 1984; Janssen et al., 1985, Schlegel et al., 1986). Halvorson and Black (1985a) suggested that a one-time, high-rate ($>$50 kg P $ha^{-1}$) application of P fertilizer may be one way to satisfy the P needs of crops grown with reduced and no-till

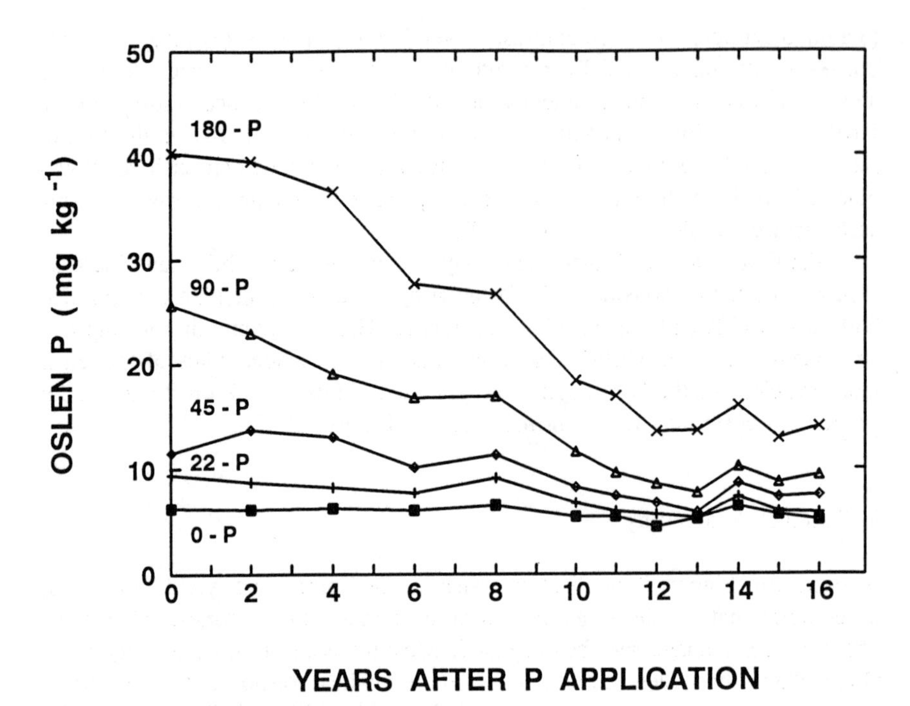


**Figure 6.** Changes in soil test P (Olsen P) levels with time after P fertilizer application (kg $ha^{-1}$) to a Williams loam in Montana. (Data adapted from Halvorson and Black, 1985a.)

systems for several years. Data from A.D. Halvorson and J.L. Havlin (1991 unpublished data) in Colorado supports this suggestion.

Phosphorus fertilization changed soil test P levels of several central Great Plains soils for several years (Hooker et al., 1980). Halvorson and Black (1985a) showed that soil test P levels were increased above the initial soil test P level for more than 16 years, by the one-time P applications on a Williams loam in Montana (Figure 6). After the initial increase, soil test P levels declined for about 12 years and then stabilized at a higher soil test level than was initially present, thus establishing what appears to be a new equilibrium level of soil available P. Fixen (1986) reported similar changes in soil test P levels with time. Crop yields reported by Halvorson and Black (1985a) were also improved by the residual P fertilizer for 16 years (Figure 7). Based on soil test P levels for the highest P rates, grain yields would have been increased for several more cropping seasons had the study been continued.

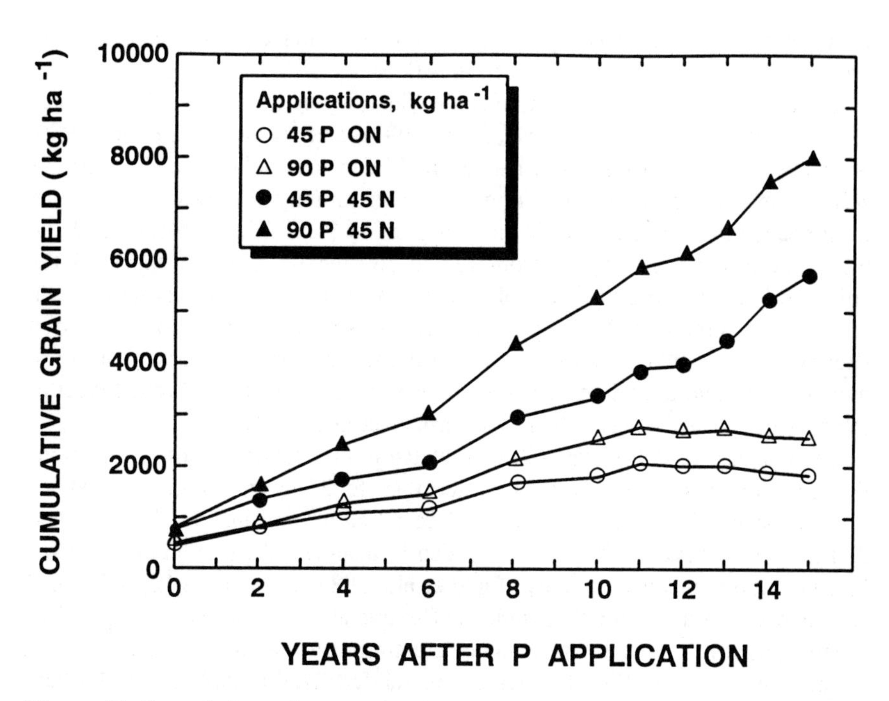


**Figure 7.** Cumulative wheat grain yield with years after initial P application with or without N applied each crop year. (Data adapted from Halvorson and Black, 1985a.)

### B. Manure Management

Manure and organic wastes can be a valuable source of P to crops, improve soil physical properties, and increase soil organic matter content. In fact, by improving vegetative ground cover, manure application can reduce runoff volume and erosion. The concentration of P in manure is highly variable, thereby introducing uncertainty into meeting crop needs. However, the fertilizer value of manures is generally inversely related to their water and carbon contents.

Animal production operations have concentrated in localized areas for economic reasons, which include the close proximity of feed supply and meat processing plants. As a result, manure production often exceeds crop P requirements of both the producing and adjacent farms. Thus, disposal of the concentrated animal waste, that accumulates in confined production systems, is an increasing problem facing the industry. A major part of this problem has arisen from the fact that manure application rates have been based on N management. In most cases, this will lead to an increase in soil P, often well in excess of levels required for maximum yields. The importance to efficient management systems of basing manure application rates on soil P rather than N, are discussed in more detail in Section VII. B.1.

The continual land application of cattle (Sharpley et al., 1984b; Vitosh et al., 1973), poultry (Field et al., 1985; Sims, 1992), and swine (King et al., 1990; Sharpley et al., 1991b) manure, has resulted in an accumulation of P in surface soil. From a survey of several farms in Sussex County, Delaware, one of the most concentrated poultry production areas in the U.S., Sims (1992) found soil test-P levels (Bray I) ranging from 123 to 369 mg kg $^{-1}$, which exceeded the "high" criteria (50 mg kg $^{-1}$) established by Delaware soil test laboratories. In addition, a decrease in P adsorption capacity of soil following manure addition (Reddy et al., 1980; Sharpley et al., 1991b), increases the potential for P movement in lateral and vertical soil water flow (Brown et al., 1989; Magette, 1988; McLeod and Hegg, 1984; Westerman et al., 1983).

The increase in soil P availability is related to the rate of manure application. For four loam soils receiving long-term (8 to 35 years) poultry or swine manure, available P content of the surface 50 cm of soil increased an average 27 kg P ha $^{-1}$ for every 100 kg P ha $^{-1}$ added in manure (Sharpley et al., 1991b). In a laboratory incubation study, Field et al. (1985) observed a 12 to 23 kg P ha $^{-1}$ increase in available P (double-acid extractable) for every 100 kg P ha $^{-1}$ added in poultry manure. These values are similar to the proportional increase in soil test P (13 to 28%) following mineral fertilizer P application (Barber, 1979; McCollum, 1991; Rehm et al., 1984).

Organic wastes that are important in localized regions include sewage sludge from municipalities, waste from livestock slaughtering facilities, and wastes from the food processing and other industries. As for fertilizer and animal manure management, the composition, rate, placement, and time of application are major factors affecting soil P availability.

## C. Crop Yield Potential

The relationship between the sodium bicarbonate P test and relative yield potential of wheat grown in a dryland wheat-fallow system is shown in Figure 8 (Halvorson, 1986a). These data indicate that a 26 mg P kg $^{-1}$ level in the surface 15 cm of soil is needed to achieve 100% of the wheat yield potential in this semiarid environment. The relationship shown in Figure 8 is useful in estimating potential yield reductions caused by inadequate available P levels. When P fertilizer is added to most soils in the Great Plains, an increase in soil test P levels can be expected. The amount of increase will depend upon soil texture and other soil characteristics. Halvorson and Kresge (1982) used this approach to estimate the amount of broadcast-incorporated P fertilizer needed to optimize yields. If less P fertilizer is applied than recommended, wheat yield potentials are reduced along with N fertilizer needs. Halvorson and Kresge (1982) estimated that 4 to 5 kg P ha $^{-1}$ was needed to raise the soil test P level 1 mg kg $^{-1}$. In Illinois, application of 9 kg P ha $^{-1}$ is estimated to increase the soil

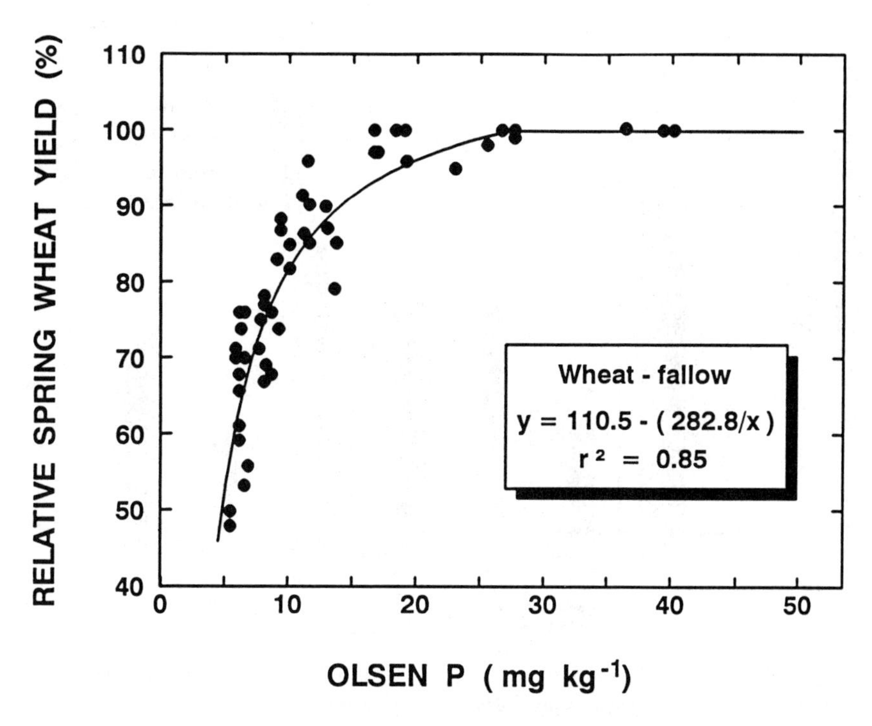


**Figure 8.** Relative spring wheat yield as a function of soil test P on a Williams loam in Montana. (Data adapted from Halvorson, 1986a.)

test P level 1 mg $kg^{-1}$ (Agronomy Staff, 1989). Based on initial soil test P level and P application rate, Halvorson et al. (1987b) developed a model to predict the change in soil test P level with time on a Williams loam after an application of P fertilizer in a wheat-fallow system. This approach could be used, along with P removal rate by the crop, to predict when future additions of P will be needed. Similar data are needed for other soil types and cropping conditions.

Application of fertilizer P to bring the soil test P level to about 21 mg P $kg^{-1}$ (Olsen P) or about 30 mg P $kg^{-1}$ (Bray-I P) followed by smaller P applications to maintain this soil test level, may result in optimum wheat yields and optimum short- and long-term profitability. This approach to P fertilization would probably provide the potential for optimum wheat yields each crop year. In dry years, a high level of soil P ($>20$ mg $kg^{-1}$) will enhance yield potential and in the wet years, a high level of soil P will provide the opportunity to more efficiently utilize available water supplies, provided N is not limiting. For example, Black (1982) showed that in dry years spring wheat yields were increased an average of 417 kg $ha^{-1}$ while in wet years yields were increased by 712 kg $ha^{-1}$ with 180 kg P $ha^{-1}$.

In summary, P should not be a yield limiting factor for crop production. Phosphorus is a relatively immobile nutrient, not readily subject to leaching

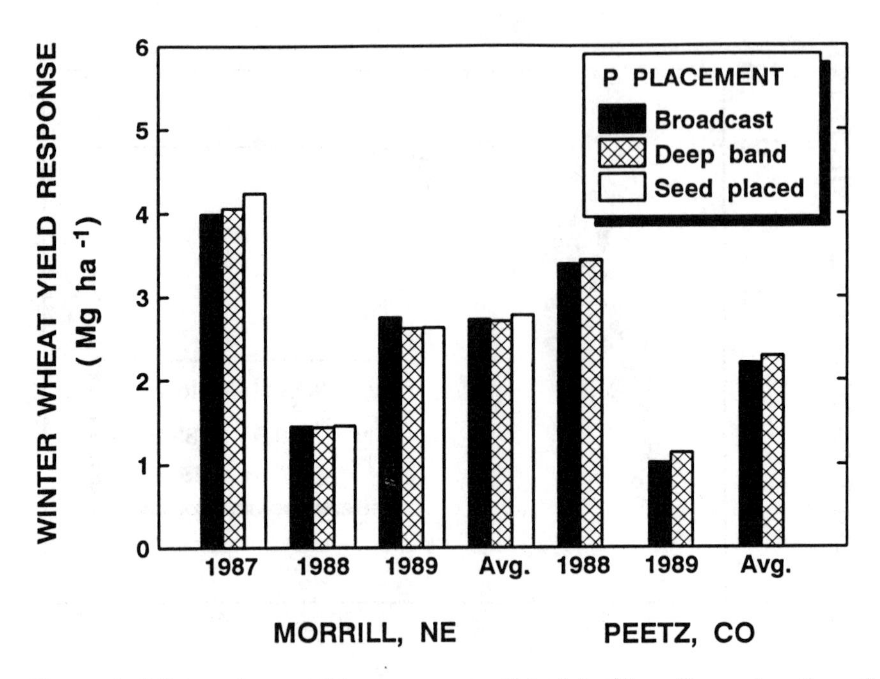


**Figure 9.** Winter wheat yield response to residual fertilizer P as a function of P application method on loam soils at two locations.

losses. The loss mechanisms are mainly through soil erosion and that removed in the harvested portion of the crop. Phosphorus fertilization is an investment that will pay dividends for several years and should be considered a capital improvement to the land. Therefore, a program to build soil P to a level adequate for maximum crop yield potential and maintain it at this level will probably be the most profitable in the long-term and environmentally sound. Establishing a soil P level adequate to eliminate P as a deficient crop nutrient can be accomplished by one of two methods: 1) by applying a one-time application of P, either broadcast or band, that is sufficient to raise the soil test P level to an optimum level; or 2) by applying smaller rates of P, either broadcast or band, for several crop years.

### D. Nitrogen Required for Efficient Residual P Use

Adequate levels of N are essential to get full benefit from residual P fertilizer and efficient P utilization, regardless of the method of P application. Halvorson and Havlin (unpublished data) found that the addition of 45 kg N $ha^{-1}$ increased winter wheat response to residual fertilizer P at Morrill, Nebraska. They also found that initial P placement method had no effect on winter wheat response to residual P (Figure 9). Yield data from Montana (Black, 1982; Halvorson and

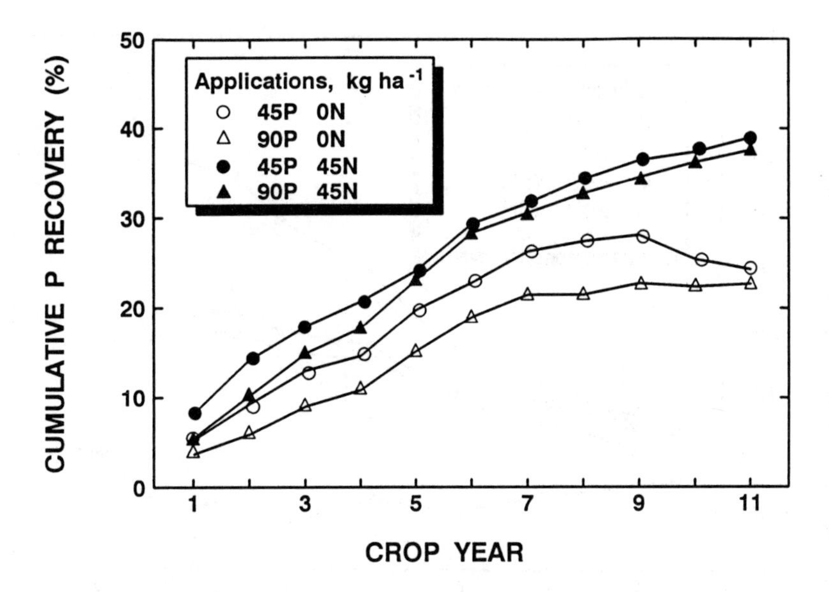


**Figure 10.** Cumulative P fertilizer recovery in wheat grain with harvest of each additional crop from a single P application with or without N applied each crop year. (Data adapted from Halvorson and Black, 1985b.)

Black, 1985a) also shows that N fertilization was needed to get optimum response of spring wheat to residual P fertilizer (Figure 7). Fertilizer P recovery also improved with each additional crop year (Figure 10). Thus, by having adequate P present and balancing the N needs of the crop based on yield potential, optimum yield and profit potentials can be realized.

Halvorson (1989) found that the presence of adequate levels of P also improved N uptake by irrigated winter wheat. Residual soil $NO_3$-N levels in the soil profile were significantly less where adequate P was present to optimize yield, thus, reducing the quantity of potentially leachable $NO_3$-N and ground water quality concerns. Phosphorus uptake and removal with the harvested grain generally increased as the soil $NO_3$-N plus fertilizer N level increased to an adequate level for maximum wheat yields. Estimated fertilizer P recovery of a single 67 kg P $ha^{-1}$ application in the fall of 1983 in the harvested grain of three winter wheat crops was 7.2, 22.4, 27.6, 26.0, and 23.3% for the 0, 34, 67, 134, and 268 kg N $ha^{-1}$ treatments, respectively. For the single 67 kg $ha^{-1}$ P application, cumulative P fertilizer recovery was 2.1%, 4.9%, and 7.3% without N and 7.6%, 19.1%, and 27.6% with 67 kg N $ha^{-1}$ added for 1984, 1985, and 1986, respectively. Thus, time and N fertilization significantly improved the recovery of fertilizer P in the harvested grain. The positive benefits of residual P fertilizer availability on irrigated and dryland crop yields demonstrate that P fertilizer-use efficiency needs to be evaluated over a longer period than just one

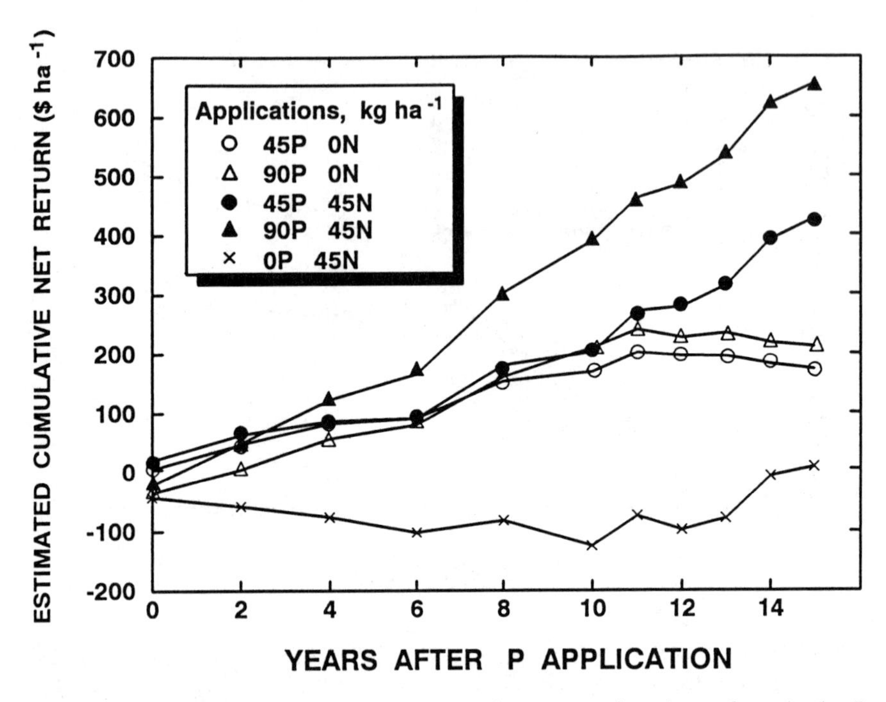


**Figure 11.** Cumulative net return with time as a function of a single P application with or without N applied each crop year. (Data adapted from Halvorson et al., 1986.)

crop year. It may need to be evaluated for more than 20 years, depending on P rate, soil type, and cropping system.

### E. Economics of Fertilizer Phosphorus Management

Many farmers today consider themselves economically stressed as a result of rising production costs while crop prices have remained relatively constant, despite federal price support programs. Current farm management emphasis is on increasing input-use efficiency. Adequate levels of plant nutrients are essential for obtaining optimum economic yields while protecting the environment. By soil testing, more accurate fertilizer recommendations can be made by giving credit for residual N and P in the soil profile, thus helping farmers achieve the required nutrient balance without over- or under-investing in fertilizer. This will require that soils previously receiving banded P fertilizer applications be properly sampled to insure that the soil test accurately reflects the true P status of the soil.

The short- (1 to 2 crop yr) and long-term (>2 crop yr) economics of P fertilization need to be considered (Figure 11). The long-term economics of a

**Table 4.** Percent of soils testing medium or less in available P in north-central states

| State | 1984 | 1986 |
|---|---|---|
| Illinois | 59/70 | 38/50 |
| Indiana | 31 | 22 |
| Iowa | 66 | 44 |
| Kansas | 61 | 54 |
| Kentucky | 75 | 58 |
| Michigan | 31 | 28 |
| Minnesota | 33 | 33 |
| Missouri | 55 | 55 |
| Nebraska | 69 | 69 |
| North Dakota | 72 | 75 |
| Ohio | 42 | 38 |
| South Dakota | 76 | 76 |
| Wisconsin | 38 | 34 |
| Average | 54 | 48 |

Data adapted from Potash and Phosphate Institute, 1985 and 1987a.

large one-time application of P fertilizer can be profitable on some soils (Jose, 1981; Halvorson et al., 1986; Wagar et al., 1986). However, the short-term profitability may be marginal for a one-time large P application (>90 kg P $ha^{-1}$). Jose (1981) concluded there was a long-term economic advantage of a single high rate broadcast P application over an equal quantity of P band applied to several crops at a lower rate at dryland sites in Canada. Crop price, fertilizer cost, and current soil test P level will govern how much P can be profitably applied in a given year. The investment in P fertilizer needs to be amortized over several years, similar to machinery, in order to maximize wheat yields and optimize responses to N fertilization. Fixen and Halvorson (1991) point out that length of land tenure, initial soil P level, expected response to P, and expected lifetime of the investment, influence the level of P fertilizer that can economically be applied.

## IV. Soil Testing for Phosphorus Availability

Soil testing is the best tool available to assess the need for fertilizer P. Soil test summaries for 1984 show that on an average, 54% of the soils in the north-central region tested medium or less in available P (Table 4). In 1986, the number had dropped to 48% indicating that some progress has occurred in reducing P deficiency. These data indicate that there are many soils, that if fertilized with P, would probably result in significant increases in crop yields.

In contrast, there are many soils in several eastern U.S. states that would not respond to P for many years (Table 1). Accurately assessing soil P availability and the quantity of P fertilizer required to alleviate any P deficiency becomes very important if maximum economic yields (MEY) are to be obtained. The current emphasis on the need for precise fertilizer rates to optimize grain yields necessitates that the short- and long-term economic and environmental impact of P fertilizer applications be evaluated.

Soil testing is essential if accurate, profitable-fertilizer recommendations are to be made. While soil testing doesn't directly tell us how much fertilizer P will be required for profitable crop production, proper interpretation of the analytical results will guide producers as to the amount of P required. Though questions about the validity of soil testing in predicting fertilizer needs have been raised, the basic problem probably stems not from the analytical results but from the interpretation of these results for making P recommendations (Leikam, 1987). Interpretation of these results and the development of recommended rates of P application should be based on research data (public and private), but the recommendations may need to be refined locally for each situation to obtain maximum benefit from the soil testing/recommendation process.

Soil testing success depends on collection of a representative soil sample and an accurate laboratory analysis. Many soil samples, from the appropriate soil depth, need to be collected from representative areas of the field. A field should be divided, avoiding odd or problem areas, into about 8 to 16 ha lots (40 ha lots may be more practical but less desirable) and 10 to 15 subsamples collected from each lot for an adequate composite sample. Different or special problem areas should be sampled separately. Clearly, a non-representative sample can be misleading and may be worse than no sample at all.

Soil test procedures to estimate plant available P commonly use a variety of chemical extraction methods, which are closely related to plant growth and uptake of P under certain climatic and soil conditions (Fixen and Grove, 1990). Alternative procedures, using anion exchange resin, isotopic $P^{32}$, buffer capacities, adsorption-desorption relationships, and quantity-intensity curves have been developed to assess plant available P. However, none has found widespread routine application, thus, the basic dilute solution extraction techniques used by soil test laboratories have changed little over the last 25 years. The most commonly used procedures are Bray (Bray and Kurtz, 1945), Mehlich (Mehlich, 1984), and Olsen (Olsen et al., 1954) extractants (Fixen and Grove, 1990).

Despite this lack of change in methodology, improvements in soil characterization, sampling, and fertilizer application techniques have made soil test P methodology and interpretation a weak link in the P recommendation process. Thus, further test development is necessary, particularly in assessing the sustainability of soil P fertility for different management strategies, which rely more heavily on precise fertilizer placement, non-incorporation of crop residues, and utilization of residual inorganic and organic P. These developments should assess both positional and chemical availability of P, soil test P-crop yield relationships, and the use of nonconventional soil test procedures.

### A. Positional Availability

The accumulation of P in specific soil horizons with fertilizer P banding, continuous manure applications, and reduced tillage, will present sampling problems to determine subsequent fertilizer P requirements. For example, if location of the fertilizer band is known, what portion of samples should be collected on and off the band and if the band's location is not known, is a random sampling strategy adequate? Collection of 15 (Ward and Leikam, 1986; Shapiro, 1988) to 30 random samples (Hooker, 1976) have been reported to adequately reflect P availability in fields where P bands exist.

Sampling strategies for minimum and no-till conditions would be similar to conventional tillage situations when P has been broadcast applied. However, the sampling strategies described by Kitchen et al. (1990) should be followed for reduced- and no-till situations where P has been banded. When location of the P bands are known, sampling involves one-in-the-band soil sample for every 20 or 8 between-the-band samples for 76- and 30-cm band spacing, respectively (Kitchen et al., 1990). When band location is unknown, paired sampling (approximately 10) where a first random sample and a second sample 50% of the band-spacing distance from the first sample perpendicular to band direction, reduces soil test P variability over completely random sampling.

### B. Chemical Availability

As shown in an earlier section, the mineralization of soil organic P can be an important process supplying available P in certain soils. Consequently, it is possible that soil P tests may be improved in certain situations by accounting for or giving credit to mineralizable organic P as well as inorganic P. Several studies have reported that potential soil P supply, as reflected by crop yields, was more closely estimated by including extractable organic P (Abbott, 1978; Adepetu and Corey, 1976; Bowman and Cole, 1978; Daughtrey et al., 1973). Bowman and Cole (1978) used a modification of the Olsen method (Olsen et al., 1954), which measured the total amount of P (inorganic plus organic) extracted by the reagent. Where other soil test P test methods are recommended, a similar adaptation may be used. As conditions of organic P extraction may not replicate the dynamic field conditions influencing organic P mineralization, caution must be used in relating amounts of extractable organic P to expected crop response.

With an increase in residual P levels and potential formation of less soluble P-rich compounds (Adepoju et al., 1982; Pierzynski et al., 1990), it is possible that current soil test P extractants and their interpretation may not adequately reflect residual P availability. It may also be necessary to develop procedures to credit soil test P levels for residual P fertilizer that had been previously banded when making P fertilizer recommendations.

### C. Soil Test-Crop Yield Relationships

The relationship between the sodium bicarbonate P test and relative yield potential of wheat grown in a dryland wheat-fallow system (Figure 8; Halvorson, 1986a) was discussed in section III.C. Application of adequate P fertilizer to bring the soil test P level to about 21 mg P $kg^{-1}$ (Olsen P test) on calcareous soils should optimize wheat yield potentials (Halvorson, 1986a, 1986b; Fiedler et al., 1987). The Bray-I P soil test is widely used for acid soils (pH <6.5) with similar soil test P levels (about 30 mg $kg^{-1}$) required to achieve maximum wheat yield potentials in the north-central Region (Agronomy Staff, 1989; Fiedler et al., 1987; Fixen, 1986; Oplinger et al., 1985). Bray-I soil test values considered adequate for wheat on acid soils range from about 25 to as high as 50 mg $kg^{-1}$, depending on yield goal and location. Usually, higher soil test values are desirable in northern areas where spring soil temperatures are cooler (Murphy, 1988). As more intensive crop management systems are adapted, particularly under reduced or no-till practices, higher soil test P levels may be required to optimize yields unless stratified P is redistributed within the root zone. In terms of efficient residual P utilization, it may be necessary to develop soil test-yield relationships for different soils by establishing different soil test P levels with one-time applications of variable P rates up to at least 200 kgP $ha^{-1}$, and then measuring crop response to each residual P level, using optimum farm management practices such as those used with the MEY concept.

### D. Development of Nonconventional Procedures

The benefits of several nonconventional soil test methods should be evaluated further. For example, quantity (concentration of sorbed P) and intensity (solution P concentration) factors in soil P test procedures (Kuo, 1990; Moody et al., 1988), as well as the use of resin accumulators (Skogley et al., 1990; Yang et al., 1991) or iron oxide-impregnated paper strips ($Fe_2O_3$ strips) as a sink for plant available P (Menon et al., 1989a,b) have been proposed. Anion exchange resins more closely simulated soil P removal by plant roots and their action is generally independent of soil type; however, their widespread use in routine soil testing has been limited by slow and cumbersome methodology. Although exchange resins in membrane and spherical forms have been used in biochemical and medical research for some time, they have only recently received attention in soil testing (Sagger et al., 1990; Schoenau and Huang, 1991).

The amount of P extracted by $Fe_2O_3$ strips (strip P), was more closely related to both dry matter yield and P uptake of maize than Bray-I P for four soils ranging in pH from 4.5 to 8.2 (Menon et al., 1989a,b,c). Sharpley (1991) reported the $Fe_2O_3$ strips removed primarily physically-bound P (anion exchange resin P) from 203 soils representing all soil orders. As strip P was closely related to different P tests for soils on which the use of the test is recommended and with resin P (Sharpley, 1991), it is possible that the $Fe_2O_3$

strip may extract amounts of P closely related to plant availability for soils ranging widely in physical and chemical properties.

The close correlation between strip P and soil test P does not in itself justify adoption of the procedure to quantify P uptake. However, it emphasizes the potentially wide applicability of strips to estimate plant available soil P, including residual P, and suggests further evaluation of the method is warranted. For example, P extracted by $Fe_2O_3$ strips embedded in soil columns, was closely related to Bray-I P for acid and Olsen P for alkaline and calcareous soils (Menon et al., 1990). The potential use of the strips as a nondestructive method to measure *in-situ* soil P availability, may be of value in estimating banded residual P availability. In addition, dry matter yield and P uptake by maize from soils treated with rock P (RP) and partially acidulated RPs, was more closely correlated with strip P ($r^2$ = 0.83 and 0.88, respectively) than Bray-I P($r^2$ = 0.53 and 0.45, respectively) (Menon et al., 1989a). Acid extractants like Bray-I reagent can overestimate P from such soils by dissolving more P than would be available for plant use, whereas Olsen could underestimate P (Chien, 1978; Mackay et al., 1984). Consequently, strip P was more effective than other soil P tests in evaluating P availability from different RPs applied to soil.

## V. Transport of P in Runoff

### A. Forms

The transport of P in runoff can occur in soluble (SP) and particulate (PP) forms. Particulate P encompasses all solid phase forms, which includes P sorbed by soil particles and organic matter eroded during runoff and constitutes the major portion of P transported from conventionally tilled land (75-90%). Runoff from grass or forest land carries little sediment, and is, therefore, generally dominated by the soluble form. While SP is, for the most part, immediately available for biological uptake (Nurnberg and Peters, 1984; Peters, 1977; Walton and Lee, 1972), PP can provide a long-term source of P for aquatic plant growth (Bjork, 1972; Carignan and Kalff, 1980; Wildung et al., 1974).

In the past, most studies have measured only SP and total P (TP) transport in runoff. However, estimation of bioavailable P (BAP) transport in runoff is needed to estimate more accurately the impact of agricultural management practices on the biological productivity of surface waters. Bioavailable P represents P that is potentially available for algal uptake and is comprised of SP plus bioavailable PP (BPP). Although BPP can be quantified by algal culture tests (US EPA, 1971), these assays generally involve long-term incubations (100 days) and, thus, do not lend themselves to routine analysis. Hence, more rapid chemical extraction procedures have been used to simulate utilization of PP by algae (Hegemann et al., 1983; Sonzogni et al., 1982). Chemical extractants that have been used to measure the BPP content of eroded soil material are NaOH (Butkus et al., 1988; Logan et al., 1979; Sagher et al., 1975), $NH_4F$ (Dorich et

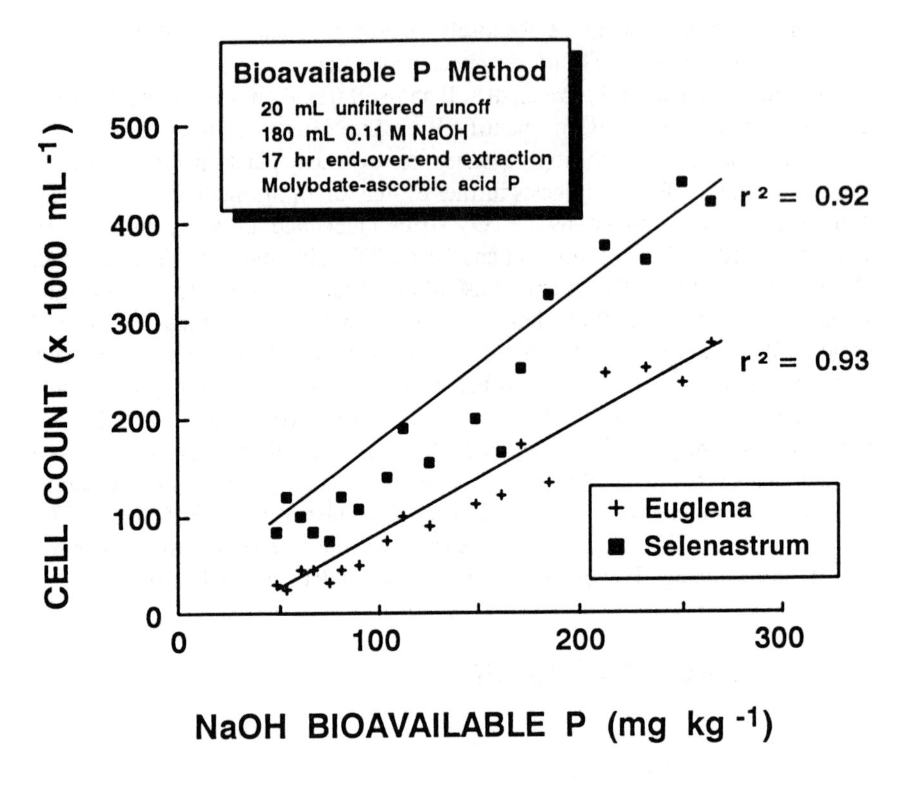


**Figure 12.** Relationship between the bioavailable P content of runoff sediment, extraction by NaOH and growth of P-starved algae during a 29-day incubation.

al., 1980; Porcella et al., 1970), and ion exchange resins (Armstrong et al., 1979; Hanna, 1989; Huettl et al., 1979).

Use of these methods has shown a wide range in percent bioavailability of P in suspended and deposited lake sediments (Sharpley and Menzel, 1987). However, these extractions require collection of a large volume of runoff to provide an adequate amount of sediment for analysis and are, thus, not applicable to the routine measurement of BPP transport in runoff. As a result there is a lack of information on the effect of agricultural management on BAP transport in runoff. To fill this gap, a chemical extraction procedure to estimate the BPP content of sediment transported in runoff was proposed by Sharpley et al (1991a). This procedure is a modification of the method of Dorich et al. (1985) and involves extraction of unfiltered runoff with 0.1 *M* NaOH for 17 hours (Sharpley et al., 1991a). Using this method, BPP contents of runoff from 9 Oklahoma watersheds, were closely related to the growth of P-starved algae (*Selenastrum capricornutum*) incubated with the same runoff sediments (Figure 12). Consequently, it is suggested that the extraction of P from unfiltered runoff by 0.1 *M* NaOH can be used as a rapid and interference-free method to routinely

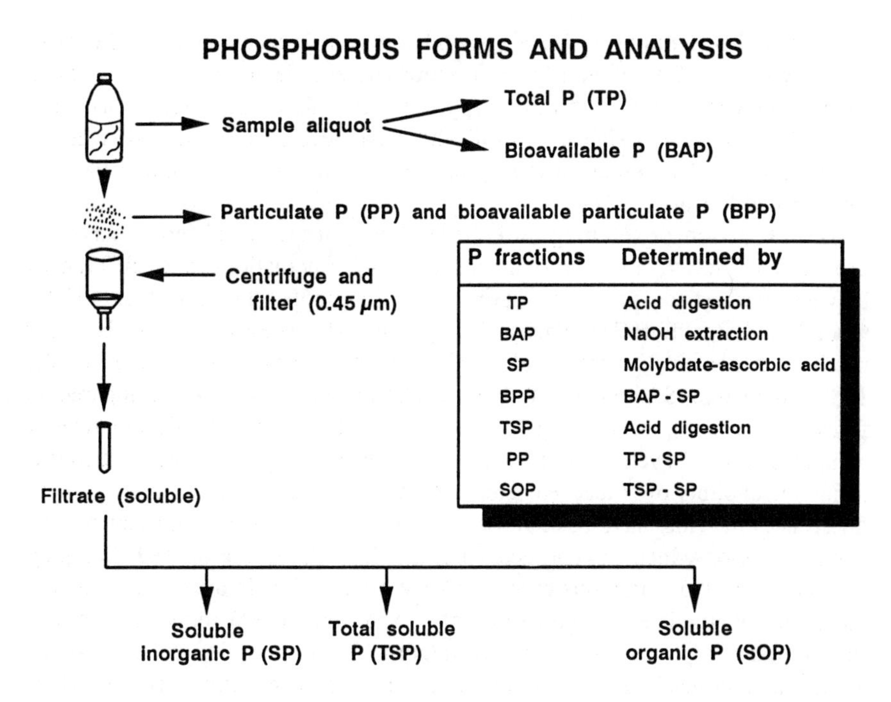


| P fractions | Determined by |
|---|---|
| TP | Acid digestion |
| BAP | NaOH extraction |
| SP | Molybdate-ascorbic acid |
| BPP | BAP - SP |
| TSP | Acid digestion |
| PP | TP - SP |
| SOP | TSP - SP |

**Figure 13.** Phosphorus forms and their analysis commonly used in agricultural runoff studies.

estimate the BPP and BAP (BPP plus SP) in runoff. This method has been included in a schematic summary of P forms and analysis commonly used in runoff studies (Figure 13).

## B. Processes

### 1. Soluble P

The transport of SP in runoff is initiated by the desorption, dissolution, and extraction of P from soil and plant material (Figure 2). These processes occur as a portion of rainfall interacts with a thin layer of surface soil before leaving the field as runoff. Simulated rainfall studies have shown this layer to range from 1 to 5 mm in depth (Sharpley 1985b). Although this depth has not been measured under field conditions, it is expected to be highly dynamic due to variations in rainfall intensity, soil tilth, and vegetative cover. The remaining rainfall percolates through the soil profile where sorption of P by P-deficient subsoils generally results in low concentrations of soluble P in subsurface flow. Exceptions may occur in organic or peaty soils, where organic matter may accelerate the downward movement of P together with organic acids and Fe and

Al (Fox and Kamprath, 1971; Singh and Jones, 1976; Duxbury and Peverly, 1978; Miller, 1979). Similarly, P is more susceptible to movement through sandy soils with low P adsorption capacities (Ozanne et al., 1961; Adriano et al., 1975; Sawhney, 1977) and in soils that have become waterlogged, where a decrease in Fe (III) content occurs (Ponnamperuma, 1972; Gotoh and Patrick, 1974; Khalid et al., 1977).

The extraction or leaching of P from plant material in different stages of growth and decay, may account for seasonal fluctuations and differences between watersheds in the transport of SP in runoff (Burwell et al., 1975; Gburek and Heald, 1974). This will be particularly true for watersheds under no till management (Barisas et al., 1978; Langdale et al., 1985; Pesant et al., 1987). Increased SP loss in runoff from alfalfa plots (33 g $ha^{-1}$) compared to forested (4 g $ha^{-1}$), corn (11 g $ha^{-1}$) and oat plots (16 g $ha^{-1}$), during several simulated rainfall events (7.4 to 12.2 $cm^{-1}$) over a 2-yr period, were attributed to larger amounts of P leached from alfalfa (Wendt and Corey, 1980). These differences in P loss have been partially explained by studies of nutrient release from vegetation which was cut, decaying, dried, and/or freeze-thawed (Sharpley and Smith, 1989a; Timmons et al., 1970; White, 1973). In a study of growing plants under simulated rainfall (6.4 $cm^{-1}$), Sharpley (1981) found that cotton, sorghum, and soybean plants could maintain a SP concentration of 0.02 to 0.13 mg $L^{-1}$ in plant leachate (i.e., rainfall intercepted by vegetation). The contribution of plant leachate P to runoff losses was subsequently calculated from the difference in SP concentration of planted and bare soil. For mature plants (40 days after planting), receiving 50 or 100 kgP $ha^{-1}$, leached SP accounted for approximately 20% of SP transported in runoff for each crop (0.036 to 0.087 mg $L^{-1}$). However, when the plants were P deficient (no fertilizer P applied) or reached senescence (80 days after planting), plant leachate was the major source of P transported in runoff (up to 90% of 0.024 to 0.154 mg $L^{-1}$).

Many studies have investigated soil P desorption in relation to soil fertility and water quality, using a wide range in ionic strength and species of extracting medium (Sharpley and Menzel, 1987). However, few have used filtered runoff or lake water (Bahnick, 1977; Barlow and Glase, 1982) as the extracting medium, due to difficulties in preparing large volumes of filtrate of constant chemical composition. As P desorption is a function of the type of extracting medium (Ryden and Syers, 1977) and solution:soil ratio used (Hope and Syers, 1976; Barrow and Shaw, 1979), a standardized method is needed to relate P desorption to its transport and potential bioavailability in runoff. One such method may involve the use of $Fe_2O_3$ strips as a sink for algal available P. These strips have been used to investigate the desorption kinetics of reversibly adsorbed P in Dutch soils (Van der Zee et al., 1987) and more recently, successfully applied to the estimation of plant available soil P (Menon et al., 1989a, b).

## 2. Particulate P

As the sources of PP in streams include eroding surface soil, streambanks, and channel beds, processes determining soil erosion also control PP transport (Figure 2). In general, the P content and desorption-adsorption potential of eroded particulate material is greater than that of source soil, due to preferential transport of clay-sized material ($<2$ $\mu$m). This has led to the determination of enrichment ratios (ER) for P, calculated as the ratio of the concentration of P in the sediment (eroded soil) to that in the source soil (Knoblauch et al., 1942; Neal, 1944; Rogers, 1941; Stoltenberg and White, 1953). More recently, Sharpley (1985c) observed that the enrichment of available P (Bray-l - 2.45 and labile P - 2.89) was greater than for other P forms (total, inorganic, and organic) (1.48) for 6 soils using simulated rainfall. The relatively greater enrichment of available than total P forms was attributed to less aggregation of runoff sediment compared to source soil reducing the physical protection of P. Phosphorus reactivity, in terms of the desorption-adsorption characteristics; buffer capacity (1.49), sorption index (1.56), and equilibrium P concentration ($EPC_0$) (1.80), were also enriched in runoff sediment compared to source soil. The $EPC_0$ is the SP concentration of water due to sediment sorption and desorption processes.

Processes controlling PP bioavailability are more complex than for PP transport, due to particle size sorting during transport and the contribution of less dense organic material to bioavailable P. Thus, the selective transport and enrichment of organic and fine material in runoff will increase PP bioavailability. This will be of particular significance to situations where manure has been applied. Despite a lack of information, increases in P bioavailability in runoff will be expected following manure application due to the increased transport of low density organic material and high solubility of manure P. However, the magnitude of these increases may differ between types of manure as a function of its density, form (wet or dry), organic matter content, and P content. Clearly, the dynamic nature of these interactions must be considered in determining both the short- and long-term potential of runoff to increase the biological productivity of receiving water bodies.

## 3. Changes in Bioavailablity during Transport

Transformations between SP and PP, occurring during transport in stream flow, can alter both the amount and bioavailability of P entering a lake, compared to edge-of-field losses (Figure 14). These transformations are accentuated by the selective transport of fine materials, which have a greater capacity to sorb or desorb P and will thus be important in determining the bioavailability of P transported. In addition, P may be taken up by aquatic biota and PP deposited or eroded from the stream bed with a change in stream flow (Meyer, 1979; Vincent and Downes, 1980). The direction and extent of P exchange between

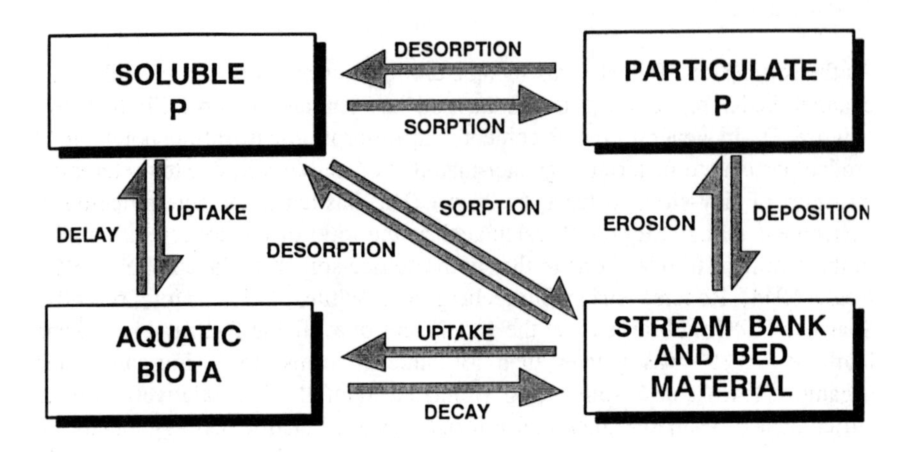


**Figure 14.** Phosphorus transformations during transfer from terrestrial to aquatic ecosystems.

SP and PP will depend on their relative concentration in stream flow, $EPC_0$ of the sediments contacted, and rate of stream flow. Several studies have shown that sediments rapidly sorb added P and concluded that the close agreement between $EPC_0$ and the SP concentration of runoff and stream flow indicated that sediments may determine SP concentration (Klotz, 1988; Meyer, 1979; Taylor and Kunishi, 1971). For example, if the SP concentration of stream flow falls below the $EPC_0$ of the particulate material contacted, P will be desorbed. However, if SP concentration increases above, the $EPC_0$, P may be sorbed. Soluble P concentrations of 0.10 to 0.13 mg $L^{-1}$ of runoff from fertilized fields were reduced to 0.009 mg $L^{-1}$ by sorption during movement downstream (Kunishi et al., 1972).

Clearly, changes in P bioavailability can occur between the point where it leaves a field to where it enters a water body. Consequently, the extent to which transformations between SP and PP occur during stream flow must be considered in assessing the impact of P transported in runoff as a function of agricultural management on the potential biological productivity of a receiving lake.

### C. Amounts Transported

Amounts of P transported in runoff from uncultivated or pristine land is considered the background loading, which cannot be reduced. These inputs determine the natural status of a lake and, as will be seen later, may be sufficient to cause eutrophication. As we try to assess the impact of agricultural management on P loss in runoff, it becomes clear that little quantitative

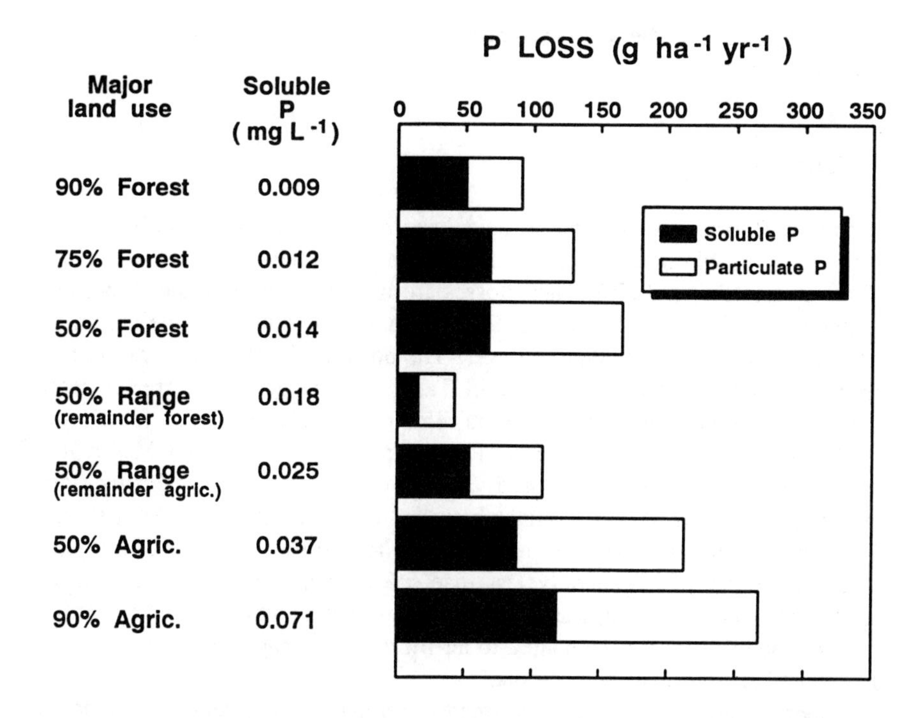


**Figure 15.** Phosphorus loss in runoff as a function of land use in the U.S. (Data adapted from Omernik, 1977.)

information is available on background losses of P from a given location before cultivation. Consequently, it is still difficult to quantify any increase in P loss following cultivation. These problems result mainly from the expensive and labor intensive nature of water quality monitoring studies, which are site-specific and impossible to replicate, due to spatial and temporal variations in climatic, edaphic, and agronomic conditions. Despite these problems, an investigation of published studies enables generalizations about the effect of agricultural management on P transport in runoff. Several surveys of U.S. watersheds (Omernik, 1977; Rast and Lee, 1978), have clearly shown that P loss in runoff increases as the portion of the watershed under forest decreases and agriculture increases (Figure 15). The loss of P from forested land tends to be similar to that found in subsurface or base flow from agricultural land (Ryden et al., 1973; House and Casey, 1988). In general, forested watersheds conserve P, with P input in rainfall usually exceeding outputs in stream flow (Hobbie and Likens, 1973; Schreiber et al., 1976; Taylor and Kunishi, 1971). As a result forested areas are often utilized as buffer or riparian zones around streams or water bodies to reduce P inputs from agricultural land (Lowrance et al., 1984a, b; 1985). However, the potential loss of P from agricultural land is to a large extent dependent on the relative importance of surface and subsurface runoff in the watershed.

### 1. Surface Runoff

Increased P loss in surface runoff have been measured after the application of fertilizer P (Table 5) and manure (Gilbertson et al., 1970; Holt et al., 1970; Sharpley and Syers, 1976). These losses are influenced by the rate, time, and method of application; form of fertilizer or manure; amount and time of rainfall after application; and vegetative cover. The portion of fertilizer P transported in runoff for the studies reported in Table 5, was generally greater from conventional compared to conservation tilled watersheds. However, fertilizer P application to no till corn reduced PP transport (McDowell and McGregor, 1984), probably due to an increased vegetative cover afforded by fertilization. Although it is difficult to distinguish between losses of fertilizer and native soil P, without the use of expensive and hazardous radiotracers, the loss of fertilizer P in surface runoff is generally less than 5% of that applied. Often, manure application rates are so large as to cause concern that a relatively greater portion of P will be lost in runoff compared to fertilizer and the capacity of soil to retain P against leaching may be exceeded.

A more detailed evaluation of the impact of tillage management on the amount and bioavailability of P transported in surface runoff is given for an intensive monitoring program of 28 watersheds in Oklahoma and Texas (Sharpley et al., 1992; Smith et al., 1991). These watersheds are representative of agricultural practices in the Southern Plains, with fertilizer P applications recommended by soil test P levels. Mean annual soil and P losses from sorghum, wheat, and native grass are summarized for a 5-yr period (1984-1989) in Figure 16. Soil loss from sorghum and wheat and P loss from wheat where similar amounts of P were added were appreciably lower from conservation than conventional tillage practices (Figure 16). However, P bioavailability increased. For example, mean annual SP concentrations were significantly greater from conservation compared to conventionally tilled watersheds, even though similar amounts of fertilizer P were applied to wheat. Further, the portion of PP that was bioavailable increased with an increase in vegetative soil cover. As a result, a greater portion of P transported from native grass and no till practices was bioavailable (SP plus BPP, 46 to 85%) than that from conventionally tilled practices (21 to 32%) (Figure 16). This difference results from a greater portion of finer-sized particles transported in runoff with reduced soil loss, leaching of crop residue P, and a build up of P in the surface soil with reduced mechanical mixing of the top soil under conservation tillage. Thus, it should be recognized that an increase in SP and BPP transport as a result of conservation tillage practices, may not bring about as great a reduction in the trophic status of a water body as may be expected from inspection of total loads only.

**Table 5.** Effect of fertilizer P on the loss of P in surface and subsurface runoff

| Land use | P applied | Concentration Sol. | Concentration Partic. | Amount Sol. | Amount Partic. | Fertilizer loss Sol. | Fertilizer loss Partic. | Reference |
|---|---|---|---|---|---|---|---|---|
| | kg ha$^{-1}$ yr$^{-1}$ | mg L$^{-1}$ | | kg ha$^{-1}$ yr$^{-1}$ | | % | | |
| ***Surface Runoff*** | | | | | | | | |
| Grass | 0 | 0.18 | 0.24 | 0.50 | 0.67 | | | Sharpley and Syers |
| | 50 | 0.98 | 0.96 | 2.80 | 2.74 | 4.6 | 4.1 | (1979), New Zealand |
| Grass | 0 | 0.01 | 0.06 | 0.01 | 0.20 | | | McColl et al. (1977), |
| | 75 | 0.03 | 0.14 | 0.04 | 0.29 | 0.04 | 0.1 | New Zealand |
| No-till corn | 0 | 0.23 | 0.46 | 1.10 | 2.20 | | | McDowell and |
| grain | 30 | 0.57 | 0.51 | 1.80 | 1.60 | 2.3 | 27.3[a] | McGregor (1984), |
| | | | | | | | | Mississippi |
| Conventional | 15 | 0.07 | 3.57 | 0.30 | 15.10 | | | |
| corn | 30 | 0.11 | 9.71 | 0.20 | 17.50 | +3.3[a] | 16.0 | |
| Contour corn | 40 | 0.19 | 0.71 | 0.12 | 0.45 | | | Burwell et al. (1977), |
| | 66 | 0.25 | 1.27 | 0.15 | 0.76 | 0.1 | 1.2 | Minnesota |
| Wheat-summer | 0 | 0.30 | 1.80 | 0.20 | 1.40 | | | Nicholaichuk and |
| fallow | 54 | 3.70 | 7.40 | 1.20 | 2.90 | 1.9 | 2.8 | Read (1978), |
| | | | | | | | | Western Canada |
| ***Subsurface Runoff*** | | | | | | | | |
| Alfalfa | 0 | 0.18 | - | 0.12 | - | | | Bolton et al. (1970), |
| (tile drainage) | 29 | 0.21 | - | 0.19 | - | 1.0 | - | Canada |
| Continuous corn | 0 | 0.02 | 0.10 | 0.13 | 0.29 | | | Culley and Bolton |
| | | | | | | | | (1983), Canada |
| (tile drainage) | 30 | 0.11 | 0.36 | 0.20 | 0.42 | 0.2 | 0.4 | |
| Bluegrass sod | 0 | 0.02 | 0.15 | 0.06 | 0.09 | | | |
| | 30 | 1.01 | 3.29 | 0.16 | 0.21 | 0.3 | 0.4 | |
| Oats | 0 | 0.02 | 0.09 | 0.10 | 0.19 | | | |
| | 30 | 0.42 | 1.10 | 0.20 | 0.30 | 0.3 | 0.4 | |
| Alfalfa | 0 | 0.02 | 0.11 | 0.12 | 0.20 | | | |
| | 30 | 0.37 | 1.03 | 0.20 | 0.31 | 0.3 | 0.3 | |

[a] Percent decrease in P loss from fertilized compared to check treatment.

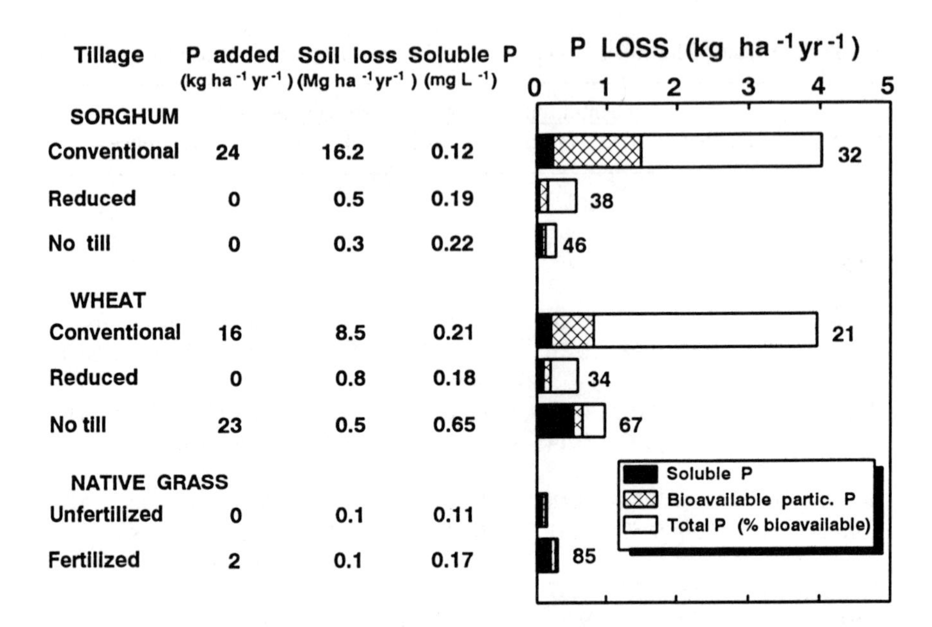


**Figure 16.** Phosphorus loss in runoff as a function of tillage management of several Southern Plains watersheds, averaged over a 5-year period.

2. Subsurface Runoff

The loss of P in subsurface runoff is appreciably lower than that in surface runoff, because of sorption of P from infiltrating water as it moves through the soil profile (Table 5). Subsurface runoff includes tile drainage and natural subsurface runoff, where tile drainage is percolating water intercepted by artificial drainage systems, such as tile or mole drains, thus accelerating its movement into streams. In general, P concentrations and losses in natural subsurface runoff are lower than in tile drainage (Table 5), due to the longer contact time between subsoil and natural subsurface flow than tile drainage, enhancing SP removal. Increased sorption of P from percolating water also accounted for lower TP loads from 1.0 to 0.6 m deep tiles draining a Brookston clay soil under alfalfa (Figure 17, from Culley et al., 1983). For the shallower drains, TP loads were about 1% of fertilizer P applied, whereas 1 m deep tiles exported about 0.6% of that applied (Figure 17).

Subsurface drainage of a soil can reduce P loss in runoff, via enhanced infiltration and thereby decreased runoff volumes. For example, Bengtson et al. (1988) found annual TP loss in runoff from a Commerce clay loam under corn in the lower Mississippi Valley, was reduced an average 36% in 6 yr following subsurface drainage with 1 m deep tiles (5.0 kg $ha^{-1}$ $yr^{-1}$) compared to undrained soil (7.8 kg $ha^{-1}$ $yr^{-1}$) (Figure 18). Of this TP loss reduction, only 8% was exported in tile drainage (0.3 kg $ha^{-1}$ $yr^{-1}$) (Bengtson et al., 1988). The reduction

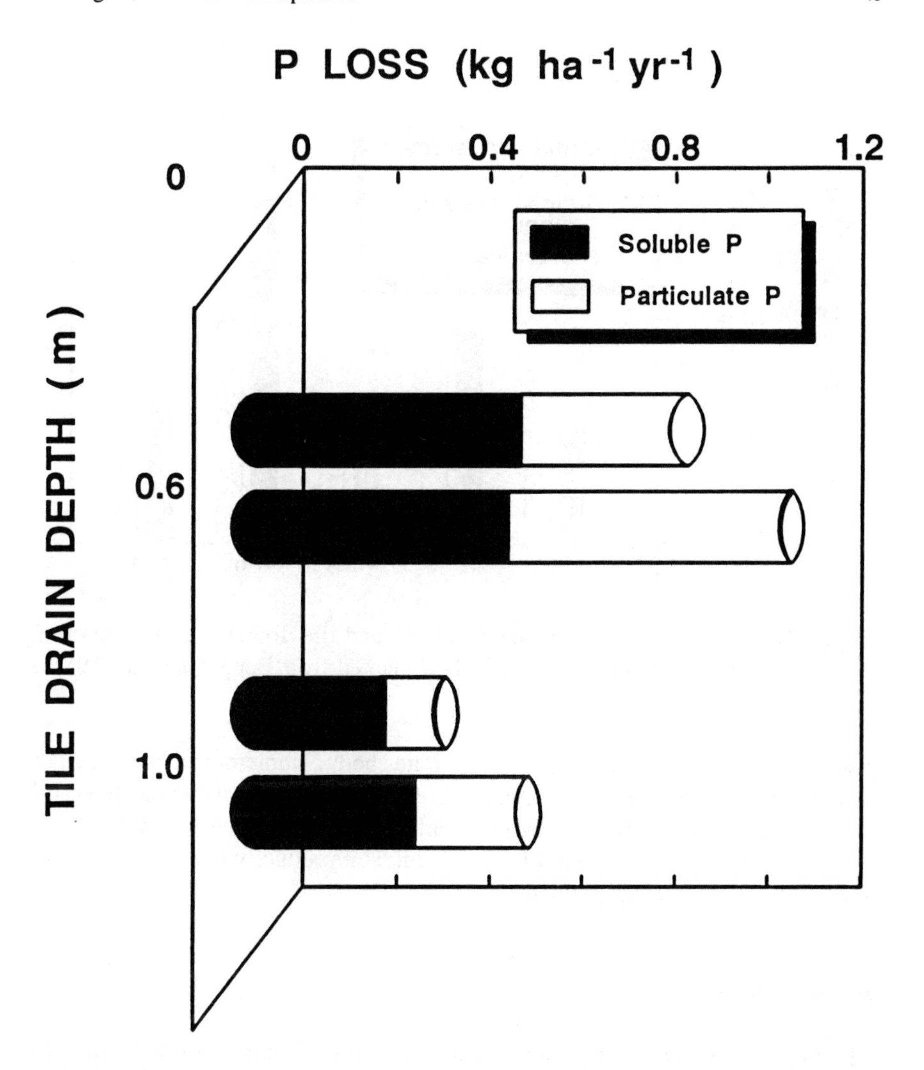


**Figure 17.** Phosphorus loss in tile drainage from a Brookston clay loam under alfalfa in Ontario, Canada as a function of tile drain depth. (Data adapted from Culley et al., 1983.)

in P loss must be weighed in terms of a potential increase in $NO_3$-N loss in tile drained fields (Bengtson et al., 1988).

## VI. Predicting Phosphorus Availability and Transport

Accurate predictions of soil P availability and its transport in runoff are required to evaluate the relative effects of current and proposed agricultural management

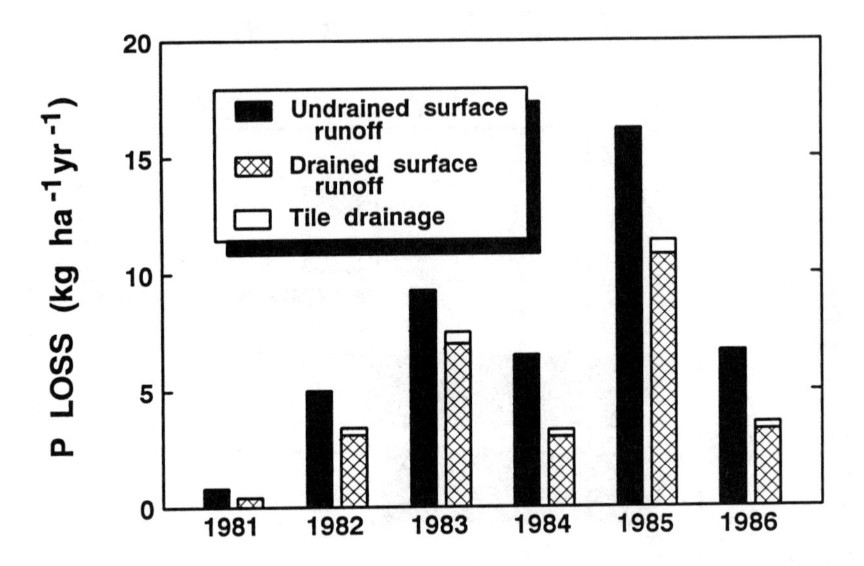


**Figure 18.** Phosphorus loss in surface runoff and tile drainage from a drained and undrained Commerce clay loam. (Data adapted from Bengston et al., 1988.)

practices on the biological response of a water body. Numerous comprehensive models have been developed to simulate the fate of agricultural chemicals in soil and their transport to surface waters, with the purpose of aiding selection of management practices capable of maximizing soil productivity and minimizing associated water quality problems.

### A. Soil Availability

Claassen and Barber (1974) developed a conceptual model to simulate soil nutrient flux to plant roots, which has been used to predict P uptake by corn (Schenk and Barber, 1979a) and soybean (Silberbush and Barber, 1983). Although the model does not simulate P cycling in inorganic and organic pools, it has provided a quantitative assessment of the importance of several factors, such as root morphology (Schenk and Barber, 1979a, b), distribution (Kuchenbuch and Barber, 1987), soil temperature and water content (Barber et al., 1988; Mackay and Barber, 1984), liming (Alder and Barber, 1991), and fertilizer placement (Anghioni and Barber, 1980; Kovar and Barber, 1987) on the availability of P to several crops.

A detailed model of P cycling in grassland soils was developed by Cole et al. (1977). Plant uptake of soil solution P is assumed to be limited by diffusion. The solution P pool is replenished from a labile P pool, which in turn is supplied by slightly soluble P minerals, adsorbed P, and organic P mineralization. Many of the concepts used in this and soil organic matter models (Parton et al., 1983) were used in developing the Century model (Parton et al., 1987; 1988), which simulates the dynamics of C, N, P, and S cycling in the soil plant system using a monthly time step. These complex, long-time frame models, incorporate the effects of moisture, temperature, soil properties, plant phenology, and organic matter decomposition on nutrient flows and simulate many soil and plant processes. However, their use is limited by the availability of detailed soil and plant data. Even so, they have revealed gaps in our knowledge of processes such as the contribution of organic matter cycling to long-term soil productivity as a function of management (Parton et al., 1988). Thus, these models provide valuable direction for future research.

A user-oriented model, "Decide," was developed by Bennett and Ozanne (1972) and Helyar and Godden (1976) to simulate residual fertilizer-P recovery and used to give advice on fertilizer use to farmers. The model constructs a fertilizer response curve for each farming situation by combining research information with the farmers knowledge of past soil fertility and future yield goals. The model was developed using highly weathered and leached soils of western Australia and is presently being modified for use on slightly weathered soils of eastern Australia. Thus, the model only estimates fertilizer P recommendations for a limited range of soil types.

A more comprehensive model simulating soil P availability and fertilizer requirements (Jones et al., 1984b), is included in the EPIC model, which is composed of physically based components for simulating erosion, plant growth and related processes, and economic components to assess the cost of erosion and determining optimum management strategies (Erosion-Productivity Impact Calculator, Sharpley and Williams, 1990; Williams et al., 1984). The P model simulates uptake and transformations between several inorganic and organic pools in up to 10 soil layers of variable soil thickness. Fertilizer P is added to the labile inorganic P pool which rapidly achieves equilibrium with active inorganic P. Crop uptake from a soil layer is sensitive to crop P demand and the amounts of labile P, soil water, and root in the layer. Stover and root P are added to the fresh organic P pool upon their death and/or incorporation into the soil. Decomposition of fresh and stable organic matter may result in net immobilization of labile P or net mineralization of organic P. Regression equations were developed to estimate labile P from soil test P, organic P from total N or organic C, and fertilizer availability index from soil chemical and taxonomic characteristics (Sharpley et al., 1984a). Thus, except for soil test P, the minimum data set required to run the P model can be obtained from soil survey information.

Simulation of soil P transformations and availability under various long-term management strategies have been evaluated (Jones et al., 1984a). Changes in surface soil labile (i.e., Olsen P) and organic P contents of several Great Plains soils observed by Haas et al. (1961), were in most cases, accurately simulated by the EPIC model (Table 6). Measured and simulated values after the period of cultivation were not significantly different (at the 0.10 probability level). The

**Table 6.** Measured changes in surface soil organic and labile P within the Great Plains and changes simulated by the EPIC model

| Location | Duration of study | Rotation | Organic P | | | Labile P | | |
|---|---|---|---|---|---|---|---|---|
| | | | | Cultivated | | | Cultivated | |
| | | | Virgin | Meas. | Sim. | Virgin | Meas. | Sim. |
| | years | | mg kg$^{-1}$ | | | | | |
| Havre, MT | 31 | SWF[a] | 157 | 102 | 108 | 11 | 13 | 15 |
| Moccasin, MT | 39 | " | 308 | 183 | 169 | 14 | 14 | 20 |
| Dickinson, ND | 41 | " | 292 | 148 | 174 | 10 | 12 | 14 |
| Mandan, ND | 31 | " | 139 | 132 | 97 | 9 | 12 | 7 |
| Sheridan, WY | 30 | " | 120 | 93 | 86 | 12 | 14 | 9 |
| Laramie, WY | 34 | " | 142 | 91 | 96 | 13 | 24 | 9 |
| Akron, CO | 39 | WWF[a] | 115 | 82 | 81 | 26 | 45 | 19 |
| Colby, KS | 31 | " | 158 | 61 | 92 | 34 | 30 | 27 |
| Hays, KS | 30 | W. Wheat | 174 | 97 | 108 | 11 | 40 | 8 |
| Lawton, OK | 28 | " | 128 | 71 | 73 | 8 | 9 | 8 |
| Dalhart, TX | 29 | Maize | 84 | 39 | 53 | 17 | 13 | 9 |
| Big Spring, TX | 41 | W. Wheat | 55 | 30 | 29 | 12 | 12 | 6 |
| Mean | 34 | | 156 | 94 | 97 | 15 | 20 | 13 |

[a] SWF and WWF represent spring wheat-fallow and winter wheat-fallow, respectively.
Data adapted from Haas et al. (1961).

slight overestimation of P forms may be due in part to the fact that soil erosion by wind and water, was kept minimal during the simulation. The decline in residual soil P availability (Olsen P) in 12 years following a single broadcast P application to a Williams loam under spring wheat in the Great Plains (Black, 1982), was also reliably simulated by the EPIC model (Figure 19). These examples illustrate the potential use of the model in the effect of long-term soil management of processes influencing soil P availability.

### B. Transport in Runoff

Numerous comprehensive mathematical models have been developed to simulate the transport of agricultural chemicals in runoff, with the purpose of evaluating the relative effectiveness of management practices to minimize transport. These include, but are not limited to, ANSWERS (Areal Nonpoint Source Watershed Environment Response Simulation, Beasley et al., 1985), AGNPS (Agricultural Nonpoint Pollution Model, Young et al., 1989), GAMES (Guelph Model for Evaluating the Effects of Agricultural Management Systems on Erosion and

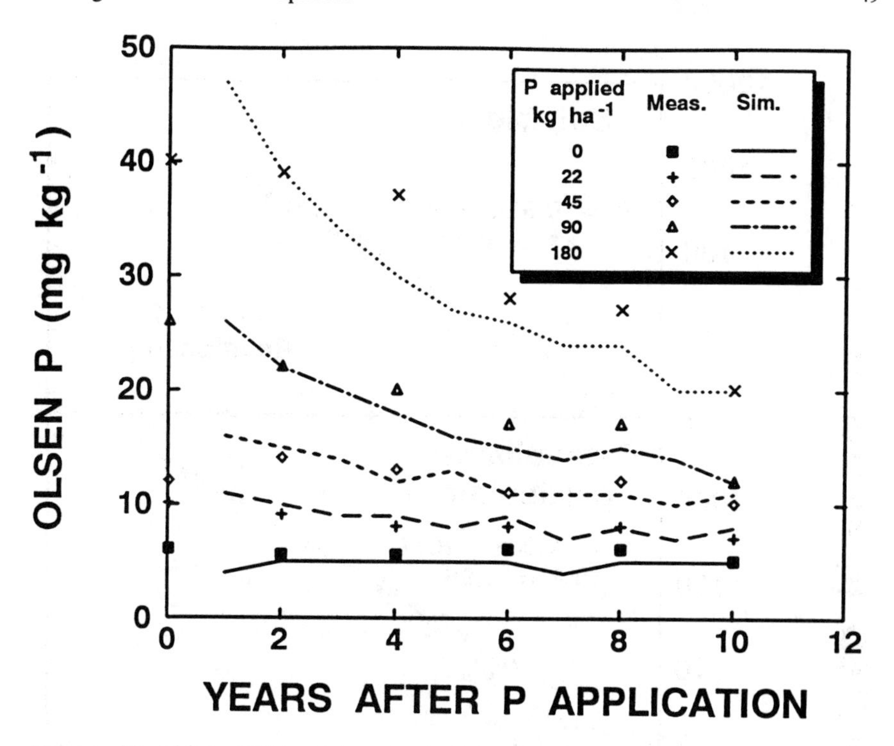


**Figure 19.** Effect of a single broadcast fertilizer P application on Olsen P content of a Williams loam under spring wheat at Culbertson, MT. (Data adapted from Black, 1982.)

Sedimentation, Cook et al., 1985), ARM (Agricultural Runoff Model, Donigian et al., 1977), and EPIC (Erosion-Productivity Impact Calculator, Sharpley and Williams, 1990).

Although physically-based descriptions of the various transport processes are used, the lack of data to drive the models and limited field data for testing has resulted in an over-simplified representation of P transport processes. In particular, equilibrium extraction coefficients have generally been used to predict SP, BPP has been assumed to be a constant proportion of total P, and no attempts have been made to predict BPP. Conceptually-based equations have been developed to describe the chemical and physical processes involved in the release from soil and transport of SP and BPP in runoff; they are described below. Soluble P, PP, BPP, and BAP concentrations were predicted using these equations and compared with values measured in runoff from 28 watersheds in Oklahoma and Texas, over a 5-yr period (Sharpley et al., 1992). These comparisons are presented (Figure 20) on a mean annual loss basis, which were calculated from concentrations and volumes of each runoff event.

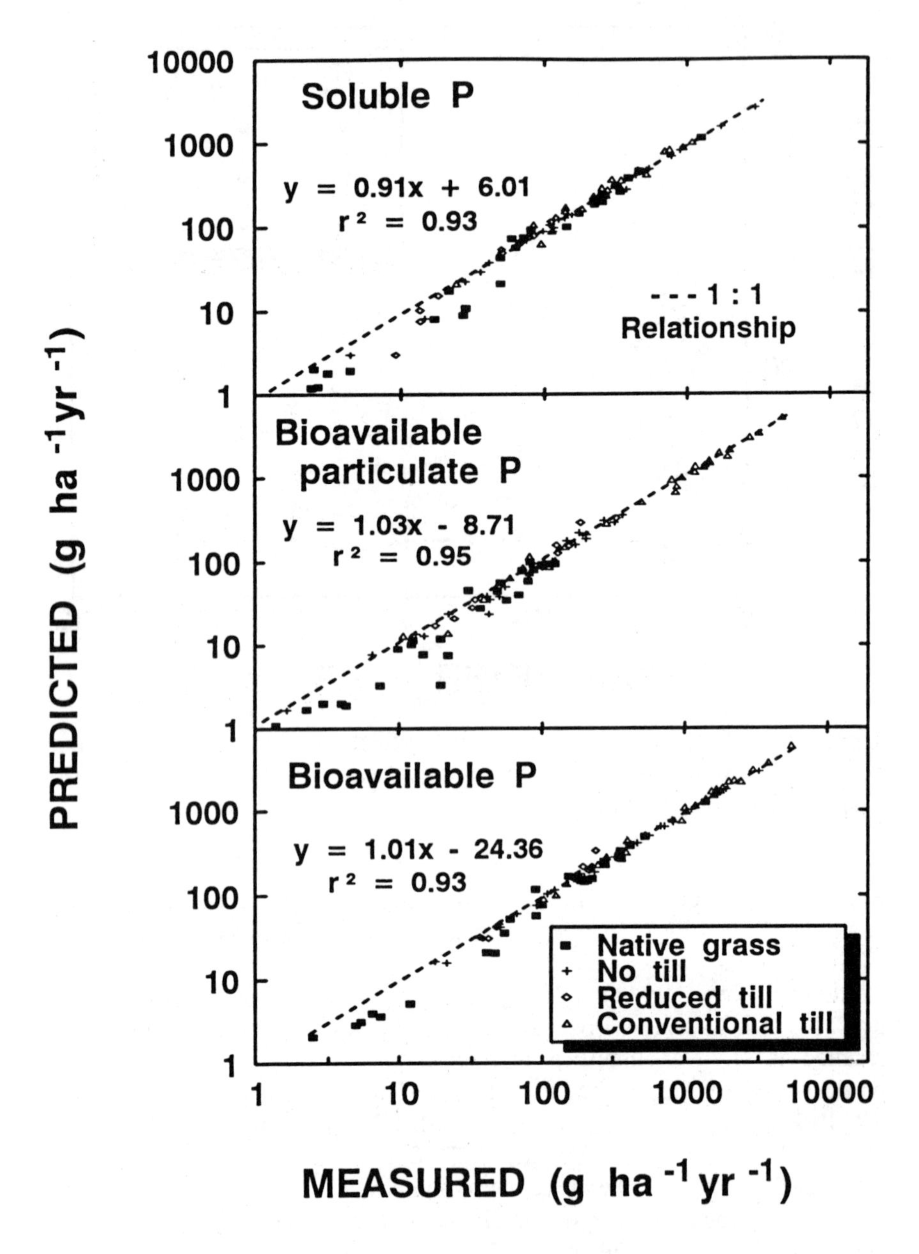


**Figure 20.** Relationship between measured and predicted P loss in runoff from the Oklahoma and Texas watersheds.

### 1. Soluble Phosphorus

The SP concentration of runoff is predicted by the following equation describing the kinetics of soil P desorption (Sharpley and Smith, 1989b):

$$P_r = \frac{K\, P_a\, B\, D\, t^{\alpha}\, W^{\beta}}{V} \qquad [1]$$

where $P_r$ is average SP concentration of an individual runoff event (mg $L^{-1}$), $P_a$ soil test (Bray-I) P (mg $kg^{-1}$) of surface soil (0-50 mm depth) before each runoff event, D effective depth of interaction between surface soil and runoff (mm), B bulk density of soil (Mg $m^{-3}$), t runoff event duration (min), W runoff water/soil (suspended sediment) ratio, V total runoff during the event (mm), and K, $\alpha$, and $\beta$ constants for a given soil. Values of D and Eq. [1] constants were estimated from soil loss (Sharpley, 1985b) and surface soil clay/organic C content (Sharpley, 1983), respectively. Accurate SP predictions were obtained for each management practice, covering a wide range in measured losses (1 to 2980 g $ha^{-1}$ $yr^{-1}$, Figure 20). A general underestimation of SP loss in runoff from native grass and no till compared to the other watersheds was observed (Figure 20). This may result from an inadequate representation of the contribution of P release from vegetative material to P loss in runoff.

The extent of surface-soil vegetative cover will influence the degree of interaction between surface soil and runoff (D, Eqs. [1]), initiating P extraction and transport and the differential release of P from vegetation. It is well established that the release of P from vegetation can be an important source of P in runoff, which is influenced by several soil and crop factors such as soil nutrient status, soil water content, crop type, and growth stage (Burwell et al., 1975; Klausner et al., 1974; Sharpley, 1981; Timmons et al., 1970; Wendt and Corey, 1980). However, there has been limited success in simulating these processes, particularly for growing plants (Schrieber, 1990).

### 2. Particulate Phosphorus

The selective transport of clay-sized particles in runoff has led to the concept of enrichment ratios (ER) for P, defined as the ratio of the P content of runoff sediment to that of surface soil, used to predict particulate P transport (Menzel, 1980; Sharpley et al., 1985). Particulate P concentrations of runoff are calculated from total and bioavailable P contents of surface soil, respectively, using ER:

$$\text{Particulate P} = (\text{Soil total P}) * (\text{Sediment concentration}) * (\text{PER}) \qquad [2]$$

$$\text{Bioavailable particulate P} = (\text{Soil bioavailable P}) * (\text{Sediment concentration}) * (\text{BIOER}) \qquad [3]$$

where the units of soil total and bioavailable P are mg $kg^{-1}$ and those for sediment concentrations in runoff are g $L^{-1}$. The enrichment ratios were predicted from soil loss (kg $ha^{-1}$) using the following equation developed by Sharpley (1985c):

$$\text{Ln (ER)} = 1.21 - 0.16 \text{ Ln (Soil loss)} \qquad [4]$$

Accurate predictions of PP and BPP loss from each management practice were obtained, for a wide range in measured values (Figure 20). The enrichment ratio approach has also been successfully used for other U.S. regions and soils in AGNPS, CREAMS, GAMES, and ARM models to predict PP transport in runoff (Beasley et al., 1985; Cook et al., 1985; Donigian et al., 1977; Knisel et al., 1988). As for SP, BPP predictions were generally underestimated for the native grass and no till compared to reduced and conventionally tilled watersheds, particularly at low concentrations (Figure 20). This discrepancy may result in part from a lower soil loss from the grassed and no till watersheds (Figure 16).

As the predicted relationship between ER and soil loss (Eq. [3]) is logarithmic, predicted values of ER will be affected more than a unit quantity of soil loss at low values of loss ($< 50$ kg $ha^{-1}$ $yr^{-1}$) than higher values ($> 500$ kg $ha^{-1}$ $yr^{-1}$). Additionally, preliminary testing of Eq. [3], showed that this relationship varied between watersheds (Sharpley et al., 1985). It may, thus, be inappropriate to use constant slope and intercept values for different management practices, where the runoff-surface soil interaction may differ. For example, in grassed and no till practices, organic matter may contribute a greater proportion of particulate material transported. Consequently, making slope and intercept of Eq. [4] a function of factors affecting soil loss or runoff, such as rainfall intensity, vegetative cover, and management practice, should improve predictions. This may involve use of specific surface area and density of eroded material and enrichment of particle size fractions, rather than total sediment loss.

### 3. Bioavailable Phosphorus

Bioavailable P loss was calculated as the sum of predicted SP and BPP losses. Accurate BAP predictions were obtained (Figure 20). As both SP and BPP losses in runoff from native grass and no-till watersheds were slightly underestimated, predicted BAP losses were lower than measured values for these watersheds (Figure 20). However, above BAP losses of about 100 g $ha^{-1}$ $yr^{-1}$, predictions closely followed a 1:1 relationship.

### C. Model Limitations

There are many models available which simulate soil nutrient cycling and transport in runoff, facilitating identification of alternative management strategies and basic research needs. However, a major limitation is often the lack of input data to run a model, as many models require detailed information on soil physical, chemical, and biological properties as well as crop and tillage operations. As model output will only be as reliable as the data input, use of these models to provide quantitative estimates of P availability and transport in runoff under specific environmental conditions is limited. Thus, their use is recommended for comparison of relative effects of P management on soil and water productivity.

Some limitations to model use and application may be overcome by development and generation of adaptable data bases, which may fill information gaps. Weather generators and soil data bases are being developed to provide needed inputs, so that all a user needs to know is the soil name, location, and management practice to be imposed. Linkage of soil productivity and water quality models may be necessary to evaluate the effect of agricultural management strategies on water quality. Such an effort is underway to combine the soil productivity and water quality models of EPIC and GLEAMS. In addition, it is still difficult to relate P loss in runoff as a function of watershed management to the biological productivity of a receiving water body. Models simulating watershed (AGNPS) and lake (FARMPND) processes were linked by Summer et al. (1990). However, a lack of adequate field data limits rigorous testing of their ability to simulate a lake's response to changes in agricultural management and weather conditions.

### D. Indexing Phosphorus Availability

Because of the above limitations, indices which identify soil and management practices that may enrich the P content of surface waters should be developed. The basic principle of water quality indices is to synthesize data such as analytical results and experimental information, by means of a simple quality vector or algorithm. Hence, the index will be a simplified expression of a complex combination of several factors or parameters, that make information more easily and rapidly interpretable than a list of numerical values. Such an index would be helpful for field personnel with limited resources working with land owners, to identify sensitive areas and suggest management alternatives to reduce the risk of water quality problems associated with P.

For initial development of such an index, critical factors influencing soil P availability, transport in runoff, and susceptibility of receiving water bodies to P inputs, have been identified in Figure 21. These are: erosion potential; runoff

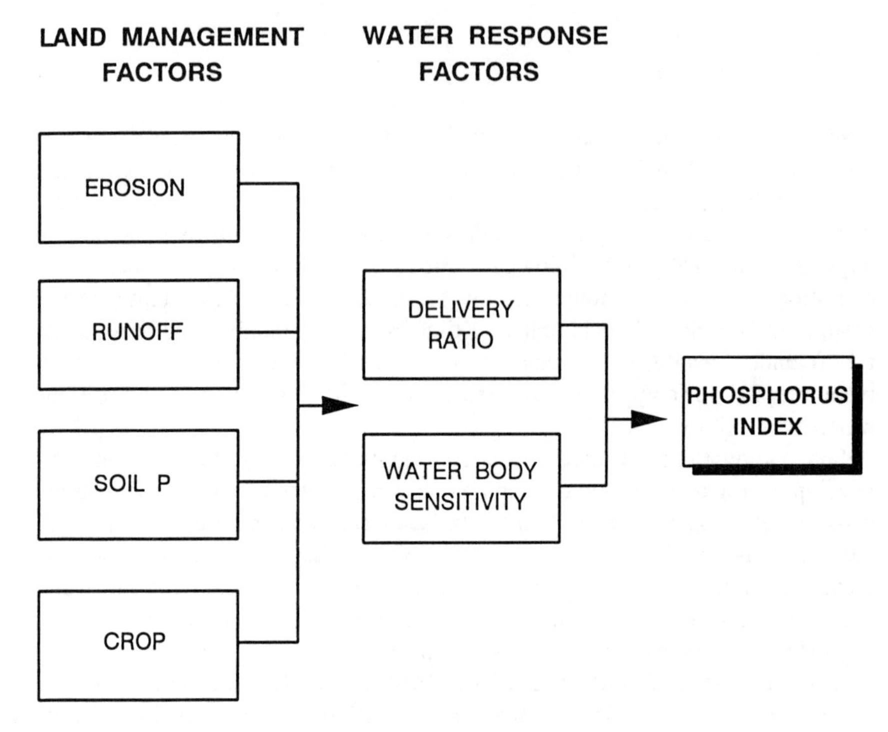


**Figure 21.** Components of a soil and water P availability index.

potential; soil P; crop type and yield goals; water delivery ratio; and waterbody sensitivity.

As the major portion of P transported in runoff from cropland is associated with eroded soil material, parameters describing ***soil erosion and runoff potential*** will be primary components of the index. An estimate of soil erosion and runoff potential may be obtained from soil survey and site information, such as soil texture, permeability, and cover; slope length and gradient; crop management and tillage; rainfall; and erosion control practices. As a first approximation, this information is used with equations predicting erosion such as the Universal Soil Loss Equation (USLE, Wischmeier and Smith, 1965; 1978), Modified USLE (MUSLE, Williams, 1975), Rangeland USLE (RUSLE, Renard et al., 1991), or Water Erosion Potential Predictor (WEPP, Laflen et al., 1991).

The ***soil P factor*** (Figure 21), includes available soil P content and P sorption capacity. Because of the importance in available soil P in determining P loss in runoff and its dynamic nature in surface soil, an estimate of its content will be required for the index. Soil test P may be used as a first approximation. However, an estimate of the bioavailable P content of surface soil ($Fe_2O_3$ strip or 0.1 *M* NaOH extractable) may be more appropriate for indexing the potential impact of P transport on accelerated eutrophication. Because of the ease of bioavailable soil P determination and a lack of general correlation with soil test P (Sharpley and Smith, 1992; Wolf et al., 1985), its measurement should not be substituted by soil test P. The P sorption capacity of a soil will determine the disposition of added P into available and unavailable forms, which will ultimately influence the portion of P transferred to runoff in soluble and particulate forms. For example, in a soil of low sorption capacity, more added P will remain in an available form and be more susceptible to loss as SP than an equivalent addition to a soil of high P sorption capacity. Phosphorus sorption may be determined by a single point isotherm (Bache and Williams, 1971) or from regression equations relating it to associated soil properties, such as pH, clay, organic C, and $CaCO_3$ content (Sharpley et al., 1984a).

***Crop factors*** such as type and yield will influence the degree of vegetative soil cover and amount of P removed from the soil system, if the crop is harvested. In addition to influencing soil erosion, type of crop cover can affect the amount (Sharpley, 1981; Burwell et al., 1975; Timmons et al., 1970) and relative bioavailability of P transported (Sharpley et al., 1992; McDowell and McGregor, 1984; Wendt and Corey, 1980). For example, Muir et al., (1973) found a significant correlation between SP concentration in major streams of Nebraska and legume acreage statewide. They suggested that SP concentration in the Platte River, Nebraska, may reflect P leached from alfalfa residues, carried in runoff during the growing season. Soil P utilization also varies from crop to crop, which will influence the amount of available soil P that could be transported in runoff.

The effect of in-stream physical processes and chemical transformations on the amount and form of P transferred from edge-of-field to lake, can be summarized by a ***P delivery ratio***. Physical processes involve deposition and erosion of stream bank and bed material with a change in stream flow velocity. These processes accentuate P transformations by the overall selective transport of fine materials, which have a greater capacity to sorb or desorb P. The relationship between the form of P transported in runoff and P sorption capacity was demonstrated by data from several unfertilized watersheds in the Southern Plains (Sharpley and Smith, 1990). A decrease in SP concentration with an increase in suspended sediment concentration of runoff was found to be a function of P sorption capacity of the dominant soil type at each watershed (Sharpley and Smith, 1990). Thus, SP concentration can be modified to a great extent during transport in runoff by suspended sediment from source soil having a higher sorption capacity. Clearly, in-stream changes in the relative amounts of SP and PP transported, approximated by the P delivery ratio, will influence the short- and long-term impact on lake productivity.

The ***sensitivity of a water body*** to either SP or PP inputs may determine management control strategies on source areas. For example, depth of photic zone, degree of surface mixing, development of reducing conditions at the water-sediment interface, and water residence time of a water body will influence its sensitivity to P inputs. If properties of the water body are such that particulate material and associated P rapidly settle from the water column (i.e.,

**Table 7.** An indexing to rate the potential P loss in runoff from given site characteristics

| Site characteristic (weight) | Phosphorus Loss Potential (Value) | | | | |
|---|---|---|---|---|---|
| | None (0) | Low (1) | Medium (2) | High (4) | Very High (8) |
| ***Transport Factors*** | | | | | |
| Soil erosion (1.5)[a] | Negligible | <10 | 10 - 20 | 20 - 30 | >30 |
| Irrigation erosion[b] (1.5) | Negligible | Tailwater recovery for QS <6 for very erodible soils or QS <10 for others | QS >10 for erosion resistant soils | QS >10 for erodible soils | QS >6 for very erodible soils |
| Runoff class (0.5) | Negligible | Very low or low | Medium | High | Very high |
| ***Phosphorus Source Factors*** | | | | | |
| Soil P test (1.0) | Negligible (0) | Low (1) | Medium (2) | High (4) | Excessive (8) |
| P fertilizer application rate (0.75)[c] | None applied | 1 - 15 | 16 - 45 | 46 - 75 | >76 |

| | | | | | |
|---|---|---|---|---|---|
| P fertilizer application method (0.5) | None applied | Placed with planter deeper than 5 cm | Incorporated immediately before crop | Incorporated >3 months before crop or surface applied <3 months before crop | Surface applied >3 months before crop |
| Organic P source application rate (0.%)[c] | None applied | 1 - 15 | 16 - 30 | 30 - 45 | >45 |
| Organic P source application method (1.0) | None | Injected deeper than 5 cm | Incorporated immediately before crop | Incorporated >3 months before crop or surface applied <3 months before crop | Surface applied >3 months before crop |

[a] Units for soil erosion are Mg $ha^{-1}$.
[b] Q is irrigation furrow flow rate (gal $min^{-1}$) and S is furrow slope (%).
[c] Units for P application are kg P $ha^{-1}$.

a deep, stratified lake of long water residence time), then control of SP inputs may be of greater short-term benefit in reducing biological productivity. The trophic response or sensitivity of a given water body to P input may be approximated by an empirical equation, such as those developed by OECD (1982), Vollenweider and Kerekes (1980), and Jones and Lee (1982).

The main land management and water response factors that should be considered in developing an index to assess the risk of agricultural P loss in runoff to degrade water quality are described above. These factors are incorporated into a P index being developed by a team of scientists (Phosphorus Indexing Core Team - PICT), led by the USDA-SCS, National Water Quality Technology Development Staff. The index is outlined in Tables 7 and 8. Each site characteristic has been arbitrarily assigned a weighting, assuming that certain characteristics have a relatively greater effect on potential P loss than others. Each site characteristic is given a rating value (Table 7), although each user must establish a range of values for different geographic areas.

As discussed earlier, erosion and runoff class can be obtained from soil loss equations (i.e., USLE, RUSLE) and soil survey data (i.e., estimated soil saturated hydraulic conductivity and slope), respectively. Irrigated erosion is calculated as the product of flow rate in the furrow (Q) and furrow slope (S). Soil test P is represented by the recommended procedure for each state. At the present time, the user must categorize soil test P into low, medium, high, and very high based on regional experience. Categories for rate and method of fertilizer and organic P application are self-explanatory (Table 7). In general, increasing from low to very high category values for application method, depict longer surface exposure time between P application, incorporation, and crop utilization.

An assessment of site vulnerability to P loss in runoff is made by selecting the rating value for each site characteristic from the P index (Table 7). Each rating value is multiplied by the appropriate site characteristic weight factor. Weighted values of all site characteristics are summed and site vulnerability obtained from Table 8. A hypothetical site is used as an example, where soil erosion is 25 Mg $ha^{-1}$ (weighting $\times$ value; $1.5 \times 4 = 6$), irrigated erosion is not applicable (0), runoff class is medium ($0.5 \times 2 = 1$), soil test P is medium ($1 \times 2 = 2$), 20 kg P $ha^{-1}$ of fertilizer ($0.75 \times 2 = 1.5$), and 50 kg P $ha^{-1}$ of animal manure ($0.5 \times 8 = 4$) are broadcast in early spring prior to planting ($1 \times 4 = 4$). The sum of these weighted values (6, 0, 1, 2, 4, 1.5, 4, and 4) is 22.5, which has a high site vulnerability (Table 8). In this hypothetical situation, conservation measures to minimize erosion and runoff as well as a P management plan should be implemented to reduce the risk of P movement and probable water quality degradation.

The index is intended for use as a tool for field personnel to easily identify agricultural areas or practices that have the greatest potential to accelerate eutrophication. It is intended that the index will identify management options available to land users that may allow them flexibility in developing control strategies.

**Table 8.** Site vulnerability to P loss as a function of total weighted rating values from the index matrix

| Site vulnerability | Total index rating value |
|---|---|
| Low | <10 |
| Medium | 10 - 18 |
| High | 19 - 36 |
| Very High | >36 |

# VII. Management Challenges

## A. Minimizing Phosphorus Transport

Efforts to minimize P transport in runoff can be divided into fertilizer management and erosion control measures. Emphasis should be placed on minimizing the potential sources of P for transport.

### 1. Fertilizer Management

Efforts to reduce P losses to surface runoff by fertilizer management, should include subsurface placement, credit for alternative P sources such as manures, and utilization of accumulated residual P.

#### a. Subsurface Placement

With the increasing adoption of conservation tillage and more efficient agricultural systems, where P fertilizer and/or manure is broadcast on land which may not be tilled for several years, it will be necessary from a water quality perspective to minimize excessive surface soil accumulations of P. Where possible, this may be achieved by subsurface placement of fertilizer and manure away from the zone of removal by surface runoff. Preliminary data on fertilizer P management under conservation tillage systems in the Midwest region of the U.S. (Rehm, 1991), show that without this redistribution of stratified P, higher soil test P levels may be required for optimum production compared to conventional tillage practices.

#### b. Alternative Sources

Recent economic pressures have resulted in a diversification of agricultural systems, which in Alabama, Arkansas, Delaware, and Oklahoma for example, have been a rapid increase in poultry and swine operations. Although manure is a valuable source of nutrients, utilization of large amounts of accumulated animal manure is an increasing problem facing the industry. Consequently, manure can be utilized as a supplement or alternative to fertilizer P, preferably by subsurface placement or injection. However, it may be necessary in the future to establish cost sharing programs or subsidies, involving the consumer, producer, and farmer to transport excess manure from areas of high to low intensity production.

Because of the rapid fixation of fertilizer P by soil and conversion to unavailable forms, application rates usually exceed starter requirements, in all soils except those with very high P test values. A lack of agronomically effective slow release P fertilizers has limited the reduction in application rates of traditional soluble P fertilizers. However, the use of new sources of more reactive North Carolina RP material, adoption of more efficient systems using naturally occurring amendments, and inclusion of potentially soil acidifying forage legumes in crop rotations, may enhance the potential agronomic and economic benefit of RP as a slow release fertilizer on acid soils. The use of RP may decrease the potential loss of P in runoff, via banded or subsurface application in combination with smaller amounts of soluble fertilizer P. Research is needed, however, to evaluate the relative agronomic, economic, and environmental benefit of manure and RP as alternative sources of P as a function of different climatic conditions, soil types, and management systems. This should determine under what conditions RP can compete with high analysis fertilizers due to greater transportation costs of the former.

### c. Utilization of Residual Soil P

In some cases, continued manure application has resulted in an excessive accumulation in surface soil P. For example, soil test P contents (Bray-I) in the surface 10 cm of soil receiving long-term (8 to 35 years) poultry and swine manure (Sharpley et al., 1991b) and cattle feedlot waste (Sharpley et al., 1984b), were increased up to 38-fold compared to untreated soils (Figure 22). We estimated the P concentration of a 2.5 cm runoff event of 20 kg $ha^{-1}$ soil loss from the treated and untreated soils, using the predictive equations presented earlier (Eqs. [1] to [4]) with measured soil P content. These runoff and soil loss values are means for all events occurring on grassed watersheds in the Southern Plains during the last 15 years. Predicted P concentrations of runoff from treated soils were dramatically higher than from untreated soils, if runoff occurs (Figure 22). Under grass, erosion is minimal and, thus, most of the transported P will be in a bioavailable form (57 to 98%). Although this is a hypothetical runoff situation, it clearly emphasizes the need to carefully manage repeated applications of manure over a long period of time to minimize the potential for runoff to occur from these soils.

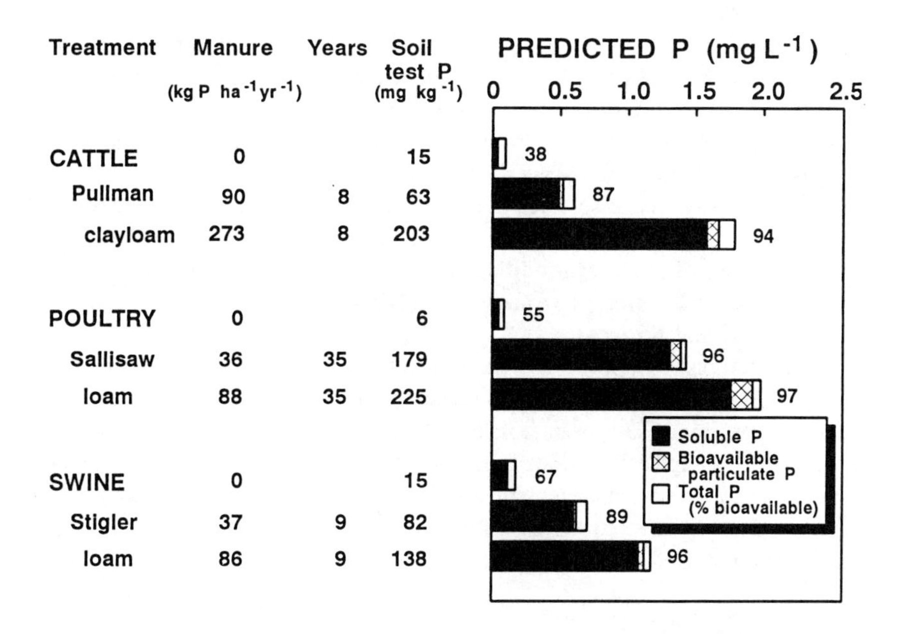


**Figure 22.** Effect of animal manure application on soil test P (Bray-I, 0-10 cm) and predicted P concentration of runoff from three soils in Oklahoma and Texas.

Brown et al. (1989) investigated the effect of various animal manure control practices on nonpoint source P loading to the eutrophic Cannonsville Reservoir in the West Branch of the Delaware River watershed, New York. Reductions in P loss from dairy operations of 50 to 90% were shown to be achievable using practices that reduced the volume of runoff from these areas (Brown et al., 1989). However, the contribution of P in runoff from dairy operations to P loss from the West Branch watershed, was substantially less than that from corn, on which manure had been spread. Hence, manure spreading schedules that guide the location and timing of spreading had the potential to reduce P loading from the area studied by as much as 35%. However, demands on farmer's daily schedules often limit the practicality of precise timing of manure applications. As a result, application timing is possibly the greatest practical obstacle to better manure management, with many best management practices needing to be done at the busiest times of the year for farmers.

## 2. Erosion Control

### a. Conservation Tillage

Because of the fixation of P by soil, practices that reduce erosion will also reduce total P losses in runoff. Soil erosion and associated P transport may be reduced by increasing vegetative and residue ground cover through reduced tillage practices (see Figure 16). As a trade-off to this benefit, the concentration of SP in runoff from conservation tillage wheat and sorghum was greater than from conventional practices, even though similar or less P fertilizer, respectively, was added (Figure 16). Several other studies have suggested that conservation tillage may increase the bioavailability of P transported in runoff (McDowell and McGregor, 1980; Wendt and Corey, 1980). Consequently, implementation of conservation tillage alone may not reduce the potential for P-associated eutrophication of surface waters, as great as may be expected from inspection of total P loads.

In an attempt to improve the trophic status of Lake Erie, the U.S. Army Corps of Engineers (1982) showed that nonpoint source P load needed to be reduced by about 26% to reduce lake eutrophication significantly. From field experimentation and calculation, Forster et al. (1985) concluded that accelerated implementation of best management systems, which includes conservation tillage on soils suited to those practices on the U.S. side of the Lake Erie basin, can achieve the required P load reduction in 20 years.

#### b. Buffer Strips/Riparian Zones

In addition to conservation tillage, filter strips or zones can effectively reduce sediment and PP transport in runoff from cropland or animal production facilities. Vegetative filter strips are bands of planted or indigenous vegetation and riparian zones, are normally areas of native forest or woodland.

Dillaha et al. (1989) reported that a vegetative filter strip of orchardgrass (*Dactylis glomerata*) effectively reduced sediment and PP loss in runoff from bare Groseclose silt loam plots (11% slope and 5.5 by 18.3 m) in Virginia (Figure 23). The loss of soil and PP in two 60 min and four 30 min rainfalls of 50 mm $hr^{-1}$ was reduced a respective 82 and 76% for a 4.6 m strip and 97 and 93% for a 9.1 m strip compared to no strip. Although SP loss in runoff was reduced 44% by the 9.1 m strip, SP increased 78% with the 4.6 m strip. Dillaha et al. (1989) attributed the SP increase to lower removal efficiencies for soluble nutrients and the release of previously trapped PP in the filters.

Because of the low installation and maintenance costs and effectiveness, filter strips are approved cost-share practices in many states and were recently incorporated into the USDA's Conservation Reserve Program (Alder, 1988). However, inadequate information on their effectiveness in reducing SP losses and on the dynamics of sediment and nutrient transport in the strips has resulted in strip installation in areas where they are inappropriate because of topographic limitations. Detailed reviews of filter strips are given by Dillaha et al. (1987) and Magette et al. (1987).

Riparian forests adjacent to surface waters have also been shown to reduce P inputs from agricultural runoff (Lowrance et al., 1985; Schlosser and Karr, 1981; Yates and Sheridan, 1983). For example, Peterjohn and Correll (1984)

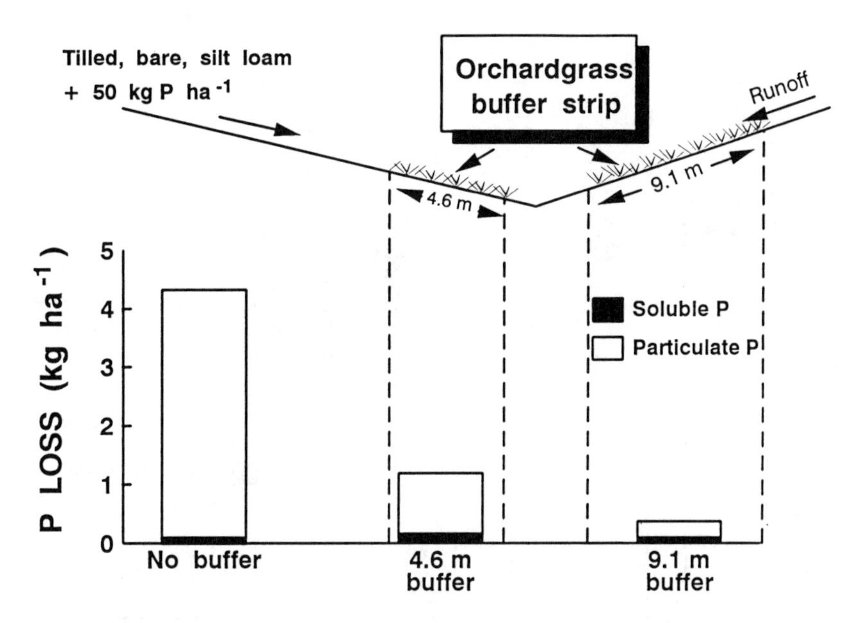


**Figure 23.** Phosphorus loss in runoff from Groseclose silt loam of 11% slope as a function of length of orchardgrass buffer strip. (Data adapted from Dillaha et al., 1989.)

conducted a detailed study of nutrient dynamics in a 16.3 ha agricultural watershed adjacent to Chesapeake Bay, Maryland. In the watershed, 10.4 ha was planted with corn and 5.9 ha as a riparian forest of broadleaved, deciduous trees. Peterjohn and Correll (1984) found that the total P input from cropland to riparian forest in surface runoff and ground water was 3.4 and 0.091 kg $ha^{-1}$ $yr^{-1}$, respectively (Figure 24). Phosphorus export in stream flow from the riparian forest was reduced 80%, with similar amounts exported in surface runoff (0.43 kg $ha^{-1}$ $yr^{-1}$) and ground water (0.30 kg $ha^{-1}$ $yr^{-1}$). Clearly, riparian forests can reduce the transfer of P in agricultural runoff to receiving surface waters. Thus, coupled natural and agricultural systems within a watershed may reduce the occurrence of P-associated eutrophication. Even so, vegetative filter strips or riparian zones should not be relied upon as the sole or primary means to reduce P losses in runoff from agriculture (Magette et al., 1989).

## B. Management Implications

It is apparent that several questions regarding the management of P availability in soil and water systems, must be answered to minimize P transport in runoff and bring about a further improvement in both soil productivity and surface water quality. These questions involve manure management based on P or N;

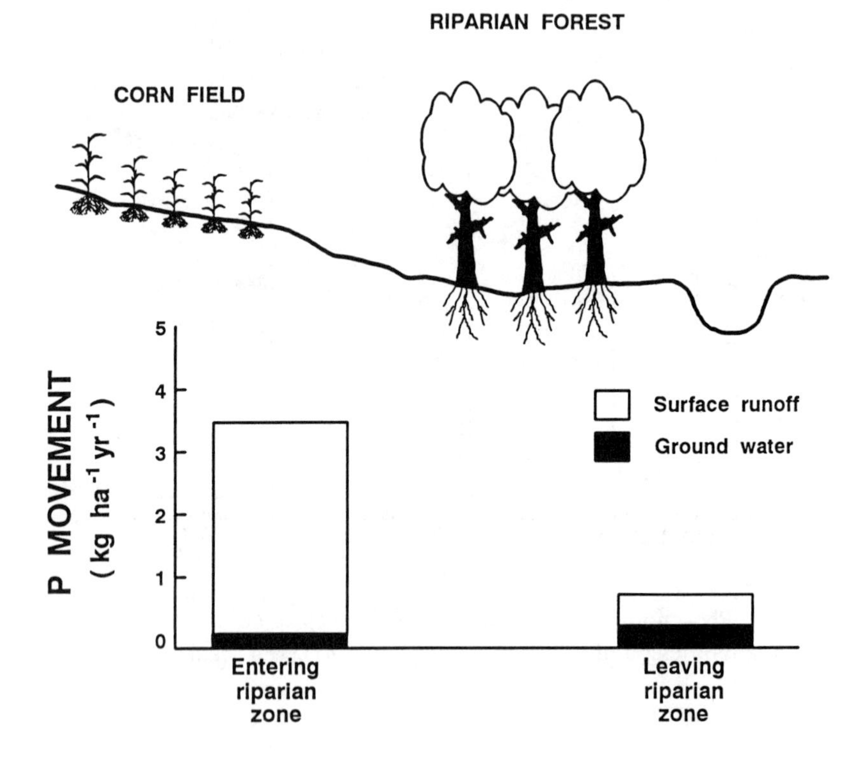


**Figure 24.** Amounts of total P entering a riparian forest from a corn field and leaving the forest in stream flow for a Maryland watershed. (Data adapted from Peterjohn and Correll, 1984.)

management effects on the dynamics of soil P cycling; tillage, crop, and residue management; soil test methodology and interpretation; and impact assessment.

### 1. Manure Management

Manure is a valuable source of both N and P for crop uptake in efficient management systems. However, continual long-term applications of manure have increased $NO_3$-N leaching (Cooper et al., 1984; Ritter and Chirnside, 1984) and the potential for P loss in runoff. For example, Liebhardt et al. (1979) applied poultry manure at rates of 0 to 179 Mg ha$^{-1}$ (wet wt. basis) to sandy soils in Delaware and found a direct relationship between application rate and $NO_3$-N concentrations in ground water in excess of the recommended 10 mgL$^{-1}$ limit (U.S. EPA, 1976). In addition, the buildup of P at the surface of soils receiving continual long-term manure applications has increased the potential for P enrichment of associated surface waters (Magette, 1988; Sharpley, 1991; Sims, 1992). Consequently, should manure application rates be based on soil N or P?

This question may be partially answered by determining if the soil, to which manure is to be applied, is susceptible to runoff and erosion or leaching as by the P indexing procedure (Tables 6 and 7). If the potential for runoff or erosion from an application site exists, then P should be a priority management consideration.

Currently, most manure application rates are based primarily on the management of N to minimize nitrate-N losses by leaching. In most cases, this has led to an increase in soil P levels in excess of crop requirements due to the generally lower ratio of N:P added in manure than taken up by crops. For example, dairy, beef, swine, sheep, and poultry manure has an average N:P ratio of 4.1 (Gilbertson et al., 1979), while the N:P requirement of major grain and hay crops is 7.3 (Fertilizer Handbook, 1982).

Sims (1992) emphasized that basing manure applications on P rather than N management, would present several problems to many farmers. For many crops and soils, manure application rates would be low if based on soil P, requiring farmers to identify larger areas of land to dispose of generated manure. In addition, farmers presently relying on manure to supply most of their crop N requirements may be forced to buy commercial fertilizer N, instead of using their own manure N. Clearly the development of environmentally sound management systems for the use of P originating from manure is a challenge from both research and environmental standpoints. For example, basing recommendations on soil P levels may resolve the issues of P enrichment of surface water, but place unacceptable economic burdens on farmers.

## 2. Soil Management

In conservation tillage systems, surface crop residues minimize soil water evaporation and erosion. Similarly, in other sustainable systems where cover crops are being increasingly included, the cover crop is killed before maturity and left in the soil to minimize water and light competition with the subsequent cash crop. What effect will this have on the amount and bioavailability of P in runoff? If the crop residue is left on the surface or occasionally plowed into the soil, is the potential loss of residue P in runoff affected? Crop residue management can affect P cycling and availability as a function of residue amount, type, and degree of incorporation with tillage. Sharpley and Smith (1989a) found that mineralization and leaching of residue P was greater when the residue of several crop types was surface applied compared to incorporated. A greater amount of residue will increase the amount of P being cycled and, particularly if left on the surface of the soil, will reduce evaporation losses and keep surface soil moist for more days during the growing season, thereby enhancing microbial activity and mineralization.

Under conservation tillage practices or continuous heavy fertilizer, manure or sludge applications, P may accumulate in certain soil horizons. Thus, is it possible to select a "scavenger" crop that may have a higher affinity or requirement for P and thereby reduce soil nutrient stratification? Alfalfa for example, has reduced subsoil nitrate accumulations (Mathers et al., 1975). May the same be true for surface soil accumulations of P? It is possible that by utilizing residual soil P, careful crop selection will reduce the amount of nutrients potentially available to be transferred to surface waters. However to be successful, the "scavenger" crop and P must be economically sustainable and leave the farm and move to a P-deficient site.

The benefits of improved technology and information on fertilizer and residue management on the fate of P in soil and water, are to a large extent dependent upon reliable soil test procedures. However, with an increase in soil P stratification and amounts in organic forms under no-till systems, are soil test procedures adequate to determine positional and chemical availability? This may be of particular importance to conservation tillage systems and where a cover crop is returned to the soil, contributing to an increase in organic matter content of the surface soil. In these situations, mineralizable organic P may be an important source of P to crops (Table 2) and subsequently runoff. In tropical soils, more frequent soil wetting and drying cycles and greater amounts of soil organic P, increase potential mineralization of organic P compared to temperate region soils (Table 2). Considering maximum organic P mineralization rate, crop uptake of P and runoff potential frequently occur at about the same time of year (i.e., spring) which emphasizes the need to quantify the contribution of organic P to P transfer within terrestrial ecosystems.

Several studies have shown that soil P supply, represented by crop yields, were more closely estimated by including extractable organic P (Abbott, 1978; Adepetu and Corey, 1976; Bowman and Cole, 1978; Daughtrey et al., 1973). Bowman and Cole (1978) used a modification of the Olsen P test (16 hr instead of 30-min extraction), and measured the total amount of P (inorganic plus organic) extracted by the reagent. Such modified soil P tests may be conducted on soils identified as having a potentially large contribution of organic P mineralization to the plant available P pool. Caution must be exercised, however, in relating amounts of organic P extracted to crop response in the field, as the conditions of P extraction may not duplicate conditions for organic P mineralization in the field and sorption of mineralized organic P could affect its availability. Even so, soil test methods that estimate or give credit for mineralizable organic P may avoid potentially excessive fertilizer P applications.

## 3. Water Management

It is apparent from the above discussion that conservation tillage can reduce soil and P transfer in surface runoff, although the proportion that is bioavailable both in soluble and particulate forms may increase. Is this increase in bioavailability sufficient to increase the short- and long-term biological productivity of receiving water bodies? In all examples given, SP and TP concentrations of runoff were consistently above the critical values associated with accelerated eutrophication of a water body (0.01 and 0.02 mg $L^{-1}$, respectively; Sawyer, 1947; Vollenwieder and Kerekes, 1980). This is true even for unfertilized native grass watersheds, where native soil fertility is apparently high enough in some cases to enrich the SP concentration of runoff. As a result, sustainable or conservation tillage practices may not lower P concentrations of runoff from cultivated or grazed land to critical values. Consequently, the critical P level approach should not be used as the sole criterion in quantifying permissible tolerance levels of P in surface runoff as a result of differing management practices.

Furthermore, as conservation tillage systems reduce P loss in surface runoff, but not its bioavailability, should eutrophication-agricultural management decisions be based on total loss or bioavailability? Several studies have indicated little decrease in lake productivity with reduced P inputs and have attributed this to an increased bioavailability of P entering lakes, as well as internal recycling of P (Gray and Kirkland, 1986; Logan, 1982; Young and DePinto, 1982). Consequently, the measurement of P bioavailability, as both SP and BPP, is essential to more accurately estimate the impact of agricultural management practices on the biological productivity of surface waters.

A detailed procedure to estimate the amounts and form of P in water was presented earlier (Figure 13). However, does the complexity of this scheme limit its use, particularly in the estimation of bioavailable P as recommended? In many situations, the answer is often yes, especially in developing countries where limited resources restrict even simple analytical capabilities. Consequently, there is a need for methodology that will facilitate P estimation. One such method may be the use of $Fe_2O_3$ strips, which have been developed (Menon et al., 1989a, b) and successfully applied to the estimation of plant available P in a wide range of soils and cropping situations (Menon et al., 1989c; Sharpley, 1991). The strips are made by soaking filter paper in a 10% $FeCl_3$ solution, conversion to $Fe(OH)_2$ by ammonia vapor, and cutting the paper into 10 × 2 cm strips. One strip is shaken with soil (1 g soil in 40 mL 0.01 *M* $CaCl_2$) for 16 h and P removed from the strip by 40 mL 0.1 *M* $H_2SO_4$.

It is proposed that one $Fe_2O_3$ strip be shaken with 50 mL of unfiltered runoff with the amount of P removed from the strip representing potentially bioavailable P (subsequently referred to as strip BAP). The strip BAP content of runoff sediment from several Southern Plains watersheds, is related to the growth of P-starved *Euglena* and *Selenastum* incubated for 29 days with runoff sediment as the sole source of P (Figure 25). This relationship is similar to that shown earlier for the frequently used NaOH extractable BAP (Figure 12). In fact, the BAP concentration of runoff from the Southern Plains watersheds, determined by $Fe_2O_3$ strip and NaOH methods were not significantly different (at the 0.1% level, Figure 26).

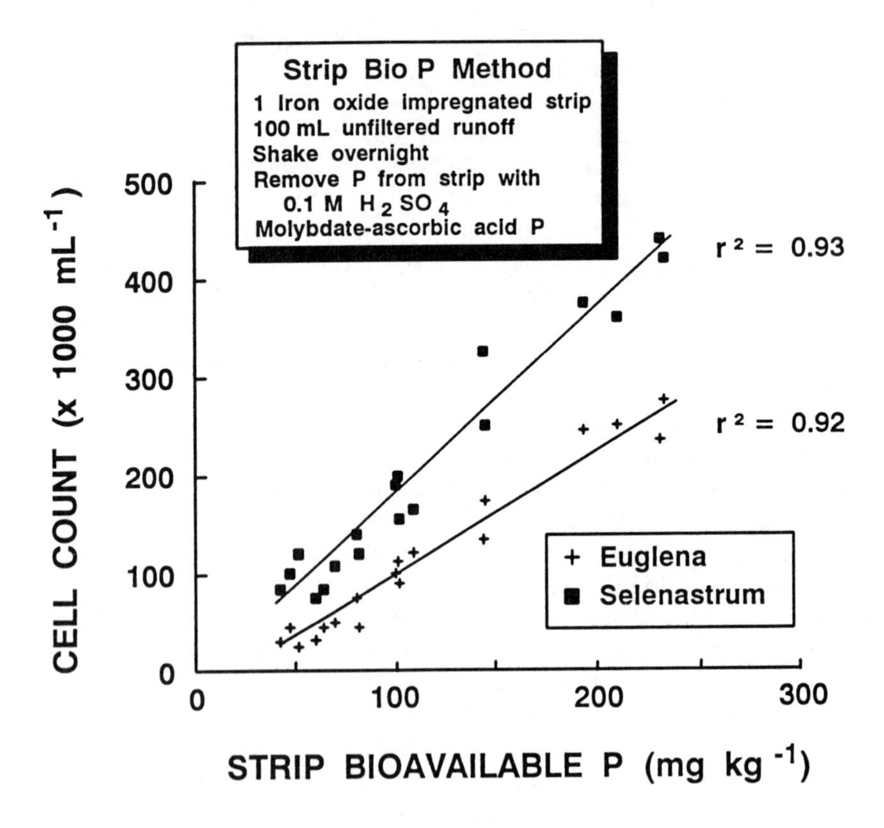


**Figure 25.** Relationship between the bioavailable P content of runoff sediment, extraction by iron oxide-impregnated paper strip, and growth of P-starved algae during a 29-day incubation.

It is suggested that the $Fe_2O_3$ strips act as a P-sink and thereby more closely simulate P removal from sediment-water systems by algae. In addition, the validity of relating the form or availability of P extracted by chemicals differing by more than 1 pH unit from runoff or sediment, to *in-situ* bioavailability may be questioned. In as much, the strip method has a stronger theoretical justification for its use over chemical extractants in estimating BAP. In addition, prepared paper strips may be sent to a field location of limited resources and strip BAP measured using only a 100 to 500 mL bottle in which a strip and unfiltered runoff sample is shaken overnight. The strip may then be dried and returned to an analytical laboratory for P removal and measurement. This would also avoid potential problems with P transformations during sample storage and shipping (U.S. EPA, 1979). Future strip development with the use of prepackaged color reagents may allow P determination in the field by comparison with a color chart, similar to measurement of pH with litmus paper.

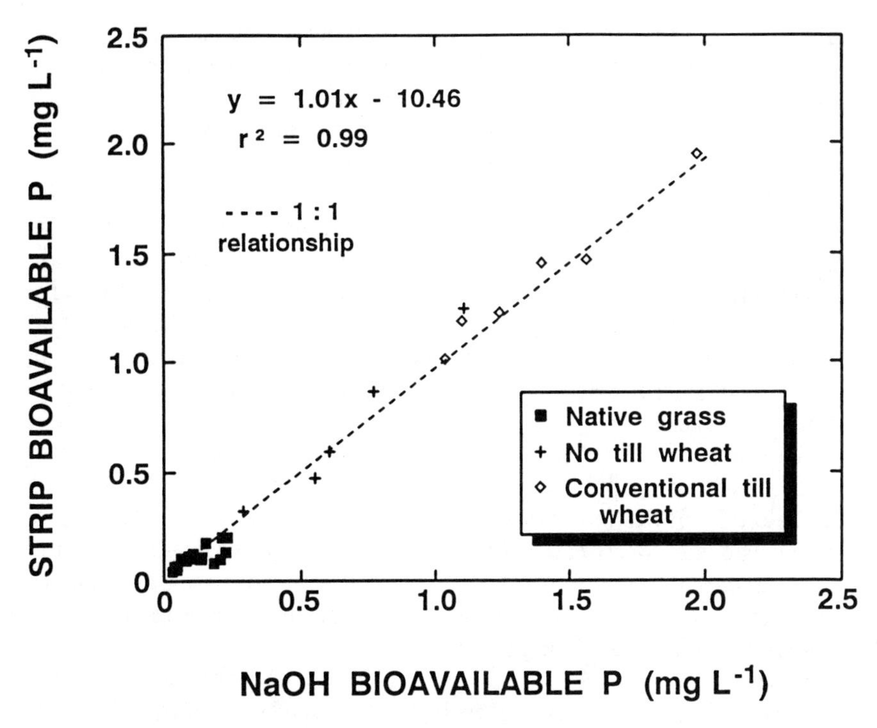


**Figure 26.** Relationship between the bioavailable P concentration of runoff from the Oklahoma and Texas watersheds, as determined by NaOH and iron oxide-impregnated paper strips.

## VIII. Research Needs

The proper management of soil P fertility is essential in maintaining crop yields for an ever increasing world population and to keep unit cost of production low. While its use has been associated with accelerated eutrophication of surface waters, judicious P amendments can reduce P enrichment of agricultural runoff via increased crop uptake and vegetative cover. Yet it is clear that natural soil fertility and P release from plant residues, can result in SP concentrations of runoff from unfertilized pristine land, great enough to stimulate increased biological growth of surface waters. Nevertheless, it is of vital importance that we implement management practices that minimize surface P accumulations, utilize alternative P sources and residual soil P levels, and improve methods estimating P bioavailability in soil and runoff, to affect a decrease in agricultural P inputs to surface waters. Otherwise, the perception by the public that agriculture cannot manage itself for the good of the environment will increase. This may lead to legislation, such as that limiting sewage sludge application to soil, based on soil test P levels to regulate P fertilizer and manure use. Unfortunately, the benefit from implementation of control measures on water quality improvement, will not be immediately visible to a concerned public. Consequently, future research and policy should emphasize the long-term economic and environmental benefits of these measures.

From this discussion, several general areas of future research are suggested to minimize agricultural P loss to surface waters.

***Systems Research***:

Future research regarding residual fertilizer P in soils needs to consider how a farmer might apply high enough rates of P fertilizer to eliminate P as a deficient nutrient in crop production systems while maintaining an economically sustainable and environmentally sound system in the short-term. Other residual P factors that need to be considered include evaluating one-time, high rate applications of P fertilizer as one method of managing P needs of crops grown with reduced- and no-till systems for several years. On soils that already have excessively high soil P levels, reliable methods are needed to determine how long these soils can supply adequate levels of available P for optimum crop yields before P fertilization beyond low starter rates is required. In addition, we must evaluate the effectiveness/efficiency of residual P availability for crop production versus recently applied ($<$ 2 months) P fertilizer and the need for higher than normal P rates on more soil types, especially when high yield, crop management practices are being used.

Information is also needed on the long-term effects of conservation and low-input systems on the dynamics of P cycling, in terms of the buildup of residual soil levels, fertilizer use, and transfer of bioavailable forms to runoff. This will involve the development of methods to credit residual and organic P effectiveness when considering P fertilization needs, economics, and P fertilizer-use efficiency. In addition, a better understanding of the effect of subsurface placement or injection of P fertilizers and manures under various agricultural systems on soil productivity and surface water quality is warranted. We must also develop more accurate methods of metering and applying manure.

***Interdisciplinary Research***:

More emphasis should be placed on interdisciplinary research, crossing agricultural and limnological boundaries. Innovative methods must be developed to eliminate P as a deficient nutrient for crop production, which give credit for residual fertilizer and organic P effects, while maintaining short-term economic sustainability and low labor-machinery requirements. Considerable research has been conducted to quantify soil and fertilizer P losses in runoff as a function of management practice. Yet it is still difficult to relate P inputs to lakes to a quantitative description of water quality. While the effect of P concentration on algal growth receives continued attention, little information is available on how lake macrophytes are affected, even though macrophytes present a more serious economic problem than algae in many lakes.

***Modeling Research***:

Computer simulation of soil productivity and impact of agricultural practices on water quality have improved dramatically over the last decade and are expected to have a major positive impact on farmers and agricultural-eutrophication management decisions. Research is needed on the development of mathematical models simulating changes in residual soil and organic P availability with time. Although first order desorption kinetics have explained soil P release over short-time periods, they have not adequately described long-term changes. As the emphasis on input efficiency increases, accurate predictions of changes in P availability will be required to determine the sustainability and potential transport of bioavailable P in runoff from systems having minimal external P inputs. Considering the need to fertilize the soil rather than the field, models or expert systems should be formulated to variably apply P fertilizer by soil type and need, rather than on a field basis. Credit should be given for previously applied (residual) P fertilizer, such as for example, a budget-type systems approach.

Future research should also be directed towards improving partitioning of soluble, particulate, and especially bioavailable P transported in runoff and their dynamics in lakes. This should focus on the mechanisms of exchange between sediment and solution P and standardization of routine methods to quantify bioavailable P. With the accumulation of fertilizer and residual P at the soil surface under conservation tillage practices, the relative importance of the partitioning processes outlined above may need to be reevaluated. In particular, more accurate simulations of P release from residual soil P under long-term cropping are needed. With the move to more efficient systems, models will enable evaluation of P transport in runoff from soils with high residual P levels with minimal external P inputs.

Although many models are available, it is difficult to select the most appropriate model to obtain the level of detailed information required. Once the appropriate model is chosen, a major limitation is often the lack of input data to drive the model. This most frequently limits model use and the output will only be as reliable as the data input. Consequently, use of these models to provide quantitative estimates of P loss under specific environmental conditions is limited, and the use of many models is recommended to be restricted to the relative comparison of the impact of management options. Because of these limitations more research should be directed to development of a soil index, to identify soil and management practices that may increase the bioavailable P content of surface waters. Agronomic and economic studies have shown that measures to control nonpoint source pollution are much more effective if concentrated on specific source areas rather than on a general basis over large areas (Heatwole et al., 1987; Prato and Wu, 1991). Consequently, index procedures that will specify source areas vulnerable to P loss in runoff, will aid implementation of nonpoint source abatement programs.

It is hoped that answers to these questions will result in agricultural management practices that are both economically and environmentally viable, efficiently utilizing fertilizer and on-farm waste materials, and thereby improving water quality for the next century.

## Acknowledgement

The help of Elaine Mead in preparing the text and graphs for this manuscript is gratefully acknowledged.

## References

Abbott, J.L. 1978. Importance of the organic phosphorus fraction in extracts of calcareous soils. *Soil Sci. Soc. Am. J.* 42:81-85.

Adepetu, J.A., and R.B. Corey. 1976. Organic phosphorus as a predictor of plant available phosphorus in soils of Southern Nigeria. *Soil Sci.* 122:159-164.

Adepoju, A.Y., P.F. Pratt, and S.V. Mattigod. 1982. Availability and extractability of phosphorus from soils having high residual phosphorus. *Soil Sci. Soc. Am. J.* 46:583-588.

Adriano, D.C., L.T. Novak, A.E. Erickson, A.R. Woolcoot, and E. Ellis. 1975. Effect of long-term disposal by spray irrigation of food processing wastes on some chemical properties of the soil and subsurface water. *J. Environ. Qual.* 4:242-248.

Agboola, A.A., and B. Oko. 1976. An attempt to evaluate plant available P in western Nigerian soils under shifting cultivation. *Agron. J.* 68:798-801.

Agronomy Staff. 1989. Illinois Agronomy Handbook - 1989-1990. Univ. IL. at Urbana-Champaign. Coop. Ext. Ser. Cir. No. 1290. p. 53-60.

Alder, K. 1988. CRP water quality report: Signing-up for filter strips. National Water Quality Evaluation Project Notes, No. 31. North Carolina Agric. Ext. Ser., North Carolina State Univ., Raleigh, NC.

Alder, P.R., and S.A. Barber. 1991. Prediction of the calcium oxide effect on phosphorus availability with a mechanistic model. *Commun. Soil Sci. Plant Anal.* 22:147-158.

Alessi, J., and J.F. Power. 1980. Effects of banded and residual fertilizer phosphorus on dryland spring wheat yield in the Northern Plains. *Soil Sci. Soc. Am. J.* 44:792-796.

Alston, A.M. 1980. Response of wheat to deep placement of nitrogen and phosphorus fertilizers on a soil high in phosphorus in the surface layer. *Aust. J. Agric. Res.* 31:13-24.

Anghioni, I., and S.A. Barber. 1980. Predicting the most efficient phosphorus placement for corn. *Soil Sci. Soc. Am. J.* 44:1016-1020.

Armstrong, D.E., J.R. Perry, and D. Flatness. 1979. Availability of pollutants associated with suspended or settled river sediments, which gain access to the Great Lakes. Final Report, Wisconsin Water Res. Cent., Madison, WI.

Bache, B.W., and E.G. Williams. 1971. A phosphate sorption index for soils. *J. Soil Sci.* 22:289-301.

Bahnick, D.A. 1977. The contribution of red clay erosion to orthophosphate loadings into southwestern Lake Superior. *J. Environ. Qual.* 6:217-222.

Bailey, L.D., and C.A. Grant. 1989. Fertilizer phosphorus placement studies on calcareous and noncalcareous chernozemic soils: Growth, P-uptake and yield of flax. *Commun. Soil Sci. Plant Anal.* 20:635-654.

Bailey, L.D., E.D. Spratt, D.W.L. Read, F.G. Warder, and W.S. Ferguson. 1977. Residual effects of phosphorus fertilizer: For wheat and flax grown on chernozemic soils in Manitoba. *Can. J. Soil Sci.* 57:263-270.

Barber, S.A. 1979. Soil phosphorus after 25 years of cropping with five rates of phosphorus application. *Commun. Soil Sci. Plant Anal.* 10:1459-1468.

Barber, S.A., A.D. Mackay, R.O. Kuchenbuch, and P.B. Barraclough. 1988. Effects of soil temperature and water on maize root growth. *Plant Soil* 111:267-269.

Barisas, S.G., J.L. Baker, H.D. Johnson, and J.M. Laflen. 1978. Effect of tillage systems on runoff losses of nutrients, a simulation rainfall study. *Trans. ASAE* 21:893-897.

Barlow, J.P., and M.S. Glase. 1982. Partitioning of phosphorus between particles and water in a river outflow. *Hydrobiologia* 91:253-260.

Barrow, N.J. 1980. Differences amongst a wide-ranging collection of soils in the rate of reaction of phosphate. *Aust. J. Soil Res.* 18:215-224.

Barrow, N.J., and T.C. Shaw. 1979. Effects of solution/soil ratio and vigor of shaking on the rate of phosphate adsorption by soil. *J. Soil Sci.* 30:67-76.

Beasley, D.B., E.J. Monke, E.R. Miller, and L.F. Huggins. 1985. Using simulation to assess the impacts of conservation tillage on movement of sediment and phosphorus into Lake Erie. *J. Soil Water Conserv.* 40:233-237.

Bengtson, R.L., C.E. Carter, H.F. Morris, and S.A. Bartkiewics. 1988. The influence of subsurface drainage practices on nitrogen and phosphorus losses in a warm, humid climate. *Trans. ASAE* 31:729-733.

Bennett, D., and P.G. Ozanne. 1972. Australia, CSIRO Division of Plant Industry. Annual Report. pp. 45.

Berry, J.T., and N.L. Hargett. 1989. Fertilizer summary data, 1988. TVA/NFDC-89/3, Bulletin Y-209. Natl. Fert. Develop. Center, TVA, Muscle Shoals, AL, pp 130.

Bjork, S. 1972. Swedish lake restoration program gets results. *Ambio* 1:153-165.

Black, A.L. 1982. Long-term N-P fertilizer and climate influences on morphology and yield components of spring wheat. *Agron. J.* 74:651-657.

Bolton, E.F., J.W. Aylesworth, and F.R. Hove. 1970. Nutrient losses through tile drainage under three cropping systems and two fertility levels on a Brookston clay soil. *Can. J. Soil Sci.* 50:272-279.

Borkert, C.M., and S.A. Barber. 1985. Soybean shoot and root growth and phosphorus concentration as affected by phosphorus placement. *Soil Sci. Soc. Am. J.* 49:152-155.

Bowman, R.A. 1989. A sequential extraction procedure with concentrated sulfuric acid and dilute base for soil organic phosphorus. *Soil Sci. Soc. Am. J.* 53:362-366.

Bowman, R.A., and C.V. Cole. 1978. An exploratory method for fractionation of organic phosphorus from grassland soils. *Soil Sci.* 125:95-101.

Bray, R.H., and L.T. Kurtz. 1945. Determination of total, organic, and available forms of phosphorus in soils. *Soil Sci.* 59:39-45.

Brookes, P.C., D.S. Powlson, and D.S. Jenkinson. 1982. Measurement of microbial biomass phosphorus in soil. *Soil Biol. Biochem.* 14:319-329.

Brookes, P.C., D.S. Powlson, D.S. Jenkinson. 1984. Phosphorus in the soil microbial biomass. *Soil Biol. Biochem.* 16:169-175.

Brown, M.P., P. Longabucco, M.R. Rafferty, P.D. Robillard, M.F. Walter, and D.A. Haith. 1989. Effects of animal waste control practices on nonpoint-source phosphorus loading in the West Branch of the Delaware River watershed. *J. Soil Water Conserv.* 44:67-70.

Burwell, R.E., D.R. Timmons, and R.F. Holt. 1975. Nutrient transport in surface runoff as influenced by soil cover and seasonal periods. *Soil Sci. Soc. Am. Proc.* 39:523-528.

Burwell, R.E., G.E. Schuman, H.G. Heinemann, and R.G. Spomer. 1977. Nitrogen and phosphorus movement from agricultural watersheds. *J. Soil and Water Conserv.* 32:226-230.

Butkus, S.R., E.B. Welch, R.R. Horner, and D.E. Spyridakis. 1988. Lake response modeling using biologically available phosphorus. *J. Water Pollut. Cont. Fed.* 60:1663-1669.

Carignan, R., and J. Kalff. 1980. Phosphorus sources for aquatic weeds: Water or sediments? *Science* 207:987-989.

Chien, S.H. 1978. Interpretation of Bray I extractable phosphorus from acid soils treated with phosphate rocks. *Soil Sci.* 126:34-39.

Claassen, N., and S.A. Barber. 1974. A method for characterizing the relation between nutrient concentration and flux into roots of intact plants. *Plant Physiol.* 54:564-568.

Cole, C.V., G.S. Innis, and J.W.B. Stewart. 1977. Simulation of phoshorus cycling in semiarid grasslands. *Ecology* 58:1-15.

Cook, D.J., W.T. Dickinson, and R.P. Rudra. 1985. GAMES - The Guelph Model for Evaluating the Effects of Agricultural Management Systems in Erosion and Sedimentation. User's Manual. Tech. Rep. 126-71. School of Engineering, Univ. of Guelph, Guelph, Ontario, Canada.

Cooper, J.R., R.B. Reneau, Jr., W. Kroontje, and G.D. Jones. 1984. Distribution of nitrogenous compounds in a Rhodic Paleudult following heavy manure application. *J. Environ. Qual.* 13:189-193.

Crowder, B., and C.E. Young. 1988. Managing farm nutrients: Tradeoffs for surface- and ground-water quality. Agric. Econ. Rpt. No. 583. U.S. Dept. Agric., U.S. Govt. Print. Off., Washington, DC.

Culley, J.L.B., and E.F. Bolton. 1983. Suspended solids and phosphorus loads from a clay soil: II. Watershed study. *J. Environ. Qual.* 12:498-503.

Culley, J.L.B., E.F. Bolton, and V. Bernyk. 1983. Suspended solids and phosphorus loads from a clay soil: I. Plot studies. *J. Environ. Qual.* 12:493-498.

Dahnke, W.C., L.J. Swenson, and A. Johnson. 1986. Fertilizer placement for small grains. *North Dakota Farm Research* 43(4):36-38.

Dalal, R.C. 1979. Mineralization of carbon and phosphorus from carbon-14 and phosphorus-32 labelled plant material added to soil. *Soil Sci. Soc. Am. J.* 43:913-916.

Daughtrey, Z.W., J.W. Gilliam, and E.J. Kamprath. 1973. Soil test parameters for assessing plant-available P of acid organic soils. *Soil Sci.* 115:438-446.

Dillaha, T.A., R.B. Reneau, S. Mostaghimi, and D. Lee. 1989. Vegetative filter strips for agricultural nonpoint source pollution control. *Trans. ASAE* 32:513-519.

Dillaha, T.A., R.B. Reneau, S. Mostaghimi, V.O. Shanholtz, and W.L. Magette. 1987. Evaluating nutrient and sediment losses from agricultural lands: Vegetative filter strips. U.S. Environ. Prot. Agency, Report No. CBP/TRS 4/87. Washington, DC.

Doerge, T., and E.H. Gardner. 1978. Soil testing for available P in southwest Oregon. p. 143-152. In: Proc. 29th Ann. Northwest Fertilizer Conf., Beaverton, OR.

Donigian, A.S., Jr., D.C. Beyerlein, H.H. Davis, Jr., and N.H. Crawford. 1977. Agricultural Runoff Management (ARM) Model Version II: Refinement and testing. 294 pp. U.S. Environ. Prot. Agency. EPA 600/3-77-098. Environ. Res. Lab., Athens, GA.

Dorich, R.A., D.W. Nelson, and L.E. Sommers. 1980. Algal availability of sediment phosphorus in drainage water of the Black Creek watershed. *J. Environ. Qual.* 9:557-563.

Dorich, R.A., D.W. Nelson, and L.E. Sommers. 1985. Estimating algal available phosphorus in suspended sediments by chemical extraction. *J. Environ. Qual.* 14:400-405.

Duxbury, J.M., and J.H. Peverly. 1978. Nitrogen and phosphorus losses from organic soils. *J. Environ. Qual.* 7:566-570.

Edwards, C.A., R. Lal, P. Madden, R.H. Miller, and G. House. 1988. (eds.) Sustainable agricultural systems. Pub. Soil and Water Conserv. Soc., Ankeny, IA.

Fertilizer Handbook. 1982. The fertilizer handbook. pp. 274. *In* W.C. White and D.N. Collins (eds.), Pub. The Fertilizer Institute, Washington, DC.

Fiedler, R.J., D.H. Sander, and G.A. Peterson. 1987. Predicting winter wheat grain yield response to applied P with different soil P tests and sampling depths. *J. Fert. Issues* 4:19-28.

Field, J.A., R.B. Reneau, and W. Kroontje. 1985. Effects of anaerobically digested poultry manure on soil phosphorus adsorption and extractability. *J. Environ. Qual.* 14:105-107.

Fixen, P.E. 1986. Residual effects of P fertilization: Lessons for the 80's. p.1-8. *In* Proc. of the Sixteenth North Central Extension-Industry Soil Fertility Workshop. Oct. 29-30, 1986, St. Louis, MO. Potash and Phosphate Institute, Atlanta, GA.

Fixen, P.E., and J.N. Grove. 1990. Testing soils for phosphorus. p. 141-180. *In* R.L. Westerman (ed.), Soil Testing and Plant Analysis, 3rd ed. SSSA Book Series No. 3. Soil Sci. Soc. Am., Madison, WI.

Fixen, P.E., and A.D. Halvorson. 1991. Optimum phosphorus management for small grain production. *Better Crops with Plant Food.* 75(3):26-29.

Fixen, P.E. and D.F. Leikam. 1989. Understanding phosphorus placement. *Better Crops with Plant Food.* 73:18-21.

Follett, R.H., D.G. Westfall, J.W. Echols, R.L. Croissant, and J.S. Quick. 1987. Integration of soil fertility trials into on-going cultivar testing programs. *J. Agron. Educ.* 16:81-84.

Forster, D.L., T.J. Logan, S.M. Yaksich, and J.R. Adams. 1985. An accelerated implementation program for reducing the diffuse-source phosphorus load to Lake Erie. *J. Soil and Water Conserv.* 40:136-141.

Fox, R.L., and E.J. Kamprath. 1971. Adsorption and leaching of P in acid organic soils and high organic matter sand. *Soil Sci. Soc. Am. Proc.,* 35:154-156.

Francis, C.A., C.B. Flora, and L.D. King. 1990. (eds.) Sustainable agriculture in temperate zones. J. Wiley and Sons, New York, NY.

Gburek, W.J., and W.R. Heald. 1974. Soluble phosphate output of an agricultural watershed in Pennsylvania. *Water Resour. Res.* 10:113-118.

Gilbertson, C.B., T.M. McCalla, J.R. Ellis, E.O. Cross, and W.R. Woods. 1970. The effects of animal density and surface slope on the characteristics of runoff, soil wastes, and nitrate movement on unpaired beef feedlots. *Nebr. Agric. Exp. Sta. Bull.* SB508.

Gilbertson, C.B., F.A. Norstadt, A.C. Mathers, R.F. Holt, A.P. Barnett, T.M. McCalla, C.A. Onstad, and R.A. Young. 1979. Animal waste utilization on cropland and pastureland. USDA Utilization Research Report No. 6. 135 p.

Gotoh, S., and W.H. Patrick, Jr. 1974. Transformations of iron in a waterlogged soil as influenced by redox potential and pH. *Soil Sci. Soc. Am. Proc.* 38:66-71.

Gray, C.B.J., and R.A. Kirkland. 1986. Suspended sediment phosphorus composition in tributaries of the Okanagan Lakes, BC. *Water Res.* 20:1193-1196.

Haas, H.J., D.L. Grunes, and G.A. Reichman. 1961. Phosphorus changes in Great Plains soils as influenced by cropping and manure applications. *Soil Sci. Soc. Am. Proc.* 24:214-218.

Halvorson, A.D. 1986a. Soil test and P rate relationships to maximum yield: West. *In* Proc. Maximum Wheat Yield Systems Workshop, Potash and Phosphate Institute, Denver, CO, March 5-7, 1986.

Halvorson, A.D. 1986b. Phosphorus management for MEY and quality. *In* Implementing Maximum Economic Wheat Yield Systems Workshop Proceedings, July 8-11, 1986. Bismarck, ND. Dept. Soil Science, North Dakota State University.

Halvorson, A.D. 1987. Utilizing soil testing for greater wheat profitability. *In* Proceedings of the National Wheat Research Conference, Feb. 24-26, 1987, Kansas City, MO. National Assoc. of Wheat Growers Foundation, Washington, DC.

Halvorson, A.D. 1989. Multiple-year response of winter wheat to a single application of phosphorus fertilizer. *Soil Sci. Soc. Am. J.* 53:1862-1868.

Halvorson, A.D., and A.L. Black. 1985a. Long-term dryland crop responses to residual phosphorus fertilizer. *Soil Sci. Soc. Am. J.* 49:928-933.

Halvorson, A.D., and A.L. Black. 1985b. Fertilizer phosphorus recovery after seventeen years of dryland cropping. *Soil Sci. Soc. Am. J.* 49:933-937.

Halvorson, A.D. and P.O. Kresge. 1982. FLEXCROP: A dryland cropping systems model. U.S. Dept. of Agric. Production Research Report No. 180.

Halvorson, A.D., M.M. Alley, and L.S. Murphy. 1987a. Management of the wheat crop: Nutrient requirements and fertilizer use. *In* E. G. Heyne (ed.), Wheat and Wheat Improvement. ASA Monograph No.13, 2nd Edition, Madison, WI.

Halvorson, A.D., E.H. Vasey, and D.L. Watt. 1987b. PHOSECON: A computer economics program to evaluate phosphorus fertilization of wheat. *Applied Agric. Res.* 2:207-212.

Halvorson, A.D., A.L. Black, D.L. Watt, and A.G. Leholm. 1986. Economics of a one-time phosphorus application in the northern Great Plains. *Applied Agric. Res.* 1:137-144.

Hanna, M. 1989. Biologically available phosphorus: Estimation and prediction using an anion-exchange resin. *Can. J. Fish. Aquat. Sci.* 46:638-643.

Harapiak, J.T. and N.A. Flore. 1984. Preplant banding of N and N-P fertilizers: Western Canadian research. p. 43-51. *In* Proceedings of 35th Annual Northwest Fertilizer Conference, Pasco, WA, July 17-18, 1984.

Harrison, A.F. 1978. Phosphorus cycles of forest and upland grassland systems and some effects of land management practices. pp. 175-195. *In* Phosphorus in the Environment: Its Chemistry and Biochemistry. CIBA Foundation Symp. 57, Amsterdam, Elsevier/North-Holland.

Harrison, A.F. 1982. P-32 method to compare rates of mineralization of labile organic phosphorus in woodland soils. *Soil Biol. Biochem.* 15:93-99.

Havlin, J.L., D.G. Westfall, and H.M. Golus. 1984. Six years of phosphorus and potassium fertilization of irrigated alfalfa on calcareous soils. *Soil Sci. Soc. Am. J.* 48:331-336.

Heatwole, C.D., A.B. Bottcher, and L.B. Baldwin. 1987. Modeling cost-effectiveness of agricultural nonpoint pollution abatement programs on two Florida basins. *Water Res. Bull.* 23:127-131.

Hedley, M.J., and J.W.B. Stewart. 1982. Method to measure microbial phosphate in soils. *Soil Biol. Biochem.* 14:377-385.

Hedley, M.J., J.W.B. Stewart, and B.S. Chanhan. 1982. Changes in inorganic and organic soil phosphorus fraction induced by cultivation practices and by laboratory incubations. *Soil Sci. Soc. Am. J.* 46:970-976.

Hedley, M.J., R.W. Tillman, and G. Wallace. 1989. The use of nitrogen fertilizers for increasing the suitability of reactive phosphate rocks for use in intensive agriculture. *In* R.W. White and L.D. Currie (eds.), Nitrogen Fertilizer Use in New Zealand Agriculture and Horticulture. Occasional Report No. 3, Fertilizer and Lime Research Centre, Massey University, New Zealand.

Hegemann, D.A., A.H. Johnson, and J.D. Keenan. 1983. Determination of algae-available phosphorus on soil and sediment: A review and analysis. *J. Environ. Qual.* 12:12-16.

Helyar, K.R., and D.P. Godden. 1976. The phosphorus cycle - What are the sensitive areas? pp. 23-30. *In* G.J. Blair (ed.), Prospects for Improving Efficiency of Phosphorus Utilization. Proc. of Symp. at Univ. of New England, Armidale, N.S.W. Australia Reviews in Rural Sci. III.

Hobbie, J.E., and G.E. Likens. 1973. Output of phosphorus, organic carbon, and fine particulate carbon from Hubbard Brook watersheds. *Limnol. Oceanog.* 18:734-742.

Holford, I.C.R. 1989. Efficacy of different phosphate application methods in relation to phosphate sorptivity in soils. *Aust. J. Soil Res.* 27:123-133.

Holt, R.F., D.R. Timmons, and J.J. Latterell. 1970. Accumulation of phosphates in water. *J. Agric. Food Chem.* 18:781-784.

Hooker, M.L. 1976. Soil Sampling intensities required to estimate available N and P in five Nebraska soil types. MS thesis, Univ. Nebraska, Lincoln, NE. (Cat. No. LD 3656 H665X 1976).

Hooker, M.L., G.A. Peterson, D.H. Sander, and L.A. Daigger. 1980. Phosphate fractions in calcareous soils as altered by time and amounts of added phosphate. *Soil Sci. Soc. Am. J.* 44:269-277.

Hope, G.D., and J.K. Syers. 1976. Effects of solution to soil ratio on phosphate sorption by soils. *J. Soil Sci.* 27:301-306.

House, W.A., and H. Casey. 1988. Transport of phosphorus in rivers. p. 253-282. *In* H. Tiessen (ed.), Phosphorus Cycles in Terrestrial and Aquatic Ecosystems. 1. Europe. SCOPE, UNEP, Univ. Saskatchewan, Saskatoon, Canada.

Huettl, P.J., R.C. Wendt, and R.B. Corey. 1979. Prediction of algal available phosphorus in runoff suspension. *J. Environ. Qual.* 4:541-548.

Janssen, K.A., D.A. Whitney, and D.E. Kissel. 1985. Phosphorus application frequency and sources for grain sorghum. *Soil Sci. Soc. Am. J.* 49:754-758.

Jones, C.A., A.N. Sharpley, and J.R. Williams. 1984a. A simplified soil and plant phosphorus model: III. Testing. *Soil Sci. Soc. Am. J.* 48:810-813.

Jones, C.A., C.V. Cole, A.N. Sharpley, and J.R. Williams. 1984b. A simplified soil and plant phosphorus model: I. Documentation. *Soil Sci. Soc. Am. J.* 48:800-805.

Jones, R.A., and G.F. Lee. 1982. Recent advances in assessing eutrophication for water quality management. *J. Water Res.* 16:503-515.

Jose, H.D. 1981. An economic comparison of batch and annual phosphorus fertilizer application in wheat production in western Canada. *Can. J. Soil Sci.* 61:47-54.

Khalid, R.A., W.H. Patrick, Jr., and R.D. Delaune. 1977. Phosphorus sorption characteristics of flooded soils. *Soil Sci. Soc. Am. J.* 41:301-305.

Khasawneh, F.E., E.C. Sample, and E.J. Kamprath. 1988. The role of phosphorus in agriculture. ASA, CSSA, and SSSA, Madison, WI.

King, L.D., J.C. Burns, and P.W. Westerman. 1990. Long-term swine lagoon effluent applications on "Coastal" Bermudagrass: II. Effects on nutrient accumulations in soil. *J. Environ. Qual.* 19:756-760.

Kitchen, N.R., J.L. Havlin, and D.G. Westfall. 1990. Soil sampling under no-till banded phosphorus. *Soil Sci. Soc. Am. J.* 54:1661-1665.

Klausner, S.D., P.J. Zwerman, and D.F. Ellis. 1974. Surface runoff losses of soluble nitrogen and phosphorus under two systems of soil management. *J. Environ. Qual.* 3:42-46.

Klotz, R.L. 1988. Sediment control of soluble reactive phosphorus in Hoxie Gorge Creek, New York. *Can. J. Fish. Aquat. Sci.* 45:2026-2034.

Knisel, W.G., Jr. Editor. 1980. CREAMS: A field scale model for chemicals, runoff, and erosion from agricultural management systems. Cons. Res. Rept. No. 26. U.S. Dept. Agric., U.S. Govt. Printing Office, Washington, DC, U.S. 640 p.

Knoblauch, H.C., L. Koloday, and G.D. Brill. 1942. Erosion losses of major plant nutrients and organic matter from Collington sandy loam. *Soil Sci.* 53:369-378.

Kovar, J.L., and S.A. Barber. 1987. Placing phosphorus and potassium for greatest recovery. *J. Fert. Issues* 4:1-6.

Kuchenbuch, R.O., and S.A. Barber. 1987. Yearly variation in root distribution with depth in relation to nutrient uptake and corn yield. *Commun. Soil Sci. Plant Anal.* 18:255-263.

Kunishi, H.M., A.W. Taylor, W.R. Heald, W.J. Gburek, and R.N. Weaver. 1972. Phosphate movement from an agricultural watershed during two rainfall periods. *J. Agric. Food Chem.* 20:900-905.

Kuo, S. 1990. Phosphate sorption implications on phosphate soil tests and uptake by corn. *Soil Sci. Soc. Am. J.* 54:131-135.

Laflen, J.M., L.J. Lane, and G.R. Foster. 1991. WEPP: A new generation of erosion prediction technology. *J. Soil Water Conserv.* 46:34-38.

Lammond, R.E. 1987. Comparison of fertilizer solution placement methods for grain sorghum under two tillage systems. *J. Fert. Issues* 4:43-47.

Langdale, G.W., R.A. Leonard, and A.N. Thomas. 1985. Conservation practice effects on phosphorus losses from Southern Piedmont watersheds. *J. Soil Water Conserv.* 40:157-160.

Larsen, S., D. Gunnary, and C.D. Sutton. 1965. The rate of immobilization of applied phosphate in relation to soil properties. *J. Soil Sci.* 16:141-148.

Leikam, D.F. 1987. Phosphorus fertility management - A key to profitability. *In* Proc. Central Great Plains Profitable Wheat Management Workshop, Wichita, KS, August 18-20, 1987. Potash and Phosphate Institute, Atlanta, GA.

Leikam, D.F., L.S. Murphy, D.E. Kissel, D.A. Whitney, and H.C. Moser. 1983. Effects of nitrogen and phosphorus application method and nitrogen source on winter wheat yield and leaf tissue phosphorus. *Soil Sci. Soc. Am. J.* 47:530-535.

Liebhardt, W.C., C. Golt, and J. Tupin. 1979. Nitrate and ammonium concentrations of ground water resulting from poultry manure applications. *J. Environ. Qual.* 5:211-215.

Logan, T.J. 1982. Mechanisms for release of sediment-bound phosphate to water and the effects of agricultural land management on fluvial transport of particulate and dissolved phosphates. *Hydrobiologia* 92:519-530.

Logan, T.J., T.O. Oloya, and S.M. Yaksich. 1979. Phosphate characteristics and bioavailability of suspended sediments from streams draining into Lake Erie. *J. Great Lakes. Res.* 5:112-123.

Lopez-Hernandez, D., and C.P. Burnham. 1974. The covariance of phosphate sorption with other soil properties in some British and tropical soils. *J. Soil Sci.* 25:196-206.

Lowrance, R.R., R.A. Leonard, and J.M. Sheridan. 1985. Managing riparian ecosystems to control non-point pollution. *J. Soil and Water Conserv.* 40:87-91.

Lowrance, R.R., R.L. Todd, and L.E. Asmussen. 1984a. Nutrient cycling in an agricultural watershed: II. Stream flow and artificial drainage. *J. Environ. Qual.* 13:27-32.

Lowrance, R.R., R.L. Todd, J. Fail, Jr., O. Hendrickson, Jr., R. Leonard, and L. Asmussen. 1984b. Riparian forests as nutrient filters in agricultural watersheds. *BioScience* 34:374-377.

McColl, R.H.S., E. White, and A.R. Gibson. 1977. Phosphorus and nitrate runoff in hill pasture and forest catchments, Taita, New Zealand. *N.Z. J. Mar. Freshwater Res.* 11:729-744.

McCollum, R.E. 1991. Buildup and decline in soil phosphorus: 30-year trends on a Typic Umprabuult. *Agron. J.* 83:77-85.

McDowell, L.L., and K.C. McGregor. 1980. Nitrogen and phosphorus losses in runoff from no-till soybeans. *Trans. ASAE* 23:643-648.

McDowell, L.L., and K.C. McGregor. 1984. Plant nutrient losses in runoff from conservation tillage corn. *Soil Tillage Res.* 4:79-91.

McLaughlin, M.J., and A.M. Alston. 1986. The relative contribution of plant residues and fertilizer to the phosphorus nutrition of wheat in a pasture/cereal rotation. *Aust. J. Soil Res.* 24:517-526.

McLaughlin, M.J., A.M. Alston, and J.K. Martin. 1988. Phosphorus cycling in wheat-pasture rotations. III. Organic phosphorus turnover and phosphorus cycling. *Aust. J. Soil Res.* 26:343-353.

McLeod, R.V., and R.O. Hegg. 1984. Pasture runoff water quality from application of inorganic and organic nitrogen sources. *J. Environ. Qual.* 13:122-126.

Mackay, A.D., and S.A. Barber. 1984. Soil temperature effects on root growth and phosphorus uptake by corn. *Soil Sci. Soc. Am. J.* 48:818-823.

Mackay, A.D., J.K. Syers. P.E.H. Gregg, and R.W. Tillman. 1984. A comparison of three soil testing procedures for estimating plant-available phosphorus in soils using either super phosphates or phosphate rock. *N.Z. J. Agric. Res.* 27:231-245.

Magette, W.L. 1988. Runoff potential from poultry manure applications. p. 102-106. *In* E.C. Naber (ed.), Proc. Natl. Poultry Waste Management Symp., Columbus, OH 1988. Ohio State Univ. Press, Columbus, OH.

Magette, W.L., R.B. Brinsfield, R.E. Palmer, and J.D. Wood. 1989. Nutrient and sediment removal by vegetated filter strips. *Trans. ASA E.* 32:663-667.

Magette, W.L., R.B. Brinsfield, R.E. Palmer, J.D. Wood, T.A. Dillaha, and R.B. Reneau. 1987. Vegetative filter strips for agricultural runoff treatment. U.S. Environ. Prot. Agency, Report No. CBP/TRS2/87. Washington, DC.

Mathers, A.C., B.A. Stewart, and B. Blair. 1975. Nitrate removal from soil profiles by alfalfa. *J. Environ. Qual.* 4:403-405.

Mehlich, A. 1984. Mehlich 3 soil test extractant: A modification of Mehlich 2 extractant. *Commun. Soil Sci. Plant Anal.* 15:1409-1416.

Menon, R.G., S.H. Chien, and L.L. Hammond. 1989a. Comparison of Bray 1 and $P_i$ tests for evaluating plant-available phosphorus from soils treated with different partially acidulated phosphate rocks. *Plant Soil* 114:211-216.

Menon, R.G., L.L. Hammond, and H.A. Sissingh. 1989b. Determination of plant-available phosphorus by the iron hydroxide-impregnated filter paper ($P_i$) soil test. *Soil Sci. Soc. Am. J.* 52:110-115.

Menon, R.G., S.H. Chien, L.L. Hammond, and J. Henoa. 1989c. Modified techniques for preparing paper strips for the new $P_i$ soil test for phosphorus. *Fert. Res.* 19:85-91.

Menon, R.G., S.H. Chien, L.L. Hammond, and B.R. Arora. 1990. Sorption of phosphorus by the iron oxide-impregnated filter paper ($P_i$ soil test) embedded in soils. *Plant Soil* 126:287-294.

Menzel, R.G. 1980. Enrichment ratios for water quality modeling. p. 486-492. *In* W. Knisel (ed.), CREAMS - A Field Scale Model for Chemicals, Runoff and Erosion from Agricultural Management Systems. Vol. III. Supporting Documentation, USDA, Cons. Res. Rep. 26. U.S. Govt. Printing Office, Washington, DC.

Meyer, J.L. 1979. The role of sediments and bryophites in phosphorus dynamics in a headwater stream ecosystem. *Limnol. Oceanog.* 24:365-375.

Miller, M.H. 1979. Contribution of nitrogen and phosphorus to subsurface drainage water from intensively cropped mineral and organic soils in Ontario. *J. Environ. Qual.* 8:42-48.

Moody, P.W., R.L. Aitken, B.L. Compton, and S. Hunt. 1988. Soil phosphorus parameters affecting phosphorus availability to, and fertilizer requirements of, Maize (*Zea mays*). *Aust. J. Soil Res.* 26:611-622.

Muchovej, R.M.C., J.J. Muchovej, and V.H. Alvarez. 1989. Temporal relations of phosphorus fractions in a Oxisol amended with rock phosphate and *Thiobacillus thioxidans*. *Soil Sci. Soc. Am. J.* 53:1096-1100.

Muir, J., E.C. Seim, and R.A. Olson. 1973. A study of factors influencing the nitrogen and phosphorus contents of Nebraska waters. *J. Environ. Qual.* 2:466-470.

Munson, R.D. and L.S. Murphy. 1986. Factors affecting crop response to phosphorus. p. 9-24. *In* Phosphorus for Agriculture, A Situation Analysis. Potash and Phosphate Institute, Atlanta, GA.

Murphy, L.S. 1988. Phosphorus management strategies for MEY of spring wheat. *In* Proc. Profitable Spring Wheat Production Workshop, Jan. 6-7, 1988, Fargo, ND.

Murphy, L.S. and D.W. Dibb. 1986. Phosphorus and placement. p. 35-48. *In* Phosphorus for Agriculture, A Situation Analysis. Potash and Phosphate Institute, Atlanta, Ga.

Neal, O.R. 1944. Removal of nutrients from the soil by plants and crops. *Agron. J.* 36:601-607.

Nicholaichuk, W., and D.W.L. Read. 1978. Nutrient runoff from fertilized and unfertilized fields in western Canada. *J. Environ. Qual.* 7:542-544.

Nurnberg, G.K., and R.H. Peters. 1984. Biological availability of soluble reactive phosphorus in anoxic and oxic freshwaters. *Can. J. Fish. Aquat. Sci.* 41:757-765.

Olsen, S.R., and L.E. Sommers. 1982. Phosphorus. *In* A.L. Page et al. (eds.), Methods of Soil Analysis, Part 2, 2nd edition. *Agronomy* 9:403-429.

Olsen, S.R., C.V. Cole, F.S. Watanabe, and L.A. Dean. 1954. Estimation of available phosphorus in soils by extraction with sodium bicarbonate. USDA Circ. 939. U.S. Govt. Print. Office, Washington, DC.

Omernik, J.M. 1977. Nonpoint source - stream nutrient level relationships: A nationwide study. EPA-600/3-77-105. Corvallis, OR.

Oplinger, E.S., D.W. Wiersma, C.R. Grau, and K.A. Kelling. 1985. Intensive wheat management. *Univ. Wisconsin, Coop. Ext. Serv. Bull.* A3337, Madison, WI.

Organization for Economic Cooperation and Development. 1982. Eutrophication of waters: Monitoring, assessment, and control. O.E.C.D. Paris, France.

Ozanne, P.G., D.J. Kirton, and T.C. Shaw. 1961. The loss of phosphorus from sandy soils. *Aust. J. Agric. Res.* 12:409-423.

Parton, W.J., J.W.B. Stewart, and C.V. Cole. 1988. Dynamics of C, N, P and S in grassland soils: A model. *Biogeochemistry* 5:109-131.

Parton, W.J., D.W. Anderson, C.V. Cole, and J.W.B. Stewart. 1983. Simulation of soil organic matter formations and mineralization in semiarid agroecosystems. p. 533-550. *In* R.R. Lowrance, R.L. Todd, L.E. Asmussen, and R.A. Leonard (eds.), Nutrient Cycling in Agricultural Ecosystems. The Univ. of Georgia, College of Agric. Expt. Sta., Special Publ. 23. Athens, GA.

Parton, W.J., D.S. Schimel, C.V. Cole, and D.S. Ojima. 1987. Analysis of factors controlling soil organic matter levels in Great Plains grasslands. *Soil Sci. Soc. Am. J.* 51:1173-1179.

Pesant, A.R., J.L. Dionne, and J. Genest. 1987. Soil and nutrient losses in surface runoff from conventional and no-till corn systems. *Can. J. Soil Sci.* 67:835-843.

Peterjohn, W.T., and D.L. Correll. 1984. Nutrient dynamics in an agricultural watershed: Observations on the role of a riparian forest. *Ecology* 65:1466-1475.

Peters, R.H. 1977. Availability of atmospheric orthophosphate. *J. Fish. Res. Board. Can.* 34:918-924.

Peterson, G.A., D.H. Sander, P.H. Grabouski, and M.L. Hooker. 1981. A new look at row and broadcast phosphate recommendations for winter wheat. *Agron. J.* 73:13-17.

Peterson, L.A., and A.R. Kreuger. 1980. Variation in content of available P and K (Bray I) in soil samples from a cropped N, P, and K fertility experiment over 8 years. *Commun. Soil Sci. Plant Anal.* 11:993-1004.

Pierzynski, G.M., T.J. Logan, and S.J. Traina. 1990. Phosphorus chemistry and mineralogy in excessively fertilized soils: Solubility equilibria. *Soil Sci. Soc. Am. J.* 54:1589-1595.

Ponnamperuma, F.N. 1972. The chemistry of submerged soils. *Adv. Agron.* 24:29-96.

Porcella, D.B., J.S. Kumazar, and E.J. Middlebrooks. 1970. Biological effects on sediment-water nutrient interchange. *J. Sanit. Eng. Div., Proc. Am. Soc. Civil Eng.* 96:911-926.

Potash and Phosphate Institute. 1985. Soil test summary for phosphorus and potassium. *Better Crops with Plant Food* 69:16-17.

Potash and Phosphate Institute. 1987a. Soil test summary for phosphorus and potassium. *Better Crops with Plant Food.* 71:12-13.

Potash and Phosphate Institute. 1987b. The vital role of phosphorus in our environment. Potash and Phosph. Inst. Pub. 11-87-A, Atlanta, GA.

Potash and Phosphate Institute. 1989. Conventional and low-input agriculture. Potash and Phosphate Inst., Atlanta, GA.

Potter, R.L., C.F. Jordan, R.M. Guedes, G.J. Batmanian, and X.G. Han. 1991. Assessment of a phosphorus fractionation method for soils: Problems for further investigation. *Agric. Ecosystems and Environ.* 34:453-463.

Power, J.F., W.W. Wilhelm, and J.W. Doran. 1986. Crop residue effects on soil environment and dryland maize and soybean production. *Soil Tillage Res.* 8:101-111.

Prato, T., and S. Wu. 1991. Erosion, sediment and economic effects of conservation compliance in an agricultural watershed. *J. Soil Water Conserv.* 46:211-214.

Rast, W., and G.F. Lee. 1978. Summary analysis of the North American (U.S. Portion) OECD eutrophication project: Nutrient loading - lake response relationships and trophic state indices. EPA 600/3-78-008, U.S. EPA, Corvallis, OR.

Read, D.W.L., E.D. Spratt, L.D. Bailey, and F.G. Warder. 1977. Residual effects of phosphorus fertilizer. I. For wheat grown on four chernozemic soil types in Saskatchewan and Manitoba. *Can. J. Soil Sci.* 57:255-262.

Reddy, K.R., M.R. Overcash, R. Kahled, and P.W. Westerman. 1980. Phosphorus absorption-desorption characteristics of two soils utilized for disposal of animal manures. *J. Environ. Qual.* 9:86-92.

Rehm, G. 1991. Management of phosphate fertilizers in conservation tillage production systems: The midwest. p. 108-111. *In* F.J. Sikora (ed.), Future directions for agricultural phosphorus research. National Fert. and Environ. Res. Cent., TVA, Muscle Shoals, AL.

Rehm, G.W., R.C. Sorensen, and R.A. Wiese. 1984. Soil test values for phosphorus, potassium and zinc as affected by rate applied to corn. *Soil Sci. Soc. Am. J.* 48:814-818.

Renard, K.G., G.R. Foster, G.A. Weesies, and J.P. Porter. 1991. RUSLE: Revised universal soil loss equation. *J. Soil Water Conserv.* 46:30-33.

Ritter, W.F., and A.E.M. Chirnside. 1984. Impact of land use on groundwater quality in Southern Delaware. *Ground Water* 22:39-47.

Roberts, T.L., and J.W.B. Stewart. 1987. Update of residual fertilizer phosphorus in western Canadian soils. Saskatchewan Institute of Pedology Publication No. R523.

Rogers, M.T. 1941. Plant nutrient losses by erosion from a corn, wheat, clover rotation on Dunmore silt loam. *Soil Sci. Soc. Am. Proc.* 6:263-271.

Ryden, J.C., and J.K. Syers. 1977. Desorption and isotopic exchange relationships of phosphate sorbed by soils and hydrous ferric oxide gel. *J. Soil Sci.* 28:596-609.

Ryden, J.C., J.K. Syers, and R.F. Harris. 1973. Phosphorus in runoff and streams. *Adv. Agron.* 25:1-45.

Ryding, S.O., M. Enell, and L. Wennberg. 1990. Swedish agricultural nonpoint source pollution: A summary of research and findings. *Lake and Reserv. Mgt.* 6:207-217.

Sagger, S., M.J. Hedley, and R.E. White. 1990. A simplified resin membrane technique for extracting phosphorus from soils. *Fert. Res.* 24:173-180.

Sagher, A., R.F. Harris, and D.E. Armstrong. 1975. Availability of sediment phosphorus to microorganisms. Univ. of Wisc. Water Resour. Center, Tech. Report. WIC WRC 75-01. Madison, WI. pp. 56.

Sawhney, B.L. 1977. Predicting phosphate movement through soil columns. *J. Environ. Qual.* 6:86-89.

Sawyer, C.N. 1947. Fertilization of lakes by agricultural and urban drainage. *J. New England Water Works Assoc.* 61:109-127.

Schenk, M.K., and S.A. Barber. 1979a. Phosphate uptake by corn as affected by soil characteristics and root morphology. *Soil Sci. Soc. Am. J.* 43:880-883.

Schenk, M.K., and S.A. Barber. 1979b. Root characteristics of corn genotypes as related to P uptake. *Agron. J.* 71:921-924.

Schlegel, A., R.E. Gwin, and W.A. Conrad. 1986. Effect of nitrogen, phosphorus, and potassium on irrigated corn and sorghum. Kansas State University, Agric. Experiment Sta. Report of Progress 509, p 45-48.

Schlosser, I.J., and J.R. Karr. 1981. Water quality in agricultural watersheds: Impact of riparian vegetation during base flow. *Water Resour. Bull.* 17:233-240.

Schoenau, J.J., and W.Z. Huang. 1991. Assessing P, N, S and K availability in soil using anion and cation exchange membranes. p. 131-136. *In* Proc. Western Phosphate and Sulfur Workgroup. Ft. Collins, CO.

Schreiber, J.D. 1990. Estimating soluble phosphorus from green crops and their residues in agricultural runoff. p. 77-95. *In* D.G. DeCoursey (ed.), Small watershed model (SWAM) for water, sediment and chemical movement: Supporting documentation U.S. Dept. Agric., U.S. Govt. Printing Office, Washington, DC. ARS-80.

Schreiber, J.D., P.D. Duffy, and D.C. McClurkin. 1976. Dissolved nutrient losses in storm runoff from five southern pine watersheds. *J. Environ. Qual.* 5:201-205.

Shapiro, C.A. 1988. Soil sampling fields with a history of fertilizer bands. *In* Soil Science New - Nebraska Cooperative Extension Service. Vol. 10, No. 5.

Sharpley, A.N. 1981. The contribution of phosphorus leached from crop canopy to losses in surface runoff. *J. Environ. Qual.* 10:160-165.

Sharpley, A.N. 1983. Effect of soil properties on the kinetics of phosphorus desorption. *Soil Sci. Soc. Am. J.* 47:805-809.

Sharpley, A.N. 1985a. Phosphorus cycling in unfertilized and fertilized agricultural soils. *Soil Sci. Soc. Am. J.* 49:905-911.

Sharpley, A.N. 1985b. Depth of surface soil-runoff interaction as affected by rainfall, soil slope and management. *Soil Sci. Soc. Am. J.* 49:1010-1015.

Sharpley, A.N. 1985c. The selective erosion of plant nutrients in runoff. *Soil Sci. Soc. Am. J.* 49:1527-1534.

Sharpley, A.N. 1991. Soil phosphorus extracted by iron-aluminum-oxide-impregnated filter paper. *Soil Sci. Soc. Am. J.* 55:1038-1041.

Sharpley, A.N., and R.G. Menzel. 1987. The impact of soil and fertilizer phosphorus on the environment. *Adv. Agron.* 41:297-324.

Sharpley, A.N., and S.J. Smith. 1983. Distribution of phosphorus forms in virgin and cultivated soil and potential erosion losses. *Soil Sci. Soc. Am. J.* 47:581-586.

Sharpley, A.N., and S.J. Smith. 1989a. Mineralization and leaching of phosphorus from soil incubated with surface-applied and incorporated crop residues. *J. Environ. Qual.* 18:101-105.

Sharpley, A.N., and S.J. Smith. 1989b. Prediction of soluble phosphorus transport in agricultural runoff. *J. Environ. Qual.* 18:313-316.

Sharpley, A.N., and S.J. Smith. 1990. Phosphorus transport in agricultural runoff: The role of soil erosion. p. 351-366. *In* J. Boardman, I.D.L. Foster, and J.A. Dearing (eds.), Soil Erosion on Agricultural Land. J. Wiley and Sons, London, UK.

Sharpley, A.N., and S.J. Smith. 1992. Application of phosphorus bioavailability indices to agricultural runoff and soils. p. 43-57. *In* K. Hoddinott (ed.), Application of agricultural analysis in environmental studies. Am. Soc. Techniques and Materials, Philadelphia, PA.

Sharpley, A.N., and J.K. Syers. 1976. Phosphorus transport in surface runoff as influenced by fertilizer and grazing cattle. *N.Z. J. Sci.* 19:277-282.

Sharpley, A.N., and J.K. Syers. 1979. Phosphorus inputs into a stream draining an agricultural watershed: II. Amounts and relative significance of runoff types. *Water, Air and Soil Pollut.* 11:417-428.

Sharpley, A.N., and J.R. Williams (eds.). 1990. EPIC-Erosion/Productivity Impact Calculator. 1. Model documentation. USDA Technical Bull. 1768. 235 pp. U.S. Govt. Print. Office, Washington, DC.

Sharpley, A.N., J.K. Syers, and P.E.H. Gregg. 1978. Transport in surface runoff of phosphate derived from dicalcium phosphate and superphosphate. *N.Z. J. Sci.* 21:301-310.

Sharpley, A.N., W.W. Troeger, and S.J. Smith. 1991a. The measurement of bioavailable phosphorus in agricultural runoff. *J. Environ. Qual.* 20:235-238.

Sharpley, A.N., C.A. Jones, C. Gray, and C.V. Cole. 1984a. A simplified soil and plant phosphorus model: II. Prediction of labile, organic, and sorbed phosphorus. *Soil Sci. Soc. Am. J.* 48:805-809.

Sharpley, A.N., S.J. Smith, W.A. Berg, and J.R. Williams. 1985. Nutrient runoff losses as predicted by annual and monthly soil sampling. *J. Environ. Qual.* 14:354-360.

Sharpley, A.N., S.J. Smith, B.A. Stewart, and A.C. Mathers. 1984b. Forms of phosphorus in soil receiving cattle feedlot waste. *J. Environ. Qual.* 13:211-215.

Sharpley, A.N., U. Singh, G. Uehara, and J. Kimble. 1989. Modeling soil and plant phosphorus dynamics in calcareous and highly weathered soils. *Soil Sci. Soc. Am. J.* 53:153-158.

Sharpley, A.N., S.J. Smith, O.R. Jones, W.A. Berg, and G.A. Coleman. 1992. The transport of bioavailable phosphorus in agricultural runoff. *J. Environ. Qual.* 21:30-35.

Sharpley, A.N., B.J. Carter, B.J. Wagner, S.J. Smith, E.L. Cole, and G.A. Sample. 1991b. Impact of long-term swine and poultry manure applications on soil and water resources in eastern Oklahoma. *Okla. State Univ., Tech. Bull.* T169, 51 pp.

Silberbush, M., and S.A. Barber. 1983. Prediction of phosphorus and potassium uptake by soybean with a mechanistic mathematical model. *Soil Sci. Soc. Am. J.* 47:262-265.

Sims, J.T. 1992. Environmental management of phosphorus in agriculture and municipal wastes. p. 59-64. *In* F.J. Sikora (ed.), Future Directions for Agricultural Phosphorus Research. Nat. Fert. Environ. Res. Cent., TVA Muscle Shoals, AL.

Singh, B.B., and J.P. Jones. 1976. Phosphorus sorption and desorption characteristics of soil as affected by organic residues. *Soil Sci. Soc. Am. J.* 40:389-394.

Skogley, E.O., S.J. Georgitis, J.E. Yang, and B.F. Schaff. 1990. The phytoavailability soil test - PST. *Commun. Soil Sci. Plant Anal.* 21:1229-1243.

Sleight, D.M., D.H. Sander, and G.A. Peterson. 1984. Effect of fertilizer phosphorus placement on the availability of phosphorus. *Soil Sci. Soc. Am. J.* 48:336.

Smith, S.J., A.N. Sharpley, J.W. Naney, W.A. Berg, and O.R. Jones. 1991. Water quality impacts associated with wheat culture in the Southern Plains. *J. Environ. Qual.* 20:244-249.

Soltanpour, P.N., R.L. Fox, and R.C. Jones. 1987. A quick method to extract organic phosphorus from soils. *Soil Sci. Soc. Am. J.* 51:255-256.

Sonzogni, W.C., S.C. Chapra, D.E. Armstrong, and T.J. Logan. 1982. Bioavailability of phosphorus input to lakes. *J. Environ. Qual.* 11:555-563.

Soper, R.J., and Y.P. Kalra. 1969. Effect of mode of application and source of fertilizer on phosphorus utilization by buckwheat, rape, oats, and flax. *Can. J. Soil Sci.* 49:319-326.

Stewart, J.W.B., and A.N. Sharpley. 1987. Controls on dynamics of soil and fertilizer phosphorus and sulfur. p. 101-121. *In* R.F. Follett, J.W.B. Stewart, and C.V. Cole (eds.), Soil fertility and organic matter as critical components of production. SSSA Spec. Pub. 19, Am. Soc. Agron., Madison, WI.

Stewart, J.W.B., and H. Tiessen. 1987. Dynamics of soil organic phosphorus. *Biogeochemistry* 4:41-60.

Stoltenberg, N.L., and J.L. White. 1953. Selective loss of plant nutrients by erosion. *Soil Sci. Soc. Am. Proc.* 17:405-410.

Strong, W.M., and R.J. Soper. 1974. Utilization of pelleted phosphorus by flax, wheat, rape and buckwheat from a calcareous soil. *Agron. J.* 65:18-21.

Summer, R.M., C.V. Alonso, and R.A. Young. 1990. Modeling linked watershed and lake processes for water quality management decisions. *J. Environ. Qual.* 19:421-427.

Syers, J.K., and D. Curtin. 1988. Inorganic reactions controlling phosphorus cycling. p. 17-29. *In* H. Tiessen (ed.), Phosphorus cycles in terrestrial and aquatic ecosystems. UNDP, Pub. by Saskatchewan Inst. Pedology, Saskatoon, Canada.

Tate, K.R., T.W. Spier, D.J. Ross, R.L. Parfitt, K.N. Whale, and J.C. Cowling. 1991. Temporal variations in some plant and soil P pools in two pasture soils of different P fertility status. *Plant Soil* 132:219-232.

Taylor, A.W., and H.M. Kunishi. 1971. Phosphate equilibria on stream sediment and soil in a watershed draining an agricultural region. *J. Agric. Food Chem.* 19:827-831.

Thibaud, M.C., C. Morel, and J.C. Fardeau. 1988. Contribution of phosphorus issued from crop residues to plant nutrition. *Soil Sci. Plant Nutr.* 34:481-491.

Thompson, L.F. 1990. A new rock phosphate. Is there a place for it in our agriculture? Am. Soc. Agron. Abstract. Southern Section Meeting, Little Rock, AK. Feb., 1990.

Tiessen, H., J.W.B. Stewart, and J.O. Muir. 1983. Changes in organic and inorganic P composition of two grassland soils and their particle size fractions during 60-90 years of cultivation. *J. Soil Sci.* 34:815-823.

Timmons, D.R., R.F. Holt, and J.J. Latterell. 1970. Leaching of crop residues as a source of nutrients in surface runoff water. *Water Resour. Res.* 6:1367-1375.

U.S. Corps of Engineers. 1982. Lake Erie wastewater management study. Final Report. Buffalo, NY. 225 pp.

U.S. Environmental Protection Agency. 1971. Algal assay procedure - Bottle test. Nat. Eutrophication Res. Program, Pacific Northwest Laboratory, Corvallis, OR. 82 pp.

U.S. Environmental Protection Agency. 1976. Quality criteria for water. U.S. Govt. Print. Office, Washington, DC.

U.S. Environmental Protection Agency. 1979. Methods for chemical analysis of water and wastes. US-EPA-600-4-79-020. Environmental Monitoring Support Lab., Cincinnati, OH.

U.S. Environmental Protection Agency. 1984. Report to Congress: Nonpoint source pollution in the U.S. U.S. Govt. Printing Office, Washington, DC.

Van der Zee, S.E.A.T.M., L.G.J. Fokkink, and W.H. van Riemsdjkik. 1987. A new technique for assessment of reversibly adsorbed phosphate. *Soil Sci. Soc. Am. J.* 51:599-604.

Vincent, W.F., and M.T. Downes. 1980. Variation in nutrient removal from a stream by water cress (*Nasturtium Officinale* R. BR.). *Aquatic Bot.* 9:221-235.

Vitosh, M.L., J.F. Davis, and B.D. Knezek. 1973. Long-term effects of manure, fertilizer, and plow depth on chemical properties of soils and nutrient movement in a monoculture corn system. *J. Environ. Qual.* 2:296-299.

Vollenwieder, R.A., and J. Kerekes. 1980. The loading concept as a basis for controlling eutrophication: Philosophy and preliminary results of the OECD program on eutrophication. *Progr. Water Technol.* 12:5-38.

Wagar, B.I., J.W.B. Stewart, and J.L. Henry. 1986. Comparison of single large broadcast and small annual seed-placed phosphorus treatments on yield and phosphorus and zinc content of wheat on chernozemic soils. *Can. J. Soil Sci.* 66:237-248.

Walbridge, M.R., and P.M. Vitousek. 1987. Phosphorus mineralization potentials in acid organic soils: Processes affecting $^{32}PO_4^{3-}$ isotope dilution measurements. *Soil Biol. Biochem.* 19:709-717.

Walton, C.P., and G.F. Lee. 1972. A biological evaluation of the molybdenum blue method for orthophosphate analysis. *Tech. Int. Ver. Limnol.* 18:676-684.

Ward, R., and D.F. Leikam. 1986. Soil sampling techniques for reduced tillage and band fertilizer application. *In* Proc. Great Plains Soil Fertility Workshop. March 4-5, 1986, Denver, CO.

Welch, L.F., D.O.L. Mulvaney, L.V. Boone, C.E. McKibben, and J.W. Pendleton. 1966. Relative efficiency of broadcast versus banded phosphorus for corn. *Agron. J.* 58:283-287.

Wendt, R.C., and R.B. Corey. 1980. Phosphorus variations in surface runoff from agricultural lands as a function of land use. *J. Environ. Qual.* 9:130-136.

Westerman, P.W., T.L. Donnely, and M.R. Overcash. 1983. Erosion of soil and poultry manure - A laboratory study. *Trans. ASAE* 26:1070-1078 and 1084.

Westfall, D.G., R.H. Follett, and J.W. Echols. 1986. Fertilization of dryland winter wheat. Colorado State University Service-In-Action No. 114. Fort Collins, CO.

White, E.M. 1973. Water leachable nutrients from frozen or dried prairie vegetation. *J. Environ. Qual.* 2:104-107.

Wildung, R.E., R.L. Schmidt, and A.R. Gahler. 1974. The phosphorus status of eutrophic lake sediments as related to changes in limnological conditions - total, inorganic, and organic phosphorus. *J. Environ. Qual.* 3:133-138.

Williams, J.R. 1975. Sediment yield prediction with universal equation using runoff energy factor. p. 244-252. *In* Present and prospective technology for predicting sediment yields and sources. USDA-SEA, ARS-40. U.S. Govt. Printing Office, Washington, DC.

Williams, J.R., C.A. Jones, and P.T. Dyke. 1984. A modeling approach to determining the relationship between erosion and soil productivity. *Trans. ASAE* 27:129-144.

Wischmeier, W.H., and D.D. Smith. 1965. Predicting rainfall-erosion losses from cropland east of the Rocky Mountains - Guide for selection of practices for soil and water conservation. USDA-Agr. Handbook No. 282. U.S. Govt. Printing Office, Washington, DC.

Wischmeier, W.H., and D.D. Smith. 1978. Predicting rainfall erosion losses - A guide to conservation planning. USDA-Agr. Handbook No. 537. U.S. Govt. Printing Office, Washington, DC.

Wolf, A.M., D.E. Baker, H.B. Pionke, and H.M. Kunishi. 1985. Soil tests for estimating labile, soluble, and algae-available phosphorus in agricultural soils. *J. Environ. Qual.* 14:341-348.

Yang, J.E., E.O. Skogley, S.S. Georgitis, B.E. Schaff, and A.H. Ferguson. 1991. Phytoavailability soil test: Development and verification of theory. *Soil Sci. Soc. Am. J.* 55:1358-1365.

Yao, J., and S.A. Barber. 1986. Effect of one phosphorus rate placed in different soil volumes on P uptake and growth of wheat. *Commun. Soil Sci. Plant Anal.* 17:819-827.

Yates, P., and J.M. Sheridan. 1983. Estimating the effectiveness of vegetated floodplains/wetlands as nitrate-nitrite and orthophosphorus filters. *Agric. Ecosyst. and Environ.* 9:303-314.

Yerokum, O.A., and D.R. Christenson. 1990. Relating high soil test phosphorus concentrations to plant phosphorus uptake. *Soil Sci. Soc. Am. J.* 54:796-799.

Yost, R.S., E.J. Kamprath, G.C. Naderman, and E. Lobato. 1981. Residual effects of phosphorus applications on a high phosphorus adsorbing oxisol of central Brazil. *Soil Sci. Soc. Am. J.* 45:540-543.

Young, R.A., C.A. Onstad, D.D. Bosch, and W.P. Anderson. 1989. AGNP S: A nonpoint-source pollution model for evaluating agricultural watersheds. *J. Soil Water Conserv.* 44:168-173.

Young, T.C., and J.V. DePinto. 1982. Algal - availability of particulate phosphorus from diffuse and point sources in the lower Great Lakes basin. *Hydrobiologia* 91:111-119.

# Impact of Nitrogen Fertilization of Pastures and Turfgrasses on Water Quality

Rosa M.C. Muchovej and Jack E. Rechcigl


I. Nitrogen Fertilizers and Their Uses . . . 91
A. Current and Future Use of Nitrogen Fertilizers . . . 91
B. Sources, Forms and Application . . . 92
C. Nitrogen Requirements of Pastures and Turfgrasses . . . 94
II. Soil Nitrogen Processes . . . 97
A. The Nitrogen Cycle . . . 97
B. Transformations of Nitrogen Fertilizers in Soil . . . 98
C. Nitrate Leaching from Pastures and Turfgrasses . . . 99
III. Nitrogen Fertilizer Effects on Water Quality . . . 111
A. Groundwater . . . 111
B. Surface and Subsurface Water . . . 112
C. Health Effects of Nitrates . . . 113
IV. Ameliorating Adverse Effects . . . 114
A. Best Management Practices . . . 114
B. Evaluation of Potential Nitrate Leaching . . . 118
C. Legislation for Water Protection . . . 119
D. Removal of Nitrate from Water Systems . . . 120
V. Conclusions . . . 121
Acknowledgements . . . 123
References . . . 123


## I. Nitrogen Fertilizers and Their Uses

### A. Current and Future Use of Nitrogen Fertilizers

Nitrogen is the most limiting nutrient for plant growth in most of the world's agricultural soils, and hence, crop production worldwide relies heavily on inputs of N from organic or inorganic sources. Plants may utilize some forms of organic N, such as amino acids and amines, although most of the N assimilated by plants from inorganic ammonium ($NH_4$) and nitrate ($NO_3$) salts.

0-87371-980-8/94/$0.00+$.50

Consumption of N fertilizers has been increasing on a global scale. From 1967 to 1987, there was an increase in world N fertilizer utilization from 22.1 to 72.7 million metric tons (FAO, 1962-1987) and by 1991, this amount had increased to 77 million metric tons (Fertilizer Institute, 1993). It has been estimated that by the end of 1993, world consumption of N fertilizer will exceed 83.5 million metric tons (FAO, 1988), and that by the year 2000 between 111 and 134 million metric tons will be produced and consumed (Newbould, 1989).

### B. Sources, Forms and Application

The ultimate source of N is inert dinitrogen ($N_2$) gas, which comprises approximately 78% of the earth's atmosphere. Dinitrogen can be assimilated directly only by a few microorganisms. It is "fixed" or converted to forms usable by higher plants and animals by combination with $H_2$ or $O_2$ in a number of processes: in biological $N_2$ fixation by bacteria (either free living or associated with plants), in atmospheric discharges which produce oxides of N, and industrially, as $NH_3$, $NO_3$ or cyanamide for the manufacture of commercial N fertilizers. More detailed descriptions on N fertilizer types, sources and forms of application have been published (Follett et al., 1981; Boswell et al., 1985; Peterson and Frye, 1989). There are several rapid- and slow-release N fertilizers commonly used for grasses, especially turfgrasses (Table 1). Readily available N sources include the inorganic $NH_4$ and $NO_3$ salts (including several ammonium phosphates) and urea. Slow-release N compounds, such as urea-formaldehyde complexes (ureaform), crotonylidene diurea (CDU), isobutylidene urea (IBDU) and magnesium ammonium sulfate are sometimes used to extend the residence time of the N fertilizer in soil. Urea can be lost as free ammonia after hydrolysis by the enzyme urease, or it can undergo microbially-mediated changes with the formation of $NH_4$ and $NO_3$. Sulfur-coated urea (SCU) was developed by the Tennessee Valley Authority to provide slow release of N to crops (Blowin and Rindt, 1967). Microorganisms degrade ureaform N, whereas N release from IBDU is resultant of dissolution and hydrolysis (Turner and Hummel, 1992).

Nitrification inhibitors are mixed with the N fertilizer or applied as a surface coating on the fertilizer particles so that N remains as $NH_4$ in soils. This decreases the rate of $NO_3$ leaching and losses due to denitrification. Several compounds have been shown to inhibit nitrification but only two are commercially available in the U.S.A., nitrapyrin [2-chloro-6-(trichloromethyl)pyridine] and etridiazol [5-ethoxy-3-(trichloromethyl)-1,2,4-thiadiazole] (Peterson and Frye, 1989).

Organic materials such as animal wastes, composts, and sludges are also applied to supply nutrients to grasses. The low available-N content and the high cost per unit of N supplied limit extensive utilization of organic-N fertilizers for plant production. However, release of the N present in organic fertilizers is

**Table 1.** Nitrogen fertilizer sources used for grasses

| Nitrogen source | Nitrogen content (%) | Nutrient release | Comments |
|---|---|---|---|
| Ammonium nitrate | 33 | Rapid | High N content |
| Ammonium sulfate | 21 | Rapid | Acidifying, supplies S |
| Calcium nitrate | 16 | Rapid | Deliquescent, supplies Ca |
| Diammonium phosphate | 20 | Rapid | Supplies P |
| Monoammonium phosphate | 11 | Rapid | Supplies P |
| Potassium nitrate | 13 | Rapid | Supplies K |
| Sodium nitrate | 16 | Rapid | Naturally occurring inorganic |
| Urea | 45 | Rapid | Very high N content, water-soluble organic |
| Isobutylidene diurea (IBDU) | 31 | Slow | Release controlled by hydrolysis, largely temperature independent |
| Resin-coated (e.g., Osmocote) | Variable | Slow | Release by diffusion through the resin coat |
| Sewage sludge (e.g., Milorganite) | 6 | Slow | Natural organic, release by microbial action, very low N content |
| Sulfur-coated urea (SCU) | 32 | Slow | Release by weathering of S coating |
| Ureaformaldehyde | 38 | Slow | Release by microbial action, temperature dependent |

(From Cisar et al., 1991.)

usually slow, providing a continuous supply of the element, which is absorbed readily and, thus, is less subject to losses.

The method of application and placement of N fertilizers influences the efficiency of its use, and the delivery of the fertilizer will depend on the amount to be applied and on the plant species. When large quantities of fertilizer are used, it is common to broadcast part of it and work it into the soil before planting. If the rate of fertilizer applied is high, seed injury may occur. Normally pastures and lawns are fertilized at the time of seeding and a top-dressing is performed in succeeding years.

The optimum time of the year to apply N fertilizer depends on the amount and distribution of rainfall, type of grass, and kind of fertilizer to be applied. Most often, fertilizers are applied in the spring or fall, except in the southwestern U.S.A. Summer application of fertilizer results in reduced yield responses, particularly by cool-season grasses. In winters with high precipitation, N fertilizer applied in fall may be leached before the spring growing season begins. Nitrogen fertilizer applied in the fall to seeded grasses on mountain rangelands in northeastern Utah and southeastern Idaho was leached away by 76 to 102 cm of late fall and winter precipitation (Hull, 1963). All of the N applied had been leached from the top 107 cm of the soil by the end of the following spring. On irrigated meadows in Nevada, there appeared to be little difference between spring and fall for N fertilizers if the fall application was made after all the growth and irrigation had stopped, provided the soils were not kept wet during the winter by flooding (Hackett et al., 1969). Nitrogen should not be applied in the fall to lowlands which are subject to flooding in the winter or spring. However, on dry foothill range in the intermountain region, fall application allows for N to be moved into the root zone for early spring growth (Cook, 1965). Spring application may be too late to move the N into the root zone, particularly when the weather is dry, unless fertilization is done early.

Application of N fertilizers for turfgrasses varies with the annual N rate, the N source and grass use. For instance, N is applied more often to intensive-use bermudagrass than to roadside bahiagrass. Increased frequency of application demands increased labor; the practice of fertigation permits delivery of small amounts of fertilizer as needed, so that leaching losses are decreased and labor is reduced.

## C. Nitrogen Requirements of Pastures and Turfgrasses

### 1. General Plant Responses

The requirement of all plants for N was established in the last century. Nitrogen is a structural component of amino acids, and thus is required for protein synthesis. Nitrogen is a vital part of many compounds essential for plant growth, including chlorophyll and many coenzymes. Nitrogen is a component of metabolites such as nucleic acids, adenosine nucleotides (ATP, ADP, AMP), nicotinamide- or flavin-containing dinucleotides (NADH, NADPH, FAD), porphyrins, diamines, and various secondary metabolites, and N is even found in cell walls.

Nitrogen is essential for carbohydrate utilization, enhances root growth and development, and increases the absorption of other nutrients. In most plants, it regulates the uptake and utilization of K, P and other constituents. Nitrogen is present in greater concentrations in plants than all other essential and nonessential elements, except O, C and sometimes K.

Plants with an insufficient or inadequate supply of N are generally stunted, have restricted root systems and many chlorotic leaves. Such chlorotic leaves become necrotic prematurely and tend to drop off. When applied as a fertilizer, inorganic N appears to have the quickest and most pronounced effect on plants, as compared with other mineral nutrients. Excess N will disturb the nutrient balance and disrupt the physiology of the plant; leaves become dark green and there is excessive vegetative growth and succulence in plants provided with high levels of N. In many cases, luxury consumption of N results in lodging and a weakening of fibers, reduced sugar content, delayed flowering, delayed crop maturity and enhanced disease and insect susceptibility.

Yield-N response curves are normally curvilinear, for a given crop, area, climate, year, with a gradual decrease in yield increment per unit N (Bock and Hergert, 1991). Addition of N fertilizer beyond the requirement for maximum yield usually does not increase crop yields, but this practice will likely produce plants with undesirable characteristics as well as increases in potential hazards to the environment.

## 2. Pasture Grasses

Nitrogen fertilizer increases palatability, protein content, and vigor of grasses, through increases in height, basal area per plant, number of seedstalks per plant, seed production, size of stems and leaves, greater number and weight of roots (Blaser, 1964; Cook, 1965). Nitrogen also increases the depth and surface area of root growth, resulting in increased subsoil moisture extraction (Black, 1968). The increased root growth aids in drought resistance.

In terms of requirement and plant response N is the most important nutrient for bahiagrass (*Paspalum notatum* Flugge). Mata and Blue (1973) reported a positive response in forage yield of a bahiagrass pasture to fertilization with $NH_4NO_3$. Nitrogen fertilization of bahiagrass was shown to increase carrying capacity but failed to enhance individual animal performance, despite increases in forage quality (Pitman et al., 1992).

In grassland systems, it is estimated that approximately 50% of the productivity is due to N-fertilizer input (Garrett, 1991). Crude protein concentration is increased substantially by addition of N to small grains and grass forages, such as annual ryegrasses (*Lolium* spp.). Addition of N at 56 and 112 kg $ha^{-1}$, with or without S, significantly increased protein content of early and late-season wheat (*Triticum* sp.) forage, used by many Great Plains agricultural producers for grazing by livestock (Freyh and Lamond, 1992).

Usually native and extensively managed grasslands are N deficient, and thus, there is practically no $NO_3$ available for leaching. Grassland systems recover between 50 and 65% of the fertilizer N as harvested yield in the year of application (Schepers and Mosier, 1991). Large inputs of N fertilizer and concentrates contribute significantly to current increased productivity of grasslands and the livestock raised on intensively managed grassland farms (van der Meer and van Uum-van Lohuyzen, 1986; Steenvoorden et al., 1986).

High soil-$NO_3$ levels, reduced light intensity during maturation, soil moisture deficiency and plant nutrient deficiencies can promote $NO_3$ concentrations as high as 3000 mg $kg^{-1}$ in some forage crops (Stevenson, 1986). These $NO_3$ levels may constitute health hazards to humans and animals.

3. Turfgrasses

Turfgrasses are established for several purposes including aesthetics, soil stabilization and recreation. The major turfgrasses adapted to the cool humid regions and irrigated areas of the cool arid regions are species of *Agrostis*, *Poa*, *Festuca* and *Lolium*; for warm humid and irrigated areas, the best adapted are *Cynodon*, *Zoysia*, *Stenotaphrum*, *Eremochloa*, *Paspalum*, *Festuca*, and *Agropyron*; in the non-irrigated warm arid regions, species of *Buchloe* and *Bouteloua* are the best turfgrasses.

Nitrogen fertilization is one of the most important management practices in turfgrass production and maintenance. Turfgrasses grown on sandy soils such as those in the southeastern U.S.A., require large quantities of N fertilizers to maintain the desirable dark green color. Turfgrasses respond markedly to N fertilization with increased growth, better competition with weeds and the characteristic dark-green color. Fertilizer application to sports turf has a significant effect on the playing characteristics of the surface, as well as its aesthetic appearance (Lawson, 1989). Nutrient availability affects leaf and root density, surface hardness, infestation by weed species, resistance to fungal pathogens, tolerance to wear, drought or cold in turfgrasses. The practice of increased sand content to promote free-draining root zones for sports areas is leading to a larger demand for N, and also is leading to an increased interest in the use of slow-release N fertilizers. Ureaformaldehyde followed by IBDU and S-coated ureas (SCU) were developed to provide fertilizers composed specifically for turfgrass (Watson et al., 1992).

On British golf greens, *Festuca* and *Agrostis* species are best suited to the management practices used. Ammonium sulfate is the best source of N for these genera and the optimum rate is about 200 to 250 kg $ha^{-1}$ annually (Isaac and Canaway, 1987). Kissel and Smith (1978) studied the fate of 560 kg N $ha^{-1}$ applied as $Ca(^{15}NO_3)_2$ to coastal bermudagrass grown in plastic cylinders. The authors determined in this field study that 49% of the fertilizer was present in plant tops; 34% was in organic forms (soil organic matter and roots) but only 6% was in mineral forms. The form of N was also shown to affect root and shoot growth of creeping bentgrass (*Agrostis palustris* Huds.) "Penncross" plants (McCrimmon and Karnok, 1992). Ammonium (as urea) and $NO_3$ ($Ca(NO_3)_2$) were applied at various $NH_4$:$NO_3$ ratios (100:0, 75:25, 50:50, 25:75, and 0:100). The 100% $NO_3$ treatment resulted in 30% more roots during the fall than did the 100% $NH_4$. However, in the spring the 100% $NH_4$ promoted 26% greater root length when compared with the two highest $NO_3$ treatments. The authors concluded that a fertilizer program that contains a portion of its N as $NO_3$ would be more beneficial certain times of the year than one containing $NH_4$ or $NO_3$ alone.

Utilization of slow-release materials reduces the risk of N leaching and/or volatilization, as well as producing a more uniform growth and color (Sartain, 1992). Sartain (1988) tested five thermo-plastic coated (Meister) ureas with release rates varying from 70 to 270 days on "Pennant" ryegrass (*Lolium perenne* L.) and "Tifway" bermudagrass [*Cynodon dactylon* (L.) Pers] × *Cynodon transvaalinsis* Burtt Davy. The author determined that during the cool season an extended-release Meister-coated urea should be applied according to its release rate and not on a 90 day cycle. The influence of 8 selected natural and synthetic organic N sources was evaluated on "Tifway" bermudagrass (*Cynodon dactylon* × *Cynodon transvaalinsis* Burtt Davy), overseeded during the cool season with "Penant" ryegrass (*Lolium perenne* L.) (Sartain, 1992). Application of 5 g N $m^{-2}$ of formulated natural organic product "Restore" every 90 days was sufficient to maintain good visual quality of ryegrass, but application of the same rate every 45 days failed to maintain quality of bermudagrass. Turf restore produced higher growth rate and improved visual quality responses on ryegrass when compared with other N sources.

Excessive N levels reduce turfgrass cold hardiness and heat tolerance, and increases luxury N consumption and susceptibility to disease and thatch accumulation (Turner and Hummel, 1992). Excessive N also results in decreased carbohydrate levels which are needed for energy generation, thus restriction and death of roots, rhizomes and stolons may occur.

## II. Soil Nitrogen Processes

### A. The Nitrogen Cycle

The N supply in the atmosphere is vast, yet it is present mainly as the free, inert dinitrogen ($N_2$) gas, which is useless for most life systems and industry. Most of the soil N is in the form of organic matter, derived largely from inputs of animal manures, decomposition of crop residues remaining after harvest, seasonal renewal of grass root systems, and soil fauna and microbes. The remaining, significantly smaller portion is in the form of inorganic N (largely $NH_3$-N, $NH_4$-N, and $NO_3$-N), and is directly available to plants.

Nitrogen inputs in the soil are in the form of aerial deposition, biological $N_2$ fixation, animal and green manures, municipal sludges, and synthetic N fertilizers. The transformations of N from one form to another, including the movement of N into the soil and its loss, some of which may be controlled by humans, constitute the nitrogen cycle (Figure 1). The major divisions of the soil-N cycle are immobilization (removal of N) and mineralization (release of N). Direct removal of N by crops, animals and microbial populations are forms of N immobilization. The immobilization of N by grasses is higher than that of field crops (Legg and Meisinger, 1982). Products of mineralization are susceptible to further transformations, resulting in losses or outputs of inorganic N such as volatilization, leaching or denitrification, all dependent on the conditions prevailing in the specific environment. For more details on the N cycle, several reviews may be consulted (Stevenson, 1982; Keeney, 1983, 1986,

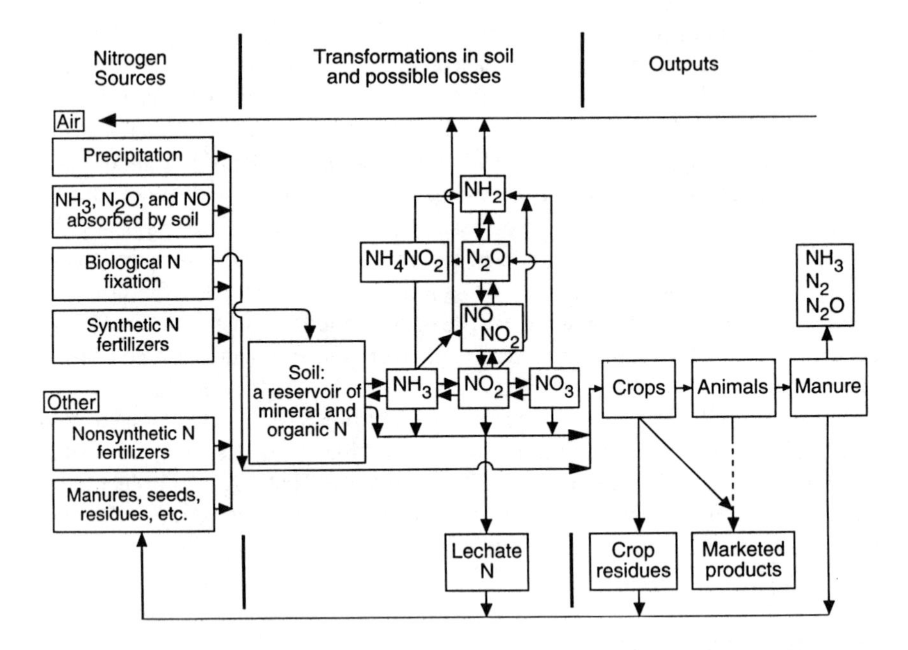


**Figure 1.** The nitrogen cycle in soil. (Reprinted with permission of Macmillan College Publishing Company from *Soil Fertility and Fertilizers* by Tisdale et al., 1985.)

1989; Petrovic, 1990). Most of the environmental impacts of N fertilizers concern losses of N to water and/or the atmosphere.

## B. Transformations of Nitrogen Fertilizers in Soil

The fate of N fertilizers in the soil-plant system is governed by several physical, chemical, and biological factors which interact with each other and with the environment (Boswell et al., 1985). Ammonia reacts readily with soil water and H ions to form the positively-charged $NH_4$ ion which can be absorbed, adsorbed to cation exchange sites, transformed further by biological processes, or enter into fixation reactions with certain clay minerals, most frequently vermiculite and illite (Stevenson, 1982; 1986). Ammonia ($NH_3$) can also be fixed by reactions with lignin and organic matter fractions, and may not be immediately usable by plants, although eventually this fixed N will be released by mineralization (Stevenson, 1982). Substantial losses of anhydrous gaseous $NH_3$ can occur in the environment through volatilization when quantities applied exceed the sorption capacity of the soil and the gas is injected superficially. Ammonia volatilization from fertilizers may reach 70% and can be reduced by application at depths below 10 cm (Legg and Meisinger, 1982). Ammonia can also be lost rapidly from calcareous soils (pH > 7.0), under conditions of increasing soil and air temperatures, in soils with low cation exchange capacity, such as sands, or when high organic-N substances, such as manures, are placed on the soil surface (Stevenson, 1982). Control of volatilization loss from urea fertilizers includes utilization of slow-release coated urea, urease inhibitors, acidifying substances and deeper placement. Urea may become the dominant fertilizer for temperate grassland production if effective measures to control volatilization are developed (Garrett, 1991). Surface application of animal manures may result in $NH_3$ losses by volatilization in the magnitude of 10 to 15 kg N $ha^{-1}$ $yr^{-1}$ (OECD, 1986). From an application of manure at the rates of approximately 70, 140, and 210 kg N $ha^{-1}$ $yr^{-1}$, losses from volatilization of $NH_3$ accounted for as much as 35% of the applied N (Kelly et al., 1991). Direct utilization of animal wastes for land application and irrigation purposes increases the risk of groundwater pollution by $NO_3$. In order to prevent and minimize pollution, it is necessary to apply geological-hydrogeological evidence and concepts to wastewater irrigation for groundwater protection, thereby predicting changes in groundwater quality as a result of infiltrating wastes. Procedures for evaluation and quality prediction have been described (Goldberg, 1989).

Ammonium-based fertilizers may be transformed biologically by populations of *Nitrosomonas* and *Nitrobacter* to $NO_3$ (nitrification), under proper environmental conditions of pH, moisture, aeration and temperature. The $NO_3$ forms of N, whether supplied as fertilizers or produced by nitrification of $NH_4$, are readily soluble in soil solution. Thus, $NO_3$ normally moves with soil water, upward due to capillary forces during extremely dry weather, and downward, under conditions of excessive precipitation or irrigation. However, on positively charged soils, such as andisols, which have an anion exchange capacity rather than a low cation exchange capacity, $NO_3$ is retained. Soil texture and organic matter (OM) content can have a major influence on leaching losses of $NO_3$. In Sweden, Bergstrom and Johansson (1991) verified that leaching was greater in a sandy soil with low OM content, intermediate losses occurred in loamy soils and the smallest losses were observed in a clay soil or a sandy soil rich in OM. Nitrate may be transformed in soils biologically by bacteria to dinitrogen ($N_2$) or nitrous oxide ($N_2O$), in a process known as denitrification. Although the denitrifying prokaryotes are aerobic, they are capable to use $NO_3$ as an electron acceptor in the absence of $O_2$.

### C. Nitrate Leaching from Pastures and Turfgrasses

#### 1. Pastures

Permanent grass swards generally have a high absorption capacity for $NO_3$-N, even when annual fertilization is performed (Larson et al., 1971; Jaakkola, 1984; Barraclough et al., 1984; Stanley et al., 1990). Nitrate does not generally accumulate under native grasslands during the growing season because it is utilized as it is formed by nitrification (Fuleky, 1991). The highly branched root mass of perennial grasses is capable of absorption of N present in excess of crop demands (Garrett, 1991; Schepers and Mosier, 1991). In non-fertilized permanent pastures, $NO_3$ levels are usually very low and losses from leaching are not very serious (Porter, 1969; Henry and Menely, 1993). Thus, leaching from soils under grassland is considerably lower than from arable land. Power (1970) determined that little leaching of N fertilizer applied to bromegrass (*Bromus inermis* Leyss.) occurs as compared with that from corn (*Zea mays*). Kolenbrander (1972) calculated the loss of fertilizer N to be 1% on grassland at an application level of 250 kg N $ha^{-1}$ and 3.5% at 60 kg N $ha^{-1}$ on cultivated land.

Grasslands have not been regarded as a major contributor of $NO_3$ to water supplies except after plowing and cultivation when much of the organic N is mineralized (Garwood and Ryden, 1986). A study conducted by Thomas and Crutchfield (1974) in central Kentucky evidenced annual losses in the range of 25 kg $ha^{-1}$ which are attributable to native N loss since fertilizer application in the area is low. Plowing enhances mineralization of OM increasing the release of N, and in many instances, in excess of crop requirements (Juergens-Gschwind, 1989). In Great Britain, N release from an old grassland containing 8,100 kg humus N $ha^{-1}$ to a depth of 25 cm was calculated as 490 kg $ha^{-1}$ $yr^{-1}$ on average for three years (Triboi and Gachon, 1985). Croll and Hayes (1988) reported a release of 500 kg N $ha^{-1}$ $yr^{-1}$ from plowed grassland.

In a grass system, the amount of N lost by leaching depends on: the form and amount of soluble N present or added; amount and time of rainfall; infiltration and percolation rate of water; water-holding capacity; evapotranspiration, rate of removal of N by crop and whether the N is leached below the root zone (Allison, 1961). Usually 50-80% of N applied to forage grasses is recovered in the harvested herbage, when applied N enhances soil available N, recovery may exceed 100% (Dougherty and Rhykerd, 1985). Recovery is higher at N rates between 200 and 300 kg $ha^{-1}$ $yr^{-1}$ since at the lower rates N is retained in roots and lower stems (Dougherty and Rhykerd, 1985). At rates higher than 300 kg N $ha^{-1}$ $yr^{-1}$, not all N is absorbed, and part is lost through leaching and volatilization. When heavier dosages, in the order of 500 kg N $ha^{-1}$ or more, are applied to grasslands, $NO_3$-N may reach the subsoil (Barraclough et al., 1984).

Cultivation of legumes with grasses may result in lower losses of N. Nitrate losses were markedly lower from grazed grass/clover pastures than those from grazed grass receiving large amounts of N fertilizer (Garwood and Ryden, 1986). Nitrate concentrations were measured in soil solution at 60 or 90 cm below grazed perennial ryegrass and/or white clover (*Trifolium repens*) swards on freely draining soil in the UK receiving different fertilizer N rates over two years, and the annual leaching losses for each was calculated (Macduff et al., 1990). Only unfertilized grass and mixed grass/clover swards gave mean $NO_3$ concentrations below the European 11.3 mg N $L^{-1}$ limit. Leguminous crops are able to absorb large quantities of soil $NO_3$, applied to previous crops. However, generally, for soils under legumes, $NO_3$-N contents in drain water are higher than under grass (Low, 1973).

Nitrate leaching losses are usually greatest during the winter months, and the highest concentrations in the seepage water occurs in periods when maximum precipitation exceeds evapotranspiration (Allison, 1961; Hood, 1976). Nitrate leaching is especially important during the winter months in humid climates (Jeffrey, 1988). Nitrogen is taken up as $NO_3$ and $NH_4$ by most grasses. However, at lower temperatures (<14°C), $NO_3$ uptake decreases, while $NH_4$ uptake may continue even to temperatures as low as 3°C (Lazenby, 1983).

High $NO_3$-N levels in drainage water from intensively managed forage and pasture crops have been reported. An average concentration of 8.4 mg $L^{-1}$ (11 kg $ha^{-1}$ $yr^{-1}$) of $NO_3$-N was verified in the lines draining a pasture in Great Britain which had received 250 kg N $ha^{-1}$ (Hood, 1976). When N application rate was increased to 750 kg $ha^{-1}$, $NO_3$-N loss increased to 54 kg $ha^{-1}$ with the average concentration of 33 mg $L^{-1}$. Nitrogen is commonly applied to pasture crops at around 400 kg $ha^{-1}$ in Great Britain. Kolenbrander (1981) compiled results from leaching experiments at various rates of N fertilization. In general, $NO_3$ leaching below cut grassland on sandy soils is minimal when N is applied in quantities below 200 kg $ha^{-1}$ $yr^{-1}$ and leaching losses increase rapidly at higher rates. On cut grassland, the European maximum $NO_3$ concentration for drinking water (11.3 mg $NO_3$-N) was reached under pastures receiving fertilization rate between 350 to 400 kg N $ha^{-1}$ $yr^{-1}$. In the Netherlands, optimal total N supply for intensive grassland management is approximately 400 kg N $ha^{-1}$ $yr^{-1}$. A long-term study revealed that up to 420 kg $ha^{-1}$ $yr^{-1}$ cause little risk of high $NO_3$ accumulation in either herbage or soil (van Burg, 1986). A great deal of information on N losses from grassland obtained from cut swards has indicated that at a recommended application rate of approximately 400 kg $ha^{-1}$ $yr^{-1}$ the risk of loss to the environment, especially $NO_3$ leaching, is not very significant (van der Meer and van Uum-van Lohuyzen, 1986). This appears to be the case for ungrazed systems.

Results from a field investigation into $NO_3$ leaching from grassland plots suggested that application rates at or below the calculated optimum, $NO_3$ leaching from cut, permanent grass was low, but as rates surpassed optimum levels, leaching losses also increased sharply (Barraclough et al., 1983). A three year investigation into $NO_3$ leaching from grassed ryegrass (*L. perenne* var. Mella) in monolith lysimeters treated with doubly ($^{15}NH_4{}^{15}NO_3$) or singly ($^{15}NH_4NO_3$) labelled N fertilizer at 250, 500, and 900 kg $ha^{-1}$ $yr^{-1}$ was undertaken by Barraclough et al. (1984). Over the 3 yrs, a portion of the fertilizer leached as $NO_3$ at all application rates was derived via nitrification, from the fertilizer $NH_4$ (Table 2). Increasing fertilizer applications caused a rise in the leaching of soil and fertilizer-derived N.

McLeod and Hegg (1984) evaluated surface runoff water quality from a fescue (*Festuca arundinacea* L. Schreb.) pasture receiving organic wastes and commercial fertilizer in South Carolina (rate 112 kg N $ha^{-1}$). The authors determined that $NO_3$-N concentration in the runoff from plots receiving the

**Table 2.** Nitrate leaching from monolith lysimeters grassed with perennial ryegrass (average of 3 years)

| Fertilizer N rate (kg $ha^{-1}$ $yr^{-1}$) | Lysimeter[a] | Flow (mm) | $NO_3$-N leached (kg $ha^{-1}$) | % leached $NO_3$ derived from $NH_4$ | % leached $NO_3$ derived from $NH_4$ + $NO_3$ |
|---|---|---|---|---|---|
| 250 | a | 233.0 | 3.08 | 7.03 | --- |
| | b | 232.0 | 2.77 | --- | 8.93 |
| 500 | a | 230.1 | 47.48 | 18.07 | --- |
| | b | 230.2 | 37.08 | --- | 39.25 |
| 900 | a | 249.7 | 270.48 | 26.05 | --- |
| | b | 232.4 | 222.92 | --- | 74.65 |

[a] Lysimeter a = average of two lysimeters receiving $^{15}NH_4{}^{15}NO_3$; b = average two lysimeters receiving $^{15}NH_4NO_3$.
(Adapted from Barraclough et al., 1984.)

$NH_4NO_3$ exceeded the permissible standard during the first runoff event (which occurred 1 day after application), but nutrient concentrations were more dependent on the number of rainfalls since application than on the quantity of rainfall or runoff.

Stout and Jung (1992) verified that the biomass accumulation rate of orchardgrass (*Dactylis glomerata* L.) in the fall was controlled by N fertilization, and spring total N uptake rates were controlled by soil-N levels and N fertilization. Fertilizer-N recovery was about 42% for the spring growth period but dropped to only about 15% for the fall growth period. The authors concluded that low fertilizer-recovery rates in orchardgrass during the fall growing period may result in unused N being leached into the groundwater under this species.

Higher amounts of nutrients are being applied to grazed grassland as a result of increased livestock numbers per unit of cropland and the use of supplementary feeds (Juergens-Gschwind, 1989). Currently animal manure previously used as fertilizer on grasslands and arable lands is treated as a waste product (Juergens-Gschwind, 1989). Thus, an oversupply of N may result, and consequently undesirable N losses may occur to the environment.

Grazing animals have a large impact on the leaching losses of $NO_3$ from grazed pastures (Haynes and Williams, 1993). The extent of pollution due to livestock on pasture depends on the stocking density, length of the grazing period, average manure loading rate, distribution of manure by grazing livestock, decomposition of manure with time and proximity to a water body (Garrett, 1991). The major part of these losses is from urine (Haynes and Williams, 1993) and feces patches that may return up to 90% of the N ingested by grazing livestock to the soil/plant system (Cooke, 1985). A portion of the returned N is quickly lost as $NH_3$ through volatilization, but a great part is transformed into $NO_3$, an amount that usually greatly exceeds requirements of the pasture crop (Juergens-Gschwind, 1989).

The intensification of grassland farms can lead to potentially large emissions of N to the environment, predominantly as leaching and surface runoff loss of N. The impact of agricultural intensification and fertilizer inputs on stream water quality were examined at different catchments in areas of mixed land use (Heathwaite et al., 1989). Ammonium and suspended sediment were mobilized in surface runoff, and the $NO_3$ transferred in through flow from heavily grazed pasture was 53% of the total rainfall input, which delivered 1.4 kg N $ha^{-1}$. Surface runoff from ungrazed land was only 7% of rainfall input.

A beef cattle-pasturing system involving four rotationally grazed summer pastures (SG) and four pastures used rotationally, for winter grazing/feeding (WGF) was studied on sloping upland watersheds in Ohio to determine effects of livestock management on N levels in water (Owens et al., 1983). Both summer and winter areas annually received 224 kg N $ha^{-1}$ as $NH_4NO_3$ fertilizer. Surface runoff was collected automatically during runoff events, and subsurface flow was sampled from spring developments on a weekly basis. Although seasonal N concentration and transport in surface runoff tended to be greater in the area occupied by the cattle, N concentration and transport in runoff from the two areas were quite similar and did not significantly impair water quality, based on U.S. Public Health Standards. The $NO_3$-N concentration in the subsurface flow from the WGF area was higher than in the subsurface flow from the SG area. The $NO_3$-N concentration in the subsurface flow from both areas increased progressively throughout the study period, and $NO_3$-N reached levels as high as 18 mg $L^{-1}$. The subsurface flow provided the main pathway for N transport, with the surface transport being approximately 20 and 14% of the total N transport from the SG and WGF areas, respectively. The amount of sediment-N transported was very small because of low soil loss.

Leaching losses from cut swards are generally small. Losses were estimated to be 5.6 to 10.2 times greater under grazed swards as compared with cut swards at the same N-fertilizer application rate (420 kg N $ha^{-1}$ $yr^{-1}$) (Ryden et al., 1984). Kolenbrander (1981) reported similar results. Grazed swards leach more $NO_3$ due to uncontrolled and uneven deposition of excreta by animals. Liberation of N in certain areas is in excess of plant demand. No grazing systems with proper N-fertilizer (including slurry) management, may be adopted to control leaching losses in problem areas. Use of nitrification inhibitors may also be recommended with slurry application (Garrett, 1991).

Ungrazed ryegrass pasture receiving an application rate of 420 kg N $ha^{-1}$ lost only 29 kg N $ha^{-1}$ (Garwood and Ryden, 1986). For ungrazed pastures with annual fertilizer inputs of 200 to 400 kg N $ha^{-1}$, combined N losses due to leaching and denitrification were determined as no more than 5 to 15% (Ball and Ryden, 1984). These authors indicated that grazing increases losses through leaching and volatilization, thus, reducing N fertilizer-use efficiency. Other reports have indicated $NO_3$-N concentrations above the legal standard from grazed European grasslands (Haigh and White, 1986; Roberts, 1987). A strong correlation between $NO_3$-N leaching to groundwater and rate of N fertilizer was observed for intensively managed grassland systems in the Netherlands, with the highest $NO_3$-N concentrations being observed for grazed systems (Steenvoorden et al., 1986).

Data obtained from studies on $NO_3$ leaching from grasslands in UK suggest that substantial loss of N as $NO_3$ can occur in leachate and surface runoff, particularly in swards grazed throughout the year (Garwood and Ryden, 1986). On grazed grasslands, studies have indicated that at the same level of N fertilizer applied, $NO_3$ leaching is higher than that from cut grassland. Nitrate-N concentrations in leachate are approximately 10-20 mg $L^{-1}$ greater for cattle farms than for cut grassland on sandy soils amended with N fertilizer varying from 200 to 450 kg $ha^{-1}$ $yr^{-1}$ (Steenvoorden et al., 1986). On studying the effects of fertilizer-N input to grazed swards, Jarvis and Barraclough (1991) reported that measured losses of $NO_3$ during the drainage period increased with increasing soil $NO_3$ profiles at 3 of 5 sites. As the ratio of $NH_4$ to $NO_3$ concentrations increased, leaching decreased. Intensive grazing of fertilized pasture grasses (4 cattle $ha^{-1}$ fed supplementary concentrates) can cause excess N input (Juergens-Gschwind, 1989). On coarse, sandy soils, $NO_3$ leaching may then approach levels observed under arable land (Cooke, 1985; Kolenbrander, 1981; Ryden et al., 1984).

Leaching losses from grazed pastures have also been reported to be low. Comparisons of $NO_3$-N concentrations in runoff from grazed watersheds receiving 112 or 448 kg N $ha^{-1}$ over 4 years indicated that only once during the entire study did $NO_3$-N concentrations exceed 10 mg $L^{-1}$ (Kilmer et al., 1974). For a Nebraska pasture fertilized with 74 kg N $ha^{-1}$ $yr^{-1}$, Doran et al. (1981) verified that the $NO_3$-N concentration in the runoff exceeded 10 mg $L^{-1}$ only once, and the event was immediately after the pasture was fertilized. Also in Nebraska, greater $NO_3$-N concentrations in runoff from grazed pastureland than in runoff from adjacent nongrazed pasture were reported (Schepers and Francis, 1982).

The practice of winter-feeding on land used as summer pasture is common in beef cattle production. The effects of this management practice on water quality with regards to surface and subsurface runoff were investigated on pastures receiving 56 kg N $ha^{-1}$ $yr^{-1}$ (Owens et al., 1982). Results from the five year study on sloping uplands in Ohio indicated that although the $NO_3$-N concentration was higher in the surface runoff from summer-grazing/winter-feeding systems than from summer grazing system only, no significant degradation of water quality occurred. Reduction of vegetative cover and increased soil disturbance on the winter-feeding area increased surface runoff and soil erosion, and consequently there was more N transport and loss. In a subsequent study on pastures receiving a much higher rate (224 kg N $ha^{-1}$), the authors (Owens et al., 1983) verified that although seasonal N concentration and transport in surface runoff tended to be greater in the area occupied by cattle, N concentration and transport in runoff did not impair water quality. However, unlike the seasonal variations of N content in the surface runoff, $NO_3$-N concentrations in the

**Table 3.** Number of runoff events and amounts of runoff occurring in nine watersheds grouped according to the $NO_3^-$-N in the runoff (data are for events >0.1 mm runoff)

| $NO_3$-N concentration | Number of runoff events | Amount of runoff |
|---|---|---|
| mg $L^{-1}$ | | mm |
| 0-5 | 732 | 2891 |
| 5-10 | 94 | 326 |
| 10-15 | 31 | 72 |
| 15-20 | 9 | 52 |
| >20 | 24 | 71 |
| Total | 890 | 3412 |

(From Owens et al., 1984.)

subsurface flow increased progressively throughout the study period and exceeded 10 mg $L^{-1}$ in 8 of the 10 seasons, reaching levels as high as 18 mg $L^{-1}$.

Nitrate-N in runoff events from nine watersheds were monitored over a 5-year period (Owens et al., 1984). Sixty-four of the 890 events had concentrations above 10 mg $L^{-1}$ and fertilizer application had preceded all surface runoff with high $NO_3$-N (Table 3). Although there can be high $NO_3$ concentrations in surface runoff, especially when fertilizer application occurs within a few days before the event, $NO_3$-N concentrations in surface runoff from fertilized grasslands are usually lower than 10 mg $L^{-1}$. Values reported from pastures receiving between 56 to 224 kg N $ha^{-1}$ $yr^{-1}$ ranged from less than 1 to 5 mg $L^{-1}$ (Owens et al., 1992).

In the U.S.A., very little research has been done on $NO_3$ leaching from under pasture grasses (Keeney, 1989). Normally, only limited amounts of $NO_3$ are available for leaching in native and extensively managed grasslands since they are N deficient. However, intensively managed forages and grazed grasslands may be a source of considerable losses of $NO_3$ to groundwater (Keeney, 1989). Few pastures in the U.S.A. have both high rates of N-fertilizer application and subsurface drainage systems (Gilliam et al., 1985) and losses of N are usually quite small. From a fertilized Kentucky bluegrass (*Poa pratensis* L.) watershed in North Carolina, the average loss was determined as 12 kg N $ha^{-1}$, whereas 3 kg N $ha^{-1}$ were lost from a similar unfertilized watershed (Kilmer et al., 1974). In these watersheds, most of the N lost was as $NO_3$ and $NO_3$ contents in drainage water were 4 and 1 mg $kg^{-1}$, respectively, for the fertilized and unfertilized bluegrass. Nitrogen uptake rate for Pensacola bahiagrass can be as high as 6 kg $ha^{-1}$ $day^{-1}$ (Blue and Graetz, 1977). Thus, N fertilizer is not expected to remain for long periods of time.

A three year field plot in which $^{15}N$ fertilizer was applied at approximately 112 kg N $ha^{-1}$ $yr^{-1}$ to 8 area soils representative of 5 soil orders was conducted by Smith et al. (1982). More than half of the fertilizer was absorbed by the sorghum (*Sorghum bicolor*)-sudangrass (*Sorghum sudanense*) plants (tops and roots) and about one third was incorporated in soil OM. Less than 5% remained in an inorganic leachable form and about 7% was unaccounted for. The unaccounted N was considered to be lost to denitrification since no evidence of fertilizer-N leaching below the root zone was verified through soil profile sampling. Pre-plant soil $NO_3$ was found to be a better indicator of total soil-N availability the first 2 years, whereas mineralization potential was better the third year (Smith et al., 1982). The authors concluded that from an environmental standpoint the use of N fertilizer at the rate described and under prevailing farming conditions appeared to pose no potential pollution problems.

In the southeastern U.S.A., $NO_3$ movement into groundwater is of great concern since the area is characterized by sandy soils requiring high N inputs, high average annual rainfall (125 cm), and expanded irrigated land due to availability of ground water (Hubbard and Sheridan, 1989). Nitrate contamination of groundwater in Florida is a potential problem due to increased N-fertilization rates (over the last few years), abundant rainfall, extremely sandy soils, frequent shallow groundwater tables, and reliance on groundwater as a source of drinking water for most urban and rural areas (Sveda et al., 1992). Conditions in the state also include rapid water percolation and high levels of nitrification in properly limed soils and N leaching may be a major problem, both economically and environmentally (Blue, 1974). With Florida's spodosols, N application in July, during the normal wet period, results in a lower absorption efficiency than when it is applied in March. Thus, in the July-August period, N is more subject to leaching or denitrification or both due to the high soil water content (Blue and Graetz, 1977).

Bahiagrass is the most prevalent grass grown in the state of Florida for cattle production, occupying over 1.0 million ha, and N is the most important nutrient for dry matter production. Elevated $NO_3$-N levels have been reported in water from several counties in Florida where bahiagrass grows (Jones et al., 1990). During the first 4 years of application of five different N sources at 112 and 224 kg $ha^{-1}$, N-uptake efficiency of bahiagrass was in the 40 to 50% range, and it increased to 65 to 75% from the fourth to the tenth year (Blue, 1974). The high percentage of N recovery in forage grass and the large amounts of N in stolon-root material and soil strongly suggested that N leaching was not important when N was applied to perennial grass pastures at biologically usable rates during the growing season (Blue, 1974). Three field experiments with $^{15}N$-enriched fertilizer were superimposed on nontracer N experiments on coarse Florida soils planted to Pensacola bahiagrass (Impithuksa et al., 1984). The solutions prepared from soil samples collected at 60 and 120 cm showed no evidence of N leaching at the rates tested, although forage recovery rates were only approximately 53%.

Results from field studies with 3 warm-season grasses [Ona stargrass (*Cynodon nlemfuensis* Vanderyst), Transvala digitgrass (*Digitaria decumbens* Stent), and Pensacola bahiagrass] fertilized with up to 200 kg $^{15}N$ $ha^{-1}$ ($NH_4NO_3$) indicated that total recovery of applied N fertilizer was 70%, 66% and 75%, for Ona stargrass, Transvala digitgrass and Pensacola bahiagrass, respectively.

There was a relatively large retention of labeled N in the sandy soil (Impithuksa and Blue, 1985). Nitrate-N only exceeded 1 mg $L^{-1}$ in the soil solution from the 200 kg N $ha^{-1}$ treatment. In a previous experiment, in which N fertilizer was applied in rates up to 336 kg $ha^{-1}$ $yr^{-1}$ it was verified that at the beginning of the growing season, the predominant N source for the three warm season grasses was the labelled fertilizer, but later, non-fertilizer sources became dominant (Impithuksa et al., 1979).

A greenhouse study was established to evaluate the influence of various rates and sources of N fertilizer on N movement in soil and on effluent water quality (Sveda et al., 1992). The highest yields and N recovery by bahiagrass were obtained with $(NH_4)_2SO_4$; however, no difference between rates or sources was obtained for N uptake. There was no $NO_3$ present in the soil columns at the end of the 90 day study, and regardless of N rate or source, effluent water $NO_3$-N concentrations were well below the 10 mg $NO_3$-N $L^{-1}$ limit established for drinking water by the Environmental Protection Agency. A field study was initiated in 1991 to evaluate the influence of the various rates and sources of N fertilizer on N movement in soil and on water quality. Preliminary results indicate that water $NO_3$-N levels averaged 0.2 mg $L^{-1}$ on plots receiving up to 168 kg N $ha^{-1}$ and that the N fertilizers applied have not leached below the root zone (Muchovej, Rechcigl and Nkedi-Kizza, unpublished data).

Depth of rooting forage influences $NO_3$ leaching. Pensacola bahiagrass is able to store nutrients in large stolon-root mass that it produces. Leaching losses from $NH_4NO_3$ at the rate of 538 kg N $ha^{-1}$ were 2.1 kg N $ha^{-1}$ from Pensacola bahiagrass, 8.6 from Pangola digitgrass, 15.6 from Coastal bermudagrass and 90.8 from carpetgrass (*Axonopus affinis* Chase) (Volk, 1956). In semi-arid regions, soil depth influences soil N uptake, but not fertilizer N uptake. Soil depth had a much greater effect on total N uptake by tall fescue and switchgrass (*Panicum virgatum* L.) than on fertilizer N uptake by these species (Stout et al., 1991). The authors concluded that there is a potential for fertilizer [rates of 0, 90 and 180 kg N $ha^{-1}$ as $(NH_4)_2SO_4$)] loss from spring growth of grasslands in humid regions where soil water is less than 15 cm.

Soil texture has been shown to affect $NO_3$ leaching in grasslands. Applications of 380 kg $ha^{-1}$ $yr^{-1}$ N fertilizer to grasslands leads to losses of approximately 20 kg $ha^{-1}$ $yr^{-1}$ and 70 kg $ha^{-1}$ $yr^{-1}$, respectively, for clay and sandy soils (Steenvoorden et al., 1986). The difference in $NO_3$ leaching between the soils is attributed to a higher crop uptake, finer structure that favors denitrification and a higher soil water content at field capacity for the clay soils. Leaching of N from heavy clay soils, especially in the winter was also verified on all-grass perennial ryegrass swards, fertilized with 480 kg N $ha^{-1}$, although part of the N remained in the soil (Prins, 1983).

Irrigation on sandy soils greatly enhances the potential for $NO_3$ contamination of underlying groundwater. Irrigation of cool-season grasses may complement forage of warm season grasses. With irrigated pastures, as with any heavily fertilized crop, there is a large potential for groundwater contamination by $NO_3-N$ leached from the root zone. Research performed at the Bet Dagan

Experimental Station, Israel, investigated the effects of different irrigation and $(NH_4)_2SO_4$-fertilizer regimes on yield of Rhodes grass (*Chloris gayana*) and movement of $NO_3$ residues through the soil. The results indicated that in finer-textured soils, leaching did not occur during the irrigation season, while on sandy soils, some leaching occurred even when no deliberate water excess was applied (Rawitz et al., 1980). Reducing the first irrigation by half has been shown to lower annual N flux below 195 cm depth by more than 50% (Hubbard and Sheridan, 1989). A three year experiment in West Central Nebraska evaluated water and $NO_3$ leaching losses from orchardgrass subjected to three irrigation levels and four N application rates (0, 122, 224, and 336 kg $ha^{-1}$) (Watts et al., 1991). Nitrate-N leaching loss ranged from 6 to 228 kg $ha^{-1}$, depending on N and irrigation regime. Winter and early spring leaching losses were a significant part of the total N loss. The authors concluded that under reduced in-season drainage and a N rate to attain 80 to 85% of the maximum production, a minimum annual leaching loss of 35 kg $ha^{-1}$ can be expected and that irrigation and N must be controlled together to minimize the $NO_3$-leaching problem. Nitrogen leaching losses may be large from dry grassland (>20 kg $ha^{-1}$ out of 50 kg applied) also, but prewetting helped to decrease them (5.7 kg N $ha^{-1}$) (Smaling and Bouma, 1992).

Owens et al. (1992) examined the impact of conventional ($NH_4NO_3$) and slow-release (Methylene Urea) N fertilizer at 168 kg N $ha^{-1}$ $yr^{-1}$ to small, grazed watersheds on groundwater quality in eastern Ohio. After a 5-year pre-study, $NO_3$-N levels in the groundwater were in the range of 3 to 5 mg $L^{-1}$, but increased more sharply during the last 5 years of the study. During the ninth and tenth year of high fertilizer application, groundwater-$NO_3$ levels varied from 10 to 16 and 7 to 14 mg $L^{-1}$ for the conventional and slow-release fertilizer, respectively. In that study, conducted on well-drained residual silt loams, it was concluded that the rate of 168 kg N $ha^{-1}$ was excessive for the predominantly orchardgrass and Kentucky bluegrass pasture, regardless of N source.

## 2. Turfgrasses

Turfgrasses are considered as excellent biological filters that can minimize surface runoff and leaching of groundwater contaminants due to their total coverage of the soil surface and highly branched root systems (Cisar et al., 1991). Runoff losses of sediments and nutrients and leaching of $NO_3$ are very low in turfgrasses when compared with agronomic crops. Turfgrass is a dense, thick, and frequently thatchy crop. The dense nature of turfgrass should prevent the loss of large amounts of sediment. According to Cisar et al. (1991), it is necessary to maintain turfgrasses in a healthy, active condition to maximize their efficiency as a biological filter.

Optimum-N fertilizer requirements of turfgrass emphasizing top and root growth and other physiological effects have been researched considerably (Beard, 1973). In turfgrass, N is applied in the largest quantity of any of the essential nutrients (Beard, 1973). The potential exists for turfgrass to contribute significantly as a source to pollution to local water supplies, especially through leaching of $NO_3$-N into groundwater and runoff of N to surface waters (Gross et al., 1990; Petrovic, 1990). In Maryland, intensively managed turfgrass receives N applications of up to 200 kg $ha^{-1}$ $yr^{-1}$ (Gross et al., 1990). Two studies conducted at separate locations investigated nutrient and sediment losses via leaching and runoff from turfgrass (Gross et al., 1990). The studies demonstrated that $NO_3$-N concentrations generally decreased with depth and that losses from the sodded tall fescue/Kentucky bluegrass plots, fertilized with 200 kg N $ha^{-1}$ $yr^{-1}$ as granular or liquid urea, were very low. The results led the authors to conclude that "judiciously" fertilized turf is not a significant source of nutrients or sediment in surface or groundwater.

Some researchers have verified small increases in $NO_3$-N leaching from fertilized turfgrass (Starr and De Roo, 1981; Snyder et al., 1984) when moderate rates are applied. Other studies have shown that soluble N fertilizers, high rates of N fertilizers, heavy irrigation, and sandy soils can cause increased $NO_3$-N losses in turfgrass (Morton et al., 1988; Snyder et al., 1984).

A review of the effects of N applied to turfgrass on groundwater was presented by Petrovic (1990). The proportion of the N fertilizer taken up by turfgrass plants appears to vary from 5 to 74% of the applied N, depending on N release rate, N application rate, and species of grass. Losses by $NH_3$ volatilization can be reduced substantially by irrigation after application, and those due to denitrification are only significant on fine-textured, saturated, warm soils. The amount of fertilizer N present in the soil plus thatch pool varies as a function of N source, release rate, age of site, and clipping management. With a soluble N source, fertilizer N found in the soil and thatch was 15 to 21% or 21 to 26% of applied N, respectively, with the higher values reflecting clippings being returned (Petrovic, 1990).

Exner et al. (1991) recorded deep $NO_3$ movement under fixed irrigation (64 cm during 34 days) with variable $NH_4NO_3$ rates applied to a Nebraska municipal sports turf consisting of Kentucky bluegrass and creeping red fescue. Results indicated that 95% of the $NO_3$ applied in late August leached below the turfgrass root zone, and the average $NO_3$ concentrations in the pulse ranged from 34 to 70 mg $NO_3$-N $L^{-1}$. The relatively high and uniform $NO_3$ concentrations in the control plot led the authors to conclude that excessive irrigation supplied $NO_3$ in excess of the turfgrass need. They also pointed out that since most lawns are not fertilized or irrigated by professionals, the potential exists for excessive applications of both fertilizer and irrigation water. Recently Geron et al. (1993) studied the effects of turfgrass establishment method, late-season fertilization and different N fertilizers applied at 218.2 kg N $ha^{-1}$ $yr^{-1}$ on N leaching on a silt loam soil under Kentucky bluegrass turfgrass. Nitrate-N leaching losses exceeded the maximum concentration limit in the first year but were below 10 mg $L^{-1}$ during the second year of the experiment. The authors attributed the high concentrations to soil disturbance during turfgrass establishment and concluded that the N sources and fertilization programs did not result in greater $NO_3$-N losses than the unfertilized plots.

Fertilizer use on homelawns has increased continuously since 1970 (Watschke, 1983). This increase suggests a potential for an increase in off-site losses and environmental pollution. Furthermore, homelawns are irrigated with little regard for soil moisture status or soil water holding capacity. Irrigation in excess of evapotranspiration demands increases $NO_3$-N leaching (Snyder et al., 1984; Exner et al., 1991). A study conducted with Kentucky bluegrass turf subjected to three levels of fertilization (0, 97, and 224 kg N $ha^{-1}$ $yr^{-1}$ as urea and methylene urea) and two irrigation regimes demonstrated that mean annual flow concentrations of inorganic N in the soil-water percolate were below the U.S.A. drinking water standard on all treatments. However, overwatering in conjunction with fertilization resulted in higher flow concentrations than unfertilized controls.

Soil texture has a marked effect on N leaching from turfgrass due to its influence on the rate and amount of percolating water, on denitrification, and on $NH_4$-retention capacity (Petrovic, 1990). Application of soluble N at higher than normal rates to highly irrigated sandy turfgrass system resulted in considerable $NO_3$ leaching (Brown et al., 1977). The use of slow-release N sources (Brown et al., 1982; Snyder et al., 1984), finer textured soils (Brown et al., 1977) and limiting irrigation to replace only the moisture removed from evapotranspiration (Morton et al., 1988; Snyder et al., 1984) will reduce or limit $NO_3$ leaching from turf. Fertigation, or application of fertilizer by sprinkler irrigation, and irrigation based on soil moisture content resulted in low $NO_3$ leaching, less than 1% of the applied N, from bermudagrass turf (Snyder et al., 1984). The authors also verified that regardless of the irrigation method employed, occurrence of heavy rain shortly after application of water-soluble N caused leaching.

Leaching losses for fertilizer N are highly influenced by fertilizer management practices (N rate, source, and timing of application), irrigation, and soil texture. The degree of $NO_3$ leaching from fertilized turfgrass is variable, and reports range from little or no leaching to values as high as 80% of applied N fertilizer, but in general these values for leaching loss are far less than 10% (Petrovic, 1990). Where turfgrass fertilization poses a threat to groundwater quality, management strategies can allow the turfgrass manager to minimize or avoid $NO_3$ leaching.

Short-term studies on sandy loam soil cores have indicated that nitrification inhibitors may play an important role in helping to control non-point source pollution problems by reducing the rate of entry and the total quantity of $NO_3$-N leached into the subsurface layer (Owens, 1981; 1987) particularly during the growing season when N is utilized by plants (Timmons, 1984). Two nitrification inhibitors, nitrapyrin (N-serve) and dicyandiamide (DCD) tested on turf failed to inhibit $NO_3$ formation (Turner and Hummel, 1992). Some urease inhibitors such as phenylphosphorodiamidate (PPD) and cationic materials have been tested on turfgrass with some degree of success (Turner and Hummel, 1992). Moderate applications of slow-release N sources to golf-greens on a regular basis provided minimum $NO_3$-N loss and a continuous N supply to golf-greens (Brown et al., 1982).

## III. Nitrogen Fertilizer Effects on Water Quality

Although the benefits of N for plant production have been well recognized, excessive use of N fertilizers has often been indicated as a major source of $NO_3$ in ground and surface water in rural and suburban areas (De Roo, 1980; Spalding et al., 1982; Schepers et al., 1984). Sources other than soluble fertilizers and soil N also contribute to high levels of $NO_3$ encountered in certain areas. Livestock feeding, barnyards, septic tanks, animal and human contamination are frequently associated with elevated $NO_3$ in well waters (Stevenson, 1986). Nitrate leaching is a potential problem in urban and rural areas that are characterized by the predominance of sandy soils, abundant rainfall, frequent shallow water tables, high fertilization rates and reliance on groundwater as a source of drinking water. Reports from Florida, dating as early as 1916 (Collison and Walker, 1916) mentioned that $(NH_4)_2SO_4$, rather than $NaNO_3$, should be the preferred N fertilizer form in the wet season to prevent leaching losses. Excessive irrigation (>10 cm) or heavy rainfall is likely to move $NO_3$ below the root zone of many crops grown on sandy soils in Florida (Graetz et al., 1973; Endelman et al., 1974). In humid regions and under irrigation, leaching loss can account for up to 50% of the N input (Legg and Meisinger, 1982; Keeney, 1986).

### A. Groundwater

Groundwater is the major source of drinking water for a large percentage of the urban and the rural population in the U.S.A. (CAST, 1985). Nitrate-N rarely occurs in groundwater in concentrations greater than 3 mg $L^{-1}$ in natural systems (Bachman, 1984). Elevated $NO_3$ levels, greater than 10 mg $NO_3$-N $L^{-1}$, have been found in areas of recharge, but in areas where the aquifers move under clay cover, levels were lower (Croll and Hayes, 1988). Application of excessive amounts of N fertilizers has frequently been suggested as a primary source of $NO_3$ contamination in groundwater (Spalding et al., 1982; Schepers et al., 1984).

A recent review of federal, state and local surveys totalling more than 200,000 $NO_3$-N data points in U.S.A. groundwaters was published by Spalding and Exner (1993). It appears that $NO_3$-N levels in water decrease with depth, shallow wells (<8m) presenting the highest levels. Regions under irrigated cropland and with well-drained soils are more prone to have large areas that exceed the maximum level of 10 mg $NO_3$-N $L^{-1}$. Nitrate pollution of groundwater is not a serious problem in poorly drained soils, especially if seepage to deep aquifers is small. Denitrification in the subsoil is normally an intense process of N removal under these conditions (Martin et al., 1991).

Concerns with fertilizer contamination of groundwater are growing nationally and internationally, although most of the surveys indicate that currently $NO_3$ contaminated areas are small in proportion to the total land use (Spalding and Exner, 1993). Undoubtedly protective measures can and must be taken so that irremediable problems are avoided in the future. Remediation of surface water pollution may be aided by naturally occurring environmental processes, such as heating and cooling, exposure to sunlight, microbial transformations, and oxidation that chemically transform pollutants. Groundwater is not subject to these processes, and thus, recovery from $NO_3$ pollution, if it occurs at all, is expected to be slow.

### B. Surface and Subsurface Water

Nutrients and sediments are the major non-point pollutants of surface waters, but groundwater may add to this contamination (Hallberg, 1989). Excellent reviews on $NO_3$ movement and transformation in aquifers have been published (Lowrance and Pionke, 1989; Pionke and Lowrance, 1991), and the reader should refer to them for better understanding of these systems.

Movement of $NO_3$ with drainage water or runoff from cropped lands may contribute significantly to pollution of surface waters (Martin et al., 1991; CAST, 1992). However, soluble-N losses through surface runoff are generally small, unless high rates of N fertilizer have been applied at the surface just before heavy rains (Gilliam et al., 1974; Legg and Meisinger, 1982). Poorly drained soils with high water tables normally lose less $NO_3$ to drainage water than well-drained soils (Gambrell et al., 1975a). Upon drainage, $NO_3$ movement in soils is enhanced, even in the absence of additional fertilizer N. Water draining from soils with high N-mineralization rates, in excess of crop uptake, may have a high $NO_3$-N content, even if no fertilization is done (Duxbury and Peverly, 1978; Miller, 1979).

Whenever N fertilizer application greatly exceeds N absorption by the crop, large losses may occur through tile outlets, and thus, excess N may enter streams or simply seep into nearby lakes (Gilliam et al., 1985). Nitrogen losses from pasture crops to surface water are quite small and application of N fertilizers does not usually have a large influence on N movement from pastures (Gilliam et al., 1985). A major fraction of the N lost by surface runoff is organic N, which is associated with silt-sized particles and sediments (Schuman et al., 1973; Kissel et al., 1976). The intensification of grassland farms can lead to potentially large emissions of N to the environment (van der Meer and Uum-van Lohuyzen, 1986) and leaching and surface runoff loss of N (Garwood and Ryden, 1986; Steenvoorden et al., 1986; Keeney, 1989). Efficiency of N use for intensive grassland production may be improved through management practices that will be discussed in a later section.

Water management practices that enhance denitrification and reduce the amount of $NO_3$ entrance to surface and groundwater have been suggested since the early 1970s (Raveh and Avnimelich, 1973). Surface water contamination may be minimized by numerous management practices (Novotny and Chesters, 1981). Modification of the landscape, contouring, and use of structures such as terraces, grass waterways, tile drains have been used successfully to control sediment discharge. Restoring aquifers has not been attempted to date. Techniques proposed for aquifer restoration include physical containment and *in situ* chemical/biological treatment (Lowrance and Pionke, 1989). Physical containment would have a lower potential of utilization due to the high structural and energy costs.

### C. Health Effects of Nitrates

In 1971, the World Health Organization (WHO) International Standards established 10 mg $NO_3$-N $L^{-1}$ or 44.5 mg $NO_3$ $L^{-1}$ as the limit for $NO_3$ content in drinking water, upon consideration of higher human fluid intake and bacteria concentration in some non-European tropical countries (OECD, 1986). The European Community (EC) Directive on the Quality of Water for Human Consumption (1980, effective 1985) established 11.3 mg $NO_3$-N $L^{-1}$ as the maximum admissible concentration and suggested 5.7 mg $NO_3$-N $L^{-1}$ as a guide level, due to the increasing intake from other dietary sources and potential risk of cancer (Bockman and Granli, 1991). The U.S. Public Health Service Standards specify that $NO_3$-N content of drinking water should not exceed 10 mg $L^{-1}$ (USEPA, 1976).

A number of comprehensive reviews have been published on health effects of nitrates (Aldrich, 1984; Brezonik, 1978; CAST, 1985; Magee, 1982). High $NO_3$ concentrations in drinking water are recognized as a health risk. Reduction of $NO_3$ to $NO_2$ leads to methemoglobinemia, or "blue baby syndrome", a disease associated with a reduction in oxygen-carrying capacity of the blood. The condition appears to be reversible in normal adults, but death may occur even in adults, if more than 40% of the hemoglobin is converted (Bockman and Granli, 1991). Young infants, in particular, are more susceptible to methemoglobinemia. Usually the symptoms of methemoglobinemia are easily recognized and treated (Keeney and Follett, 1991) by injection of methylene blue (Fedkiw, 1991).

Nitrites can react with amines and amides to form nitrosamines and nitrosamides, and 80% of these compounds tested have proved to be potent carcinogens in animal experiments (OECD, 1986). Certain epidemiological studies indicated a positive correlation between $NO_3$ consumption in drinking water and the incidence of gastric cancer (Hill et al., 1973), although several authors (Addiscott et al., 1991; Bockman and Granli, 1991) state that they found very limited evidence linking stomach cancer with $NO_3$ in water. Some evidence exists of other health disorders, such as non-Hodgkin's lymphoma (Weisenburger, 1991), increased infant mortality (Super et al., 1981), central nervous system birth defects (Dorsche et al., 1984), and hypertension (Malberg et al., 1978).

Nitrate poisoning of animals from water is not common but was reported in Saskatchewan where well water contained 626 mg $L^{-1}$ $NO_3$-N (Campbell et al., 1954). In cattle, various clinical effects have been related to high $NO_3$ ingestion through feed (forages) and/or water. The disorders include deficient growth/health in young animals, increased abortion rates, depressed yields and milk quality (OECD, 1986). No details regarding $NO_3$ levels in grass, feed and water were given. Since N fertilization of forages is expected to increase in European countries and since there is higher $NO_3$ present in the water, $NO_3$ intake by farm animals is likely to increase (OECD, 1986). Oat (*Avena sativa*) hay or oat straw containing high $NO_3$ levels and fed to animals have caused a loss of cattle in Canada (Henry and Menely, 1993). Excess $NO_3$ accumulation in plants is enhanced under conditions where a sudden drought occurs.

## IV. Ameliorating Adverse Effects

### A. Best Management Practices

Best Management Practices (BMPs) have been recognized by the United States Congress as the standard for controlling non-point pollution sources (NPS) (CAST, 1992). Often $NO_3$ increases in groundwater are a direct result of the use of low efficiency N fertilizer application and the main goal in a BMP approach is the utilization of fertilizer N in a most effective manner. The quantity of $NO_3$ derived from fertilizer that leaches below the root zone is subject to some degree of control by fertilizer management practices (Boswell et al., 1985). Various management practices may be adopted to improve N use efficiency, minimizing $NO_3$ leaching into the groundwater, both in humid and drier regions (Russelle and Hargrove, 1989; Newbould, 1989; Smith and Cassel, 1991). These include selection of optimum: 1) fertilization (rate, type, form and time of application); 2) irrigation and drainage; 3) surface cover/residues; 4) plant species; and 5) waste disposal.

#### 1. Fertilization

Selection of rate, type and placement of a N fertilizer is a function of the crop, soil characteristics, and climate. Consideration of the desired crop yield, fertilizer cost, and impact of the fertilizer upon the environment must also be included. There are recommended fertilizer rates for each crop for a range of typical soils and climates, and these rates should not be exceeded (Ritter, 1988). If rates are at or slightly below crop assimilative capacity, N will be absorbed readily, reducing the risk of loss. Researchers in Europe agree that for many

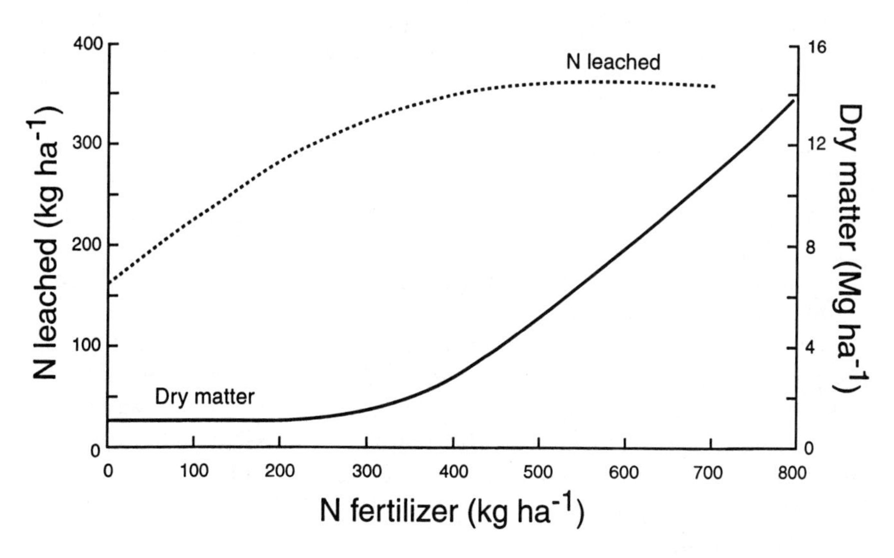


**Figure 2.** Nitrogen leached from grass and dry matter production of grass, as a function of increasing levels of N fertilizer. (After Addiscott et al., 1991.)

arable crops approximately 50% of the applied N is recovered in the harvested crop (Kolenbrander, 1977). Grasses, however, are better competitors for mineral N than field crops (Legg and Meisinger, 1982). As N fertilizer is applied at rates exceeding the requirement, crop recovery decreases and leaching losses of N as $NO_3$ become most significant (Figure 2). When making N fertilizer recommendations, non-fertilizer N inputs should be accounted for, such as contributions from residual soil N, from mineralization of OM, from salts in irrigation water, contributions of $N_2$-fixing legume crops grown in mixture with grasses and contributions from animal manures (Schepers and Mosier, 1991; Bock and Hergert, 1991).

Nitrate leaching is controlled to a certain extent by the form of N fertilizer. Nitrate sources are subject to immediate loss, whereas $NH_4$ sources must undergo nitrification first. However, nitrification will probably occur within a few weeks, and whether the source is $NO_3$ or $NH_4$ the effect is likely to be the same eventually. Nitrification inhibitors reduce leaching and denitrification losses of fertilizer N from the root zone but the use of these materials requires the synchronization of soil-N availability and crop-N uptake to prevent N deficiencies (Boswell et al., 1985). The use of slow-release materials with irrigation based on replacement of water lost to evapotranspiration has been proven to be an effective practice to minimize $NO_3$-N leaching from turfgrasses during dry periods (Snyder et al., 1984).

Surface or broadcast application of N fertilizers increases the risk of N loss with surface runoff water, especially if intense rain occurs shortly after application. Incorporation of the fertilizer into the soil minimizes this risk (Smith and Cassel, 1991). Injection of liquid or gaseous forms of $NH_3$ is usually the most efficient form of N application. This procedure may also reduce leaching by increasing $NO_3$ availability in the root zone and encouraging early crop uptake.

An effective control of $NO_3$ leaching may be achieved if water uptake and N assimilation by the crop are high at the time when leaching is likely to occur, or if depletion of most of the stored water and N occurred earlier (Peterson and Power, 1991). For N applied in the spring, losses in drainage water can be small, but these losses can be considerable if N is applied in the period before crop uptake (fall application).

Split applications of the recommended rate in order to satisfy immediate crop needs and reduce surplus in the early stages of growth is an important principle of BMP for N fertilization. Split dressings are desirable in all soils, particularly on well-drained, light-textured soils which are more susceptible to $NO_3$ leaching. Chemigation or fertigation may be a more efficient mode of application, since the N fertilizer is readily available, but will be applied when needed by the crop (Smith and Cassel, 1991).

Nitrate leaching to ground water can be reduced by applying just enough fertilizer to meet yield goals, timing the application (no fall application on sandy soils) and the use of nitrification inhibitors and slow-release fertilizers (Ritter, 1989; Martin et al., 1991).

### 2. Irrigation and Drainage

Irrigation and drainage are the major forms of soil water management (Martin et al., 1991). For non-irrigated conditions, water availability dictates crop production more than any other factor (Hargrove et al., 1988). Since water is also the carrier of $NO_3$ to groundwater, efficient water and N use is an important step in development of sustainable production. Martin et al. (1991) pointed out that combining water and N management will be more effective on reducing $NO_3$ accumulation than either approach alone.The soil water balance can be altered to discourage leaching. Mechanisms for this alteration include maintenance of or an increase in the OM content and preventing over-irrigation which will promote downward movement of water and $NO_3$ beyond the root system, especially if irrigation follows fertilization. Irrigation to remove excess salts from the root zone will also leach $NO_3$. Irrigation water should also be tested for N content, and N fertilizer rates should be adjusted accordingly.

Proper irrigation design and scheduling minimizes $NO_3$ leaching. Irrigation water should be applied when needed by the crop and in amounts just sufficient for storage and use within the crop's rooting zone (Smith and Cassel, 1991). Over-irrigation will invariably lead to $NO_3$-leaching losses (Pratt, 1984). For adequate irrigation of turfgrasses, the root-zone depth, root-zone soil water holding capacity, and irrigation system output must be known (Cisar et al., 1991). Soil moisture sensors, such as tensiometers, may be used to automatically schedule irrigations on the basis of soil moisture content (Cisar et al., 1991). Deep and active root systems with greater capacity to recover applied N may be promoted by several cultural practices including height of cut and aeration (Beard, 1973).

### 3. Surface Cover/residues

Practices which slow the water runoff in soils will normally increase infiltration. When runoff is minimum and infiltration is maximum, there is a greater chance for water to move through the root zone (Peterson and Power, 1991). Maintenance of surface residues without incorporation may enhance the potential for denitrification, particularly in humid climates due to increased soil water content, increased C availability, and decreased air-filled pore spaces, thus decreasing $O_2$ diffusion (Power and Doran, 1988). Under denitrifying conditions, $NO_3$ leaching is decreased. However, this would be a very inefficient way of controlling N leaching in fertilized crops (Thomas et al., 1989).

### 4. Plant Species

Plant variety or hybrid, plant parts removed, root morphological and physiological characteristics need to be considered when managing N to prevent $NO_3$ losses to the environment. Grasses present variations in depth of rooting, as mentioned in a previous section. Deeper-rooted plant species, such as alfalfa (*Medicago sativa* L.) can be used to intercept and remove $NO_3$ from great depths in the soil (Peterson and Russelle, 1991) or $NO_3$ accumulated after several years of residual N in soils under abandoned feedlots (Schepers, 1988).

### 5. Waste Disposal

Water contamination from animal agriculture may result from livestock production on pastureland, animals confinement facilities and waste disposal areas (Ritter, 1988). The pollutants may be leached to ground-water or carried by surface runoff. The amount of N leached is dependent upon the stocking density, length of grazing period, average manure loading rate, manure spreading uniformity by grazing livestock, degradation of manure with time, and distance from the water body. Reduction of N losses via leaching and surface runoff can be achieved by measures which influence the amount of mineral N in and on the soil and the hydrological situation, i.e., soil water content and flow (Steenvoorden et al., 1986).

Animal operations require proper management of wastes to reduce $NO_3$ pollution. These operations include management of feedlots to minimize nitrification and leaching and adequate application of wastes to land. Some of the BMP practices that may be adopted for unconfined animal production include (USEPA, 1984): 1) adoption of an effective erosion program; 2) adjusting grazing program and stocking rate to the climate, topography, geology and vegetation of the particular area; 3) locating holding pens or high density grazing away from streams; 4) maintaining grass cover on slopes and along stream banks as a vegetative filter. For feedlots, some of the BMPs are to locate them away from streams or drainage channels, divert runoff away from feedlot, collect solids carried off the feedlot and install a grass filter strip at least twice as large as the feedlot, provided it is a small feedlot (Johnson et al., 1982; USEPA, 1984).

For confined animals (feedlots), several management practices are suggested (Johnson et al., 1982): 1) locate the facility away from a stream or drainage channel; 2) divert outside runoff from the feedlot surface using diversion terraces and roof gutters; 3) collect solids carried off the feedlots surface by runoff water. Solids should be settled out in channels, debris basins, or grass waterways where they can be removed and disposed of properly on land; 4) on small feedlots install a grass filter strip at least twice as large as the feedlot if the site is close to a waterbody, to improve the quality of runoff before it enters a body of water; 5) install a runoff holding pond on large feedlots if the water quality risk is high and on small feedlots if the location prevents the use of vegetative filter. The holding pond should have a settling basin installed before it, and the collected runoff should be disposed of by irrigation on nearby crops or pastureland; 6) make best use of the nutrients in the manure to improve the soil's physical properties by applying manure to cropland.

### B. Evaluation of Potential Nitrate Leaching

Currently substantial interest exists by the agricultural community in use of models to guide the application of chemicals and water to soils and crops and to predict the fate of these in the environment. A variety of approaches has been developed to describe the movement of water and solutes in field soils. Several new models have been proposed and they vary widely in their conceptual approach, complexity and resolution. These models are also strongly influenced by the environment and many have been produced as the result of research into the basic physics and chemistry of N transport and transformation in agricultural soils. More simple, computer-based approaches such as those involving the use of a Geographic Information System (GIS) can be used for regional assessment. The use of computer simulation models for evaluating agricultural management strategies with respect to $NO_3$ leaching and movement is promising.

A number of simulation models have been developed and reviewed in the literature (Frissel and van Veen, 1981; Follett, 1989; Follett et al., 1991; Muchovej and Rechcigl, 1994). Such models aim to develop agricultural BMPs to prevent degradation of environmental quality. For instance, the model described by Addiscott and Whitmore (1991) was found to simulate the leaching characteristics of three soil types, loamy sand, sandy loam and sandy silt loam. The DRASTIC Index model, developed by the EPA (Allen et al., 1985), measures $NO_3$ leaching risk, by taking into account and assigning values to each of seven different factors (**D**epth to water, net **R**echarge, **A**quifer media, **S**oil media, **T**opography, **I**mpact on the unsaturated zone, and **C**onductivity of the aquifer). The use of detailed, accurate and local hydrologic data can provide an indication of the relative susceptibility of a site to contamination (Knox and Moody, 1991). The quantity of water percolating through the rooting zone is also important for determining $NO_3$ leaching (Williams and Kissel, 1991). The soil, crop, and climatic factors that collectively affect water percolation are well discussed by these authors. The Erosion-Productivity Impact Calculator (EPIC) (Williams et al., 1984) and the N-leaching index (LI) (Williams and Kissel, 1991) models appear to be useful tools for evaluation of leaching potential at various agricultural regions in the U.S.A.

A major contribution toward improving N-use efficiency can be obtained by using a soil-N test (Meisinger et al., 1992a). The *Presidedress Soil $NO_3$ Test* (PSNT) (Magdoff et al., 1984) aims to help conserve fertilizer N and reduce $NO_3$-N losses to the environment by measuring $NO_3$ at the time before significant N uptake has begun (Magdoff, 1991). This assay can successfully identify N-sufficient sites across a range of textures, drainage classes, and years (Meisinger et al., 1992b). The qualitative N-screening model *Long-Term Potentially Leachable Nitrogen* (LPLN) estimates a field-scale N budget (Meisinger and Randall, 1991). Based on the calculated excess N, a given field can be classified into a potential N-leaching risk group. It is suggested that such a classification would lead to a field sampling program to verify the existence of a potential environmental N problem.

There has been little work on the development of dynamic models of nutrient cycling under pastures (Haynes and Williams, 1993). These models may be prove to be very important for both pasture and turfgrass management regarding N fertilization so as to develop strategies to maximize cycling, minimize losses and produce more sustainable grassed areas.

### C. Legislation for Water Protection

Programs and legislation at all government levels have been established to promote the protection of ground and surface water from agricultural contaminants (CAST, 1992). The Federal Water Quality Legislation includes: 1) **The Clean Water Act**, under Section 303 of 101 (a) United States Code, Vol. 331251 (a), The Federal Water Pollution Control Act Amendments of 1972 (PL 92-500) with subsequent changes in 1977, 1982 and 1987, has been the framework for water pollution control in the U.S. The Clean Water Act is directed at maintaining and restoring the quality of waters of the U.S. in a broad sense, including wetlands. The Act is enforced by the EPA, which has established standards for maximum amounts of pollutants that may be released.

Individual states are authorized to establish their own standards for allowable levels of N as long as these are at least as stringent as those established by the EPA; 2) **The Coastal Zone Management Act**, originally passed in 1972, amended in 1980 and re-authorized in 1990, which requires that each state, with a Federal Coastal Zone Management Program, develop a non-point pollution control protection program to curb physical and biological degradation of the coastal environment; and 3) **The Safe Drinking Water Act** (SDWA), initially passed in 1974 and has undergone several amendments. The SDWA regulates drinking water standards for all public water supplies with the establishment of Maximum Contaminant Levels (MCL) for contaminants that may adversely affect health.

Various Federal Programs have been established by the U.S.A. Government Agencies **United States Environmental Protection Agency** (USEPA), the **United States Department of Agriculture** (USDA) and **United States Geological Survey** (USGS). There are several **State/Regional/Local Programs** which consist of efforts initiated solely at the state, regional or local levels and at the federal level, and some of these are reported in CAST (1992). The public is becoming more aware of the necessity of protecting the environment. Surveys conducted in the U.S. and other countries on $NO_3$ in ground water have been useful for detection of potential problem areas. With knowledge of vulnerable regions, prevention of $NO_3$ pollution, using various strategies, as mentioned in previous sections, we can reduce the source of contamination and reduce the risks to human health and the environment.

### D. Removal of Nitrate from Water Systems

Since $NO_3$ has a very low affinity, or ability to bind, to either soil particles or organic material once it is dissolved in water, $NO_3$ will remain dissolved until physically or biologically removed either by plants, by anaerobic microbial conversion to $N_2$ gas or by some physical process. Conventional treatment for drinking water removes virtually no $NO_3$. The most feasible physical processes for $NO_3$ removal from contaminated waters are ion exchange and reverse osmosis (Goodrich et al., 1991). Ion exchange is rather costly, especially when costs involved in disposal of spent regenerant are added to the basic costs of the exchange resins. The cost of treatment will vary with the amount of $NO_3$ that must be removed. Rajagopal and Tobin (1989) calculated that for membrane systems designed to treat 378,000 liters per day, the cost to the consumer could run from \$1.00 to \$1.50 for 3,780 liters. Therefore, a family of three which consumes 37,800 liters of water a month and is located in a community that treats 378,000 liters a day should expect to pay \$10 to \$15 per month for $NO_3$ removal.

Biological denitrification is performed by naturally occurring microorganisms, that reduce $NO_3$ under anoxic conditions. The extent of N removal by natural denitrification will depend on the amount of C either present in the subsoil or transported from the surface soil, $NO_3$ concentration, content of ferrous material in the environment, soil physical properties, such as moisture level and temperature (Gambrell et al., 1975b; Rice and Rogers, 1991; Obenhuber and Lowrance, 1991). The role of denitrification has shifted from a negative process to a positive process with potential to prevent contamination of aquifers by $NO_3$ (Henry and Menely, 1993). Reduction of nitrates in groundwater in North Carolina (Gilliam et al., 1974), in Ontario (Egboka, 1984), and in clay-confined aquifers (Foster et al., 1985) have been attributed to denitrification. Denitrification is a rather inexpensive process, but unfortunately cannot be used at every site. It may be unsuitable at borehole sites because it requires expert attention and cannot be brought into use at a short notice (Gauntlet, 1975). Denitrification by storage is sometimes the option for $NO_3$ removal from large bodies of water, but a significant effect is only observed on a long-term basis. In riparian zones denitrification has been reported to be responsible for removal of up to 98% of the $NO_3$ from groundwater (Schipper et al., 1991).

Nitrogen removal associated with buried sand filter/greywater systems was assessed in a field laboratory and two full-scale systems in Rhode Island (Lamb et al., 1991). The research tested the ability of buried sand filters to support nitrification and the suitability of domestic greywater as a carbon source for denitrification. The buried sand filter/greywater systems removed about 50% of the total N in household wastewater before the latter entered the soil absorption field. Denitrification levels of 100% were observed, using greywater as a C source and a rock tank with a 3-d retention period as an anaerobic environment.

Both processes, ion exchange and denitrification, may give rise to secondary contamination of drinking water. Methanol residues, bacterial proliferation, organic metabolites, organochlorinated compounds have been associated with denitrification, while release of undesirable organic compounds may occur with the use of resins in ion exchange. Costs are high either to remove $NO_3$ from water or to prevent contamination, and both aspects should be considered for establishment of a rational long-term policy (OECD, 1986). Long term planning and preventive quality management are essential for any water supply policy, especially in the case of $NO_3$ contamination of aquifers where the lag time may be quite long, possibly decades for both detection of contamination and restoration.

## V. Conclusions

Concerns about the environment and prevention of its degradation are increasing worldwide. Nonetheless, production of high quality pastures and turfgrasses depends, to a major extent, on N fertilization. Nitrogen fertilizers are applied to pastures and turfgrasses to correct deficiencies, increase the nutrition levels, maintain soil fertility conditions and improve crop vigor, yield, and quality. On intensively managed grassland farms N inputs also originate from waste depositions. Increasing $NO_3$ levels in groundwater have generated questions regarding recommendations for grasses so as to avoid economic and environmental implications from application of excessive amounts of N fertilizers.

The fate of N fertilizers applied to grasses may be altered by several management practices, including rate, form (rapidly and slowly available) and time of N fertilizer application, plant species, and irrigation, which are manageable by man. The use of slow-release N sources, finer textured soils, and limiting irrigation to replace only the moisture removed from evapotranspiration will reduce or limit $NO_3$ leaching from grass systems. Thus, where grass fertilization might be a threat to groundwater quality, management strategies can be adopted to minimize or avoid $NO_3$ leaching. The complete elimination of $NO_3$ leaching from pastures and turfgrasses grown on coarse-textured soils may be impossible to achieve, if adequate yields and quality are still to be maintained. Also, the occurrence of heavy rain shortly after application of water-soluble N will invariably cause leaching.

Permanent grass systems generally have a high absorption capacity for $NO_3$-N, even when N fertilizer is applied annually and $NO_3$ does not generally accumulate in soil under native grasslands. Grasses are usually considered as excellent biological filters that can minimize surface runoff and leaching of groundwater contaminants due to their complete coverage of the soil surface and extensive and dense root systems, which permit assimilation of N in excess of crop demands. Usually a high percentage (50-80%) of N applied to both pasture and turfgrasses is recovered by the plant. It has been stated that from an environmental standpoint the use of N fertilizer at the rates up to 400 kg $ha^{-1}$ $yr^{-1}$ appear to pose no potential pollution problems, regarding risk of high $NO_3$ accumulation in either plant tissues, soil or water, especially for ungrazed systems. In general, $NO_3$ leaching below grass systems land on sandy soils is minimal when N is applied in quantities below 200 kg $ha^{-1}$ $yr^{-1}$, however leaching losses may increase rapidly at higher rates.

The severity of adverse health and environmental aspects of $NO_3$ contamination resultant from N fertilization of pastures and turfgrasses are still not evident. The existing literature indicates that the quantity of applied N has to be quite elevated and that management practices rather poor, for $NO_3$ leaching from pasture and turfgrass systems to be significant. However, preventive and protective practices should be adopted since restoration of a degraded environment, especially water quality, is not an easy task. The increase in fertilizer use on homelawns since the 1970s suggests a potential for an increase in off-site losses and environmental pollution. Homelawns are irrigated with little regard for soil moisture status or soil water holding capacity.

Adequate information, training and motivation of farmers, turf specialists and home "gardeners" should be promoted by various state and federal government agencies for best management of soil and fertilizer N, as well as compliance with existing laws and regulations for the specific area in question. Training should address both plant development and pollution prevention and should take into account optimum yield as well as soil, water and health protection criteria.

## Acknowledgements

The authors wish to express their gratitude to Drs. R.S. Pacovsky, A.K. Alva, F.M. Pate and K.L. Buhr, for critically reviewing this chapter. They also thank Ms. Amie Smith, for drawing the figures, and S.J. Muchovej and M. Izabel Esteves, for assistance with typing and indexing the references.

## References

Addiscott, T.M. and A.P. Whitmore. 1991. Simulation of solute leaching in soils of differing permeability. *Soil Use Management* 7:94-102.

Addiscott, T.M., A.P. Whitmore and D.S. Powlson. 1991. *Farming, Fertilizers and the Nitrate Problem*. C. A. B. International Wallingford, Oxon.

Aldrich, S.R. 1984. Nitrogen management to minimize adverse effects on the environment. In: R.D. Hauck (ed.), *Nitrogen in Crop Production*. ASA, Madison, WI.

Allen, L., T. Bennett, J.H. Lehr and R.J. Petty. 1985. DRASTIC - A standardized system for evaluating ground-water pollution potential using hydrogeologic settings. EPA 600/2-85-028. U.S. Gov. Print. Office, Washington, DC.

Allison, F.E. 1961. Nitrogen transformations in soils. *Soil Crop Sci. Soc. FL* 21:248-254.

Bachman, J.L. 1984. Nitrate in the Columbia aquifer, central Delmarva Peninsula, Maryland. U.S. Geol. Surv. Water-Resour. Invest. Rep. 84-4322. U.S. Geol. Surv., Towson, MD.

Ball, P.R. and J.C. Ryden. 1984. Nitrogen relationships in intensively managed temperate grasslands. *Plant Soil*, 76:23-33.

Barraclough, D., M.J. Hyden and G.P. Davies. 1983. Fate of fertilizer nitrogen applied to grassland: I. Field leaching results. *J. Soil. Sci.* 34:483-497.

Barraclough, D., E.L. Geens and J.M. Maggs. 1984. Fate of fertilizer nitrogen applied to grassland: II. Nitrogen-15 leaching results. *J. Soil Sci.* 35:191-199.

Beard, J.B. 1973. *Turfgrass: Science and Culture*. Prentice Hall, Englewood Cliffs, NJ.

Bergstrom, L. and R. Johansson. 1991. Leaching of nitrate from monolith lysimeters of different types of agricultural soils. *J. Environ. Qual.* 20:801-807.

Black, A.L. 1968. Nitrogen and phosphorus fertilization for production of crested wheatgrass and native grass in Northeastern Montana. *Agron. J.* 60:213-216.

Blaser, R.E. 1964. Symposium of Forage Utilization: Effects of Fertility Levels and Stage of Maturity on Forage Nutritive Value. *J. Anim. Sci.* 23:246-253.

Blowin, G.M. and D.W. Rindt. 1967. Sulfur-coated fertilizer pellet having controlled dissolution rate and inhibited against microbial decomposition. U.S. Patents 3, 342, 577.

Blue, W.G. 1974. Efficiency of five nitrogen sources for Pensacola Bahiagrass on Leon Fine Sand as affected by lime treatments. *Soil Crop Sci. Soc. FL* 33:176-180.

Blue, W.G. and D.A. Graetz. 1977. The affect of split nitrogen applications on nitrogen uptake by Pensacola Bahiagrass from an Aeric Haplaquod. *Soil Sci. Soc. Am. J.* 41:927-930.

Bock, B.R. and G.W. Hergert. 1991. Fertilizer nitrogen management. p. 139-164. In: R.F. Follett, D.R. Keeney and R.M. Cruse (eds.), *Managing Nitrogen for Groundwater Quality and Farm Profitability*. SSSA, Madison, WI.

Bockman, O.C. and T. Granli. 1991. Human health aspects of nitrate intake from food and water. p. 373. In: M.L. Richardson (ed.), *Chemistry, Agriculture, and the Environment*. The Royal Society of Chemistry, Cambridge.

Boswell, F.C., J.J. Meisinger and N.L. Case. 1985. Production, marketing, and use of nitrogen fertilizers. p. 229-292. In: O.P. Engelstad (ed.), *Fertilizer Technology and Use*. SSSA, Madison, WI.

Brezonik, P.L. 1978. *Nitrates: An Environment Assessment*. National Academy of Sciences. Washington, DC.

Brown, K.W., R.L. Duble and J.C. Thomas. 1977. Influence of management and season on fate of N applied to golf greens. *Agron. J.* 69:667-671.

Brown, K.W., J.C. Thomas and R.L. Duble. 1982. Nitrogen source effect on nitrate and ammonium leaching and runoff losses from greens. *Agron. J.* 74:947-950.

Campbell, J.B., A.N. Davis and P.J. Myhr. 1954. Methemoglobinemia of livestock caused by high nitrate contents of well water. *Can. J. Camp. Med.* 18:93-101.

CAST (Council for Agricultural Sciences and Technology). 1985. *Agriculture and Groundwater Quality*. Council for Agricultural Sciences and Technology Reports, 103, 62.

CAST (Council for Agricultural Sciences and Technology). 1992. Water quality: Agriculture's Role. Task Force Rep. 120, CAST, Ames, IA.

Cisar, J.L., G.H. Snyder and P. Nkedi-Kizza. 1991. Maintaining quality turfgrass with minimal nitrogen leaching. Florida Cooperative Extension Service, IFAS Bul. 273, University of Florida.

Collison, S.E. and S.S. Walker. 1916. Loss of fertilizers by leaching. Univ. of Florida, Agricultural Experiment Station, Bulletin 132.

Cook, C.W. 1965. Plant and Livestock Responses to Fertilized Rangelands. *Utah Agric. Expt. Sta. Bul. 455.*

Cooke, G.W. 1985. The present use and efficiency of fertilizers and their future potential in agricultural production systems. In: F.P.W. Winteringham (ed.), *Environment and Chemicals in Agriculture*. Proc. of a symposium held in Dublin, October 15-17, 1984. Elsevier Applied Science Publishers, London.

Croll, B.T. and C.R. Hayes. 1988. Nitrate and water supplies in the United Kingdom. *Environ. Poll.* 50:163-187.

De Roo, H.C. 1980. Nitrate Fluctuation in Groundwater as Influenced by Use of Fertilizer. Connecticut Agr. Exp. Sta., University of Connecticut, New Haven, CT, Bull. 779.

Doran, J.W., J.S. Schepers and N.P. Swanson. 1981. Chemical and bacteriological quality of pasture runoff. *J. Soil Water Conserv.* 36:166-171.

Dorsche, M.M., R.K.R. Scragg, A.J. McMichael, P.A. Baghurst and K.F. Dyer. 1984. Congenital malformations and maternal drinking water supply in rural south Australia: A case control study. *Am. J. Epidemiol.* 119:473-486.

Dougherty, C.T. and C.L. Rhykerd. 1985. The role of nitrogen in forage-animal production. In: M.E. Heath, R.F. Barnes and D.S. Metcalfe (eds.), *Forages: The Science of Grassland Agriculture*, 4th edition. Iowa State University Press, Ames, IA.

Duxbury, J.M. and J.H. Peverly. 1978. Nitrogen and phosphorus losses from organic soils. *J. Environ. Qual.* 7:566-570.

Egboka, B.C.E. 1984. Nitrate contamination of shallow groundwaters in Ontario, Canada. *Sci. Total Environ.* 35:53-58.

Endelman, F.J., D.R. Keeney, J.T. Gilmour and P.G. Saffira. 1974. Nitrate and chloride movement in the Plainfield loamy sand under intensive irrigation. *J. Environ. Qual.* 3:295-298.

Exner, M.E., M.E. Burbach, D.G. Watts, R.C. Shearman and R.F. Spalding. 1991. Deep nitrate movement in the unsaturated zone of a simulated urban lawn. *J. Environ. Qual.* 20:658-662.

FAO (Food and Agriculture Organization of the United Nations). 1962-1987. *FAO Fertilizer Yearbooks*. FAO, Rome.

FAO (Food and Agriculture Organization of the United Nations). 1988. *Current World Fertilizer Situation and Outlook 1986/87-1992/93*. FAO, Rome.

Fedkiw, J. 1991. Nitrate Occurrence in U.S. Waters (and Related Questions): A Reference Summary of Published Sources from an Agricultural Perspective. Prepared by J. Fedkiw, USDA Working Group on Water Quality. US Department of Agriculture, Washington, DC.

Fertilizer Institute. 1993. *Fertilizer Facts and Figures*. Washington, DC.

Follett, R.F. (ed.). 1989. *Nitrogen Management and Ground Water Protection*: Developments in Agricultural and Managed-Forest Ecology 21. Elsevier Science Publishing Co..

Follett, R.H., L.S. Murphy and R.L. Donahue. 1981. Nitrogen fertilizers. p.23-83. In: *Fertilizers and Soil Amendments*. Prentice-Hall, Inc., Englewood Cliffs, NJ.

Follett, R.F., D.R. Keeney and R.M. Cruse (eds.). 1991. *Managing Nitrogen for Ground Water Quality and Profitability*. SSSA, Madison, WI.

Foster, S.S.D., D.P. Kelley and R. James. 1985. The evidence for zones of biodenitrification in British aquifers. p. 356-369. In: D.E Caldwell, J.A. Brierly and C.L. Brierly (eds.), *Planetary Ecology*. New York, Van Nostrand Reinhold, Planetary Ecology.

Freyh, R.L. and R.E. Lamond. 1992. Sulfur and nitrogen fertilization. *J. Prod. Agric.* 5:488-491.

Frissel, M.J. and J. A. van Veen (eds.). 1981. *Simulation of Nitrogen Behaviour of Soil-Plant Systems*. Pudoc, Centre for Agricultural Publishing and Documentation, Wageningen.

Fuleky, G. 1991. Leaching of nutrient elements from fertilizers into deeper soil zones. p. 185-208. In: M.L. Richardson (ed.), *Chemistry, Agriculture, and the Environment*. The Royal Society of Chemistry, Cambridge, Great Britain.

Gambrell, R.P., J.W. Gilliam and S.B. Weed. 1975a. Nitrogen losses from soils of the North Carolina Coastal Plains. *J. Environ. Qual.* 4:317-323.

Gambrell, R.P., J.W. Gilliam and S.B. Weed. 1975b. Denitrification in subsoils of the North Carolina Coastal Plains as affected by soil drainage. *J. Environ. Qual.* 4:311-316.

Garrett, M.K. 1991. Nitrogen losses from grassland systems under temperate climatic conditions. In: M.L. Richardson (ed.), *Chemistry, Agriculture and the Environment*. The Royal Society of Chemistry, Thomas Graham House, Science Park, Cambridge.

Garwood, E.A. and J.C. Ryden. 1986. Nitrate loss through leaching and surface runoff from grassland: Effects of water supply, soil type and management. p. 99-113. In: H.G. van der Meer, J.C. Ryden and G.C. Ennik (eds.), *Nitrogen Fluxes in Intensive Grassland Systems*. Martinus Nijhoff Publishers, Dordrecht.

Gauntlet, R.B. 1975. Nitrate removal from water by ion exchange. *Water Treat. Exam.* 24:172.

Geron, C.A., T.K. Danneberger, S.J. Traina, T.J. Logan and J.R. Street. 1993. The effects of establishment methods and fertilization practices of nitrate leaching from turfgrass. *J. Environ. Qual.* 22:119-125.

Gilliam, J.W., R.B. Daniels and J.F. Lutz. 1974. Nitrogen content of shallow groundwater in the North Carolina Coastal Plain. *J. Environ. Qual.* 3:147-151.

Gilliam, J.W., T.J. Logan and F.E. Broadbent. 1985. Fertilizer use in relation to the environment. p. 561-588. In: O.P. Engelstad (ed.), *Fertilizer Technology and Use*. SSSA, Madison, WI.

Goldberg, V.M. 1989. Groundwater pollution by nitrates from livestock wastes. *Environ. Health Perspectives*, 83:25-29.

Goodrich, J.A., B.W. Lykins and R.M. Clark. 1991. Drinking water from agriculturally contaminated groundwater. *J. Environ. Qual.* 20:707-717.

Graetz, D.A., L.C. Hammond and J.M. Davidson. 1973. Nitrate movement in a Eutis fine sand planted to millet. *Soil Crop Sci. Soc. FL.* 33:157-160.

Gross, C. M., J.S. Angle and M.S. Welterlen. 1990. Nutrient and sediment losses from turfgrass. *J. Environ. Qual.* 19:663-668.

Hackett, I., B. BrooksTaylor, N. Nichols and E.H. Jensen. 1969. Fertilization of Mountain Meadows. *Nev. Ranch Home Rev.* 4:8-11.

Haigh, R.A. and R.E. White. 1986. Nitrate leaching from a small underdrained, grassland, clay catchment. *Soil Use Manage.* 2:65-70.

Hallberg, G.R. 1989. Nitrate in groundwater in the United States. p. 35-74. In: R.F. Follett (ed.), *Nitrogen Management and Groundwater Protection.* Elsevier Science Publishers, Amsterdam.

Hargrove, W.L., A.L. Black and J.V. Mannering. 1988. Cropping strategies for efficient use of water and nitrogen: introduction. p. 1-5. In: W.L. Hargrove (ed.), *Cropping Strategies for Efficient Use of Water and Nitrogen.* ASA, CSSA, SSSA, Madison, WI, Special Publication. No. 51.

Haynes, R.J. and P.H. Williams. 1993. Nutrient cycling and soil fertility in the grazed pasture ecosystem. p. 119-199. In: D.L. Sparks (ed.), *Advances in Agronomy* Vol. 49. Academic Press, Inc., San Diego, CA.

Heathwaite, A.L., T.P. Burt, S.T. Trudgill. 1989. Runoff, sediment, and solute delivery in agricultural drainage basins: a scale-dependent approach. IAHS Publication no. 182.

Henry, J.L. and W.A. Menely. 1993. Nitrates in Western Canadian Groundwater. p. 1-31. Western Canada Fertilizer Association.

Hill, M.J., G. Hawksworth, and G. Tattersall. 1973. Bacteria, nitrosamines and cancer of stomach. *Br. J. Cancer* 28:562-567.

Hood, A.E.M. 1976. The high nitrogen trial on grassland at Jealotts Hill. *Stickstoff* 7, no. 83-84:395-404.

Hubbard, R.K., and J.M. Sheridan. 1989. Nitrate movement to groundwater in the southeastern Coastal Plain. *J. Soil Water Conserv.,* 44:20-27.

Hull, A.C. Jr. 1963. Fertilization of seeded grasses on mountainous rangelands in Northeastern Utah and Southeastern Idaho. *J. Range. Mgt.* 16:306-310.

Impithuksa, V. and W.G. Blue. 1985. Fertilizer nitrogen and nitrogen-15 in three warm-season grasses grown on a Florida Spodosol. *Soil Sci. Soc. Am. J.* 49:1201-1204.

Impithuksa, V., C.L. Dantzman and W.G. Blue. 1979. Fertilizer nitrogen utilization by three warm-season grasses on an alfic haplaquod as indicated by nitrogen-15. Proc. *Soil Crop Sci. Soc. FL* 38:93-97.

Impithuksa, V., W.G. Blue and D.A. Graetz. 1984. Distribution of applied nitrogen in soil-Pensacola Bahiagrass components as indicated by nitrogen-15. *Soil Sci. Soc. Am. J.* 48:1280-1284.

Isaac, S.P., and P.M. Canaway. 1987. The mineral nutrition of Festuca-Agrostis golf greens: a review. *Journal of the Sports Turf Research Institute* 63:9-27.

Jaakkola, A. 1984. Leaching losses of nitrogen from a clay soil under grass and cereal crops in Finland. *Plant Soil* 76:59-66.

Jarvis, S.C. and D. Barraclough. 1991. Variation in mineral nitrogen under grazed grassland swards. *Plant & Soil* 138:177-188.

Jeffrey, D.W. 1988. Mineral nutrients and the soil environment. p. 179-204. In: M.B. Jones and A. Lazenby (eds.), *The Grass Crop: The physiological basis of production.* Chapman and Hall Ltd., London.

Johnson, D.D., J.M. Kreglow, S.A. Dressing, R.P. Maas, F.A. Koehler and F.J. Humenik. 1982. State-of-the-art Review of Best Management Practices for Agriculture Nonpoint Sources Control. I. Animal Wastes. North Carolina State University, Biological and Agricultural Engineering Department, Raleigh, NC.

Jones, G.W., E.C. Dehaven, L.F. Clark, J.T. Rauch, J.R. Rasmussen and C.G. Guillen. 1990. Groundwater quality sampling results from wells in Southwest Florida Water Management District. Ambient Groundwater Quality Monitoring Program, SFWMD, and Florida Dept. of Environmental Regulations.

Juergens-Gschwind, S. 1989. Ground water nitrates in other developed countries (Europe)-Relationships to land use patterns. p. 75-138. In: R.F. Follett (ed.), *Nitrogen Management and Ground Water Protection*, Elsevier Science Publishers, Amsterdam.

Keeney, D.R. 1983. Transformations and transport of nitrogen. p. 48-64. In: F.W. Shaller and G.W. Bailey (eds.), *Agricultural Management and Water Quality*. Iowa State Univ. Press, Ames, IA.

Keeney, D.R. 1986. Sources of nitrate to ground water. *CRC Critical Rev. Environ. Control*, 16:257-304.

Keeney, D.R. 1989. Sources of nitrate to ground water. p. 23-34. In: R.F. Follett (ed.), *Nitrogen Management and Ground Water Protection*. Elsevier Science Publishers, Amsterdam.

Keeney, D.R. and R.F. Follett. 1991. Managing nitrogen for groundwater quality and farm profitability: overview and introduction. p. 1-7. In: R.F. Follett, D.R. Keeney and R.M. Cruse (eds.), *Managing Nitrogen for Groundwater Quality and Farm Profitability*. SSSA, Madison, WI.

Kelly, S., J.A. Moore, M. Gamroth, D.D. Myrold, and N. Baumeister. 1991. Movement of nitrogen from land application of dairy on pasture. In: *Nonpoint Source Pollution: the unfinished agenda for the protection of our water quality*. Proceedings from the technical session of the regional conference Tacoma, Washington, March 20-21, 1991, Washington State University, Pullman, WA.

Kilmer, V.J., J.W. Gilliam, T.F. Lutz, R.T. Joyce and C.D. Eklund. 1974. Nutrient losses from grassed watersheds in western North Carolina. *J. Environ. Qual.* 3:214-219.

Kissel, D.E. and S.J. Smith. 1978. Fate of fertilizer nitrate applied to coastal bermudagrass on a swelling clay soil. *Soil Sci. Soc. Am. J.* 42:77-80.

Kissel, D.E., C.W. Richardson and E. Burnett. 1976. Losses of nitrogen in surface runoff in the blackland prairie of Texas. *J. Environ. Qual.* 5:288-292.

Knox, E. and D.W. Moody. 1991. Influence of hydrology, soil properties and agricultural land use on nitrogen in groundwater. p. 19-57. In: R.F. Follett, D.R. Keeney and R.M. Cruse (eds.), *Managing Nitrogen for Groundwater Quality and Farm Profitability*. SSSA, Madison, WI.

Kolenbrander, G.J. 1972. The eutrophication of surface water by agriculture and the urban population. *Stikstof* 15:56-67.

Kolenbrander, G.J. 1977. Nitrogen in organic matter and fertilizers as a source of pollution. *Prog. Wat Technol.* 8:67-84.

Kolenbrander, G.J. 1981. Leaching of nitrogen in agriculture. p. 199-216. In: J.C. Brogan (ed.), *Nitrogen losses and surface runoff from landspreading of manures*. Martinus Nijhoff Publishers, Dordrecht.

Lamb, B.E., A.J. Gold, G.W. Loomis and C.G. McKiel. 1991. Nitrogen Removal for On-Site Sewage Disposal: Field Evaluation of Buried Sand Filter/Greywater Systems Kingston. *Trans. Amer. Soc. Agric. Eng.* 34:883.

Larson, K.L., J.F. Carter and E.H. Vasey. 1971. Nitrate-nitrogen accumulation under bromegrass sod fertilized annually at six levels of nitrogen for fifteen years. *Agron. J.* 63:527-528.

Lawson, D.M. 1989. The principles of fertilizer use for sports turf. *Soil Use Manage.* 5:122-127.

Lazenby, A. 1983. Nitrogen relationships in grassland ecosystems. p. 56-63. In: J.A. Smith and V.W. Hays (eds.), *Proceedings of the XIV International Grassland Congress*, Lexington, KY June 15-24 1981. Westview Press, Boulder, CO.

Legg, J.O. and J.J. Meisinger. 1982. Soil nitrogen budgets. p. 503-566. In: F.J. Stevenson (ed.), *Nitrogen in Agricultural Soils*. ASA, CSSA, SSSA, Madison, WI, Agron. Monogr. no. 22.

Low, A.J. 1973. Nitrate and ammonium nitrogen concentration in water draining through soil monoliths in lysimeters cropped with grass or clover or uncropped. *J. Sci. Food Agric.*, 24:1489-1495.

Lowrance, R.R. and H.B. Pionke. 1989. Transformations and movement of nitrate in aquifer systems. p. 373-392. In: R.F. Follett (ed.), *Nitrogen Management and Groundwater Protection*. Elsevier Science Publishers, Amsterdam.

Macduff, J.H., S.C. Jarvis, and D.H. Roberts. 1990. Nitrates: leaching from grazed grassland systems. In: R. Calvet (ed.), *Nitrates Agriculture Water*. International Symposium, Paris-La Defense 7-8 November 1990.

Magdoff, F.R. 1991. Understanding the Magdoff pre-sidedress nitrate test for corn. *J. Prod. Agric.* 4:297-305.

Magdoff, F.R., D. Ross and J. Amadon. 1984. A soil test for nitrogen availability to corn. *Soil Sci. Soc. Am. J.* 48:1301-1304.

Magee, P.N. 1982. Nitrogen as a potential health hazard. *Phil. Trans. R. Soc. Land.* B296:543-550.

Malberg, J.W., E.P. Savage and J. Osteryoung. 1978. Nitrates in drinking water and the early onset of hypertension. *Environ. Pollut.* 15:155-160.

Martin, D.L., J.R. Gilley and R.W. Skaggs. 1991. Soil water balance and management. p. 199-235. In: R.F. Follett, D.R. Keeney and R.M. Cruse (eds.), *Managing Nitrogen for Groundwater Quality and Farm Profitability*. SSSA, Madison, WI.

Mata, A. and W.G. Blue. 1973. Fertilizer nitrogen distribution in Pensacola bahiagrass sod during the first year of development on an Aeric Haplaquod. *Soil Crop Sci. Soc. FL.* 33:209-211.

McCrimmon, J.N. and K.J. Karnok. 1992. Nitrate form and seasonal root and shoot response of creeping bentgrass. *Commun. Soil Sci. Plant Anal.* 23:1071-1088.

McLeod, R.V. and R.O. Hegg. 1984. Pasture runoff water quality from application of inorganic and organic nitrogen sources. *J. Environ. Qual.* 13:122-126.

Meisinger, J.J. and G.W. Randall. 1991. Estimating nitrogen budgets for soil-crop systems. p. 85-124. In: F.R. Follett, D.R. Keeney and R.M. Cruse (eds.), *Managing Nitrogen for Ground Water Quality and Farm Profitability*. SSSA, Madison, WI.

Meisinger, J.J., F.R. Magdoff and J.S. Schepers. 1992a. Predicting N fertilizer needs for corn in humid regions: underlying principles. p. 7-27. In: B.R. Bock and K.R. Kelley (eds.), *Predicting N fertilizer Needs for Corn in Humid Regions*. TVA, National Fertilizer and Environmental Research Center, Muscle Shoals, AL.

Meisinger, J.J., V.A. Bandel, J.S. Angle, B.E. O'Keefe and C.M. Reynolds. 1992b. Presidedress soil nitrate test evaluation in Maryland. *Soil Sci. Soc. Am. J.* 56:1527-1532.

Miller, M.H. 1979. Contribution of nitrogen and phosphorus to subsurface drainage water from intensively cropped mineral and organic soils in Ontario. *J. Environ. Qual.* 8:42-48.

Morton, T.G., A.J. Gold and W.M. Sullivan. 1988. Influence of overwatering and fertilization on nitrogen losses from home lawns. *J. Environ. Qual.* 17:124-130.

Muchovej, R.M.C. and J.E. Rechcigl. 1994. Nitrogen. In: J.E. Rechcigl (ed.), *Environmental Aspects of Soil Amendments*. Lewis Publishers, Boca Raton, FL (in press).

Newbould, P. 1989. The use of nitrogen fertiliser in agriculture. Where do we go practically and ecologically? *Plant Soil* 115:297-311.

Novotny, V. and G. Chesters. 1981. *Handbook of Non-Point Pollution: Sources and Management*. van Nostrand Reinhold Co., New York, NY.

Obenhuber, D.C. and R. Lowrance. 1991. Reduction of nitrate in aquifer microcosms by carbon additions. *J. Environ. Qual.* 20:255-258.

OECD (Organization for Economic Co-operation and Development). 1986. *Water Pollution by Fertilizers and Pesticides*. OECD Publications, Paris.

Owens, L.B. 1981. Effects of nitrapyrin on nitrate movement in soil columns. *J. Environ. Qual.* 10:308-310.

Owens, L.B. 1987. Nitrate leaching losses from monolith lysimeters as influenced by nitrapyrin. *J. Environ. Qual.* 16:34-38.

Owens, L.B., R.W. van Keuren and W.M. Edwards. 1982. Environmental effects of a medium-fertility 12-month pasture program: II. Nitrogen. *J. Environ. Qual.* 11:241-246.

Owens, L.B., R.W. van Keuren, and W.M. Edwards. 1983. Nitrogen loss from a high-fertility, rotational pasture program. *J. Environ. Qual.* 12:346-350.

Owens, L.B., W.M. Edwards and R.W. van Keuren. 1984. Peak nitrate-nitrogen values in surface runoff from fertilized pastures. *J. Environ. Qual.* 13:310-312.

Owens, L.B., W.M. Edwards and R.W. van Keuren. 1992. Nitrate levels in shallow groundwater under pastures receiving ammonium nitrate or slow-release nitrogen fertilizer. *J. Environ. Qual.* 21:607-613.

Peterson, G.A. and W.W. Frye. 1989. Fertilizer nitrogen management. p. 183-219. In: R.F. Follett (ed.), *Nitrogen Management and Ground Water Protection*. Elsevier Science Publishers, Amsterdam.

Peterson, G.A. and J.F. Power. 1991. Soil, crop, and water management. p. 189-198. In: R.F. Follett, D.R. Keeney and R.M. Cruse (eds.), *Managing Nitrogen for Groundwater Quality and Farm Profitability*. SSSA, Madison, WI.

Peterson, T.A. and M.P. Russelle. 1991. Alfalfa and the nitrogen cycle in the Corn Belt. *J. Soil Water Conserv.* 46:229-235.

Petrovic, A.M. 1990. The fate of nitrogenous fertilizers applied to turfgrass. *J. Environ. Qual.* 19:1-14.

Pionke, H.B. and R.R. Lowrance. 1991. Fate of nitrate in subsurface drainage water. p. 237-257. In: R.F. Follett, D.R. Keeney and R.M. Cruse (eds.), *Managing Nitrogen for Groundwater Quality and Farm Profitability*. SSSA, Madison, WI.

Pitman, W.D., K.M. Portier, C.G. Chambliss and A.E. Kretschmer, Jr. 1992. Performance of yearling steers grazing bahia grass pastures with summer annual legumes or nitrogen fertiliser in subtropical Florida. *Tropical Grasslands* 26:206-211.

Porter, L.K. 1969. Nitrogen in grassland ecosystems. p. 377-402. In: R.L. Dix and R.G. Beidleman (eds). *The Grassland Ecosystem. A Preliminary Synthesis*. Range Science Dept., Science Series No.2. Colorado State University, Ft. Collins, CO.

Power, J.F. 1970. Leaching of nitrate nitrogen under dryland agriculture in the Northern Great Plains. p. 111-122. In: *Relationship of Agriculture to Soil and Water Pollution*. Proc. First Annual Agric. Pollution Conf., Rochester, NY, Cornell University Press, Ithaca, NY.

Power, J.F. and J.W. Doran. 1988. Role of crop residue management in nitrogen cycling and use. p. 101-113. In: W.L. Hargrove (ed.), *Cropping Strategies for Efficient Use of Water and Nitrogen*. ASA, CSSA, SSSA, Madison, WI, Special Publication No. 51.

Pratt, P.F. 1984. Nitrogen use and nitrate leaching in irrigation agriculture. p. 319-333. In: R.D. Hauck (ed.), *Nitrogen in Crop Production*. ASA, Madison, WI.

Prins, W.H. 1983. Residual effects of two years of very high nitrogen applications on clay soil under grass in a humid temperate climate. p. 305-308. In: J.A. Smith and V.W. Hays (eds.), *Proceedings of the XIV International Grassland Congress*, Lexington, KY June 15-24 1981. Westview Press, Boulder, CO p.305-308.

Rajagopal, R. and G. Tobin. 1989. Expert opinion and ground-water quality protection: the case of nitrate in drinking water. *Ground Water* 27:835-847.

Raveh, A. and Y. Avnimelich. 1973. Minimizing nitrate seepage from the Hula Valley into Lake Kinneret (Sea of Galilee): I. Enhancement of nitrate reduction by sprinkling and flooding. *J. Environ. Qual.* 2:455-458.

Rawitz, E., S. Burns, H. Etkin, R. Hardiman, D. Hillel and R. Terkeltoub. 1980. Fate of fertilizer nitrogen in irrigated fields under semi-arid conditions. p. 195-235. In: *Soil Nitrogen as Fertilizer or Pollutant*. International Atomic Energy Agency, Vienna.

Rice, C.W. and K.L. Rogers. 1991. *Denitrification in the vadose zone: a mechanism for nitrate removal.* Kansas Water Resources Research Institute Report no. G1563-04, contribution no. 1287. Kansas State University, Manhattan, KS.

Ritter, W.F. 1988. Reducing impacts of nonpoint source pollution from agriculture: a review. *J. Environ. Sci. Health* A23:645-667.

Ritter, W.F. 1989. Nitrate leaching under irrigation in the United States - a review. *J. Environ. Sci. Health* A24:349-378.

Roberts, G. 1987. Nitrogen inputs and outputs in a small agricultural catchment in the eastern part of the United Kingdom. *Soil Use Manage.* 3:148-154.

Russelle, M.P. and W.L. Hargrove. 1989. Cropping systems: ecology and management. p. 277-317. In: R.F. Follett (ed.), *Nitrogen Management and Groundwater Protection*. Elsevier Science Publishers, Amsterdam.

Ryden, J.C., P.R. Ball and E.A. Garwood. 1984. Nitrate leaching from grassland. *Nature* 311:1-3.

Sartain, J.B. 1988. Turfgrass responses to new slow-release N sources. *Proc. Fla. Hort. Soc.* 101:321-322.

Sartain, J.B. 1992. Natural organic slow-release N sources for turfgrasses. *Proc. Fla. Hort. Soc.* 105:224-226.

Schepers, J.S. 1988. Role of cropping systems in environmental quality: ground water nitrogen. p. 167-178. In: W.L. Hargrove (ed.), *Cropping Strategies for Efficient Use of Water and Nitrogen*. ASA, CSSA, SSSA, Special Publication no. 51, Madison, WI.

Schepers, J.S. and D.D. Francis. 1982. Chemical water quality of runoff from grazing land in Nebraska: I. Influence of grazing livestock. *J. Environ. Qual.* 11:351-354.

Schepers, J.S. and A.R. Mosier. 1991. Accounting for nitrogen in nonequilibrium soil-crop systems. p. 125-138. In: R.F. Follett, D.R. Keeney and R.M. Cruse (eds.), *Managing Nitrogen for Groundwater Quality and Farm Profitability*. SSSA, Madison, WI.

Schepers, J.S., K.D. Frank and D.G. Watts. 1984. Influence of irrigation and nitrogen fertilization on groundwater quality. p. 21-32. Proc. Int. Union of Geodesy and Geophysics. Hamburg, West Germany, p.21.

Schipper, L.A., A.B. Cooper and W.J. Dyck. 1991. Mitigating nonpoint-source nitrate pollution by riparian-zone denitrification. p. 401-452. In: I. Bogárdi and R.D. Kuzelka (eds.), *Nitrate Contamination: Exposure Consequence and Control*. NATO ASI Series, Vol. G30, Springer-Verlag, Berlin Heidelberg.

Schuman, G.E., R.E. Burwell, R.F. Piest and R.G. Spomer. 1973. Nitrogen losses in surface runoff from agricultural watersheds on Missouri Valley loess. *J. Environ. Qual.* 2:299-302.

Smaling, E.M.A. and J. Bouma. 1992. Bypass flow and leaching of nitrogen in a Kenyan vertisol at the onset of the growing season. *Soil Use & Management* 8:44-48.

Smith, S.J. and D.K. Cassel. 1991. Estimating nitrate leaching in soil materials. p. 165-188. In: R.F. Follett, D.R. Keeney and R.M. Cruse (eds.), *Managing Nitrogen for Groundwater Quality and Farm Profitability*. SSSA, Madison, WI.

Smith, S.J., D.W. Dillow and L.B. Young. 1982. Disposition of fertilizer nitrate applied to sorghum-sudangrass in the Southern Plains. *J. Environ. Qual.* 11:341-344.

Snyder, G.H., B.J. Augustin and J.M. Davidson. 1984. Moisture sensor controlled irrigation for reducing nitrogen leaching in bermudagrass turf. *Agron. J.* 76:964-969.

Spalding, R.F. and M.E. Exner. 1993. Occurrence of nitrate in groundwater - a review. *J. Environ. Qual.* 22:392-402.

Spalding, R.F., M.E. Exner, C.W. Lindau and D.W. Eaton. 1982. Investigation of sources of groundwater nitrate contamination in the Burbank-Wallula area of Washington, U.S.A. *J. Hydrology* 58:307-324.

Stanley, J., G.A. Thomas, H.M. Hunter, A.A. Webb and S. Berthelsen. 1990. Decreases over seven years in subsoil nitrate in a vertisol with grain sorghum and grass. *Plant Soil* 125:1-6.

Starr, J.L. and H.C. De Roo. 1981. The fate of nitrogen fertilizer applied to turfgrass. *Crop Sci.* 21:531-536.

Steenvoorden, J.H.A.M., H. Fonck and H.P. Oosteron. 1986. Losses of nitrogen from intensive grassland systems by leaching and surface runoff. p. 85-97. In: H.G. van der Meer, J.C. Ryden and G.C. Ennik (eds.), *Nitrogen Fluxes in Intensive Grassland Systems*. Martinus Nijhoff Publishers, Dordrecht.

Stevenson, F.J. 1982. Origin and distribution of nitrogen in soil. p. 155-215. In: F.J. Stevenson (ed.), *Nitrogen in Agricultural Soils*. ASA, CSSA, SSSA, Madison, WI, Agron. Monogr. no. 22.

Stevenson, F.J. 1986. The internal cycle of nitrogen in soil. In: *Cycles of Soil: Carbon, Nitrogen Phosphorus, Sulfur, Micronutrients*. John Wiley & Sons, New York.

Stout, W.L. and G.A. Jung. 1992. Influences of soil environment on biomass and nitrogen accumulation rates of orchardgrass. *Agron. J.* 84:1011-1019.

Stout, W.L., T.E. Staley, J.A. Shaffer and G.A. Jung. 1991. Quantitative effects of soil depth and soil and fertilizer nitrogen on nitrogen uptake by tall fescue and switchgrass. *Commun. Soil Sci. Plant Anal.* 22:1647-1660.

Super, M., H. Heese, D. MacKenzie, W.S. Dempster, J. duPless and J.J. Ferreira. 1981. An epidemiologic study of well-water nitrates in a group of southwest African Namibian infants. *Water Res.* 15:1265-1270.

Sveda, R., J.E. Rechcigl and P. Nkedi-Kizza. 1992. Evaluation of various nitrogen sources and rates on nitrogen movement, Pensacola bahiagrass production and water quality. *Commun. Soil Sci. Plant Anal.* 23:2451-2478.

Thomas, G.W. and J.D. Crutchfield. 1974. Nitrate-nitrogen and phosphorus contents of streams draining small agricultural watersheds in Kentucky. *J. Environ. Qual.* 3:46-49.

Thomas, G.W., M.S. Smith and R.E. Phillips. 1989. Impact of soil management practices on nitrogen leaching. p. 247-276. In: R.F. Follett (ed.), *Nitrogen Management and Ground Water Protection*. Elsevier Science Publishers, Amsterdam.

Timmons, D.R. 1984. Nitrate leaching as influenced by water application level and nitrification inhibitors. *J. Environ. Qual.* 13:305-309.

Tisdale, S.L., W.L. Nelson and J.D. Beaton. 1985. *Soil Fertility and Fertilizers*. Macmillan Publishing Company, New York, NY.

Triboi, E. and L. Gachon. 1985. Transfert des nitrates dans le sol en relation avec le système de culture. Devenir de l'azote mineral apporté comme engrais. p. 1-22. Vortrag Intern. Kongress "Nitrate dans les aux", 22-24 Oktober, Paris.

Turner, T.R. and N.W. Hummel, Jr. 1992. Nutritional requirements and fertilization. p. 385-439. In: D.V. Waddington, R.N. Carrow and R.C. Shearman (eds.), *Turfgrass*. Agron. Monogr. no. 32, ASA, CSSA, SSSA, Madison, WI.

USEPA (United States Environmental Protection Agency). 1976. *Quality Criteria for Water*. U.S. Gov. Printing Office, Washington, DC.

USEPA (United States Environmental Protection Agency). 1984. Report to Congress: Nonpoint Source Pollution. Washington, D.C.: Office of Water Programs Operations.

van Burg, P.F.J. 1986. The status of agriculture in western Europe and some implications for the consumption of nitrogen. p. 1-27. The World Fertilizer Conference, Amsterdam, April 20-23.

van der Meer, H.G. and M.G. van Uum-van Lohuyzen. 1986. The relationship between inputs and outputs of nitrogen in intensive grassland systems. p. 1-18. In: H.G. van der Meer, J.C. Ryden and G.C. Ennik (eds.), *Nitrogen Fluxes in Intensive Grassland Systems*. Martinus Nijhoff Publishers, Dordrecht.

Volk, G.M. 1956. Efficiency of various N sources for pasture grasses in large lysimiters of Lakeland fine sand. *Soil Sci. Soc. Amer. Proc.* 20:41-45.

Watschke, T.L. 1983. Penn State to study the effects of runoff water. *Am. Lawn Appl.* 3:13-15.

Watson, J.R., H.E. Kaerwer and D.P. Martin. 1992. The turfgrass industry. p. 29-88. In: D.V. Waddington, R.N. Carrow and R.C. Shearman (eds.), *Turfgrass*. Agron. Monogr. no. 32, ASA, CSSA, SSSA, Madison, WI.

Watts, D.G., G.W. Hergert and J.T. Nichols. 1991. Nitrogen leaching losses from irrigated orchardgrass on sandy soils. *J. Environ. Qual.* 20:355-362.

Weisenburger, D.D. 1991. Potential health consequences of groundwater contaminations by nitrates in Nebraska. p. 309-315. In: I. Bogardi and R.D. Kuzelka (eds.), *Nitrate Contamination: Exposure, Consequence, and Control*. NATO ASI Ser. G: Ecological Sciences 30. Springer-Verlag, Berlin.

Williams, J.R. and D.E. Kissel. 1991. Water percolation: an indicator of nitrogen-leaching potential. p. 59-83. In: R.F. Follett, D.R. Keeney and R.M. Cruse (eds.), *Managing Nitrogen for Groundwater Quality and Farm Profitability*. SSSA, Madison, WI.

Williams, J.R., C.A. Jones and P.T. Dyke. 1984. A modeling approach to determining relationship between erosion and soil productivity. *Trans. Am. Soc. Agr. Eng.* 27:129-144.

# Impacts of Soil N Management on the Quality of Surface and Subsurface Water

L.B. Owens

I. Introduction . . . . . . . . . . 137
A. Health and Environmental Concerns . . . . . . . . . . 137
B. Nitrogen Budgets . . . . . . . . . . 138
II. Impacts on Surface Runoff Quality . . . . . . . . . . 139
A. Tillage Practices . . . . . . . . . . 139
B. Crop Management . . . . . . . . . . 142
C. Summary . . . . . . . . . . 143
III. Impacts on Subsurface Water Quality . . . . . . . . . . 144
A. Tillage Practices . . . . . . . . . . 145
B. Crop Management . . . . . . . . . . 148
C. Fertilizer Management . . . . . . . . . . 152
IV. Research Needs . . . . . . . . . . 154
A. Soil N Availability Tests . . . . . . . . . . 154
B. Timing of Fertilizer Application . . . . . . . . . . 154
C. Manure Disposal . . . . . . . . . . 155
D. Subsurface Water Quality - A Systems Approach . . . . . . . . . . 155
E. Urban Use . . . . . . . . . . 155
F. Education . . . . . . . . . . 156
References . . . . . . . . . . 156

## I. Introduction

### A. Health and Environmental Concerns

Nitrogen is essential for growth of both plants and animals. It is the primary limiting nutrient in the production of food and fiber. Peterson and Frye (1989) present 22 years of data showing an increasing corn yield trend with an increasing trend of N fertilizer use in the U.S. However, excessive amounts of N are not only wasteful of resources, but they can be detrimental to human and animal health.

**Table 1.** Potential adverse environmental and health impacts of N

| Impact | Causative agents |
|---|---|
| Human health | |
| Methemoglobinemia in infants | Excess $NO_3^-$ and $NO_2^-$ in waters and food. |
| Cancer | Nitrosamines from $NO_2^-$, secondary amines. |
| Respiratory illness | Peroxyacyl nitrates, alkyl nitrates, $NO_3^-$ aerosols, $NO_2^-$, $HNO_3$ vapor in urban atmospheres. |
| Animal health | |
| Environment | Excess $NO_3^-$ in feed and water. |
| Eutrophication | Inorganic and organic N in surface waters. |
| Materials and ecosystem damage | $HNO_3$ aerosols in rainfall. |
| Plant toxicity | High levels of $NO_2^-$ in soils. |
| Excessive plant growth | Excess available N. |
| Stratospheric ozone depletion | Nitrous oxide from nitrification, denitrification, stack emissions. |

Some of the potential adverse effects of excess N were listed by Keeney (1982) (Table 1). The intended scope of this paper will discuss only the causative agents related to agricultural management.

High $NO_3^-$ levels in water are a concern because the ingestion of such water has the potential to reduce the oxygen carrying capacity of the blood. Nitrate is reduced to $NO_2^-$; $NO_2^-$ oxidizes the Fe of the hemoglobin in the blood; and methemoglobin, which cannot carry oxygen, is formed (Keeney, 1982). Infants have greater potential to suffer these ill effects than adults. In the United States, the maximum $NO_3^-$-N level for potable water for humans has been set at 10 mg/L (USPHS, 1962; USEPA, 1973). The World Health Organization (1971) and European Economic Community (1980) have established levels of 10.2 and 11.3 mg/L of $NO_3^-$-N, respectively.

### B. Nitrogen Budgets

The nitrogen cycle has numerous pathways and complex interrelationships of the processes involved. Not only are there chemical, physical, and biological processes, but these processes can be driven differently depending on climate, soil properties, and land management. These complexities increase the challenge but do not diminish the need for preparing N budgets as means for evaluating ecosystem N inputs and outputs. By quantifying N in each pathway, N budgets can provide a means of assessing the impacts of tillage, fertilizer, and crop management practices on outputs such as runoff and leaching, which affect water quality.

Some early N budget estimates from Rothamsted, England were provided in the 1880's by Lawes et al. (1882, 1883). Since that time numerous efforts have been made to quantify various aspects of the nitrogen cycle, and some components have been much more difficult than others. Legg and Meisinger (1982) extensively reviewed the literature and noted that although there are limitations and inadequacies in N budget research, "soil N budget principles provide a sound basis for estimating N losses, gains, and transformations for any agricultural management system." In a more recent literature discussion, Schepers and Fox (1989) note the difficulty in precisely quantifying N losses but that "sound N management practices can be developed...to minimize $NO_3$ leaching."

As has been indicated, much has already been written about N budgets and the aspects of the nitrogen cycle which impact water quality along with the subsequent adverse effects on human health and the environment. This chapter uses this literature base and adds results of current research dealing with the impacts of various facets of soil N management on the quality of surface runoff and subsurface water. Even though the N cycle has complex interrelationships which make it difficult to separate individual components, this discussion will focus on impacts of soil N management practices on water quality. A comprehensive coverage of the N cycle or N budgets is beyond the intended scope of this chapter. The emphasis will also be placed on a field or small watershed scale (or smaller) and not on a large watershed or basin scale.

## II. Impacts on Surface Runoff Quality

Surface runoff is dependent on amount and intensity of precipitation, antecedent soil moisture, soil properties, and slope, as well as management factors such as tillage, crop residue, and cropping systems. Surface runoff and infiltration are closely related. Usually factors which reduce runoff increase infiltration. Subsurface aspects will be discussed later in the chapter.

### A. Tillage Practices

#### 1. Conventional Tillage Impacts

The term "conventional tillage" usually refers to moldboard plowing plus disking. In the typical situation, conventional tillage has been employed for many years, usually leaves very little residue on the surface, and can render the soil highly vulnerable to runoff and erosion. Over recent decades, conservation methods have been developed to reduce these losses, e.g., contour farming, strip cropping, terracing, and spring plowing instead of fall plowing.

The loss of soil and subsequent problems associated with its downstream deposition have been of concern for a longer time than problems associated with N loss. Correspondingly, there is less information concerning nutrient losses in surface runoff than is available on runoff and erosion (Sharpley and Smith, 1991). The loss of N associated with sediment is usually several times greater than soluble N loss in surface runoff (Legg and Meisinger, 1982; Sharpley et al., 1987). Therefore, as practices are developed and implemented to reduce runoff and erosion, losses of sediment associated N and soluble N in runoff are also reduced.

## 2. Conservation Tillage Impacts

Various reduced-till or "conservation" tillage practices have been or are being developed as alternatives to moldboard plowing. Conservation tillage practices cause less soil disturbance and leave greater amounts of crop residue on the surface than moldboard plowing. Both of these factors help reduce runoff and erosion, reduce transport of soluble and sediment-attached chemicals, and promote infiltration (Thomas et al., 1989).

With traditional tillage the soil is plowed and followed by several diskings, which destroy the macro structure of the soil. Thus, the upper parts of cracks, root channels, earthworm burrows, etc. are destroyed and lose their capabilities as water flow pathways.

With no-tillage, the opposite extreme, disturbance of soil macro structure is minimal, and a larger portion of rainfall can infiltrate by means of macropores. This results in reduced runoff. Various intermediate conservation tillage practices will have intermediate effects on the shift of water flow from runoff to infiltration (Thomas et al., 1989). Infiltration aspects will be discussed further in the subsurface water section.

More crop residue is left on the soil surface with reduced tillage. Thomas et al. (1989) report data of Mannering et al. (1975) which show percent residue cover on soil ranging from 1.0% for conventional fall plow to 76.0% for no-tillage. On plots in Iowa (Table 2), Erbach (1982) showed that spring tillage and planting reduced the residue cover but that conservation tillage techniques left much more plant residue than moldboard plowing. On watersheds in a corn/soybean rotation at Coshocton, Ohio, Edwards et al. (1992) reported an average yearly soil loss over a 6-year period of 0.5 Mg/ha with three conservation tillage methods. This was in contrast with 4.8 Mg/ha of measured soil loss during the corn years of an earlier corn (*Zea mays*), wheat (*Triticum aestivum*), meadow, meadow rotation with conventional tillage.

**Table 2.** Effect of tillage system on percentage of soil surface covered with plant residue following corn and following soybeans

| Tillage system | Following corn | | Following soybeans | |
|---|---|---|---|---|
| | Before spring tillage[a] | After plant[b] | Before spring tillage[a] | After plant[b] |
| | % | | | |
| Slot-plant on ridges | 78 | 53 | 73 | 37 |
| Spring disk | 78 | 31 | 71 | 12 |
| Till-plant | 75 | 40 | 76 | 19 |
| Fall, chisel plow | 45 | 32 | 23 | 12 |
| Fall, moldboard plow | 11 | 6 | 4 | 3 |

[a]Average for 1979 and 1980
[b]Average for 1978, 1979, and 1980
From Erbach, 1982

Similar tillage practices on soils with different soil properties can produce different results in terms of runoff and erosion. Soil texture is a property which can cause tillage and cropping practices to have different effects. Because data do not exist for very many textural combinations, available data must often be extrapolated beyond its textural class (Thomas et al., 1989). Subsoil properties can also impact tillage-runoff-infiltration interactions. With continuous corn grown with no-till practices, summer runoff was nearly eliminated from a watershed with a heavy, non-stony subsoil and watersheds with well-drained or stony subsoils (Edwards and Amerman, 1984; Edwards et al, 1984). However, dormant season runoff from the watershed with the heavy subsoil was much greater than from the watersheds with the well-drained or stony subsoils.

Practices which reduce runoff and erosion also reduce nitrogen losses (Legg and Meisinger, 1982; Laflen and Tabatabai, 1984; Owens and Edwards, 1993). Based on rainfall simulator studies on Iowa soils, Laflen and Tabatabai (1984) found that the greatest nutrient losses occurred when soil loss was greatest and surface runoff was least. They concluded that to control surface nutrient losses, erosion must be controlled. In a 6-year corn/soybean (*Glycine max*) rotation study on small watersheds, Owens and Edwards (1993) found soluble N losses to be greater than sediment-attached N, but sediment losses were low - approximately 0.5 Mg/ha each year. Soluble N losses were also low, 7.5 and 1.9 kg/ha each year during the corn and soybean years, respectively.

Even with low transport of N because of low runoff volumes, high concentrations can occur. Owens and Edwards (1993) reported 16.6 mg/L as the average $NO_3^-$-N in runoff during the corn years of a corn/soybean rotation. Other research found high N concentrations in runoff when major runoff events occurred soon after a surface N application (Smith et al., 1988), and when a restrictive layer (such as a fragipan) in the soil profile caused interflow to occur (Lehman and Ahuja, 1985).

### B. Crop Management

#### 1. Row Crops

Factors which influence runoff are quite interrelated and more emphasis has been placed on tillage differences than on differences due to continuous row crop versus rotational row crop systems. Legg and Meisinger (1982) review a number of studies which include data on N loss in surface runoff. In a 10-yr study by Burwell et al. (1975), soluble N and sediment N losses were measured for five soil cover conditions: 1) continuous clean-cultivated fallow, 2) continuous corn (*Zea mays*), 3) corn in a corn-oats-hay rotation, 4) oats (*Avena sativa*) in rotation, and 5) hay in rotation. Average annual soluble N losses were small, ranging from 1.0 to 3.7 kg/ha. These levels of N loss in surface runoff were lower than the amounts of N contributed by precipitation. Smith et al. (1990) also noted that soluble N concentrations in runoff are often lower than N concentration values in precipitation.

#### 2. Cover Crops

Cover crops can be effective in reducing runoff and erosion, especially during the winter period in conditions where there is little crop residue present. When crop residue is already present, the extra residue from a winter cover crop may be detrimental to crop yield. Eckert (1988) found such an effect with rye (*Secale cereale*) cover crops in no-till corn and soybeans (*Glycine max*), and attributed this yield reduction to stand reduction. Sharpley and Smith (1991) reviewed several published studies which found cover crops reduced runoff, soil loss, and N transport, but had inconsistent effects on N concentrations in runoff. Using three types of winter cover crops following soybeans, Zhu et al. (1989) reported runoff reductions of 44 to 53% and soil loss reductions of 87 to 96% compared to a fallow check. Soluble $NO_3^-$-N concentration reductions ranged from 49 to 54%. Nevertheless, soil fertility, crop type, and growth stage may cause soluble N concentrations in runoff to increase (Sharpley and Smith, 1991).

When legumes are used as the winter cover crop, fixed N in the legumes can be supplied to the subsequent crop so that less fertilizer N has to be applied. This reduces the amount of inorganic N which is readily available for removal in surface runoff. Holderbaum et al. (1990) found that fall-seeded legumes can supply significant amounts of N to subsequent corn (*Zea mays*) while providing better soil protection during winter months. Crimson clover (*Trifolium incarnatum*) and hairy vetch (*Vicia villosa*) can supply a substantial portion of the subsequent corn N requirement (Wagger, 1989; Corak et al., 1991). These studies also noted that transpiration by the legumes prior to planting the next crop reduced the moisture in the soil profile. This could be beneficial in a wet spring and detrimental in a dry spring. This soil moisture reduction would have the potential to reduce both leaching and surface runoff.

### 3. Grasslands

The percentage of vegetative cover is usually quite high for grasslands. Costin (1980) and Lang (1979) report that 70 - 75% ground cover was the critical threshold, above which surface runoff was slight, and below which runoff increased rapidly. With small amounts of runoff, the transport of soluble N is also low. Less than 1 kg/ha N was loss from a Nebraska pasture receiving an annual N fertilization of 67 kg/ha (Schepers and Francis, 1982); and less than 8 kg/ha of total N (less than 4 kg/ha of $NO_3^-$-N) was lost each year in runoff from Ohio pastures receiving fertilizer N at an annual rate of 224 kg/ha (Owens et al., 1983). From research in The Netherlands, Steenvoorden et al. (1986) concluded that nutrient losses in surface runoff from pasture were usually not economically important.

Nitrogen concentrations in surface runoff from pastures were also usually found to be below the maximum levels set for human health. Schepers and Francis (1982) reported weighted averages of $NO_3^-$-N less than 1.0 mg/L from a Nebraska pasture. Flow-weighted, seasonal averages of less than 1.0 mg/L of $NO_3^-$-N were reported in the surface runoff from Ohio pastures receiving 56 kg N/ha annually (Owens et al., 1982) and less than 5.0 mg/L of $NO_3^-$-N for pastures receiving 224 kg N/ha each year (Owens et al., 1983). In permanent pastures in New Zealand, 4.5 mg/L was reported as the maximum $NO_3^-$-N concentration in surface runoff from an area receiving 60 kg N/ha (Sharpley et al., 1983). However if runoff events occur within a few days following a nitrogen application, high $NO_3^-$-N concentrations can be observed in surface runoff (Owens et al., 1984; Steenvoorden et al., 1986).

## C. Summary

Surface runoff is dependent on non-management-controlled factors such as precipitation amount and intensity and by management practices which impact ground cover. Several management options have been utilized to reduce runoff and erosion. Conservation tillage practices, winter cover crops, and grass cover (in rotation, permanent, or in riparian zones) all help reduce surface runoff and erosion. With these reductions there are reductions in loss of soluble N. Frequently, both the mass loss and concentration of soluble N in surface runoff are less than those found in precipitation. However, there can be elevated concentrations of N in surface runoff if a sizeable runoff event occurs soon after the N fertilizer application.

## III. Impacts on Subsurface Water Quality

Subsurface flow can be another major pathway for water loss from a soil-plant system. Many of the factors which reduce surface runoff promote an increase in subsurface flow. The impacts of tillage, crop management, and fertilizer management are quite interrelated, especially their impacts on soluble N, which is moved via subsurface flow. These factors take on greater importance in humid climates where greater soluble N losses usually occur with subsurface flow than with surface runoff.

### A. Tillage Practices

Conventional tillage disrupts the surface soil structure and buries a major portion of the residue cover. Macropores, which can serve as preferential flow paths for water, are destroyed in the tilled layer. In this reduced soil structure environment, water tends to flow as a front (Thomas et al., 1989).

With no-tillage, the soil macro-structure is disturbed very little and most macropores remain intact from the soil surface to their full depth. These macropores provide flow paths which enable water to bypass much of the soil matrix. Dick et al. (1986) observed much greater water movement through lysimeters under no-tillage than under conventional tillage (Table 3). Slightly greater leachate amounts were observed from minimum-tillage corn than conventional-tillage corn on monolith lysimeters (Chichester, 1977). Long-term runoff records showed that infiltration and potential groundwater recharge could increase by more than 100 mm/yr in watersheds farmed with no-tillage practices as compared to similar fields with conventional tillage (Edwards et al., 1990). The difference in hydrologic response appeared to be due to tillage-related differences in soil physical properties, including macropores created by earthworms. Practices which leave more mulch cover on the surface than moldboard plowing not only reduce surface runoff but also reduce evapotranspiration. This leaves more moisture available for movement through the soil profile.

Water infiltrating into the soil, or moving through macropores, also has the potential to move soluble N. Increasing implementation of conservation tillage practices and subsequent increased potential infiltration has further stimulated interest in the environmental impacts of N management. Much of the early work of investigating N in subsurface flow was the analysis of tile effluent from conventional-tillage corn areas. Most of the average annual $NO_3^-$-N concentrations reported from Iowa to Vermont were in the 10 to 60 mg/L range (Table 4). However, some individual samples were much higher than this range. The N loss in these tile effluent studies was 10 to 40% of the N applied. Greater percent loss tended to occur with higher rates of N application.

Smith et al. (1990) also made the point that tile effluent represented only a portion of the infiltrating water. Therefore, the N measured is also only a

**Table 3.** Percentage of water collected as leachate from lysimeters as affected by tillage

| Year | Crop | Annual mean % of water collected[a] | |
|---|---|---|---|
| | | Conventional tillage | No-tillage |
| 1975-76 | Corn | 49 | 88 |
| 1976-77 | Corn | 52 | 83 |
| 1977-78 | Corn | 60 | 84 |
| 1983-84 | Soybeans | 59 | 94 |
| 1984-85 | Soybeans | 68 | 92 |
| 1985-86 | Soybeans | 70 | 90 |

[a]% is calculated by dividing the leachate volume by the sum of the surface runoff plus leachate volumes and multiplying by 100

portion of the N moving through the soil. In a tile spacing study, Kladivko et al. (1991) showed that on a per-area basis, more N was removed from a tilled (chisel plow plus 2 passes with a disc or field cultivator) corn field with tile on a 5-m spacing than on a 20-m spacing. This raises a question as to the fate of the soluble N bypassing the tile and continuing downward.

In a comparison of conventional-tillage with no-tillage, Kanwar et al. (1988) noted that in a year of moderate precipitation, tile flow was moderate but slightly greater under no-tillage, and $NO_3^-$-N loss was also slightly greater under no-tillage. In a year with high precipitation, $NO_3^-$-N concentration in the drain water was significantly higher under conventional-tillage, but tile flow was over 100% greater under no-tillage and $NO_3^-$-N loss was over 50% greater under no-tillage. Lysimeter data (Tyler and Thomas, 1977) have shown that $NO_3^-$ can be moved through the soil much more quickly under no-tillage than conventional tillage. In some Ohio soils, it appears that most of the macropore flow occurs through earthworm burrows (Edwards et al., 1988) rather than old root channels or structural cracks. Further evaluation of the effects of earthworm burrows indicated that water movement through such burrows was over 10 times greater than their areal distribution would suggest (Edwards et al., 1989). However, $NO_3^-$-N movement was not large, 0.7 kg/ha, and $NO_3^-$-N concentrations ranged from 0 to 12.6 mg/L. Based on experience in Kentucky, Thomas et al. (1989) credit more macropore flow to root channels and structural divisions than to earthworm burrows, but state that the overall effects on water flow paths are roughly the same. Shipitalo et al. (1990) found that movement of surface applied chemicals may be strongly impacted by the nature of the first rainfall event following application. A light rain can move solutes into the soil matrix and reduce the potential for transport through macropores with subsequent rainfall.

**Table 4.** Nitrate-N concentrations and transport in tile effluent from conventionally-tilled cropland

| Reference and study description | Year | Crop | N applied kg/ha | $NO_3^-$- Conc. mg/L | N Transport kg/ha |
|---|---|---|---|---|---|
| Benoit (1973); effluent from subsurface drains collected from plots; a 1-year study on a poorly drained silt loam in Vermont. | 1969/70 | Corn | 247 | 2.9 - 4.9 (before fert.)<br>13.2 - 37.4 (after fert.) | |
| Logan and Schwab (1976); tile effluent was collected from two fields (approx. 3 ha each) with silt loam and silty clay loam soils. Samples were collected during two growing seasons. | 1972 | Corn | 20 - 60 | 10.5 - 11.6 | |
| | 1973 | Corn | 224 - 260 | 11.9 - 23.7 | |
| Bottcher et al. (1981); water samples from 17 ha subsurface drain system. This 3-year study was conducted on a silty clay soil in Indiana. | 1976 | Wheat and soybeans | 98<br>10 | 5.6 | 0.7 |
| | 1977 | Corn | 188 | 12.0 | 14.0 |
| | 1978 | Soybeans followed by wheat | 12 | 4.9 | 4.8 |

| | | | | | |
|---|---|---|---|---|---|
| Baker and Johnson (1981); effluent from tile drains was collected from two plots (<0.5 ha) with silt loam soils. This 5-year study was conducted in Iowa. | 1974 | Corn | 250/100[a] | 17.3/14.9[a] | 31.3/32.3[a] |
| | 1975 | Soybeans | 0/0 | 49.1/24.2 | 71.6/40.3 |
| | 1976 | Corn | 240/90 | 37.9/23.0 | 31.9/21.4 |
| | 1977 | Oats | 0/0 | 61.4/20.0 | 67.2/18.0 |
| | 1978 | Corn | 90/90 | 37.0/18.6 | 37.6/21.1 |
| Logan et al. (1980); tile drainage from plots, small systems, and large systems are reported for Iowa, Ohio, and Minnesota. Only Minnesota plot data are presented here. | 1974[b] | Corn | 20 | 18.7 | 17.0 |
| | | | 112 | 24.2 | 22.0 |
| | | | 224 | 33.0 | 30.0 |
| | | | 448 | 59.3 | 54.0 |
| | 1975 | Corn | 20 | 14.8 | 19.0 |
| | | | 112 | 19.5 | 25.0 |
| | | | 224 | 46.1 | 59.0 |
| | | | 448 | 93.8 | 120.0 |
| | 1977 | Corn | 20 | 35.0 | 4.9 |
| | | | 112 | 47.9 | 6.7 |
| | | | 224 | 52.1 | 7.3 |
| | | | 448 | 140.7 | 19.7 |
| | 1978 | Corn | 20 | 23.2 | 11.6 |
| | | | 112 | 57.8 | 28.9 |
| | | | 224 | 97.0 | 48.5 |
| | | | 448 | 195.6 | 97.8 |

[a]Data for the 2 plots are presented side-by-side
[b]No tile flow in 1976

## B. Crop Management

### 1. Row Crops, Cover Crops, and Rotation Systems

Nitrogen is essential for non-legume crop production. The per area application rate of fertilizer N on agricultural land in the U.S. increased nearly three-fold between 1961 and 1981 (Anderson et al., 1989). Much of this increase has been applied to monocultures of grain crops, such as corn and wheat, which require high inputs of fertilizer N. This has led to some excessive applications of N, excessive N losses, and the need to develop N management techniques to reduce such losses and their impacts on the environment. Russelle and Hargrove (1989) discuss management options, such as increased plant populations, narrow rows, and optimum planting dates, which should increase fertilizer N use efficiency and reduce the $NO_3^-$ leaching potential. Crop residues, especially non-legume and residues with a high C:N ratio, can immobilize some fertilizer N and probably restrict $NO_3^-$ leaching (Russelle and Hargrove, 1989).

The use of winter cover crops is not new and has shown benefits for centuries. However, only in recent years has much of the focus of their use been on their potential to reduce $NO_3^-$ leaching losses. Cover crops can influence the water budget by increasing infiltration, reducing evaporation, and subsequently increasing the total amount of water moving through the soil. Cover crops can also influence the water budget by reducing the water content of the soil during their active growth stage (Russelle and Hargrove, 1989; Meisinger et al., 1991). The potential for $NO_3^-$ leaching can be either increased or decreased depending on the net effect on the water budget. This net effect will depend on site specific factors of soil and environmental conditions. Another way in which cover crops have the potential to reduce $NO_3^-$ leaching losses is that they are utilizing N during the water-recharge season when N is normally not being taken up by plants and would otherwise be available for leaching. Based on results from their thorough literature review, Meisinger et al. (1991) conclude that cover crops can reduce both the mass of N leached and the concentration of the leachate 20 to 80% compared with no cover crop. They note that grasses and brassicas are two to three times more efficient than legumes in reducing N leaching. In a winter wheat/corn rotation, Martinez and Guiraud (1990) applied 200 kg N/ha in the spring of the wheat year. During the winter following wheat harvest, $NO_3^-$-N concentrations in water draining from fallow lysimeters reached 68 mg/L with an average of 40 mg/L. Nitrate-N concentrations in drain water from lysimeters with a ryegrass cover crop declined from 41 to 0.25 mg/L. The quantities of N leached were 110 and 40 kg/ha from the fallow and cover crop lysimeters, respectively.

Rotation of crops on the same field is a practice which has long been utilized. Although there are several benefits to crop rotations (e.g., lower fertilizer N inputs required because of residual N from legume crops providing N to subsequent crops, erosion reduction, minimizing crop diseases, and reduction of insect problems), relatively little attention has been given to impacts on $NO_3^-$

**Table 5.** Nitrate-N concentrations and transport in percolate from Coshocon's monolith lysimeters

| Lysimeter | Crop and tillage | N applied | Annual $NO_3^-$-N Conc. mg/L | Annual Transport kg/ha | Reference |
|---|---|---|---|---|---|
| Y 102 | grass-alfalfa (Nov 75-Oct 77) | 0 | 20.8 | 22.5 | Owens, 1990 |
| Y 103 | grass-alfalfa (Nov 75-Oct 78) | 0 | 5.3 | 7.9 | |
| Y102 | no-till corn (Nov 77-Oct 83) | 336 | | | |
| | w/o N-serve | | 28.6 | 151.9 | |
| | w N-serve | | 22.0 | 114.4 | |
| Y 103 | conv. corn (Nov 78-Oct 83) | 336 | 27.2 | 101.8 | |
| Y 102 | corn/soybean chisel plow (Nov 85-Oct 89) | 224 in corn year | 14.0 | <2.0 | Owens et al., 1990a |
| Y 103 | corn/soybean chisel plow (Nov 85-Oct 89) | 224 in corn year | 9.7 | <2.0 | |

leaching by most crop sequences. Considerable data are available for corn-soybean systems (Keeney, 1982), and a conclusion has been drawn that including soybeans in a crop rotation can result in a net deficit of N (Heichel, 1985). Owens et al. (1990a) measured less $NO_3^-$ transport and lower concentrations in lysimeter percolate under a corn-soybean rotation than had been noted previously under continuous corn (Owens, 1990) (Table 5). Forage legumes have been shown to provide a net increase of N, frequently as soil organic N.

Alfalfa (*Medicago sativa*), a deeply-rooted plant, not only fixes N, but can also be a scavenger of N in the soil profile. It has been shown to be effective in supplying N to a following corn crop, thus allowing for a reduction in fertilizer N without a reduction of yields (Smith et al., 1990; Peterson and Russelle, 1991). Climate and management practices can play major roles in the amount and timing of N losses from alfalfa (Russelle and Hargrove, 1989). For example, in the Corn Belt, tilling alfalfa stands in the spring instead of the fall would minimize N mineralization and $NO_3^-$ accumulation in the soil profile, as well as allowing alfalfa to use soil water in late fall and early spring which might be available for leaching (Peterson and Russelle, 1991).

## 2. Grasslands and Turf

Grassland ecosystems are a major part of animal product agriculture throughout the world. They include low management and/or low productivity areas, which are frequently but not exclusively found in arid or semi-arid areas; moderate productivity areas, such as the hill country pastures of Scotland and New Zealand, which receive moderate N inputs from fertilizer or legumes; and the high productivity areas of England and northern Europe, which are heavily fertilized (200 - 600 kg N/ha annually). The potential for large amounts of $NO_3^-$ leaching from low and moderate productivity systems is minimal (Keeney, 1982), and in some ungrazed systems, $NO_3^-$ leaching from high productivity systems is low, e.g., 29 kg N/ha lost from ryegrass (*Lolium perenne*) receiving 420 kg N/ha (Garwood and Ryden, 1986). Ball and Ryden (1984) note that no more than 5 to 15% of annual fertilizer inputs of 200 to 400 kg N/ha is lost through leaching and denitrification, but the inclusion of grazing ruminants greatly reduces this efficiency. Much of the ingested N is excreted in the urine and in a form much more labile than in the associated forage (Ball and Keeney, 1981). This allows for greater losses of N through volatization and leaching. Several studies have reported greater N leaching with a grazing system (Table 6). Heng et al. (1991) found lower amounts of $NO_3^-$ being leached with moderate N fertilization and grazing sheep, 9-19 kg N/ha, than has been found with intensive cattle-grazed pastures. However, annual $NO_3^-$ leaching losses from a grazed ryegrass/clover pasture were less than from a cut ryegrass area receiving 420 kg N/ha and 6.6 times less than from a grazed ryegrass pasture receiving 420 kg N/ha each year (Garwood and Ryden, 1986).

Although grasslands in the U.S. have not been the recipients of the majority of fertilizer N applied or the subsequent attention to impacts of fertilization on groundwater quality, grasslands are important to livestock based agriculture and the influence of grassland management on $NO_3^-$ leaching should not be ignored.

A minimum annual $NO_3^-$-N leaching loss of 35 kg/ha can be expected from fertilized orchardgrass (*Dactylis glomerata*) grown on irrigated sandy soils in Nebraska, even with reduced inseason drainage and an N rate sufficient for 80 to 85% of maximum production (Watts et al., 1991). On steeply sloping grassed watersheds in North Carolina, monthly $NO_3^-$-N concentrations in subsurface throughflow were below 4 mg/L when a total of 112 kg N/ha was applied over four years, and monthly $NO_3^-$-N concentrations only exceeded 10 mg/L when 112 kg N/ha was applied each year (Kilmer et al., 1974).

In small pastured watersheds in eastern Ohio receiving 56 and 224 kg N/ha annually, subsurface $NO_3^-$-N levels ranged from 3 to 5 mg/L (Owens et al., 1982) and 8 to 18 mg/L (Owens et al., 1983), respectively. The small pastured watersheds which had been receiving 224 kg N/ha annually were interseeded with alfalfa, and the non-N mineral fertilizer treatments were based on soil test recommendations. With legume supplied N instead of fertilizer N, subsurface $NO_3^-$-N levels in these areas decreased to a range of 3 to 10 mg/L (Owens et al., 1990b). Fertilized turfgrass represents a potential source for $NO_3^-$-N

**Table 6.** Nitrate-N concentrations and transport in subsurface water under grasslands

| Reference and Brief Study Description | Management | N applied kg $ha^{-1}$ | Annual $NO_3^-$-N Conc. mg/l | Annual N transport kg/ha |
|---|---|---|---|---|
| Steele et al. (1984); percolate samples collected with a suction sampler; there were cattle-grazed ryegrass pastures: a control and N treatment. Research done in New Zealand. | Grazed | 0 | 7.3 | 88.3 |
| | Grazed | 172 | 21.2 | 192.7 |
| Steenvoorden et al. (1986); data are from leachate from cut and cattle-grazed grassland receiving different N levels of fertilizer; soils are sandy. Research done in The Netherlands. | Cut grass | 250 | | 14 |
| | Grazed | 250 | | 42 |
| | Grazed | 500-530 | 36-60 | |
| | | 194 | 40% > 11.3 | 17.5 |
| Haigh and White (1986); analysis of tile drainflow from a 1.1 ha ryegrass watershed; cattle grazing in 1982, sheep grazing in 1983. Research done in England. | Grazed | (1982-83) | | 48.7 |
| | | 114 (1983-84) | 100% > 20 | |
| Ryden et al. (1984); analysis of 6 m soil profiles below cut ryegrass and grazed ryegrass (samples taken Dec. 82); 420 kg N/ha annually 1977-1982. Research done in England. | Cut ryegrass | 420 | mg/kg 100% < 4 | |
| | Grazed ryegrass | 420 | 10-17 | |

leaching into groundwater. Although a Long Island survey revealed that homeowners annually applied an average of 122 kg N/ha to their lawns, commercial lawn care companies annually applied 220 to 293 kg N/ha (Morton et al., 1988), and intensely managed turfgrass in Maryland receives up to 220 kg N/ha annually (Gross et al., 1990). In a review of the relatively limited documentation of the impacts of fertilized turfgrass on groundwater contamination, Petrovic (1990) states that leaching of fertilizer N is highly influenced by management practices, e.g., N rate, source, and timing. It was also concluded that where turfgrass fertilization may threaten groundwater quality, there are management practices which can minimize or eliminate $NO_3^-$ leaching.

## C. Fertilizer Management

The effects of time, rate, method of placement of N fertilizer on N losses from the soil vary depending on the N fertilizer source used. Much has been written and is known about management practices which increase N fertilizer use efficiency. Smith et al. (1990) give an overview of nitrogen fertilizers and considerations for developing N fertilizer management practices to reduce N leaching. Peterson and Frye (1989) present a comprehensive discussion of N fertilizers, application methods, nitrogen stabilizers, and practices to increase N efficiency. A major point made by Peterson and Frye is that a wealth of N fertilizer management knowledge exists, but this knowledge is not adequately being put into practice. Schepers et al. (1991) reported that large excesses of N fertilizer were applied in the Central Platte Natural Resource District and that "fertilizer N applied showed little relationship to fertilizer N recommended." Unfortunately, in many cases, economic considerations alone are insufficient to change the "better too much than too little" approach to N management. Major educational efforts are needed to impress the seriousness of the problem on N suppliers and farmers. Without a change in attitudes and motivation, restrictive legislation may be needed to reduce the potential negative environmental impacts of agriculture.

### 1. Amount to Apply

The first step in applying N fertilizer is to determine the amount to apply. This should be done based on realistic yield goals not merely desired yields. Accurate yield records for each field are important in setting realistic goals. Then, with a determination of the N required to produce a crop unit, the N required to produce the yield goal can be estimated.

In addition to fertilizer N being used to meet the anticipated crop need, there are other potential sources of N for the crop that need to be given appropriate credit to avoid applying excess fertilizer N. The soil can supply N to plants as mineral N, usually $NO_3^-$, and as N mineralized from organic matter. There can be substantial over-winter carryover of residual profile $NO_3^-$ (Bundy and Malone, 1988), which would allow for reduction of subsequent fertilizer N applications. The amount of N from these soil pools is very difficult to assess (Peterson and Frye, 1989; Smith et al., 1990). Efforts are being made to develop soil tests to measure soil N that is available for plant uptake (Blackmer et al., 1989; Magdoff et al., 1990). Binford et al. (1990) describe a tissue test for corn stalk $NO_3^-$ concentration to determine the degree of excess N fertilization. This end of season evaluation would allow for a fertilizer N adjustment for the next year's crop. Winter legume cover crops and legumes in rotation can provide a significant amount of N for a subsequent non-leguminous crop (El-Hout and Blackmer, 1990; McVay et al., 1989; Smith et al., 1990). Some legumes can provide as much as 120 kg N/ha, which strongly shows that credit needs to be given for legume N to reduce excessive fertilizer N applications.

Similarly, N credits need to be given for manure applications. Frequently manure is viewed as material which requires disposal rather than as a source of nutrients. Recent reports of long-term manure applications indicate that high rates of application can pose a hazard for $NO_3^-$ contamination in groundwater. A rate of 335 kg N/ha applied to 'coastal' bermudagrass (*Cynodon dactylon* L.) in North Carolina did not produce soil N accumulations which posed a groundwater quality hazard, but higher rates may create a problem (King et al., 1990). Potential pollution problems existed when manure was applied on a long-term basis at recommended rates for Alberta, Canada, 475 kg N/ha (Chang et al., 1991). Jokela (1992) found that applying manure to corn at rates equivalent, based on yield response, to 73 to 122 kg fertilizer N/ha may produce some occasional high soil solution $NO_3^-$ concentrations but "manure resulted in similar or slightly lower soil profile $NO_3^-$ than agronomically equivalent rates of fertilizer N."

## 2. Other Management Factors

The efficient use of N fertilizers can often be improved through improved placement of the fertilizer. Some traditional approaches have been surface broadcast and surface broadcast with incorporation by tillage. Recent research continues to show N efficiency benefits to subsurface placement. Howard and Tyler (1989) found significantly higher corn yields and N uptake when UAN and urea were injected in subsurface bands as contrasted to surface broadcast or surface band application methods. Maddux et al. (1991) found similar results with subsurface banding of urea-based N fertilizers on corn and suggested that there were decreases in volatization losses compared with broadcast applications, decreased immobilization of inorganic fertilizer N, and decreased $NO_3^-$ leaching losses. Subsurface N applications, with 10 cm being near optimum depth, have produced greater yield responses in cool-season grasses than surface applications (Moyer and Sweeney, 1990).

Timing of fertilizer applications is a technique which has shown mixed results for improving N fertilizer efficiency. Fox et al. (1986) showed increased corn yield and increased fertilizer N efficiency when the N fertilizer was applied a few weeks after plant emergence compared to pre-plant applications. On the other hand, Jokela and Randall (1989) found no difference in corn yield between pre-plant and sidedress applications. Sometimes timing is combined with a nitrification inhibitor (NI) to achieve greater efficiency of N fertilizer and to reduce $NO_3^-$ leaching. Bronson et al. (1991) used dicyandiamide (DCD) with a fall application of urea to winter wheat and reported that this combination can conserve fertilizer N in the soil and minimize $NO_3^-$ leaching losses. Using a NI, nitrapyrin, with urea on irrigated corn, Walters and Malzer (1990) found that the NI delayed the $NO_3^-$ leaching but not the total N loss. They emphasized that NI may reduce the potential for groundwater contamination by delaying $NO_3^-$ leaching, but they "will be most effective if coupled with proper N rates and conservative irrigation water management." Much more information about forms of N fertilizers, placement and timing of applications, and nitrification inhibitors is presented in a comprehensive discussion by Peterson and Frye (1989).

# IV. Research Needs

Much is known about how to use N for efficient crop production and to minimize its adverse impacts on surface and subsurface water quality. Nevertheless, there are areas where further knowledge is needed to fill in the gaps and to improve upon what is already known. Research priority lists will vary according to the individual(s) preparing the lists and will be influenced by the varying needs of different geographic/climatic areas.

## A. Soil N Availability Tests

One promising test for predicting soil N availability is the pre-sidedress nitrate test (PSNT). This will enable farmers to greatly reduce excess N applications. Considerable attention is currently being directed toward soil tests and tissue tests which will provide an accurate inseason assessment of N need for the crop. What works in one area may not work in another. There may be differences in accuracy of tests dependent upon the kind of crop.

## B. Timing of Fertilizer Application

Research results on the effectiveness of varying the time of application of fertilizer, especially split fertilizer applications, have been mixed. What factors promote increased N fertilizer efficiency with split applications in some areas and not in others? Form of fertilizer N applied, placement, soil type, drainage, microbial activity, and climate are some factors to consider.

### C. Manure Disposal

There have been a few long-term studies of the effect of rate of manure application on water quality. Should manure be injected instead of surface applied? Surface application may be the only reasonable alternative for some manure forms, e.g., poultry litter. (With an increasing number of large, confinement poultry operations, disposal of poultry litter is becoming more of a concern.) What pre-treatment, if any, would allow greater rates of application for disposal or promote greater N efficiency in supplying plant nutrients? What levels of organic N accumulation can be accepted without posing a potential groundwater quality problem? From the disposal aspect, are some crops more suited than others in accepting and utilizing large amounts of N? What techniques would work best for utilizing animal wastes as a nutrient source instead of only being a material which needs disposal?

### D. Subsurface Water Quality - A Systems Approach

Assessing management impacts on subsurface water quality is difficult at best. Much of the good research in this area has been conducted with soil columns and lysimeters. There is a need for more groundwater quality research on a field or watershed scale using a systems approach. It is important to collect more hard data on the quantity and concentrations of $NO_3^-$ being leached from under a number of different crops and cultural practices. These data are particularly needed for creating accurate models of nitrate leaching and verifying current models in order to make confident predictions. But there is also a major need to know impacts of the interaction of all the factors within the management system on $NO_3^-$ leaching. This is especially true of systems with crop rotations. Even though quick results are not produced, long-term studies can be critically important, e.g., a 10-year study may give different results than a 4-year study.

### E. Urban Use

Fertilization of lawns and turf areas and the subsequent impacts on groundwater quality need more attention. Research on this aspect of N management has been limited, and some of the results mixed as to whether a pollution problem is being posed. Another reason for giving this higher priority is that concern over limited landfill space may disallow continued disposal of yard wastes, e.g., grass clippings, in landfills. This may reduce the removal of grass clippings from lawns and increase the potential for organic N accumulation. Will this impact be large enough that N recommendations for lawns should be reduced?

### F. Education

The need for better education may not require research in the traditional sense; but if the scientific knowledge is not put into practice, the usefulness of the scientific research is greatly diminished. As has already been noted and stressed, knowledge about N fertilizer efficiency and techniques for minimizing $NO_3^-$ leaching are available but not consistently being used. What educational means can be used to acquaint N suppliers and users about the seriousness of N leaching and its impacts, and make the point effectively enough to cause a change in management where necessary? Even though more research needs to be conducted in some areas dealing with N management, there is too wide a gap between our current knowledge and the application of that knowledge.

## References

Anderson, J.L., G.L. Malzer, G.W. Randall, and G.W. Rehm. 1989. Nitrogenmanagement related to groundwater quality in Minnesota. *J. Minn. Acad. Sci.* 55:53-57.

Baker, J.L. and H.P. Johnson. 1981. Nitrate-nitrogen in tile drainage asaffected by fertilization. *J. Environ. Qual.* 10:519-522.

Ball, P.R. and D.R. Keeney. 1981. Nitrogen losses from urine-affected areasof a New Zealand pasture, under contrasting seasonal conditions. p. 342-344. Proceedings of the XIV International Grassland Congress. Westview Press. Boulder, CO.

Ball, P.R. and J.C. Ryden. 1984. Nitrogen relationships in intensively managed temperate grasslands. *Plant and Soil* 76:23-33.

Benoit, G.R. 1973. Effect of agricultural management of wet sloping soil onnitrate and phosphorus in surface and subsurface water. *Water Resources Res.* 9:1296-1303.

Binford, G.D., A.M. Blackmer, and N.M. El-Hout. 1990. Tissue test for excess nitrogen during corn production. *Agron. J.* 82:124-129.

Blackmer, A.M., D. Pottker, M.E. Cerrato, and J. Webb. 1989. Correlationsbetween soil nitrate concentrations in late spring and corn yields in Iowa. *J. Prod. Agric.* 2:103-109.

Bottcher, A.B., E.J. Monke, and L.F. Huggins. 1981. Nutrient and sediment loadings from a subsurface drainage system. *Trans. Am. Soc. Agric. Engr.* 24:1221-1226.

Bronson, K.F., J.T. Touchton, R.D. Hauck, and K.R. Kelley. 1991. Nitrogen-15 recovery in winter wheat as affected by application timing and dicyandiamide. *Soil Sci. Soc. Am. J.* 55:130-135.

Bundy, L.G. and E.S. Malone. 1988. Effect of residual profile nitrate oncorn response to applied nitrogen. *Soil Sci. Soc. Am. J.* 52:1377-1383.

Burwell, R.E., D.R. Timmons, and R.F. Holt. 1975. Nutrient transport insurface runoff as influenced by soil cover and seasonal periods. *Soil Sci. Soc. Am. Proc.* 39:523-528.

Chang, C., T.G. Sommerfeldt, and T. Entz. 1991. Soil chemistry after elevenannual applications of cattle feedlot manure. *J. Environ. Qual.* 20:475-480.

Chichester, F.W. 1977. Effects of increased fertilizer rates on nitrogencontent of runoff and percolate from monolith lysimeters. *J. Environ. Qual.* 6:211-217.

Corak, S.J., W.W. Frye, and M.S. Smith. 1991. Legume mulch and nitrogenfertilizer effects on soil water and corn production. *Soil Sci. Soc. Am. J.* 55:1395-1400.

Costin, A.B. 1980. Runoff and soil and nutrient losses from an improved pasture at Ginninderra, Southern Tablelands, New South Wales. Aust. J. Agric. Res. 31:533-546.

Dick, W.A., W.M. Edwards, and F. Haghiri. 1986. Water movement through soil to which no-tillage cropping practices have been continuously applied. Proc. Agricultural Impacts on Ground Water. p. 243-252. Natl. Water Well Assoc., Dublin, OH.

Eckert, D.J. 1988. Rye cover crops for no-tillage corn and soybean production. *J. Prod. Agric.* 1:207-210.

Edwards, W.M. and C.R. Amerman. 1984. Subsoil characteristics influence hydrologic response to no-tillage. *Trans. Am. Soc. Agric. Engr.* 27:1055-1058.

Edwards, W.M., P.F. Germann, L.B. Owens, and C.R. Amerman. 1984. Watershed studies of factors influencing infiltration, runoff, and erosion on stony and non-stony soils. p. 45-54. In: *Erosion and Productivity of Soils Containing Rock Fragments*, Soil Sci. Soc. Am., Madison, WI.

Edwards, W.M., M. J. Shipitalo, and L.D. Norton. 1988. Contribution ofmacroporosity to infiltration into a continuous no-tilled watershed: Implications for contaminant movement. *J. Contaminant Hydrology* 3:193-205.

Edwards, W.M., M.J. Shipitalo, L.B. Owens, and L.D. Norton. 1989. Water and nitrate movement in earthworm burrows within long-term no-till cornfields. *J. Soil Water Conserv.* 44:240-243.

Edwards, W.M., M.J. Shipitalo, L.B. Owens, and L.D. Norton. 1990. Effect of *Lumbricus Terrestris* L. burrows on hydrology of continuous no-till corn fields. *Geoderma* 46:73-84.

Edwards, W.M., G.B. Triplett, D.M. Van Doren, L.B. Owens, C.E. Redmond, and W.A. Dick. 1992. Tillage studies with a corn/soybean rotation: Hydrology and sediment loss. *Soil Sci. Soc. Am. J.* 56:52-58.

El-Hout, N.M. and A.M. Blackmer. 1990. Nitrogen status of corn after alfalfain 29 Iowa fields. *J. Soil Water Conserv.* 45:115-117.

Erbach, D.C. 1982. Tillage for continuous corn and corn-soybean rotation. Trans. Am. Soc. Agric. Engr. 25:906-911, 918.

European Economic Community. 1980. Council directive on the quality of water for human consumption. Official Journal No. 80/778, EEC L229, Luxembourg.

Fox, R.H., J.M. Kern, and W.P. Piekielek. 1986. Nitrogen fertilizer source,and method and time of application effects on no-till corn yields and nitrogen uptakes. *Agron. J.* 78:741-746.

Garwood, E.A. and J.C. Ryden. 1986. Nitrate loss through leaching and surface runoff from grassland: Effects of water supply, soil type and management. p. 99-113. In: H.G. van der Meer, J.C. Ryden, and G.C. Ennik (eds.) *Nitrogen Fluxes in Intensive Grassland Systems*. Martinus Nijhoff Publishers, Dordrecht.

Gross, C.M., J.S. Angle, and M.S. Welterlen. 1990. Nutrient and sediment losses from turfgrass. *J. Environ. Qual.* 19:663-668.

Haigh, R.A. and R.E. White. 1986. Nitrate leaching from a small,underdrained, grassland, clay catchment. *Soil Use and Management* 2:65-70.

Heichel, G.H. 1985. Nitrogen recovery by crops that follow legumes. p. 183-189. In: R.F. Barnes (ed.), *Forage Legumes for Energy-efficient Animal Production*. USDA-Agric. Res. Serv., Washington, DC.

Heng, L.K., R.E. White, N.S. Bolan, and D.R. Scotter. 1991. Leaching losses of major nutrients from a mole-drained soil under pasture. *New Zealand J. Agric. Res.* 34:325-334.

Holderbaum, J.F., A.M. Decker, J.J. Meisinger, F.R. Mulford, and L.R. Vough. 1990. Fall-seeded legume cover crop for no-tillage corn in the humid East. *Agron. J.* 82:117-124.

Howard, D.D. and D.D. Tyler. 1989. Nitrogen source, rate, and application method for no-tillage corn. *Soil Sci. Soc. Am. J.* 53:1573-1577.

Jokela, W.E. 1992. Nitrogen fertilizer and dairy manure effects on cornyield and soil nitrate. *Soil Sci. Soc. Am. J.* 56:148-154.

Jokela, W.E. and G.W. Randall. 1989. Corn yield and residual soil nitrate asaffected by time and rate of nitrogen application. *Agron. J.* 81:720-726.

Kanwar, R.S., J.L. Baker, and D.G. Baker. 1988. Tillage and split N-fertilization effects on subsurface drainage water quality and crop yields. *Trans. Am. Soc. Agric. Engr.* 31:453-461.

Keeney, D.R. 1982. Nitrogen management for maximum efficiency and minimum pollution. In: F.J. Stevenson (ed.) *Nitrogen in Agricultural Soils*, Agronomy 22:605-649. Am. Soc. Agron., Madison, WI.

Kilmer, V.J., J.W. Gilliam, J.F. Lutz, R.T. Royce, and C.D. Eklund. 1974.Nutrient loss from fertilized grassed watersheds in western North Carolina. *J. Environ. Qual.* 3:214-219.

King, L.D., J.C. Burns, and P.W. Westerman. 1990. Long-term swine lagooneffluent applications on "coastal" bermudagrass: II. Effect on nutrient accumulation in soil. *J. Environ. Qual.* 19:756-760.

Kladivko, E.J., G.E. Van Scoyoc, E.J. Monke, K.M. Oates, and W. Pask. 1991. Pesticide and nutrient movement into subsurface tile drains on a silt loam soil in Indiana. *J. Environ. Qual.* 20:264-270.

Laflen, J.M. and M.A. Tabatabai. 1984. Nitrogen and phosphorus losses from corn-soybean rotations as affected by tillage practices. *Trans. Am. Soc. Agr. Engr.* 27:58-63.

Lang, R.D. 1979. The effect of ground cover on surface runoff from experimental plots. *J. Soil Conserv. N.S.W.* 35:108-114.

Lawes, J.B., J.H. Gilbert, and R. Warington. 1882. On the amount and composition of the rain and drainage waters collected at Rothamsted. *J. Royal Agric. Soc. Engl. Ser. 2*, 17:241-279, 311-350, and 18:1-71.

Lawes, J.B., J.H. Gilbert, and R. Warington. 1883. Nitrogen as nitric acid in soils and subsoils of some of the fields at Rothamsted. *J. Royal Agric. Soc. Engl.* 19:1-39.

Legg, J.O. and J.J. Meisinger. 1982. Soil nitrogen budgets. In: F.J. Stevenson (ed.) *Nitrogen in Agricultural Soils*, Agronomy 22:503-565. Am. Soc. Agron., Madison, WI.

Lehman, O.R. and L.R. Ahuja. 1985. Interflow of water and tracer chemical on sloping field plots with exposed seepage faces. *J. Hydrology* 76:307-317.

Logan, T.J., G.W. Randell, and D.R. Timmons. 1980. Nutrient content of tiledrainage from crop land in the north central region. Research Bulletin 1119. Ohio Agric. Res. Develop. Center, Wooster, OH.

Logan, T.J. and G.O. Schwab. 1976. Nutrient and sediment characteristics oftile effluent in Ohio. *J. Soil Water Conserv.* 31:24-27.

Maddux, L.D., C.W. Raczkowski, D.E. Kissel, and P.L. Barnes. 1991. Broadcast and subsurface-banded urea nitrogen in urea ammonium nitrate applied to corn. *Soil Sci. Soc. Am. J.* 55:264-267.

Magdoff, F.R., W.E. Jokela, R.H. Fox, and G.F. Griffin. 1990. A soil testfor nitrogen availability in northeastern United States. *Commun. Soil Sci. Plant Anal.* 21:1103-1115.

Mannering, J.V., J.D. Meyer, and L.D. Meyer. 1975. Tillage for moisture conservation. Paper No. 75-2523. Am. Soc. Agr. Eng., St. Joseph, MI.

Martinez, J. and G. Guiraud. 1990. A lysimeter study of the effects of a ryegrass catch crop, during a winter wheat/maize rotation, on nitrate leaching and on the following crop. *J. Soil Sci.* 41:5-16.

McVay, K.A., D.E. Radcliffe, and W.L. Hargrove. 1989. Winter legume effects on soil properties and nitrogen fertilizer requirements. *Soil Sci. Soc. Am. J.* 53:1856-1862.

Meisinger, J.J., W.L. Hargrove, R.L. Mikkelsen, J.R. Williams, and V.W. Benson. 1991. Effects of cover crops on groundwater quality. p. 57-68. In: W.L. Hargrove (ed.), *Cover Crops for Clean Water*, Soil Water Conserv. Soc., Ankeny, IA.

Morton, T.G., A.J. Gold, and W.M. Sullivan. 1988. Influence of over watering and fertilization on nitrogen losses from home lawns. *J. Environ. Qual.* 17:124-130.

Moyer, J.L. and D.W. Sweeney. 1990. Tall fescue response to placement ofurea/ammonium nitrate solution. *Soil Sci. Soc. Am. J.* 54:1153-1156.

Owens, L.B. 1990. Nitrate-nitrogen concentrations in percolate from lysimeters planted to a legume-grass mixture. *J. Environ. Qual.* 19:131-135.

Owens, L.B. and W.M. Edwards. 1993. Tillage studies with a corn/soybean rotation: Surface runoff chemistry. *Soil Sci. Soc. Am. J.* 57:1055-1060.

Owens, L.B., W.M. Edwards, and M.J. Shipitalo. 1990a. Nitrate leaching through lysimeters in a corn/soybean rotation. p. 44. *Agronomy Abstracts* Am. Soc. Agron., Madison, WI.

Owens, L.B., W.M. Edwards, and R.W. Van Keuren. 1984. Peak nitrate-nitrogen values in surface runoff from fertilized pastures. *J. Environ. Qual.* 13:310-312.

Owens, L.B., W.M. Edwards, and R.W. Van Keuren. 1990b. Effects of grass-legume pasture systems on groundwater nitrate levels. p. 294-299. *Transactions of the 14th International Congress of Soil Science, Vol. IV* Kyoto, Japan.

Owens, L.B., R.W. Van Keuren, and W.M. Edwards. 1982. Environmental effects of a medium-fertility 12-month pasture program: Nitrogen. *J. Environ. Qual.* 11:241-246.

Owens, L.B., R.W. Van Keuren, and W.M. Edwards. 1983. Nitrogen loss from a high-fertility, rotational pasture program. *J. Environ. Qual.* 12:346-350.

Peterson, G.A. and W.W. Frye. 1989. Fertilizer nitrogen management. p. 183-219. In: R.F. Follett (ed.), *Nitrogen Management and Groundwater Protection*, Elsevier, Amsterdam.

Peterson, T.A. and M.P. Russelle. 1991. Alfalfa and the nitrogen cycle inthe Corn Belt. *J. Soil Water Conserv.* 46:229-235.

Petrovic, A.M. 1990. The fate of nitrogenous fertilizers applied toturfgrass. *J. Environ. Qual.* 19:1-14.

Russelle, M.P. and W.L. Hargrove. 1989. Cropping systems: Ecology andmanagement. p. 277-317. In: R.F. Follett (ed.), *Nitrogen Management and Groundwater Protection*, Elsevier, Amsterdam.

Ryden, J.C., P.R. Ball, and E.A. Garwood. 1984. Nitrate leaching from grassland. *Nature* 311:50-53.

Schepers, J.S. and R.H. Fox. 1989. Estimation of N budgets for crops. p. 221-246. In: R.F. Follett (ed.) *Nitrogen Management and Groundwater Protection*, Elsevier, Amsterdam.

Schepers, J.S. and D.D. Francis. 1982. Chemical water quality of runoff fromgrazing land in Nebraska: I. Influence of grazing livestock. *J. Environ. Qual.* 11:351-354.

Schepers, J.S., M.G. Moravek, E.E. Alberts, and K.D. Frank. 1991. Maizeproduction impacts ongroundwater quality. *J. Environ. Qual.* 20:12-16.

Sharpley, A.N. and S.J. Smith. 1991. Effects of cover crops on surface waterquality. p. 41-49. In: W.L. Hargrove (ed.) *Cover Crops for Clean Water*, Soil Water Conserv. Soc., Ankeny, IA.

Sharpley, A.N., S.J. Smith, and J.W. Naney. 1987. Environmental impact ofagricultural nitrogen and phosphorus use. *J. Agric. Food Chem.* 35:812-817.

Sharpley, A.N., J.K. Syers, and R.W. Tillman. 1983. Transport of ammonium- and nitrate-nitrogen in surface runoff from pasture as influenced by urea application. *Water, Air, and Soil Pollution* 20:425-430.

Shipitalo, M.J., W.M. Edwards, W.A. Dick, and L.B. Owens. 1990. Initialstorm effects on macropore transport of surface-applied chemicals in no-till soil. *Soil Sci. Soc. Am. J.* 54:1530-1536.

Smith, S.J., J.S. Schepers, and L.K. Porter. 1990. Assessing and managing agricultural nitrogen losses to the environment. p. 1-43. In: B.A. Stewart (ed.) *Advances in Soil Science*, Springer-Verlag, NY.

Smith, S.J., A.N. Sharpley, W.A. Berg, G.A. Coleman, and N.W. Welch. 1988. Nutrient losses from agricultural land runoff in Oklahoma. Proc. 22nd. Okla. Agric. Chem. Conf. 13:23-26. Oklahoma State Univ. Pub. Stillwater, OK.

Steele, K.W., M.J. Judd, and P.W. Shannon. 1984. Leaching of nitrate and other nutrients from a grazed pasture. *New Zealand J. Agric. Res.* 27:5-11.

Steenvoorden, J.H.A.M., H. Fonck, and H.P. Oosterom. 1986. Losses of nitrogen from intensive grassland systems by leaching and surface runoff. p. 85-97 In: H.G. van der Meer, J.C. Ryden, and G.C. Ennik (eds.) *Nitrogen fluxes in intensive grassland systems*, Dordrecht. The Netherlands: Martinus Nijhoff Publishers.

Thomas, G.W., M.S. Smith, and R.E. Phillips. 1989. Impact of soil manage-- ment practices on nitrogen leaching. p. 247-276. In: R.F. Follett (ed.) *Nitrogen Management and Groundwater Protection*, Elsevier, Amsterdam.

Tyler, D.D. and G.W. Thomas. 1977. Lysimeter measurements of nitrate andchloride losses from under conventional and no-tillage corn. *J. Environ. Qual.* 6:63-66.

U.S. Environmental Protection Agency. 1973. Water quality criteria. Washington, DC.

U.S. Public Health Service. 1962. Public health drinking water standards.

U.S. Public Health Publ. 956. U.S. Gov. Print. Office, Washington, DC.

Wagger, M.G. 1989. Cover crop management and nitrogen rate relation togrowth and yield of no-till corn. *Agron. J.* 81:533-538.

Walters, D.T. and G.L. Malzer. 1990. Nitrogen management and nitrificationinhibitor effects on nitrogen-15 urea: II. Nitrogen leaching and balance. *Soil Sci. Soc. Am. J.* 54:122-130.

Watts, D.G., G.W. Hergert, and J.T. Nichols. 1991. Nitrogen leaching lossesfrom irrigated orchardgrass on sandy soils. *J. Environ. Qual.* 20:355-362.

World Health Organization. 1971. International standards for drinking water.p. 36. World Health Organization, Geneva.

Zhu, J.C., C.J. Gantzer, S.H. Anderson, E.E. Alberts, and P.R. Beuselinck. 1989. Runoff, soil, and dissolved nutrient losses from no-till soybean with winter cover crops. *Soil Sci. Soc. Am. J.* 53:1210-1214.

# Animal and Municipal Organic Wastes and Water Quality

H. Kirchmann

I. Introduction . . . 164
II. Chemical Characteristics . . . 165
A. Animal Residues . . . 165
B. Sewage Sludge . . . 170
C. Municipal Compost . . . 172
III. Aerobic and Anaerobic Storage . . . 173
A. Decomposition Pathways . . . 174
B. Gaseous N Losses from Solid Manures . . . 175
C. Gaseous N Losses from Slurries . . . 183
D. Leaching Losses of N from Manures and Slurries . . . 186
IV. Reactions in Soils . . . 187
A. Ammonia Volatilization . . . 187
B. Denitrification . . . 191
C. Mineralization and Immobilization . . . 193
D. Conceptual Approaches and Models . . . 199
E. Leaching of N . . . 200
F. N Balance Studies . . . 202
V. Environmental Aspects of Organic Waste Application to Soils . 203
A. Gaseous N Losses from Organic Wastes Compared to N Fertilizers . . . 203
B. Leaching Losses of N from Organic Wastes Compared to N Fertilizers . . . 204
C. Surface Runoff . . . 205
D. Dissolved Organic Carbon . . . 206
VI. Organic Wastes and Water Quality . . . 207
A. Microbial Quality of Surface Runoff and Groundwater from Soils Amended with Organic Wastes . . . 208
B. Chemical Quality of Surface Runoff and Groundwater from Soils Amended with Organic Wastes . . . 209
C. Solutions to the Groundwater Pollution Problem . . . 209
VII. Summary . . . 212
References . . . 212

0-87371-980-8/94/$0.00+$.50

# I. Introduction

The waste products of living farm animals are feces and urine, representing metabolized remains of feed and fodder material. The main organic wastes of municipal origin are sewage sludge, derived from waste water treatment and composts derived from sorted organic household wastes. Recirculation of these materials to agricultural land usually includes three handling moments; continuous collection, storage or treatment, and spreading.

Manures and slurries have traditionally been applied to agricultural land and the addition of municipal organic wastes to soils is becoming a convenient method of disposal. However, although animal and organic wastes are regarded as "natural fertilizers" and their obvious way of recycling is the application to land, the materials are more difficult to manage than inorganic fertilizers.

Recirculation to soil of nutrients present in fresh waste materials is not complete. Nitrogen losses occur in the form of ammonia volatilization during storage and handling. In fact, on average, 81% of the ammonia emissions in Europe originate from livestock wastes and animal wastes are considered to be the main source for ammonia pollution to the atmosphere (Buijsman et al., 1987). Thus, there is an urgent need to reduce nitrogen losses from these materials to the external environment and to improve their utilization in crop production.

Manures and organic wastes are variable in composition and therefore do not form a group of fertilizers with a well-known content of plant nutrients. Municipal waste materials such as household composts and sewage sludge often contain undesirable contaminants. Through their incorporation in agricultural soil, heavy metals and xenobiotic organic compounds may be introduced into the human food chain. Thus, the composition of municipal wastes needed to be controlled and their use in soil to be monitored.

The purpose of this study is to survey current scientific literature on animal and domestic organic wastes, with the emphasis on biological changes taking place during treatment and handling. The fate of N of organic waste materials in the soil-plant system is also reviewed. Several articles dealing with various aspects of organic manures have been published earlier. Harmsen and van Schreven (1955) and Bartholomew (1965) focused on mineralization and immobilization of nitrogen. Smith and Peterson (1982) and Bouldin et al. (1984) reviewed recycling aspects and use of manure N. Jenkinson (1981) characterized the fate of plant and animal residues in soil. Terman (1979), Beauchamp (1983) and Jarvis and Pain (1990) reviewed the literature on gaseous losses from organic manures added to soil. Khaleel et al. (1980) presented an overview concerning the transport of nutrients, oxygen demanding compounds and bacteria in runoff waters from land receiving animal wastes.

# II. Chemical Characteristics

## A. Animal Residues

### 1. Metabolic Effects on Chemical Composition

The proportion of ingested nutrients utilized by farm animals is relatively small. On average, 80% of nitrogen, 78% of phosphorus and 95% of potassium present in the fodder is recovered in fresh animal residues (Sibbesen, 1990). The composition of freshly excreted animal residues depends on species, age of the animal and varies with the type of diet. Despite the variations caused by diet and the age of the animal, there are distinct differences in the composition of fresh manures from different farm animal species.

Ruminants have an equivalent nitrogen distribution in feces and urine (Smith, 1973; Tietjen, 1977). Increased N intake of ruminants results in a larger excretion of urine N (Vuuren and Meijs, 1987). Urea is the main urinary nitrogen form, amounting to 50-75% of urine N (Doak, 1952; Thomas et al., 1988). Other organic N compounds found in urine are allantoin N (3-5% of urine N), hippuric acid N (2-3% of urine N), creatine and creatinine N (1-1.8% of urine N), amino-N (10.5-15.9 % of urine N) and ammonia N (0.5-0.9% of urine N) (Doak, 1952). Cattle feces contain the main plant nutrients such as Ca, Mg, P, S and micro plant nutrients, with the exception of potassium which is mainly found in urine (Tietjen, 1977; Japenga and Harmsen, 1990). Variations in the elemental composition of dairy feces were studied by Safley et al. (1985) who estimated the nitrogen concentration in cow feces to be, on average, 2.70% $\pm$ 0.52 in dry matter.

Carbohydrates and lignin in cattle wastes accounted for 30% of the dry weight (Hrubrant and Detroy, 1979). Studies on organic fractions in cow feces show that about 1/10 of the carbohydrates originate from bacteria, amounting to 4-4.5% of the fecal dry matter (Salo, 1965). In a review by Flachowsky and Hennig (1990), the proximate composition and the mineral content of different animal wastes are compiled.

Fresh feces of cattle were found to contain acetic, propionic, butyric, valeric and isovaleric acid, constituting a total of 0.18% on a dry weight basis (Schuman and McCalla, 1976). The average proportions of the main acids were reported to be 69% acetic acid, 18% propionic acid and 13% butyric acid, respectively (Hill, 1970).

Involuntary ingestion of soil together with herbage was observed for grazing cows and sheep. The percentage of soil ingested together with the herbage could amount to 14% of the total dry matter intake (Mc Grath et al., 1982; Fleming, 1986). Feces of pasture-fed dairy cows were found to contain up to 40% soil (dry matter basis) (Healy, 1968).

Swine have about two-thirds of their excreted nitrogen in urine, of which urea is the main nitrogen form (Smith, 1973). Swine urine also contains most of the potassium taken up through the diet, whereas the other main plant nutrients are excreted with the feces (Tietjen, 1977).

Poultry excrete urine and feces together. Uric acid is the most abundant nitrogen compound in avian excreta, amounting to 52-72% of total N, while urea comprises 2-10% and ammonium N 6-23% of total excreted nitrogen (Davis, 1927; Ekman et al., 1949; Terpstra and De Hart, 1974; McNabb and McNabb, 1975; Emmanuel and Howard, 1978; Krogdahl and Dahlsgård, 1981). Total nitrogen concentrations in fresh poultry excreta varied between 7 and 9% expressed on a dry matter basis (Terpstra and De Hart, 1974) and thus were higher than in swine and cattle residues.

Phosphorus fractionation was performed on fresh manures from different farm animal species and showed that no distinct differences could be found; on average, 70% of phosphorus was present in inorganic forms and 30% in organic forms (Peperzak et al., 1959).

## 2. Changes in Chemical Composition during Storage or Treatment

Once urine and feces are excreted and collected, they become a mixture including straw or other litter material. If the mixture is solid, described by the term "manure", it may either decompose aerobically in loosely stored manure heaps or anaerobically in tightly packed heaps, pits or manure houses. If additional water is added to the mixture and the dry matter content decreases below 12%, it is called "slurry". Slurries usually undergo an anaerobic decomposition process, if not aerated. Microbial and chemical reactions during both aerobic and anaerobic treatment change the chemical composition of the residues. Table 1 shows some typical data on the elemental composition of fresh and treated animal wastes.

The initial carbon content in animal manures, ranging from 38-45%, decreased during aerobic and anaerobic decomposition. The lowest carbon concentration, 27.3%, was measured in aerobically treated poultry excreta. The percentage of nitrogen varies between 2.3 and 6.7%, being lowest in fresh cattle feces and highest in anaerobically decomposed poultry excreta. The content of non-volatile elements such as phosphorus, potassium, calcium and magnesium, increased during the decomposition process as the result of microbial carbon respiration.

Table 2 lists typical data on the nitrogen composition of fresh and treated animal wastes. In fresh animal feces, no nitrate or nitrite was found. Only after aerobic decomposition, low concentrations of nitrate could be detected in manures (Kirchmann, 1985). In slurries, nitrate was only formed if the material was aerated (Loynachan et al., 1976; Evans et al., 1986; Linke et al., 1987). Thus, the presence of nitrate in animal wastes indicates that aerobic decomposition had taken place.

Analyses of stored decomposed animal feces showed that no urea and no uric acid was present in these materials, indicating that the two compounds are

**Table 1**. Elemental analyses, based on dry matter, of fresh, aerobically and anaerobically decomposed animal dung, anaerobically digested sewage sludge and composted organic household wastes

| Type of material | C (%) | N (%) | P (%) | K (%) | Ca (%) | Mg (%) | S (%) |
|---|---|---|---|---|---|---|---|
| Cattle feces[a] | | | | | | | |
| Fresh | 43.6 | 2.33 | 0.90 | 0.73 | 1.96 | 0.60 | 0.36 |
| Aerobic | 40.6 | 3.00 | 1.04 | 1.44 | 2.20 | 0.69 | 0.44 |
| Anaerobic | 40.4 | 4.15 | 0.93 | 1.29 | 2.01 | 0.64 | 0.36 |
| Pig feces[a] | | | | | | | |
| Fresh | 44.7 | 3.08 | 2.90 | 1.14 | 2.64 | 0.98 | 0.43 |
| Aerobic | 37.3 | 4.05 | 3.61 | 3.00 | 3.82 | 1.50 | 0.72 |
| Anaerobic | 44.4 | 4.25 | 1.97 | 1.78 | 3.42 | 1.22 | 0.46 |
| Poultry excreta[a] | | | | | | | |
| Fresh | 37.7 | 5.10 | 1.93 | 1.85 | 6.65 | 0.57 | 0.50 |
| Aerobic | 27.3 | 2.38 | 3.35 | 4.07 | 11.87 | 0.95 | 0.82 |
| Anaerobic | 29.5 | 6.73 | 2.35 | 2.41 | 9.17 | 0.61 | 0.53 |
| Sewage sludge[b] | 32.0 | 4.00 | 2.50 | 0.30 | 1.50 | 0.30 | n.d. |
| Household compost[c] | 17.5 | 1.59 | 0.56 | 1.49 | 1.99 | 0.54 | n.d. |

[a] from Kirchmann and Witter, 1992
[b] from Sjöqvist and Wikander-Johansson, 1985; Mininni and Santori, 1987
[c] from Vogtmann et al., 1984

**Table 2.** Forms of nitrogen in fresh, aerobically and anaerobically decomposed animal dung, anaerobically digested sewage sludge and composted organic household wastes, all on a dry matter basis

| Type of material | Tot-N ($mg\ g^{-1}$) | $NH_4$ ($mg\ g^{-1}$) | $NO_3$ ($mg\ g^{-1}$) | Urea-N ($mg\ g^{-1}$) | Uric acid N ($mg\ g^{-1}$) | Organic-N ($mg\ g^{-1}$) | Water sol. N ($mg\ g^{-1}$) |
|---|---|---|---|---|---|---|---|
| Cattle feces[a] | | | | | | | |
| Fresh | 23.3 | 0.22 | 0.00 | trace | 0.00 | 23.08 | 5.02 |
| Aerobic | 30.0 | 0.23 | 0.02 | 0.00 | 0.00 | 29.75 | 2.82 |
| Anaerobic | 41.5 | 21.17 | 0.00 | 0.00 | 0.00 | 20.33 | 22.05 |
| Pig feces[a] | | | | | | | |
| Fresh | 30.8 | 2.55 | 0.00 | trace | 0.00 | 28.25 | 8.52 |
| Aerobic | 40.5 | 1.58 | 0.02 | 0.00 | 0.00 | 38.90 | 5.97 |
| Anaerobic | 42.5 | 21.64 | 0.00 | 0.00 | 0.00 | 20.86 | 21.28 |
| Poultry excreta[a] | | | | | | | |
| Fresh | 51.0 | 4.19 | 0.00 | 0.64 | 31.00 | 46.81 | 35.70 |
| Aerobic | 23.8 | 0.36 | 0.13 | 0.00 | 0.00 | 23.31 | 5.77 |
| Anaerobic | 67.3 | 50.85 | 0.00 | 0.00 | 0.00 | 16.45 | 47.79 |
| Sewage sludge[b] | 40.0 | 4.00 | 0.00 | 0.00 | 0.00 | 36.0 | n.d. |
| Household compost[c] | 15.9 | 3.00 | 0.30 | 0.00 | 0.00 | 12.6 | n.d. |

[a] from Kirchmann and Witter, 1992
[b] from Sjöqvist and Wikander-Johansson, 1985; Mininni and Santori, 1987
[c] from Vogtmann et al., 1984

completely decomposed under both aerobic and anaerobic conditions (see Table 1). Urea ($NH_2CONH_2$), the predominant nitrogen compound in urine from ruminants and swine, is hydrolyzed to carbon dioxide and ammonia by ureases, a group of enzymes that act on the C-N bonds (non-peptides) in linear amides. Urea may already be hydrolyzed during collection and early storage as indicated by high instantaneous losses of ammonia from fresh cow manures (Lindhard, 1954).

Uric acid ($C_5H_4N_4O_3$), the main end product of avian nitrogen metabolism, was found to be completely degraded by aerobic microorganisms during storage of poultry litter (Bachrach, 1957; Schefferle, 1965). Over 100 strains of bacteria and yeasts were capable of decomposing uric acid isolated in poultry droppings (Vuori and Nasi, 1977) and the odor of ammonia in poultry houses results from the breakdown of uric acid (Carlile, 1984).

The following degradation pathway for uric acid through pseudomonades was found (Bachrach, 1957):

→ Glyoxylic acid
Uric acid → Allantoin → Allantoic acid
→ Urea

Concentrations of ammonium N in animal residues increased during anaerobic decomposition amounting to 50-75% of the total N, whereas organic bound N was the main form in which N was present in fresh and aerobically treated feces (see results in Table 1).

The fact that organically bound nitrogen is the main nitrogen form of aerobically decomposed and composted manure is confirmed by studies of Bremner (1955), Beckwith and Parsons (1980), Castellanos and Pratt (1981) and Kirchmann (1985).

Studies on the ionic balances in the liquid and solid fraction of pig and poultry slurries by Japenga and Harmsen (1990) showed that calcium, magnesium and manganese concentration levels decreased in the liquid fraction during storage due to precipitation as carbonate compounds as pH values increased. Iron and copper concentrations in the liquid fraction increased slightly, which was attributed to changes in the dissolved organic matter. Sulfur was exclusively present as organic sulfur compounds in the liquid fraction and as inorganic sulfides in the solid fraction.

During both aerobic and anaerobic storage, a small part of the organic phosphorus present in fresh manures was converted into inorganic P (Peperzak et al., 1959). The litter present in manure served primarily to dilute the P without altering the distribution. An analysis of different P fractions in anaerobically stored pig slurry showed that, on average, 85% was present as inorganic P and 15% as organic P (Gerritse, 1981).

The inorganic phosphorus compounds found in the solid phase of slurries consisted of struvite ($NH_4MgPO_4 \cdot 6\,H_2O$), trimagnesium phosphate ($(Mg_3(PO_4)_2 \cdot 8\,H_2O$), octacalcium phosphate ($(Ca_4H(PO_4)_3 \cdot 3\,H_2O$) and dicalcium phosphate

($CaHPO_4 \cdot 2 H_2O$) (Fordham and Schwertmann, 1977a). Studies with $^{32}PO_4$ added to anaerobically stored pig slurries showed that $^{32}P$ was present in all phosphorus fractions of the slurry (Gerritse and Zugec, 1977). This showed that phosphorus was transformed into different P compounds through the microbial activity during storage.

Microbial decomposition of manure and slurry under anaerobic storage conditions resulted in the production of short-chain carboxylic acids ($C_2$-$C_5$). The production of short-chain carboxylic acids, also called volatile fatty acids, was related to the actual redox potentials during storage. The maximum production rate was found at a redox potential of -200 mV (Guenzi and Beard, 1981). Acetic acid ($CH_3COOH$), the dominant form, followed by propionic acid ($C_2H_5COOH$), butyric acid ($C_3H_7COOH$) and valeric acid ($C_4H_9COOH$) were all reported to be present in anaerobically treated residues from ruminants, swine and poultry (Bell, 1970; Cooper and Cornforth, 1978; Spoelstra, 1979; Guenzi and Beard, 1981; Patni and Jui, 1985). Isomeric forms of valeric and butyric acid were not found in anaerobically decomposed poultry litter (Bell, 1970; Kirchmann and Witter, 1989). No oxalate, citrate, tartrate, lactate, adipate, ascorbate, succinate or hippurate was found in stored slurries (Fordham and Schwertmann, 1977b). Reported concentrations of acetic acid in slurries varied between 4-12 g $kg^{-1}$, 1-4 g $kg^{-1}$ for propionic acid and 0.5-1.5 g $kg^{-1}$ for butyric acid on a wet weight basis. Expressed as the percentage of total carbon, the carbon in fatty acids amounted to ca 30% in pig slurry and 7% in cattle slurry, respectively (Cooper and Cornforth, 1978). In a study with anaerobically decomposed poultry manure, carbon of volatile fatty acids amounted to 3.8-5.5% of the total manure carbon (Kirchmann and Witter, 1989). Almost complete disappearance of short-chain carboxylic acids, present in fresh cattle manure, was measured after composting the material (Schuman and McCalla, 1976).

The pH of animal slurries was found to be mainly influenced by the concentration of volatile fatty acids and ammoniacal nitrogen (Georgacakis et al., 1982; Paul and Beauchamp, 1989c). The production of fatty acids results in the release of $H^+$, causing the slurry pH to decrease, whereas $NH_3$ mineralization of organic N results in increased slurry pH due to the combination of $H^+$ with $NH_3$ to form $NH_4^+$. Aeration of slurry resulted in a decrease in volatile fatty acid concentration and a relatively smaller decrease in total ammonia concentrations. As a result of the oxidation of fatty acids, an increase in slurry pH was found (Paul and Beauchamp, 1989c).

### B. Sewage Sludge

Sewage sludge is the residual solid material from waste water treatment plants consisting of a mixture of sedimented, microbiologically produced and phosphorus-precipitated sludge. Different forms of sewage sludge can be distinguished. Raw or primary sludge, that has not undergone any further treatment, are biologically very active due to the high content of easily decomposable carbon. Activated or anaerobically digested sewage sludges have been stabilized by the removal of readily decomposable carbon. Composted sludges, which are usually anaerobically digested before composting, are the most stabilized materials.

The elemental analyses of anaerobically digested sewage sludge (see Table 1) show that the contents of carbon (32%), nitrogen (4%) and phosphorus (2.5%) are similar to animal manures, whereas the concentration of potassium, calcium and magnesium is lower than in animal manures. The low concentrations of potassium, calcium and magnesium are caused by the low retention during waste water treatment. Analysis of the nitrogen composition (see Table 2) shows that, on average, 10% of the total N is present as ammonium, but no nitrate is found in anaerobically treated sewage sludge. A survey of the sewage sludge composition showed that N, P, and K concentrations vary within a relatively narrow range (Sommers, 1977).

In addition to valuable plant nutrients, sewage sludge contains variable concentrations of heavy metals and synthetic organic compounds. The contents of heavy metals in sewage sludge from different countries were compiled by Mininni and Santori (1987). Heavy metal concentrations in Swedish sewage sludge (see Table 3) have decreased over the last 20 years due to several measures (SNV, 1989).

Parallel to the concern over heavy metals, but more recent, is concern over the presence of synthetic organic compounds in sewage sludge. A review of 219 organic chemicals found in sewage sludge by Jacobs et al. (1987) showed that about 70 compounds were present in concentrations close to the detection limit. About 90% of the organic compounds were found at concentrations below 10 ppm and about 10% had a concentration between 10-100 ppm. Still, data by Jacobs et al. (1987) showed that sewage sludges can be highly contaminated with organic chemicals, with concentrations amounting to a few percent of the dry weight. The following eleven groups of organic chemical compounds found in sewage sludge were listed by Jacobs et al. (1987):

Aromatic and alkyl amines
Chlorinated pesticides
Dioxins and furans
Halogenated aliphatics
Miscellaneous compounds
Monocyclic aromatics
Phenols
Phthalate esters
Polyaromatic hydrocarbons (PAH's)
Polychlorinated biphenyls (PCB's)
Triaryl phosphate esters

**Table 3.** Trends in the average concentrations of heavy metals in Swedish sewage sludge since 1968

| Element (mg kg$^{-1}$) | Year 1968-70[a] | Year 1971[b] | Year 1972[b] | Year 1973[b] | Year 1989[c] |
|---|---|---|---|---|---|
| Cd | 16 | 13.2 | 10.6 | 9.7 | 2 |
| Co | 30 | 12 | 6.1 | 6.6 | n.d. |
| Cr | 1300 | 116 | 118 | 88 | 80 |
| Cu | 1000 | 1625 | 2250 | 1640 | 450 |
| Hg | 9 | 11.5 | 8.5 | 7.6 | 5 |
| Mn | 500 | 242 | 410 | 425 | n.d. |
| Ni | 100 | 48 | 47 | 35 | 40 |
| Pb | 300 | 237 | 221 | 140 | 80 |
| Zn | 2500 | 3230 | 2530 | 2250 | 800 |

[a] from Jansson, 1977; [b] Andersson and Nilsson, 1976; [c] SNV, 1989.

The contamination of sewage sludge with heavy metals may lead to an accumulation in agricultural soils, and synthetic organic compounds present in very high concentrations may have a significant impact on the soil-plant system. As a consequence, safeguards such as disposal guidelines and maximum permitted metal concentration levels were formulated (CEC, 1986; EPA, 1989) to restrict the addition of contaminants to soil.

### C. Municipal Compost

Solid municipal wastes consist of a mixture of organic and inorganic waste materials. Comparative analysis of solid wastes between countries showed that the percentages of food wastes were 36-70% for developing countries but only 15-18% for developed countries. On the other hand proportions of paper and cardboard was found to be higher in the solid wastes of developed countries (Polprasert, 1989).

Only the decomposable organic part of the waste material can be returned to land. Thus, sorting or separation of a "biological waste" fraction from the other materials is inevitable. The separate collection of "biological wastes" is the most promising way of obtaining a pure organic fraction free of contaminating inert constituents and metals (Krauss et al., 1987; Vogtmann et al., 1984)( Table 4). Analyses of municipal compost derived from a source-separated "biological waste" fraction showed by far the lowest heavy metal contents (Vogtmann et al., 1984; Franke, 1987) and concentrations were close to those measured in animal manures (Andersson, 1977).

Analyses of the elemental composition of source-separated municipal compost show that the concentrations of all macro plant nutrients, but also of carbon, are

**Table 4.** Heavy metal concentrations in municipal composts, reported on dry matter basis, as influenced by the separation technique of the organic fraction and compared to animal manure

| Element | Municipal waste | | | |
|---|---|---|---|---|
| | Produced from mixed wastes[a] (n=13) | Produced from source separated organic wastes[b,c] | | Farmyard manure[d] (n=77) |
| Cd ($mg\ kg^{-1}$) | 3.8 | 1.0 | 0.5 | 0.32 |
| Cr ($mg\ kg^{-1}$) | 48 | n.d. | 55 | 5.2 |
| Cu ($mg\ kg^{-1}$) | 340 | 33 | 47 | 41 |
| Hg ($mg\ kg^{-1}$) | 2.9 | n.d. | 0.5 | 0.09 |
| Ni ($mg\ kg^{-1}$) | 40 | 59 | 14 | 7.8 |
| Pb ($mg\ kg^{-1}$) | 560 | 21 | 62 | 6.6 |
| Zn ($mg\ kg^{-1}$) | 1370 | 166 | 198 | 215 |

[a] from Montin et al., 1984; [b] from Vogtmann et al., 1984; [c] from Franke, 1987; [d] from Andersson, 1977

lower than in animal manures (see Table 1). Nitrogen in municipal compost is mainly present in organic form and contents of ammonium and nitrate are low, as in aerobically decomposed manures (see Table 2). The chemical composition of municipal composts was reviewed by Xin-Tao et al. (1992).

## III. Aerobic and Anaerobic Storage

Only fresh residues from grazing animals are directly returned to the soil and decompose under soil environmental conditions. The major part of farm animal wastes and municipal organic wastes are treated or stored before addition to the soil. During storage and treatment, two groups of decomposition pathways can be distinguished, aerobic and anaerobic forms.

Exclusively aerobic or anaerobic decomposition conditions are only achieved if the processes are controlled and steered, which is the case for waste water and sludge treatment. Concerning the storage of manures or slurries, the terms "aerobic" and "anaerobic" indicate the predominant process in the materials, but completely aerobic or anaerobic conditions may not exist.

## A. Decomposition Pathways

### 1. Composting

The term "aerobic decomposition" describes the biological, microbial and chemical conversion of organic residues with oxygen (air) as the external electron acceptor into organic matter with increased resistance to biodegradation. An equivalent expression to the term "aerobic decomposition" is humification. The organic material is partially oxidized, and as a result carbon dioxide, ammonia, water, heat and stabilized polymeric compounds are formed.

A special case of aerobic decomposition is called composting. Composting is defined as the auto-heated, aerobic decomposition allowing a succession of mesophilic and thermophilic microbial populations. In accumulated volumes of organic material, the heat produced during aerobic decomposition causes the temperature of the material to increase (Mote and Griffis, 1982). A typical temperature pattern and three corresponding discernible biological phases can be observed (Bågstam, 1979; de Bertoldi et al., 1982; Godden et al., 1983; Nakasaki et al., 1985a,b). The first phase is characterized by an increase in the temperature to 40-50°C and high metabolic activity of mesophilic bacteria. The second phase is reached when a further temperature increase to 60-70°C becomes inhibiting to the mesophilic bacteria and a thermophilic population of mainly fungi and actinomycetes becomes dominant (Nakasaki et al., 1985b). When the material starts to cool, the third phase starts, which is referred to as the maturation phase. The soil fauna colonizes the material and a food web between different species is established (Dindal, 1978). During this phase the material attains its humus-like appearance of finished compost. The compost material is defined as the stabilized and sanitized product of composting (Zucconi and de Bertoldi, 1987). In view of the definition of composting, terms like "anaerobic composting", "biogas composting" or "aerobic fermentation" are confusing and not appropriate.

The relationship between moisture content and microbial activity during manure decomposition was found to be linear up to 50% of the water-holding capacity followed by a curvilinear response between 50% of WHC and saturation (Murwira et al., 1990).

### 2. Anaerobic Fermentations

Anaerobic fermentation is the microbial transformation of organic material without external electron acceptors (oxygen) and leads to a partial oxidation and partial reduction of the organic compounds. Only a small amount of heat is released, not leading to a significant temperature increase of the materials.

The process requires a unique specialization of the involved bacterial species and results in the formation of a microbial food chain consisting of different groups of bacteria (Thiele and Zeikus, 1988). The first group hydrolyze complex organic matter (carbohydrates, lipids, proteins) into simple organic molecules (formate, ethanol, acetate, lactate, propionate, butyrate, and succinate), carbon dioxide and hydrogen gas. The second group, the acetogenic bacteria, metabolize the hydrolytic end products to acetate, formate, bicarbonate and hydrogen gas. The third group, the methanogenic bacteria, utilize the fermentation products such as acetate and other fermentation products with two carbon atoms as energy source. Usually more than 60% of the methane produced originates from acetate and 30-40 % through reduction of carbon dioxide to methane (Thiele and Zeikus, 1988). The methane produced during anaerobic digestion makes up about one-third of the ultimate energy present in the organic materials, i.e. of the amount released when burned in the presence of oxygen (Sobel and Muck, 1983).

## B. Gaseous N Losses from Solid Manures

The major pathway for nitrogen losses from manures during storage is ammonia volatilization. Losses through denitrification were found to be relatively lower (Kirchmann, 1985). Aliphatic amines volatilized together with ammonia amounted to 2-5% relative to the released ammonia (Mosier et al., 1973). Literature data on nitrogen losses during storage of animal manures are given in Table 5. Losses were determined as the difference in amounts of N before and after the storage period.

Anaerobic fermentation of manures, stored very tightly packed in heaps, pits, ponds or manure houses, resulted in the relatively lowest nitrogen losses, amounting to 5-20% during 3-7 month storage periods (Siegel, 1936; Glathe and Metzen, 1937; Maiwald and Siegel, 1937; Hesse, 1938; Maiwald and Steigmiller; Siegel and Meyer, 1938; Bucher, 1943; Köhnlein and Vetter, 1953; Rauhe and Hesse, 1957; Rauhe and Koepke, 1967; Muck et al., 1984). In incubation experiments under completely anaerobic conditions no significant losses from manures were detected (Russell and Richards, 1917; Glathe and Seidel, 1937; Kirchmann and Witter, 1989).

The highest nitrogen losses were measured in aerobically decomposing manures, i.e., loosely stored manures in which a rapid air exchange takes place, amounting to at least 20% and up to 60% of the initial amount of nitrogen present during 3-6 months of decomposition (Russell and Richards, 1917; Siegel, 1936; Glathe and Metzen, 1937; Scheffer and Zöberlein, 1937; Bucher, 1943; Franz and Repp, 1949; Nehring and Schiemann, 1951; Tinsley and Nowakowsky, 1959; Kirchmann, 1985; Kirchmann and Witter, 1989).

The principal mechanisms through which a reduction of ammonia losses during decomposition can be achieved are as follows:

- Immobilization of ammonium through addition of easily decomposable, nitrogen-poor material.
- Adsorption of ammonium and ammonia on suitable amendments.
- pH regulation of the manure solution.

**Table 5.** Nitrogen losses from solid manures during storage

| Storage conditions | Storage (months) | N loss (%) | Reference |
|---|---|---|---|
| Loose, unsheltered heap | 3 | 27.0 | Russell and Richards, 1917 |
| Loose heap under cover | 3 | 26.0 | |
| Compacted, unsheltered heap | 3 | 28.0 | |
| Compacted, heap under cover | 3 | 26.0 | |
| Laboratory incubation (aerobic) | 7 | 29.6 | |
| Laboratory incubation (anaerobic) | | 1.1 | |
| Compacted, unsheltered heap covered with 10 cm soil | 3 | 26.8 | Iversen, 1936 |
| Compacted, unsheltered heap, biological-dynamic agents | 3 | 27.3 | |
| Loose, unsheltered heap | 3.5 | 22.2 | Siegel, 1936 |
| Loose, storage house | 3.5 | 17.9 | |
| Tightly packed, storage house | 3.5 | 12.6 | |
| Loose, unsheltered heap | 3 | 22.3 | Glathe and Metzen, 1937 |
| Loose thereafter compacted, unsheltered heap | 3 | 24.3 | |
| Tightly packed, storage house | 3 | 20.0 | |
| Laboratory incubation (anaerobic) | 6.5 | 0.0 | Glathe and Seidel, 1937 |
| Loose, unsheltered heap | 3.5 | 23.2 | Maiwald and Siegel, 1937 |
| Loose, thereafter compacted, storage house | 3.5 | 14.0 | |
| Compacted, unsheltered heap | 3.5 | 13.4 | |
| Tightly packed, storage house | 3.5 | 9.0 | |

| | | | |
|---|---|---|---|
| Loose, unsheltered heap | 6 | 17.9 | Scheffer and Zöberlein, 1937 |
| Tightly packed, high straw content, storage house | 6 | 28.7 | |
| Initially loose, thereafter compacted, storage house | 6 | 31.7 | |
| Tightly packed, storage house | 4 | 15.0 | Maiwald and Steigmiller, 1938 |
| Tightly packed, storage house | 6 | 15.0 | |
| Very tightly packed, storage house | 4 | 0.0 | Hesse, 1938 |
| Compacted, storage house | 4 | 6.5 | |
| Loose, storage house | 4 | 17.4 | |
| Tightly packed, storage house | 4 | 10.0 | Siegel and Meyer, 1938 |
| Tightly packed + 5% montmorillonite, storage house | 4 | 9.0 | |
| Tightly packed, storage house, varying litter content | | | Bucher, 1943 |
| 3 kg straw per cow and day | 3 | 10.0 | |
| 6 kg straw per cow and day | 3 | 7.7 | |
| 10 kg straw per cow and day | 3 | 2.6 | |
| Loose, storage house, varying litter content | | | |
| 6 kg straw per cow and day | 3 | 14.3 | |
| 10 kg straw per cow and day | 3 | 18.7 | |
| Compacted, storage house | 9 | 19.0 | Iversen and Dorph-Petersen, 1949 |
| Compacted, storage house | 6 | 16.9 | |
| Compacted, unsheltered heap | 9 | 16.5 | |

**Table 5.** continued

| Storage conditions | Storage (months) | N loss (%) | Reference |
|---|---|---|---|
| Loose, unsheltered heap | 4.5 | 22.9 | Franz and Repp, 1949 |
| Loose, unsheltered heap | 5.5 | 33.2 | |
| Loose, unsheltered heap | 5 | 23.5 | Nehring and Schiemann, 1951 |
| Compacted, unsheltered heap | 5 | 9.5 | |
| Loose, unsheltered heap, with soil (manure:soil = 2.5:1) | 5 | 5.8 | |
| Tightly packed, storage house, varying litter content | | | Köhnlein and Vetter, 1953 |
| 2 kg per cow and day | 2.5 | 9.0 | |
| 4 kg per cow and day | 2.5 | 12.0 | |
| 6 kg per cow and day | 2.5 | 18.0 | |
| Compacted, storage house | 9 | 22.0 | Iversen, 1957 |
| Compacted, storage house, 10% extra straw addition | 9 | 22.0 | |
| Compacted, unsheltered heap | 9 | 17.0 | |
| Compacted, unsheltered heap, 10% extra straw addition | 9 | 24.0 | |
| Loose, unsheltered brick cells | 11.5 | 34.0 | Tinsley and Nowakowsky, 1959 |
| Loose, unsheltered heap | 11.5 | 53.0 | |
| Tightly packed, pit | 3.5 | 4.1 | Rauhe and Hesse, 1957 |
| Compacted, unsheltered heap | 3.5 | 11.4 | |
| Loose, unsheltered heap | 3.5 | 26.8 | |
| Loose, unsheltered heap, with soil (manure:soil = 3:1) | 3.5 | 16.4 | |

| | | | |
|---|---|---|---|
| Tightly packed, pit | 4 | 7.3 | Rauhe and Koepke, 1967 |
| Compacted, storage house | 4 | 21.0 | |
| Loose, unsheltered heap | 4 | 33.3 | |
| Loose, unsheltered heap, with soil (manure:soil = 3:1) | 4 | 22.8 | |
| Laboratory incubation (aerobic), varying straw fineness | | | Meyer and Sticher, 1983 |
| Chopped | 5 | 8.7 | |
| Pulverized | 5 | 8.0 | |
| Bottom-loaded earthen pond | 11 | 8.0 | Muck et al., 1984 |
| Top-loaded earthen pond | 10 | 39.0 | |
| Loose, unsheltered heap, varying litter material | | | Kirchmann, 1985 |
| Straw litter | 4.5 | 36.1 | |
| Peat litter | 4.5 | 43.0 | |
| Coniferous wood chips | 4.5 | 53.4 | |
| Laboratory incubation, varying straw additions | | | Kirchmann and Witter, 1989 |
| manure:straw = 1:1 (aerobic) | 4.5 | 39.0 | |
| manure:straw = 1:2.1 (aerobic) | 4.5 | 15.7 | |
| manure:straw = 1:4.3 (aerobic) | 4.5 | 9.0 | |
| manure:straw = 1:1 (anaerobic) | 4.5 | 1.0 | |

The amount of litter and the availability of its energy content determines the amount of nitrogen that can be immobilized. Materials retaining ammoniacal nitrogen can be added to the manure. The pH value determining the equilibrium between $NH_4^+$-ions and $NH_3$ gas should be kept below 7 to prevent ammonia loss.

## 1. The Effect of Litter

The amount of litter present in fresh manure, i.e., the content of easily degradable carbon sources added to nitrogen-rich materials, influences the N losses during aerobic decomposition. Fassen and Dijk (1979), Meyer and Sticher (1983), and Kirchmann and Witter (1989) showed that the larger the content of straw litter and the higher the initial C/N ratio in manure, the lower N losses were measured during aerobic decomposition. The correlation between initial C/N ratio in manures and ammonia losses during aerobic decomposition was found to be highly significant ($r=-0.884$) (Kirchmann, 1985).

In earlier studies on the influence of varying amounts of straw litter on nitrogen losses, no reduction (Köhlein and Vetter, 1953), or even increase of nitrogen losses was measured (Iversen, 1957). In the studies by Köhnlein and Vetter (1953), the collected manures with larger straw contents did not have higher C/N ratios, indicating that the use of more straw in the stable increased the absorption of urine. Thus, decomposition of manures with different C/N ratios was not compared. The use of larger amounts of litter as bedding material to animals may simply result in a larger absorption of urine. Iversen (1957) added 10% extra straw to collected manures, which were stored compacted. A slight increase of N-losses was measured, but obviously different decomposition conditions were caused. Manure with extra straw had much higher dry matter losses and decomposed aerobically, whereas manure without extra straw additions was due to an anaerobic process. Thus, losses during aerobic decomposition were compared with anaerobic fermentation.

The use of other litter materials than straw to cattle manure, such as *Spaghnum* peat and coniferous wood chips, was tested in a series of compost piles. The initial C/N ratios in the different manure-litter mixtures were very similar. The manure with straw had the relatively lowest losses (36%), followed by manure with peat (43%) and highest losses were measured from the manure containing wood chips litter (53%) during a 4-month storage period (Kirchmann, 1985). Consequently, using coniferous wood chips or *Spaghnum* peat in similar proportions to manure showed no benefit over straw.

## 2. The Effect of Adsorbents

The possibility to reduce nitrogen losses from manures during storage through the addition of materials adsorbing ammonia gas and ammonium ions has been

**Table 6.** Ammonia and ammonium adsorption capacity of materials used as amendments to animal manures

| | Adsorption capacity (mg N $g^{-1}$ dry matter) | |
|---|---|---|
| Material | Ammonia | Ammonium |
| Barley straw, long | 6.4 - 8.2 | n.d. |
| Barley straw, chafed | 6.4 - 8.4 | n.d. |
| Oat straw, long | 3.8 - 5.0 | n.d. |
| Oat straw, chafed | 5.2 - 6.6 | n.d. |
| Cutter shavings | 6.6 - 8.0 | n.d. |
| Sawing dust | 4.6 - 5.8 | n.d. |
| Spaghnum peat | 23.2 - 27.2 | 11.5 |
| Zeolite | 1.8 | 16.5 |
| Montmorillonite | n.d. | 14.4 |
| Basalt, powdered | 0.05 | 0.15 |

from Mortland and Wolcott, 1965; Kemppainen, 1987; and Witter and Kirchmann, 1989a

tested. Different organic and inorganic materials such as *Spaghnum* peat, zeolite, montmorillonite, rock powders and soil have been used. The potential adsorption capacities of different litter materials and inorganic amendments for ammonia and ammonium are given in Table 6.

It was found that the effectiveness of the materials to adsorb ammoniacal nitrogen when mixed with manures was somewhat lower than their potential adsorption capacity. This was probably due to the competition with other cations and gases for the adsorption sites. However, placing the adsorbents into the spent-air stream of an aerated incubation trial increased the effectiveness of the adsorbents compared to mixing them with the manure (Witter and Kirchmann, 1989a).

The reduction in ammonia volatilization, measured in a 1-month incubation study, was 25% for peat and 1.5% for zeolite, if 250 mg adsorbent were mixed with one gram of manure (dry weight/dry weight) (Witter and Kirchmann, 1989a). Addition of powdered basalt increased ammonia volatilization by 29%. It is likely that the increased ammonia loss was due to the pH increase in manure upon addition of the alkaline basalt powder. The application of zeolite placed in a layer on top of sewage sludge-straw composts in a proportion of 3.4 grams per gram of sewage sludge-straw reduced nitrogen losses by 90% (Witter and Lopez-Real, 1988).

## 3. The Effect of pH Regulators

A possibility to conserve ammoniacal nitrogen in manures would be through the addition of acidifying inorganic chemicals, which keep the pH value below or close to 7. Several inorganic compounds have been tested and the addition of superphosphate ($Ca(H_2PO_4)_2$) or phosphoric acid to fresh manure resulted in an efficient reduction of ammonia losses during several days (Burri et al., 1895; Ames and Richmond, 1917; Kaila, 1950; Iversen and Dorph-Petersen, 1952; Safley et al., 1983). However, the situation that the reaction of the fresh manure plus the compound is below pH 7 is not enough, the manure must also remain acid during storage. Rather large quantities are needed for this purpose (Kaila, 1950). Due to high costs and excessively high phosphorus contents in the manure, the conservation method never came into general use.

The effect of calcium chloride additions to manures reducing ammonia volatilization was first reported by Jensen (1928; 1930). The effect of calcium chloride reducing ammonia volatilization after urea application to soils was discovered by Fenn et al. (1981). The principle is that the soluble calcium salt added regulates the pH values in decomposing fresh manures by precipitation of carbonate formed during urea hydrolysis. The reaction mechanism proposed in soils by Fenn et al. (1981) may also be valid for fresh manures containing urea:

$$CO(NH_2)_2 + 3\ H_2O \rightarrow (NH_4)_2CO_3\ .\ H_2O$$

$$(NH_4)_2CO_3 \bullet H_2O + CaX \rightarrow CaCO_3 \downarrow + 2NH_4X + H_2O$$

where X = $Cl^-$ or $NO_3^-$. The alkalinity formed during urea hydrolysis ($HCO_3^-$ and $CO_3^{2-}$) is removed through carbonate precipitation. Thereby the formation of ammonia is reduced as it reduces the $(NH_4)_2CO_3$ concentration, which easily decomposes to $NH_3$, $CO_2$ and water, as follows (Fenn and Kissel, 1973):

$$(NH_4)_2CO_3 \bullet H_2O \leftrightarrow 2\ NH_4OH \rightarrow 2\ NH_3 \uparrow + 2\ H_2O + CO_2 \uparrow$$

Witter and Kirchmann (1989a) tested the addition of calcium and magnesium chloride to fresh poultry manure (0.2-0.4 mmol $g^{-1}$ manure) in a 48-day incubation experiment and found reduced ammonia losses to 85-100% during the first 2-3 weeks and to 23-52% at the end of the experiment. On a weight basis, Ca and Mg salts showed a far greater capacity (about 100 times) for ammonia retention than addition of straw acting as an energy source for nitrogen immobilization (Kirchmann and Witter, 1989) or addition of peat or zeolite as adsorbents of ammoniacal nitrogen (Witter and Kirchmann, 1989b).

### C. Gaseous N Losses from Slurries

#### 1. Anaerobic Storage

Studies on nitrogen losses from slurry stores were started later than from manure stores as slurry did not become a common form of animal wastes before the 1960s. Generally, slurries contain more than half of the total N content as ammonium nitrogen, which means that there is a large potential for ammonia losses during storage.

Ammonia losses from slurry storage tanks have either been estimated by the difference in N content before and after storage or by direct measurements of the $NH_3$ amounts present in air drawn over the storage facilities. Losses reported from slurries during anaerobic storage (see Table 7) varied between 3 and 60% (Wedekind and Kuehn, 1971; Besson et al., 1982; Muck and Steenhuis, 1982; Dewes et al., 1990; de Bode, 1990; Sommer et al., 1992), depending on the storage conditions and the storage time. However, the very high loss values given by Muck and Steenhuis (1982) and Dewes et al. (1990) do not seem to be representative for anaerobic slurry stores.

Dewes et al. (1990) mixed the slurry completely on each sampling occasion, the incubation was made at a relative high temperature (20°C) and no surface crust was formed. Muck and Steenhuis (1982) obtained the high loss values using a model for top loaded stores. Dependable $NH_3$ estimates from non-aerated full scale slurry storage tanks indicate average loss values of 10-15% of the total N content throughout one year.

In the study of de Bode (1990), the seasonal temperature effect on losses was studied. Losses were 2 to 3 times higher during summer than winter. Also the importance of the formation of a surface crust layer for ammonia losses was shown, reducing losses by 0-37% and improvement of the crust layer through straw addition by 63-79%. Sommer et al. (1992) studied ammonia losses from undisturbed, stirred and covered slurry stores using a wind-tunnel system. If a natural surface crust developed on the slurry store, ammonia losses were reduced by 20% compared to those from stirred slurry. Mean loss rates of ammonia were lowest from stores covered by a wooden lid and a floating PVC-foil (0.5 - 1.0 g $NH_3$-N $m^{-2}$ $day^{-1}$) followed by sphagnum peat, rape oil and straw (1.0 - 2.0 g $NH_3$-N $m^{-2}$ $day^{-1}$). However, cracks were noticed in the rape oil layer and along the sides of the peat layer after some time, which increased loss rates compared to an intact cover.

The effect of acidification of the slurry through addition of phosphoric acid with the aim to reduce ammonia losses was tested by Lanyon et al. (1985). The addition was only found to be effective if the amount of phosphorus was larger than 2 grams P per kilogram of slurry.

**Table 7.** Nitrogen losses from liquid manures (slurries) during storage

| Type of slurry and storage conditions | Storage (days) | N loss (%) | Reference |
|---|---|---|---|
| Cattle slurry, anaerobic | 168 | 14.8 | Wedekind and Kuehn, 1971 |
| Pig slurry, anaerobic | 125 | 13.8 | |
| Pig slurry, aerobic (aerated) | 83 | 16.1 - 26.3 | Loynachan et al., 1976 |
| Cattle slurry, anaerobic | 48 | 3.2 | Besson et al., 1982 |
| Cattle slurry, aerobic (aerated) | 43 | 6.9 | |
| Pig slurry, anaerobic | 51 | 4.8 | |
| Pig slurry, aerobic (aerated) | 45 | 5.3 | |
| Cattle slurry, anaerobic, bottom-loaded | 40 | <15.0 | Muck and Steenhuis, 1982 |
| Cattle slurry, anaerobic, top-loaded | 40 | 3.0 - 60.0 | |
| Cattle slurry, anaerobic | 60 | 35.3 | Lanyon et al., 1985 |
| Cattle slurry, aerobic (aerated) | 60 | 49.9 | |
| Pig slurry, aerobic (aerated) | 2[a] | 36.0 - 68.0 | Linke et al., 1987 |

| | | | |
|---|---|---|---|
| Cattle slurry, anaerobic | 120 | 10.3-43.3 | Dewes et al., 1990 |
| Pig slurry, anaerobic | 35 | 4.6-20.5 | |
| | | | |
| Pig slurry, anaerobic | ca. 120 | 15.0 | de Bode, 1990 |
| Poultry slurry, anaerobic | ca. 120 | 3.0 | |
| | | | |
| Cattle slurry, anaerobic | 365 | 12.0 | Sommer et al., 1992 |
| Pig slurry, anaerobic | 365 | 8.0 | |

[a] average reaction time in a constant flow system

2. Aerated Storage

Reported losses of total N during aerobic storage of slurry through aeration varied between 5.3 and 68% (Loynachan et al., 1976; Besson et al., 1982; Lanyon et al., 1985; Linke et al., 1987), see Table 7. The aeration rate affects the nitrogen transformations in slurries and thereby the losses as outlined by Evans et al. (1986). At minimum aeration (dissolved oxygen < 1% of saturation) no nitrate was formed and inorganic nitrogen remained as ammonium. At a high rate of aeration (dissolved oxygen > 15% of saturation) nitrate was formed and the pH fell to about 5.5. At low aeration rates (dissolved oxygen 1-15% of saturation), simultaneous nitrification and denitrification occurred. Thus, nitrogen losses from aerated slurries can take place both through volatilization and denitrification.

## D. Leaching Losses of N from Manures and Slurries

Nitrogen losses through leaching from aerobically decomposing manure heaps were found to be smaller than gaseous N losses amounting to 1 - 14% (Siegel, 1936; Iversen, 1957; Kirchmann, 1985). Lowest losses (1%) were measured from manure heaps to which peat was added as litter material (Kirchmann, 1985). Peat has a high water holding capacity and can retain more water than straw. Protection of manure heaps against precipitation reduced leaching losses. On average 5% of the manure N were leached from rain protected heaps (Glathe and Metzen, 1937; Hesse, 1938; Köhnlein and Vetter, 1953).

The main nitrogen form in the leakage of freshly piled manure heaps was ammonium N followed by organic bound N. During progressive decomposition, organic N was the dominate form in the leakage. The total nitrogen concentration in the leakage was highest immediately after piling a heap and decreased with the time of decomposition. Ammonium N concentrations in the leakage varied between 1.5 and 0.5 mg N $l^{-1}$, nitrate N concentrations were lower amounting to 0.09 and 0.01 mg N $l^{-1}$ (Kirchmann, 1985).

The effect of repeated storage of large amounts of cattle manure on nitrate concentrations in the underlying soil was studied by Chang et al. (1973). They found higher concentrations of nitrate N (50-200 ppm) within 2.0 m depth in the soil profile compared to the untreated soil, but high contents of dissolved organic carbon seemed to stimulate denitrification and greatly reduced the amount of N subjected to leaching.

Liquid manure storage ponds were found to self-seal the soil-manure surface and thereby reduce the process of leaching even on a sandy soil (Miller et al. 1985). The mechanism of sealing was a physical blocking of pores by the manure particles which greatly reduced the infiltration rate (Rowsell et al., 1985). Results of infiltration studies indicated that ponds on clay soils become self-sealed within a few days and ponds lying upon loamy sands and sand within

100 days (Travis et al. 1971; Chang et al., 1974; Davis et al., 1973; Hills, 1976; Rowsell et al., 1985).

# IV. Reactions in Soil

The previous chapters have clearly demonstrated that storage of animal wastes before application to the soil both influences their chemical composition and the amounts of N lost, which affects the recycling efficiency. However, the composition and management practices of organic wastes will also affect their reactions in soil.

## A. Ammonia Volatilization

### 1. $NH_3$ Losses during Application of Animal Wastes

The development of the micrometeorological mass balance method enabled direct measurements to be made of $NH_3$ volatilization during and following the application of wastes to land. The work involved in spreading manures or slurries onto land is of rather short duration, probably only a few seconds. Pain et al. (1989) studied the ammonia losses occurring during this short period of time. The application of 70-160 kg N $ha^{-1}$ as slurry by means of a conventional vacuum tanker fitted with a discharge nozzle and splash plate, resulted in N losses amounting to 0.06-0.14 kg $NH_3$-N $ha^{-1}$. The authors found that the low amount of $NH_3$ lost during application seldom represented more than 1% of that occurring after spreading from slurry lying on the soil surface (Pain et al., 1989). Boxberger and Gronauer (1990) found losses between 4 and 7% when using conventional spreader equipment, but spreading through propelled irrigation resulted in losses of 10%. The ammonia losses occurring during spreading of solid wastes have not been determined.

### 2. $NH_3$ Losses after Application of Organic Wastes

Reported ammonia losses from manure, slurry and sewage sludge applied to soil are compiled in Table 8. Large quantities of ammonia clearly may be volatilized from organic wastes applied to soil. Ammonia loss data vary between a few percentages to over 90% of the $NH_4^+$-N applied and in most studies both low and high losses could be measured. The intensity of the ammonia flux from soil-applied waste material was highest immediately after spreading, the magnitude decreased with time and diurnal fluctuations occurred throughout the experimental periods (Beauchamp et al., 1978, 1982; Gordon et al., 1988; Horlacher and Marschner, 1990; Thompson et al., 1990a; Bless et al., 1991). The cumulative

**Table 8.** Ammonia volatilization from organic wastes applied to soil

| Type of material applied | $NH_4^+$-N applied ($g\ m^{-2}$) | Conditions | $NH_3$ loss (% N applied) | Time (days) | Reference |
|---|---|---|---|---|---|
| **Solid Manure** | | | | | |
| Cattle manure | 4.5 - 35.0 | Field, different climates | 61 - 99 | 5 - 25 | Lauer et al., 1976 |
| Poultry manure | 3.4 - 7.6 | Pasture | 21 - 88 | 6 | Lockyer and Pain, 1989 |
| **Liquid Manure, Urine** | | | | | |
| Pig slurry | 57.4 | Greenhouse, soil | 83 | 8 | Hoff et al., 1981 |
| Cattle slurry | 14.5 - 19.7 | Field | 24 - 30 | 7 | Beauchamp et al., 1982 |
| Cattle slurry | 11.0 | Field | 30 | 17 | Thompson et al., 1987 |
| Cattle slurry | 7.9 | Field | 51 | 7 | Stevens and Logan, 1987 |
| Cattle slurry | 6.5 - 20 | Field, different climates | 18 - 64 | 4 | van der Molen et al., 1989 |
| Pig slurry | 5.7 - 18 | Field | 7 - 62 | 2 - 3 | Pain et al., 1989 |
| Cattle slurry | 4.2 | Field | 39 | | " " |
| Cattle slurry | 8.5 - 9.3 | Grassland | 35 - 38 | 6 | Thompson et al., 1990a |
| Cattle slurry | 2.7 - 9.0 | Field | 12 - 65 | 14 | Horlacher & Marschner, 1990 |
| Cattle urine | 19.2 - 73.8 | Grassland | 4 - 27 | 15 | Lockyer and Whitehead, 1990 |
| **Sewage Sludge** | | | | | |
| Anaerobically digested | 8.9 - 15 | Field | 56 - 60 | 13 - 15 | Beauchamp et al., 1978 |
| Anaerobically digested | 2.9 | Chamber incubation | 8 | 1 | Donovan and Logan, 1983 |
| Anaerobically digested | 2.7 | Chamber incubation | 0.4 | 1 | " " |
| Lime-stabilized | 4.2 | Chamber incubation | 15.8 | 1 | " " |
| Composted | 0.35 | Chamber incubation | 0 | 1 | " " |
| Anaerobically digested | 8.5 | Container incubation | 40.3 | 14 | Adamsen and Sabey, 1987 |

pattern of $NH_3$ volatilization is represented by exponential loss curves (Sommer et al., 1991).

An understanding of the influence of environmental and management factors on $NH_3$ volatilization is important to be able to minimize losses. Consequently it is fundamental to identify the most significant parameters and establish general relationships between these and the loss of ammonia. The effect of wind speed on ammonia losses from slurries during 12 hours after spreading was studied by Thompson et al. (1990b) and Sommer et al. (1991). These authors showed that $NH_3$ loss rates increased when wind speed was increased up to 2.5 m $s^{-1}$. At wind speeds higher than 2.5 m $s^{-1}$ no further increase of loss rates was found. This indicates that a minimum wind speed of 2.5 m $s^{-1}$ is necessary if wind speed as a limiting factor shall be avoided in studies on ammonia volatilization.

A positive correlation between ammonia loss and air temperature was found to be valid in the period immediately following application (Horlacher and Marschner, 1990; Sommer et al., 1991). However, over a longer period of time, Brunke et al. (1988) did not find a high correlation between temperature and ammonia losses and the authors proposed that potential evaporation would be an appropriate prediction tool for ammonia volatilization. Increasing temperature may have a contrary effect on $NH_3$ loss by enhancing resistance to diffusion as slurry dries out and a crust is formed on the slurry surface.

It is well known that there is a greater potential for $NH_3$ volatilization from aqueous solutions at higher pH values. Pain et al. (1990b) showed how increasing pH values in slurries resulted in higher $NH_3^+$ volatilization. Aerobically treated slurries, usually having pH values above 8, were compared with anaerobically stored material. Loss percentages of applied $NH_4^+$-N were significantly higher from the aerobically treated slurries.

Mechanical separation and alteration of the particle size distribution of slurries resulted in somewhat different initial $NH_3$ loss rates but the total $NH_3$ loss after 6 days was not influenced (Thompson et al., 1990a). Practical measures such as incorporation of wastes into soil instead of surface application reduced $NH_3$ emissions drastically (Hoff et al., 1981; Adamsen and Sabey, 1987; Thompson et al., 1987; Bless et al., 1991). Generally less than 10% of the applied slurry $NH^4$-N was volatilized after injection into the soil.

The effect of soil moisture content on $NH_3$ losses from urine applied to soil was studied by Stewart (1970), showing that 90% of the applied urine-N was volatilized when the soil was dry and 25% when the soil was moistened. The extent of ammonia volatilization from urine was found to be lower than from slurry (Lockyer and Whitehead, 1990). The greater volatilization from slurry than from urine is probably due to two factors. Firstly, with slurry the urea component will have been converted to ammonium before application, and secondly, the solid components of the slurry will reduce the infiltration of the liquid into the soil and hence increase the susceptibility of the ammonium-N to volatilization.

Whitehead et al. (1989) found higher ammonia volatilization from cattle urine than from urea solutions after application to soil. The authors showed that hippuric acid, present in concentrations of 0.05-0.5 mg hippuric acid-N $l^{-1}$ in urine, increased the rate of urea hydrolysis in urine compared to urea solutions. The same observation was made by Doak (1952) also showing that hippuric acid increased the rate of hydrolysis of urea, but none of the other nitrogenous constituents of urine.

Cow dung left on the soil from grazing animals was found to lose less than 5% of the total nitrogen through ammonia volatilization. Dung beetles mixing dung with soil further reduced $NH_3$-N losses as the pH value decreased due to mixing (Yokoyama et al., 1991).

## 3. The Effect of Additives

Effective strategies for lowering ammonia volatilization from surface-applied slurries include acidification (Stevens et al., 1989; Frost et al., 1990; Pain et al., 1990a), dilution (Sommer and Olesen, 1991) and addition of $CaCl_2$ to slurries and manures containing urea or uric acid (Witter, 1991).

Acidification of the slurry to a pH value of 5.5 through addition of sulfuric acid decreased ammonia volatilization by 14-95% (Stevens et al., 1989; Frost et al., 1990; Pain et al., 1990a). However, the cost of acid and liming material needs to be balanced against the values of increased N efficiency, and at present the cost of acid is several times higher than the value of N saved. In addition to the economic disadvantage, $H_2S$ gas evolved during acidification, which constituted a serious health hazard (Frost et al., 1990).

Treatment of manure with monocalcium phosphate before spreading reduced ammonia losses from field-applied material by 17% (Gordon et al., 1988). Monocalcium phosphate produces dicalcium phosphate and phosphoric acid (Lindsay and Stephenson, 1959) and thereby reduces the pH value of the amended manure.

The influence of different dry matter contents in slurries, i.e., the effect of dilution with water, was studied by Sommer and Olesen (1991) showing that at high (>12%) and low (<4%) contents of dry matter, small changes in dry matter content had a limited influence on $NH_3$ loss. However, at dry matter contents between 4 and 12%, $NH_3$ losses increased significantly, being 2-3 times higher at dry matter contents above 10% than at dry matter contents of 2-4%. Investigations with slurries having dry matter contents ranging from 4.6 to 18.7% resulted in significantly higher losses using material with increasing dry matter content (Pain et al., 1989).

Addition of $CaCl_2$ to fresh and anaerobically stored chicken slurries applied to soils showed that $NH_3$ losses only decreased if fresh manures still containing urea or uric acid were used (Witter, 1991). The addition of $CaCl_2$ decreased the amount of $CO_2$ released through precipitation of $HCO^-_3$ as $CaCO_3$, thereby removing a source of alkalinity. The failure of the $CaCl_2$ addition to decrease ammonia losses from the anaerobic slurry, suggested that $HCO^-_3$ was an important source of alkalinity driving ammonia volatilization in the fresh slurry but not in the anaerobic slurry.

## B. Denitrification

Denitrification also causes losses of manure nitrogen and may be a potential atmospheric pollutant. One of the product gases, nitrous oxide, may cause global warming and an ozone depletion. There has been a re-evaluation of the release of $N_2O$ from biological processes compared to fossil fuel combustion. Biologically produced $N_2O$ may be as high as 95% of all emissions (Smith and Arah, 1990). Earlier estimates for $N_2O$ production from fossil fuels may be based on a sampling artifact, the $N_2O$ measured having been formed from NO after sampling by a reaction of the sampling containers involving $SO_2$ and water (Linak et al., 1990).

Studies by Burford (1976), Guenzi et al. (1978), Christensen (1983; 1985), Rice et al. (1988) and Paul and Beauchamp (1989a) have shown that manure, slurry and organic waste additions to soil increased denitrification (see Figure 1) and caused oxygen deficiency in microsites. These results are in agreement with the observation that the higher the readily decomposable organic matter in soil, the higher is the denitrification potential (Burford and Bremner, 1975). Denitrification in manure-amended soils was closely related to both water-soluble carbon and volatile fatty acid concentrations in the manures (Paul and Beauchamp, 1989b). However, addition of cow urine to soil did not affect denitrification in an experiment by Limmer and Steele (1983). Since Michaelis-Menten kinetics apply to carbon and nitrogen substrates during denitrification, the authors suggested that the denitrifying enzymes in the experimental soil were fully induced and any further additions of carbon and nitrogen supply would not increase denitrification rates.

Denitrification losses from surface-applied slurry were found to be lower than losses through ammonia volatilization and could amount to a maximum of approximately 30% of the $NH_4^+$-N of autumn/winter spread slurries (Thompson et al., 1987; van den Abbeel et al., 1989). Maag (1989) reported denitrification losses during a growth period to be as high as 50% of the applied $NH_4^+$-N in slurry. Studies by Thompson et al. (1987) showed that the injection of slurry significantly increased denitrification losses compared to surface application. In their study, denitrification losses amounted to 54% of the applied slurry $NH_4^+$-N during the non-cropped period despite low soil temperature throughout the winter and early spring. Thompson (1989) suggested that an increased population of denitrifying bacteria is likely to be the major factor contributing to the large increase in denitrification observed at low temperatures in field studies after slurry application; increases in activity appear to be relatively minor.

The predominant gas evolved during denitrification in soil after application of organic fertilizers was $N_2$, being at least 6 times greater than the $N_2O$

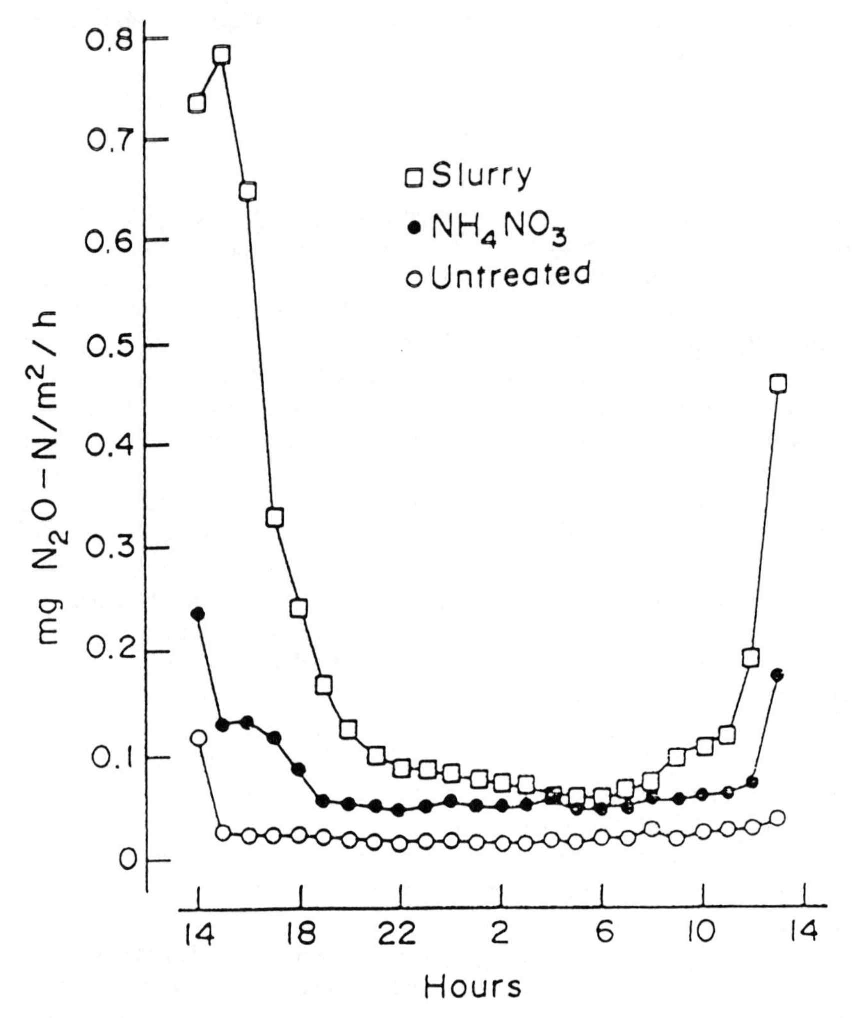


**Figure 1.** Flows of $N_2O$ from untreated soil and after application of slurry and ammonium nitrate to soil. (From Christensen, 1983.)

emissions (Rolston et al., 1978; Mosier et al., 1982). Mosier et al. (1982) concluded that although $N_2O$ emissions from agricultural soils are enhanced by the application of organic and inorganic fertilizers, only a very small fraction of the N amendments contributes to increase the $N_2O$ concentration of the troposphere.

## C. N Mineralization and Immobilization

Studies on the availability of nitrogen in organic materials has been a central research task for agronomists for several decades. The most commonly used technique today is the determination of the net N release of inorganic nitrogen from organic materials in laboratory incubation experiments after addition to soil. In this way, the potential short-term or annual release of N from organic materials can be estimated under constant temperature and moisture conditions presuming that gaseous losses through volatilization and denitrification are low and negligible. Incubation studies have shown that the net N release from organic materials can follow different patterns during the first year after the application to soil:

a) Steady, constant mineralization over time (zero-order kinetics) (Germon et al., 1979; Flowers and Arnold, 1983)
b) Steady, constant immobilization over time (zero-order kinetics) (Hébert et al., 1991; Kirchmann, 1991)
c) Mineralization that decreases over time (first-order kinetics) (Reddy et al., 1980; Lindemann and Cardenas, 1984; Chae and Tabatabai, 1986; Chescheir et al., 1986; Serna and Pomares, 1992)
d) Rapid initial mineralization followed by a slower steady mineralization (parallel first-order kinetics) (Gale and Gilmour, 1986; Boyle and Paul, 1989; Fine et al., 1989)
e) Rapid initial N immobilization followed by a steady, linear N mineralization (first/zero-order kinetics) (Kirchmann, 1991; Kirchmann and Lundvall, 1992).

The most commonly used kinetic for the description of N mineralization from organic materials rich in N is the first-order reaction. Reported kinetic parameters are in some cases calculated so that they express the decomposability of the material added, assuming no priming effect, in other cases they reflect the decomposability of the mixture, i.e., the soil amended with organic material.

### 1. Animal Manures

Manures are a variable mixture of three components, feces, urine and litter. The rate of urine separation and amount of bedding material used governs the proportions of the different components. If the urine fraction is small, fresh manures mainly consist of organically bound nitrogen. Manures can undergo an aerobic or anaerobic decomposition or remain undecomposed. Taking into account the three main species of farm animals, cows or cattle, pigs and poultry, a total of nine types of manures, three fresh-undecomposed, three aerobically and three anaerobically treated materials can be distinguished. The N mineralization of the nine types of manures not containing any litter or urine is shown in Figure 2. Despite the different types of animal manures, some general considerations can be made. Concerning fresh manures, only undecomposed

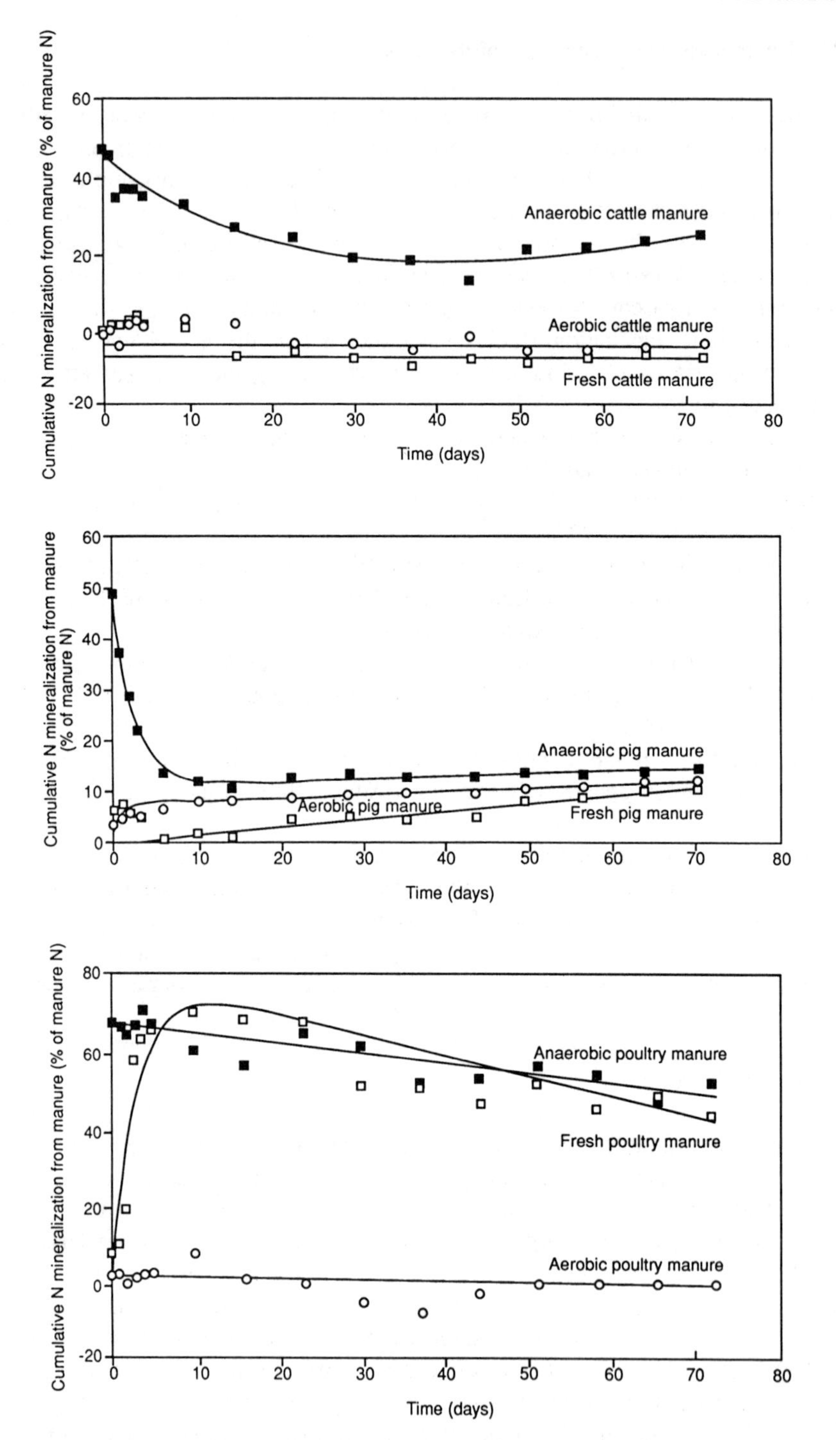


**Figure 2.** Net N mineralization/immobilization from fresh, aerobic, and anaerobic animal manures in soil. (From Kirchmann, 1991.)

poultry manure, which has been dried after excretion, may be used in agriculture. Fresh pig and cow manures are probably rarely applied to soil, disregarding grazing conditions. Manures are predominantly anaerobically decomposed during storage. Composting of manures requires a certain steering of the decomposition process to obtain aerobic conditions and aerobic manures are therefore less frequently used.

Fertilization with fresh poultry manure resulted in a rapid increase in inorganic N in soil within 2 weeks after the addition, amounting to more than 50% of the total N present (Castellanos and Pratt, 1981; Hadas et al., 1983; Gale and Gilmour, 1986; van Fassen and van Dijk, 1987; Bitzer and Sims, 1988; Kirchmann, 1991). As mentioned in Table 2, total nitrogen in fresh poultry excreta consists to more than 50% of uric acid. The fast decomposition of uric acid via urea results in the high release of N from fresh poultry excreta. In soil, the enzyme uricase (urate oxidase) and aerobic microorganisms decompose uric acid and urates (Bachrach, 1957; Durand, 1961; Martin-Smith, 1963). Thus, the fertilizer effect of this type of manure is closest to that of inorganic N. N mineralization from fresh beef and swine manure was found to be much lower than from fresh poultry manure, being lowest for fresh beef manure (Reddy et al., 1980; Kirchmann, 1991).

Composted manures have low contents of inorganic nitrogen, as shown in Table 2. Studies with composted animal manures applied to soil showed that mineralization of manure N was very low and in some experiments immobilization of N instead of mineralization was measured. Net N mineralization varied between -3.0% and 15% of total manure N applied during a period of several months and the N fertilizer value of composted manures was found to be low independent of the animal origin (Laura and Idnani, 1972; Castellanos and Pratt, 1981; Kirchmann, 1989; Hébert et al., 1991; Kirchmann, 1991). The N mineralization course of composted cattle, pig and poultry manures mainly followed zero-order kinetics (Castellanos and Pratt, 1981; Hébert et al., 1991; Kirchmann, 1991).

Use of anaerobically treated cow, pig and poultry manure showed a partial immobilization of N in soil, being highest for pig manure followed by cow manure and lowest for poultry manure (Kirchmann, 1991). Probable reasons are the high C to $N_{org}$ ratio of the organic matter of anaerobically manures compared to fresh and aerobic manures and the presence of energy-rich and easily decomposable forms of carbon such as volatile fatty acids. In spite of the immobilization, inorganic N contents were higher in soil after addition of anaerobic manures than aerobic or undecomposed material (the exception being fresh poultry manure). Therefore, anaerobic manures have the largest N effect compared to manures treated in another way.

Nitrogen interactions in soil of fresh and composted $^{15}N$ labelled poultry manure were found to be different depending on the amount of energy available in the manures. Addition of composted manure caused a larger crop uptake of unlabelled N, whereas fresh manure caused a smaller crop uptake of unlabelled N compared to the control (Kirchmann, 1990).

**Table 9.** Recovery of $^{15}N$ labelled manure nitrogen in grains and straw in a 3-year pot experiment with differently treated poultry manure

| | % harvested $^{15}N$ labelled manure N | | |
|---|---|---|---|
| Treatment | Year 1 | Year 2 | Year 3 |
| $(NH_4)_2SO_4$ | 53 | 2 | 1 |
| Aerobic manure | | | |
| C/N 10 | 7 | 5 | 2 |
| C/N 13 | 6 | 6 | 2 |
| C/N 18 | 3 | 7 | 1 |
| Anaerobic manure | | | |
| C/N 13 | 17 | 5 | 2 |
| C/N 22 | 14 | 6 | 2 |
| C/N 32 | 9 | 6 | 3 |
| Undecomposed manure | | | |
| C/N 08 | 28 | 6 | 2 |
| C/N 10 | 31 | 7 | 2 |
| C/N 15 | 23 | 8 | 2 |
| C/N 18 | 20 | 10 | 2 |
| C/N 20 | 14 | 10 | 2 |
| C/N 24 | 12 | 8 | 2 |
| C/N 30 | 6 | 11 | 2 |
| C/N 36 | 5 | 12 | 2 |

from Kirchmann, 1989.

In a 3-year N balance study with $^{15}N$-labelled manures, the long-term effect of differently treated poultry manures was measured in a pot experiment (Table 9; Kirchmann, 1989). In the third cropping year no differences in the crop uptake of manure derived $^{15}N$ was found and similar residual amounts of manure N were found in soil amounting to ca. 60%. The author concluded that the type of manure treatment before addition to the soil did not affect manure N mineralization longer than two years. Results from an early field experiment with fresh $N^{15}$-labelled cattle manure by Rauhe and Bornak (1970) also showed that more than 50% of the manure N remained in soil after two years.

## 2. Animal Slurries and Urine

Slurries, usually stored under anaerobic conditions, are mixtures of the total urine and total feces excreted, with or without addition of litter. Due to the presence of urine, the ammonium concentrations in slurries usually amount to

50% or more of the total N. Thus, soluble, inorganic N often is the major form of N in slurries and early studies by Heck (1931) showed that the amount of slurry nitrogen taken up by the crop did not exceed the amount of slurry $NH_4^+-N$ applied.

Addition of anaerobically treated pig slurry to soil resulted in a partial immobilization of up to 40% of the applied $NH_4^+$-N (Flowers and Arnold, 1983). The authors also found that the soil temperature did not affect the amount of N immobilized but soil temperature influenced the length of the immobilization period. Initial immobilization of $NH_4^+$-N from pig and cattle slurry was also observed in the incubation study by Chescheir et al. (1986) and Opperman et al. (1989). Use of $(^{15}NH_4)_2SO_4$ added to cattle slurry confirmed that the decrease in $^{15}NH_4^+$-N in soil could not be attributed to volatilization or denitrification but due to immobilization (Herbst et al., 1987). Kirchmann and Lundvall (1992) found that fatty acids present in slurries decomposed within 1-2 days at 25°C in soil. Parallel to the fatty acid decomposition, immobilization of N was measured in soil. The correlation between the initial fatty acid concentrations in the slurries and the amounts of N immobilized were found to be highly significant ($R^2$= 0.97).

Germon et al. (1979), however, found no initial immobilization of $NH_4^+$-N after the application of pig slurry to soil nor was there any additional mineralization that could be attributed to slurry organic N during 8 weeks of incubation. However, in this study any organic material larger than 5 mm was removed from the slurry by sieving. Fractionation studies of cattle slurry showed that the solid fraction had a C/N ratio of over 40 (Diaz-Fierros et al., 1987; 1988). An incubation study with the solid cattle slurry fraction clearly induced immobilization of N in soil over 130 days, whereas the liquid fraction induced a mineralization and then stabilized nitrogen levels (Diaz-Fierros et al., 1988).

Fresh animal urine-N applied to soil decomposed in the following way: a rapid hydrolysis of the urea present in urine increased the content of ammonium distinctly in soil profile within one day (Thomas et al., 1986, 1988; Spatz et al., 1992). High nitrate concentrations were found even in deeper soil layers after one week (Spatz et al., 1992).

The long-term N fertilizer value of slurries has not been evaluated as thoroughly as their initial N availability. Experiments with $^{15}N$ labelled slurry enabled Rauhe et al. (1973) and Peschke (1982) to determine the remaining amounts of slurry N in soil. On average, residual amounts of 40 to 50% of the slurry N were found in soil after two years.

### 3. Sewage Sludge

More N mineralization studies with sewage sludge applied to soil have been reported than with other organic waste materials. Generally, a net mineralization of N has been observed in sludge-amended soil (Ryan et al., 1973; Epstein et al., 1978; Magdoff and Amadon, 1980; Parker and Sommers, 1983; Lindemann and Cardenas, 1984; Chae and Tabatabai, 1986; Garau et al., 1986; Fine et al., 1989; Serna and Pomares, 1992). However, initial immobilization of N in soil has been reported for raw untreated sewage sludges (Parker and Sommers, 1983).

Comparative studies of activated, anaerobically digested and composted sludge showed that composted sludge released less nitrogen than the other sludge types, amounting to a maximum of 15% of the organic N applied (Epstein et al., 1978; Parkers and Sommers, 1983; Tester and Parr, 1983) followed by activated and anaerobically digested sludge (Garau et al., 1986; Serna and Pomares, 1992). Thus, pretreatment of the sludge influenced N mineralization, indicating less N mineralization for stabilized materials.

The proportion of N mineralized from the organic fraction of anaerobically digested sludge in soil ranges from a few percent to ca. 50%. Ryan et al. (1973) reported mineralization rates of 4-48% within 16 weeks, Epstein et al. (1978) found corresponding rates of 40-42% within 15 weeks, Lindemann and Cardenas (1984) 19-37% to be released after 32 weeks, Chae and Tabatabai (1986) found 15-23% within 26 weeks, and Serna and Pomares (1992) found variations between 17 and 46% during 16 weeks. Reported values of organic N mineralization from aerobically digested and activated sludges were found to be somewhat higher than for anaerobically treated sludge. Magdoff and Amadon (1980) found 54% to be taken up by crops during one year, Garau et al. (1986) measured mineralization rates of 23-80%, Fine et al. (1989) reported values of 41-61% within 13 weeks and Serna and Pomares (1992) values of 24-40% within 16 weeks. Previous sludge amendment had no effect on the mineralization of fresh sludge (Lindemann et al., 1988).

Increased application rates of sewage sludge to soil resulted in decreased N mineralization percentages of sludge N (Ryan et al., 1973; Garau et al., 1986). However, the inverse relationship between mineralized sludge N and sludge application rate might not have a mechanistic explanation (Fine et al., 1989). Gaseous losses of mineralized N through volatilization and denitrification may increase at higher application rates and determinations of inorganic N in soil will not be a direct measure of ammonification of the organic N added. Thus, possible gaseous losses introduce uncertainty into the interpretation.

The soil type seems to have a marked effect on N mineralization from sewage sludge (Barbarika et al., 1985; Chae and Tabatabai, 1986). A positive correlation between increased total N content of the soil and the extent of N mineralization from sewage sludge was found by Barbarika et al. (1985). However, the soil types induced significantly different gaseous losses after sludge application (Lindemann and Cardenas, 1984; Fine et al., 1989). As mentioned above, measurements of inorganic N in soil do not describe ammonification if gaseous losses occur in parallel. Thus, the large variability in N mineralization from sludge measured in different soils may not be a true effect.

The long-term dynamics of N release in sludge-amended soil was studied by Stark and Clapp (1980), Harding et al. (1985) and Boyle and Paul (1989). The authors found throughout a greater N mineralization potential in sludge-amended soils up to 5 years after the last sludge addition than in the control soils. Further, the long-term mineralization potential increased with the rate of earlier sludge applications (Boyle and Paul, 1989).

### D. Conceptual Approaches and Models

Numerous studies deal with the determination of N mineralization from organic materials in soil and it is well known that results gained from experiments with one specific material are not applicable for the prediction of the N availability of other organic material of the same kind. Potential N mineralization rates and rate constants derived from kinetic functions are specific for the organic material used and the conditions applied only. Environmental factors such as actual temperature and moisture conditions can be incorporated as variables in the first-order model enabling prediction of N mineralization of the specific substrate in the same experimental soil under field conditions (Hsieh et al., 1981a,b). Still, the approach is limited to a specific substrate and a certain soil used.

Sluijsmans and Kolenbrander (1976) distinguished three nitrogen fractions in manures and slurries: inorganic nitrogen mainly present as ammoniacal N, easily decomposable organic N and resistant organic N. The concept of dividing total N of organic wastes into three N pools has been adopted by Bhat et al. (1980) in a simulation model and suggestions to improve the approach have been discussed by Beauchamp and Paul (1989). Chemical attempts to define a labile and a resistant organic N pool have not been successful (Chescheir et al., 1986).

A mechanistic approach involving carbon decomposition and microorganisms for determination of the amounts of N released from organic materials has been described (Parnas, 1975; Knapp et al., 1983; Molina et al., 1983; van Veen et al., 1984). In these models the availability of organic C mainly controls the growth of soil biomass, which in turn controls immobilization and mineralization of N. Bjarnason (1988) has given a simple mechanistic description of the short-term N mineralization/immobilization of organic materials during decomposition.

The actual decomposition ($C_d$) of a substrate is the sum of $CO_2$-C evolved ($C_e$) plus the assimilated amount by microorganisms ($C_a$).

$$C_d = C_e + C_a$$

The relationship between actual carbon decomposition ($C_d$) and assimilated carbon ($C_a$) can be expressed by an efficiency term ($f_a$) describing the ratio of carbon assimilated to carbon decomposed, which has been found to vary between 0.4 and 0.6.

$$C_a = f_a C_d$$

Rearranging the equations given above results in the following expressions for $C_a$ and $C_d$.

$$C_a = C_e /(1/f_a - 1)$$

$$C_d = C_e /(1 - f_a)$$

If the C/N ratio of the organic material added is known, $CN_{OM}$, and the C/N ratio of the microbial biomass is assumed to be constant, $CN_B$, gross N mineralization, $N_{min}$, and gross N immobilization, $N_{im}$, can be calculated:

$$N_{min} = C_d/CN_{OM}$$

$$N_{im} = C_a/CN_B$$

The net change in amounts of inorganic N mineralized or immobilized, dN, depends on the difference between the two gross processes: the amount of N in the decomposed organic material ($C_d/CN_{OM}$) and the amount of N needed to keep the C/N ratio of the biomass unchanged ($C_a/CN_B$).

$$dN = C_d/CN_{OM} - C_a/CN_B$$

The equations consider only one pool of decomposable carbon. A number of pools, differing in stability and C/N ratio, are probably needed for a proper description. However, no fractionation technique to separate organic materials into biologically meaningful components has been found. The use of rate constants derived from kinetic C mineralization data, characterizing a very rapidly mineralizable, an intermediate and a slowly mineralizable fraction was found to be successful in a model of the same conceptual idea as described above (Gilmour et al., 1985). Thus, measurements of $CO_2$-C evolution kinetics seem to be necessary to be able to model N mineralization from organic materials.

## E. Leaching of N

The movement of nitrogen from the unsaturated into the saturated zone, i.e., from the surface soil and the root zone into the groundwater is called leaching of N. Leaching of nitrogen takes place mainly in the form of nitrate, whereas leaching of ammonium is mostly prevented by adsorption and nitrification. Significant leaching of organic N may take place under conditions of heavy fertilization with organic manures (Oosterom, 1984). Principally leaching takes place when rainfall exceeds the sum of water loss by evapotranspiration. In northwest Europe this is usually the case during autumn, winter and early spring, i.e. when temperature and crop growth are low. The water surplus during this period ranges from 0-500 mm year$^{-1}$.

At the time of application of manure, slurry or urine to soil, N is either present mainly in organic forms or distributed between ammonium and organic forms depending on the storage conditions, but no nitrate is added. That means, in the short term i.e. one to two weeks after the application, no losses of manure N via nitrate can occur and ammonia volatilization is the main loss pathway for N. When nitrate is produced both leaching and denitrifcation may occur. There is an inter-dependence of leaching and denitrification in the removal of nitrate from the soil profile. Garwood and Ryden (1986) found high leaching losses in drained plots from which low losses of denitrification were measured and vice versa. Sherwood (1980; 1986) found low leaching losses of N and high denitrification losses from cattle slurry in an impermeable clay and a moderately drained loam.

## 1. The Influence of the Time of Application

Nitrogen in animal wastes applied to arable fields during autumn and winter is much more liable to leaching than when applied in spring and during the growing season (Duthion, 1981; Vetter and Steffens, 1981; Brink and Jernelås, 1982; Herrmann et al., 1983; Torstensson et al., 1992). The amount of N leached usually decreases as the date of application of the slurry comes closer to the date of crop growth. However, in a Dutch study no different leaching of slurry N was found when applied in November and February, respectively due to low soil temperatures retarding nitrification (Steenvoorden, 1986).

In trials on arable land in north west Germany, leaching of N from 180 kg total N ha$^{-1}$ applied as pig slurry amounted to 140 and 100 kg N ha$^{-1}$ when given in autumn and spring, respectively (Vetter and Steffens, 1981). In similar trials under Swedish conditions, leaching losses on a sandy soil amounting to 239 kg N ha$^{-1}$ were found when 375 kg total N was added as slurry in August and 148 kg ha$^{-1}$ when applied in November. Spring application resulted in leaching losses of 91-93 kg N ha$^{-1}$. Torstensson et al. (1992) found 68 kg N ha$^{-1}$ being leached from 110 kg of total N applied as slurry in autumn and 46 kg N ha$^{-1}$ being leached when slurry was applied in spring. Analyses of inorganic N in soil showed that very little of the inorganic N applied through slurries in autumn was left in soil at springtime before sowing.

## 2. The Influence of the Cropping System

Generally nitrate leaching from animal wastes has been found to be higher when applied to arable land than to ungrazed, cut grasslands (Vetter and Steffens, 1981; Smith and Unwin, 1983). Cow slurry containing 200 kg total N ha$^{-1}$ was applied annually to ryegrass swards during the growing season for a period of

**Table 10.** Manure N balance sheet over 3 years showing cumulative crop uptake, gaseous losses, and remaining amounts of $^{15}N$ labelled poultry manure in soil

| | % of labelled manure N | | |
|---|---|---|---|
| Treatment | Cumulative crop uptake | Cumulative gaseous losses | Remaining in soil |
| Aerobic manure | 13.3 | 27.8 | 58.9 |
| Anaerobic manure | 21.1 | 20.6 | 58.3 |
| Fresh manure | 28.3 | 6.5 | 65.2 |
| Ammonium sulfate | 56.0 | 26.0 | 18.0 |

from Kirchmann, 1989

five years (Smith and Unwin, 1983). Only 10 kg N $ha^{-1}$ was lost each year in the leachates. When four applications of slurry were made during each season, supplying 800 kg N $ha^{-1}$, leaching losses increased to between 15 and 30 kg N $ha^{-1}$ $yr^{-1}$. However, from grazed grassland including the effect of ruminants, nitrate leached was 5.6 times greater than that leached below a comparable cut grassland amounting to 162 and 29 kg N $ha^{-1}$ $yr^{-1}$, respectively (Ryden et al., 1984). Thus the losses from the grazed grassland exceeded leaching losses normally observed from arable land. The enhanced nitrate movement below the grazed grassland can be attributed to the return in urine and dung of as much as 80-90% of the N herbage consumed by the ruminants. Urine patches, which contain most of the excreted N, form areas containing very high N concentrations (57-698 kg N $ha^{-1}$) (Richards and Wolton, 1976). Similar results were obtained by Steenvoorden et al. (1986) showing an increase in nitrate leaching of 30-60 kg N $ha^{-1}$ $yr^{-1}$ below grazed grassland compared to cut grassland.

## F. N Balance Studies

Several manure N balances in field experiments based on amounts of manure N applied, changes in soil organic N and crop removal data compared to non-manure treatments have been presented (Mattingly, 1974; Johnston, 1975; Pratt et al., 1976; Jenkinson and Johnston, 1977; Jenkinson, 1977). The duration of the experiments varied from decades to some years. Principally, the same proportional distribution of manure N in the soil-plant system was found in the experiments cited. Of the manure N applied, 15-30% was taken up by the crop, 15-45% remained in the soil, and 40-60% was lost from the soil. A 3-year pot experiment with $^{15}N$-labelled poultry manures showed that the cumulative crop uptake varied between 13 and 21% depending on the treatment of the manure, amounts remaining in soil being 58 and 65%. The uptake of labelled N by crops was lower than manure N losses, see Table 10 (Kirchmann, 1989). Treatment of soil with $(^{15}NH_4)_2SO_4$ in the same experiment showed that 56% of the labelled

N was taken up by crops, 18% remained in soil and 26% were lost after three years. Other balance studies with $^{15}N$-labelled fertilizer also showed that the efficiency of inorganic fertilizers is much higher than from organic manure ranging from 55 to 75% and losses between 10 and 28% per year (Jansson, 1963; Powlson et al., 1986).

Considering that the residual manure N in soil after 3 years amounted to 58-65% (Kirchmann, 1989), one may think that the efficiency of N in organic manures could be equivalent or even higher than from inorganic fertilizers, only the soil is cropped for a sufficient period of time. Assuming that the uptake efficiency of residual manure N and residual fertilizer N by crops would be 100% during the following years or decades, then the size of the initial losses would be a direct measure for the N efficiency of the fertilizer and of the manure. In such case, manure N would be as efficient or even more efficient than fertilizer N. However, mineralized residual manure N is exposed to gaseous and leaching losses as nitrogen from other pools in soil. Thus, a certain proportion of the residual manure N will be lost during the following years, which means that the N efficiency of manures cannot be higher. Results from long-term field experiments with manure cited above confirm that the average efficiency of manure N is lower than from inorganic fertilizers.

## V. Environmental Aspects of Organic Waste Application to Soils

### A. Gaseous N Losses from Organic Wastes Compared to N Fertilizers

The foregoing results illustrate that large gaseous losses of N from animal wastes can occur after the actual application to soil. Estimates on ammonia volatilization from surface applied fertilizers (not urea) showed that throughout less than 10% of the fertilizer N was volatilized if soil pH values are below 7 (Whitehead and Raistrick, 1990). This shows clearly that the average proportions of $NH_4^+$-N lost from surface-applied slurries or manures through volatilization are higher than ammonia losses from N fertilizers, see Table 8.

Monitoring changes in pH after application of organic manures to soil showed that there was a significant increase in pH of up to one unit within a few days after application to soil (Opperman et al., 1989; Paul and Beauchamp, 1989a; Sommer et al., 1991; Witter, 1991). The rapid increase in soil pH, which lasts a short period of time, shifts the $NH_4^+/NH_3$ equilibrium and is the main reason for higher ammonia losses from organic manures than inorganic ammonium fertilizers. Probably several reasons acting together cause the pH increase in soil after the application of organic fertilizers.

a) Anaerobically stored slurries and manures contain volatile fatty acids. The exposure of anaerobically treated organic materials to oxygen after surface application in the field causes rapid oxidation and disappearance of volatile fatty acids (Paul and Beauchamp, 1989a; Kirchmann and Lundvall, 1992). The disappearance of fatty acids shifts the $NH_4^+/NH_3$ equilibrium in favor of $NH_3$ (Paul and Beauchamp, 1989c).

b) The $CO_2$ dissolved in organic materials under anaerobic decomposition conditions is much larger than the partial $CO_2$ pressure in the atmosphere. When exposed to the atmosphere, dissolved $CO_2$ is released and as a result the pH increases (English et al., 1980).

In addition to higher ammonia losses from organic fertilization, increased denitrification was measured in soil after application of organic fertilizers compared to inorganic fertilizers. Two complementary explanations have been presented in the literature.

a) Decomposition of organic matter with readily available C leads to a high respiratory $O_2$ consumption, which may cause anaerobic zones (Rice et al.,1988). Locally, anaerobic environments are created through addition of organic materials (Stevens and Cornforth, 1974; Burford, 1976), which enhance denitrifcation.

b) Manure or slurry additions to soil provide denitrifying bacteria with water-soluble or readily mineralizable carbon sources stimulating their activity (Burford and Bremner, 1975; Paul and Beauchamp, 1989a,b). Volatile fatty acids present in animal wastes have been identified as a carbon source for denitrifying bacteria (Paul et al., 1989).

Thus, comparing inorganic fertilizers based on ammonium or nitrate ions with organic fertilizers, it can be concluded that the proportion of gaseous N losses from organic fertilizers applied to soil is throughout higher from than inorganic fertilizers. This general conclusion should be taken into consideration by the organic farming movement.

### B. Leaching Losses of N from Organic Wastes Compared to N Fertilizers

Several studies comparing leaching of N from organic manures with inorganic fertilizers have been performed (Duthion, 1981; Jarvis et al., 1987; Wermke, 1987; Powlson et al., 1989). In experiments lasting 1-3 years, nitrogen leaching from animal wastes, applied at the same total N equivalent as fertilizer N, was found to be lower than from the mineral fertilizer both under grassland and arable land (Duthion, 1981; Jarvis et al., 1987; and Wermke, 1987). However, in long-term experiments at Rothamsted fertilized with 238 kg N $ha^{-1}$ $yr^{-1}$ in form of organic manure, leaching was 124 kg $NO_3$-N $ha^{-1}$ $yr^{-1}$. Leaching from the same soil, continuously receiving 144 kg inorganic N $ha^{-1}$ $yr^{-1}$, was 25 kg $ha^{-1}$ $yr^{-1}$ (Powlson et al., 1989).

The results may be explained as follows. In the short-term experiments there was a smaller amount of inorganic N present in soil when applying animal wastes at the same total N equivalents as inorganic fertilizers. Only a part of the total N applied through animal wastes was actually present as inorganic N, the other part was present as organic bound, which was not completely mineralized within one year. Some ammonia losses are likely to occur from the organic wastes after application to soil, which further reduced the content of inorganic N in soil compared to the fertilizer treatment. Thus, lower leaching losses of N from organic manures in short-term experiments compared to inorganic fertilizers were simply due to the fact that there was less inorganic N available for leaching, when applied at the same rates of total N.

The situation is different in long-term experiments. Continuous inputs of organic manure have increased the organic matter pool in soil and thereby the amount of nitrogen that can be mineralized. Substantially larger amounts of nitrate were formed in continually manured soil even at times when crop uptake was small (Powlson et al., 1989). Thus, the reason for increased nitrate leaching from soils manured over long periods of time with animal wastes compared to fertilized soils, is the increased pool of organic matter that can be mineralized at times when the demand by the crop is low. This illustrates some of the principles involved in nitrate leaching. The production of nitrate through mineralization of soil organic matter is not necessarily synchronized with periods of crop uptake. The lack of synchronization between formation of nitrate and uptake of nitrate by crops causes large leaching losses. That means the problem is, in part, one of timing.

### C. Surface Runoff

Surface runoff or overland flow occurs at sloping sites when the infiltration rate of water into soil is lower than the rate of precipitation. Precipitation which is not absorbed by the soil is transported over the soil surface. Surface runoff is especially high for soils with a high groundwater table or badly drained and impermeable soils. Surface runoff may occur when the rainfall intensity is at least 0.5 mm $h^{-1}$ and the amount of rain is 20 mm (Sherwood and Fanning, 1981), which is usually the case during autumn and winter. Substantial amounts of animal wastes are spread at this time.

Concerning runoff from manured soils, the following principal results have been gained. The closer the rainfall event was to the date of application, the higher nutrient concentrations were measured in the runoff water (Ross et al., 1979; Sherwood and Fanning, 1981; Steenvoorden et al., 1986). The N concentrations in the runoff water decreased at each consecutive runoff event (Jarvis et al., 1987).

Several studies reported high losses of N through runoff amounting to 25-50% of the manure N applied (Hensler et al., 1970; Young and Mutchler, 1976; Uhlen, 1978; Steenhuis et al., 1981). These runoff events resulted from snowmelt or heavy rainfalls within one or several days after spreading on frozen or poorly drained soil. However, when manure was applied to frozen soil and then covered with snow, melting at a later date, low nutrient losses through runoff were measured (Klausner et al., 1976). Low losses of N from surface-applied manures amounting to less than 5% of the added manure N were found when the intervals between spreading and first runoff were longer than one week (Long, 1979; Sherwood and Fanning, 1981; Brink et al., 1983; Steenvoorden et al., 1986). Beside the interval between application and runoff event, the rate of N applied through wastes had a substantial effect on the nitrogen concentration in the runoff water. Steenvoorden et al. (1986) found that for every ton of dry matter applied per ha, the concentration of N in the runoff water increased by more than 30 g N $m^{-3}$ when a runoff event was induced 1 day after spreading.

The nitrogen in runoff water 1 to 7 days after spreading was mainly present as ammonium and soluble organic N (Uhlen, 1978; Sherwood and Fanning, 1981; Brink et al., 1983) whereas slightly higher nitrate concentrations were only found at later runoff events (Long et al., 1975; Converse et al., 1976; Long, 1979). Total nitrogen concentrations in the runoff water could exceed 150 g $m^{-3}$ when the runoff occurred within 2 days of slurry application but did not exceed 10 g $m^{-3}$ at later events (Sherwood and Fanning, 1981).

Only in the runoff water from surface-applied organic materials higher concentrations of soluble salts and nitrogen were measured compared to unmanured control plots. Injection, incorporation and application to plowed land resulted in concentrations values in the runoff water not significantly exceeding values for untreated soil (Long et al., 1975; Young and Mutchler, 1976; Uhlen, 1978; Ross et al., 1979).

### D. Dissolved Organic Carbon

The composition of water soluble organic carbon leachable from manures were investigated by Mosier et al. (1972). Free phenolics could be readily leached from manure. However, in the groundwater below manured soil only trace amounts of low-molecular weight compounds and mainly polymerized nonphenolic material were found. Further, no difference in the polymerized compounds in groundwater below manured and unmanured land was found. The authors suggested that the structure of the leachable organic material from manures is changed due to mineralization or immobilization by microbes.

The contents and distribution of dissolved organic carbon (DOC) in soil solutions of an unamended and manure amended field was studied by Rolston and Liss (1989). A mean DOC content of 40.3 mg C $kg^{-1}$ was found for the unamended soil, the bulk of samples grouped between 23.8 and 70 mg C $kg^{-1}$ soil. The addition of manure increased the mean DOC value in the soil to 71.8 mg C $kg^{-1}$ one month after addition and to 96.2 mg C $kg^{-1}$ soil four months after addition. However, after one year there was no significant difference between the DOC values of the two fields. Meek et al. (1974) found a regular increase in the DOC content ranging from 20 to 80 mg C $kg^{-1}$ in soil profiles at manure application rates from 0 to 180 tons $ha^{-1}$. DOC has been found to be readily decomposable in soil (Burford and Bremner, 1975; Jenkinson and Powlson, 1976; Cook and Allan, 1992). Analyses of soils 7 years after a heavy, single application of digested sewage sludge (150 and 330 t dry matter $ha^{-1}$) showed that no significantly higher concentrations of DOC were present in the soil profiles compared to untreated soil (Campbell and Beckett, 1988).

DOC contents in soil solutions of manured grassland amounted to 100-400 mg C $l^{-1}$ in the unsaturated zone (Oosterom, 1984). Comparable values measured in the unsaturated zone under forest soils amounted to 6-31 mg C $l^{-1}$ (Herbauts, 1980) and in leachates of peatlands to 20-63 mg C $l^{-1}$ (Urban et al., 1989). Thus, higher DOC values were generally found in the soil solution of manured fields compared to other land use.

Organic matter released from organic wastes applied to soils may not be completely decomposed during the transport through the unsaturated zone and may reach the groundwater. In the groundwater below manured grassland, DOC values ranging from 2-59 mg C $l^{-1}$ were measured when 2250-4250 kg C $ha^{-1}$ $yr^{-1}$ were added through cattle slurries (Krajenbrink et al., 1989). These DOC values were somewhat higher than in the groundwater below sandy woodlands ($< 8.5$ mg C $l^{-1}$) (Krajenbrink et al., 1988).

High concentrations of DOC (25-120 mg C $l^{-1}$) were present in a 30 m thick unsaturated zone under land irrigated with sewage effluent for over 30 years with annual applications of 140 kg C $ha^{-1}$ $yr^{-1}$ (Ronen et al., 1987). The DOC was mobile in the unsaturated zone and was oxidized when it reached the ground water resulting in an anoxic groundwater system (Ronen et al., 1987). These findings made it questionable whether irrigation with water with high DOC content should be used.

## VI. Organic Wastes and Water Quality

The quality of water is mainly determined by its microbial and chemical composition. For the microbial characterization of runoff waters from soils amended with organic wastes, the number of fecal coliforms, total coliforms and fecal streptococci were determined. The chemical characterization include a list of inorganic ions such as nitrogen forms, phosphorus, chloride, fluoride, iron, manganese, copper, zinc, boron, silver but also oxygen demanding compounds, surfactants, hydrocarbons and organochlorine compounds.

Usually the recommended standards for the quality of drinking water as outlined by the EC countries or World Health Organization are used as a reference of comparison (CEC, 1980; WHO, 1984). According to these recommendations, drinking water should not contain any pathogenic organisms. Thus, the maximum admissible concentration is 0 when determined according to the membrane filter method and less than 1 when measured according to the multiple tube method (CEC, 1980). The maximum admissible nitrate concentration in drinking water is 11.3 g $m^{-3}$ as $NO_3$-N or 50 g $m^{-3}$ as $NO_3$ and the ammonium level is 0.078 g $m^{-3}$ as $NH_4$-N or 0.1 g $m^{-3}$ as $NH_4$.

When reviewing the quality of runoff water from soils to which animal wastes were applied, only two different land areas were considered: grazed land and arable land. The influence of feedlots and storage places were excluded in this review.

### A. Microbiological Quality of Surface Runoff and Groundwater from Soils Amended with Organic Wastes

On grazed land, the amount of manure dropped per unit land and year usually is lower than amounts applied on arable land. Still, a bacterial contamination with coliforms of watersheds attached to grazing areas has been recognized in several studies (Robbins et al., 1972; Milne, 1976; Stephenson and Street, 1978; Doran and Linn, 1979). The bacteriological counts in the runoff from both ungrazed and grazed cow-calf pastures exceeded the recommended water quality standards (Doran and Linn, 1979). Further, the fecal coliform group was a better indicator of the impact of grazing than total coliforms or fecal streptococci. Thelin and Gifford (1983) found that fresh cattle feces released fecal coliform into water at a concentration of millions per 100 ml. Fecal deposits as old as 30 days produced concentrations of 40,000 per 100 ml. Buckhouse and Gifford (1976) showed that coliform bacteria within cattle feces deposited on soil and plant surfaces were able to survive intensive sunlight and heat for at least one summer.

Giddens and Barnett (1980) studied the total coliform bacterial content in the surface runoff from soils amended with poultry manure. No extremely high contents were measured and no relationship between application rate and number of total coliforms in the runoff water was found, when applied to fallow soil. However, on grassland a greater initial runoff of coliforms was measured at low manure application rates.

Culley and Phillips (1982) investigated the bacterial quality in both surface and subsurface discharge from a continuously corn cropped sandy clay loam amended with different rates of cow slurry for 6 years. With the exception of winter-applied slurry, neither rate nor time of slurry application significantly affected organism contents in the discharges in spring. Winter applications of slurry resulted in significantly higher fecal coliforms and fecal streptococci in both surface runoff and leachate compared to the unfertilized control. Fecal streptococcus populations of winter-applied slurry changed little during the first 100 days after application, while both total and fecal coliforms declined to low levels 24-40 days after spreading.

Fecal coliform bacteria in the surface runoff from sewage sludge treated soil amounted to 55,000 counts $ml^{-1}$ within the first 2 to 3 weeks after the application (Dunigan and Dick, 1980). As the soil became drier, the number of bacteria decreased rapidly to the same level present in the untreated soil. These findings are in agreement with those of van Donsel et al. (1967), who found few coliform bacteria in the runoff waters from dry manured soil.

### B. Chemical Quality of Surface Runoff and Groundwater from Soils Amended with Organic Wastes

A compilation of references describing the chemical quality of runoff and groundwater derived from manured soils is given in Table 11. Nitrate concentrations measured in percolation or groundwater were much higher than in surface runoffs. This was independent of whether the land was arable or grassland. Nitrate-N concentrations in surface runoffs usually did not exceed 10 mg $l^{-1}$ , even if large amounts of animal wastes were applied. However, in a study using sewage sludge, median concentrations ranged from 8.9 to 14 mg $l^{-1}$ and a maximum value of 65 mg $l^{-1}$ was measured (Jones and Hinesly, 1988).

Nitrate-N values measured in the groundwater below arable land were generally higher than below grassland. Values amounting to more than 100 mg $NO_3$-N $l^{-1}$ were recorded in the leachate from arable soils (Meek et al., 1974; Pratt et al., 1976; Mitchell and Gunther, 1976; Liebhardt et al., 1979; Fleige et al., 1980; Lembke and Thorne, 1980; Jones and Hinesly, 1988) but less than 100 mg $NO_3$-N in the leachate from grasslands (Adriano et al., 1971; Jordan and Smith, 1985; Steenvorden et al., 1986; Wermke, 1987).

In most studies there was an obvious relationship between the increase in the nitrogen concentration of the groundwater and the rate of application of organic wastes. Thus, the application rates for organic wastes have to be adjusted to avoid unacceptable nutrient levels in the groundwater.

### C. Solutions to the Groundwater Pollution Problem

The negative effects caused by large applications of animal wastes to soils on the quality of groundwater can be avoided if the following principles are implied:

- The density of livestock on a farm should be ruled by the number of hectares used for agricultural production. As a consequence, the application rate of manure need not exceed the need of the crop.
- The capacity for storage of animal wastes should be large enough (in Sweden 8 months) so that the application of animal wastes to soil can be made at the "right" time concerning the need of the crop.
- Animal wastes should be incorporated into soil within 4 hours after application. Beside the reduction in ammonia volatilization in this way, an enrichment of the surface runoff from manured arable land is largely avoided.

The principles outlined above are a summary of the Swedish regulations concerning livestock density, storage capacity, and spreading of livestock wastes given in the statute book of the Swedish agricultural administration coming into force 1995 (Lantbruksstyrelsen, 1990).

**Table 11.** Chemical quality of runoff and percolation water from grassland and arable land amended with organic wastes

| Conditions | Runoff (mg $l^{-1}$) | | | Percolation (mg $l^{-1}$) | | Reference |
|---|---|---|---|---|---|---|
| | $NO_3$-N | $NH_4$-N | COD | $NO_3$-N | COD | |
| **ARABLE LAND** | | | | | | |
| Field plots with liquid and solid cow manure, soil solution measured at 1.5-4.5 m depth | n.d. | n.d. | n.d. | 51-170 | n.d. | Pratt et al., 1976 |
| Laboratory study with pig slurry, 3 and 9% slope | 3-10 | n.d. | 850 | 8-221 | n.d. | Mitchell and Gunther, 1976 |
| Field plots amended with poultry manure, wells at 3, 4.5, and 6 m depth | n.d. | n.d. | n.d. | 2-174 | n.d. | Liebhardt et al., 1979 |
| Field plots irrigated with sewage sludge, soil solution measured at 0.8-1.0 m depth | n.d. | n.d. | n.d. | 35-250 | n.d. | Fleige et al., 1980 |
| Field plots, soil solution measured as 2.1 m depth | | | | | | Lembke and Thorne, 1980 |
| Beef manure applied | n.d. | n.d. | n.d. | 8.6-52 | n.d. | |
| Sewage sludge applied | n.d. | n.d. | n.d. | 96-260 | n.d. | |
| Tile-drained field plots, injection of slurry | n.d. | n.d. | n.d. | 5-30 | n.d. | Brink and Jernlås, 1982 |
| Field plots, manure applied on snow-covered soil | 2-5 | 3-45 | n.d. | n.d. | n.d. | Brink et al., 1983 |
| Field lysimeter plots, weekly irrigation of bermuda grass with pig lagoon effluents | n.d. | n.d. | n.d. | 6-27 | 26-31 | Evans et al., 1984 |

| | | | | | | |
|---|---|---|---|---|---|---|
| Field lysimeter plots, weekly irrigation of bermuda grass with pig lagoon effluents | 3-6 | n.d. | n.d. | n.d. | n.d. | Westerman et al., 1985 |
| Tile-drained field plots, with sewage sludge | 4-65 | n.d. | n.d. | 9-682 | n.d. | Jones and Hinesly, 1988 |
| **GRASSLAND** | | | | | | |
| Continuously grazed plots, soil solution measured at 3-4.5 m depth | n.d. | n.d. | n.d. | 30-50 | n.d. | Adriano et al., 1971 |
| Ungrazed plots amended with cattle manure | 2.2-7.6 | 0.4-1.3 | 38-123 | n.d. | n.d. | Long, 1979 |
| Rotationally grazed plots | 0.4-0.8 | 0.4-0.7 | 60-100 | n.d. | n.d. | Schepers and Francis, 1982 |
| Ungrazed plots, animal wastes and sewage sludge | 2-4 | 2-15 | 100-250 | n.d. | n.d. | Mc Leod and Hegg, 1984 |
| Continuously grazed plots, pig slurry added | n.d. | n.d. | n.d. | 3-7 | n.d. | Jordan and Smith, 1985 |
| Rotationally grazed plots with cow slurry, borehole water from shallow groundwater | n.d. | n.d. | n.d. | 80-137 | n.d. | Steenvoorden et al., 1986 |
| Continuously grazed plots amended with cattle slurry, soil solution measured at 0.5-2.0 m depth | n.d. | n.d. | n.d. | 20-40 | n.d. | Wermke, 1987 |

## VII. Summary

There is a large variability in the chemical composition of only one type of waste material. Thus, contrasting results can be gained in different investigations principally dealing with the same type of material. Chemical analysis and characterization of the waste material, whether treated aerobically, anaerobically or being undecomposed, will give a rational basis for the understanding of their behavior in the soil-plant system.

The pollution potential through gaseous losses volatilized from organic waste materials has been recognized. Anaerobic storage compared to aerobic treatment of manures and slurries enable ammonia volatilization to be kept at a minimum. Immediate incorporation of organic materials into soil reduces ammonia losses during the next step of manure handling.

In soil, gaseous N losses from organic wastes are higher than from inorganic fertilizers. Crop utilization of N from organic materials is lower than from fertilizers. However, innovations in management of organic materials may lead to decreased N losses and a better utilization of N through crops in the future. Concerning leaching losses, lower N losses from animal wastes have been measured when applied at the same time and N rate as inorganic fertilizers. There was simply less inorganic N available in soil when using animal wastes at the same total N equivalent as inorganic fertilizers. In soils manured over longer periods of time, however, more N was leached than from fertilized soils. The continuous inputs or organic manures increased the soil organic matter pool and thereby the amount of nitrogen that can be mineralized even at times when crop demand is low or zero.

Water pollution caused by animal wastes is due to too high applications rates to soil, which is the consequence of a surplus of manure on farms. The chemical quality of leachates from soils is highly dependent on the amount of organic wastes applied, whereas the surface runoff is mainly influenced through microbial contamination.

## REFERENCES

Abbeel, van den, R., A. Claes, and K. Vlassek. 1989. Gaseous nitrogen losses from slurry-manured land. p. 214-224. In: J.A. Hansen and K. Henriksen (eds.), *Nitrogen in organic wastes applied to soils*. Academic Press, London, England.

Adamsen, F.J., and B.R. Sabey. 1987. Ammonia volatilization from liquid digested sewage sludge as affected by placement in soil. *Soil Sci. Soc. Am. J.* 51:1080-1082.

Adriano, D.C, P.F. Pratt, and S.E. Bishop. 1971. Nitrate and salt in soils and ground waters from land disposal of dairy manure. *Soil Sci. Soc. Amer. Proc.* 35:759-762.

Ames, J.W. and T.E. Richmond. 1917. Fermentation of manure treated with sulphur and sulphates: Changes in nitrogen and phosphorus content. *Soil Sci.* 4:79-89.

Andersson, A. and K.O. Nilsson. 1976. Influence on the level of heavy metals in soil and plant from sewage sludge used as fertilizer. *Swedish J. Agric. Res.* 6:151-159.

Andersson, A. 1977. Some aspects on the significance of heavy metals in sewage sludge and related products used as fertilizers. *Swedish J. Agric. Res.* 7:1-5.

Bachrach, U. 1957. The aerobic breakdown of uric acid by certain pseudomonades. *J. Gen. Microbiol.* 17:1-11.

Barbarika, Jr., A., L. J. Sikora, and D. Colacicco. 1985. Factors affecting the mineralization of nitrogen in sewage sludge applied to soils. *J. Environ. Qual.* 49:1403-1406.

Bartholomew, W.V. 1965. Mineralization and immobilization of nitrogen in the decomposition of plant and animal residues. In: W.V. Bartholomew and F.E. Clark (eds.), Soil nitrogen. Am. Soc. Agron., *Agronomy* 10:285-306. Madison, WI.

Beauchamp, E.G. 1983. Nitrogen losses from sewage sludge and manures applied to agricultural land. p. 181-195. In: J.R. Freney and J.R. Simpson (eds.), *Gaseous loss of nitrogen from plant-soil systems.* Martinus Nijhoff/Dr W.Junk Publishers, The Hague, The Netherlands.

Beauchamp, E.G., G.E. Kidd, and G. Thurtell. 1978. Ammonia volatilization from sewage sludge applied in the field. *J. Environ. Qual.* 7:141-146.

Beauchamp, E.G., G.E. Kidd, and G. Thurtell. 1982. Ammonia volatilization from liquid dairy cattle manure in the field. *Can. J. Soil Sci.* 62:11-19.

Beauchamp, E.G. and J.W. Paul. 1989. A simple model to predict manure N availability to crops in the field. p. 140-149. In: J. A. Hansen and K. Henriksen (eds.) *Nitrogen in organic wastes applied to soils.* Academic Press, London, England.

Beckwith, C.P. and J.W. Parsons. 1980. The influence of mineral amendments on the changes in the organic nitrogen components of composts. *Plant Soil* 54:259-270.

Bell, R.G. 1970. Fatty acid content as a measure of the odor potential of stored liquid poultry manure. *Pollut. Sci.* 49:1126-1129.

Bertilsson, G. 1988. Lysimeter studies of nitrogen leaching and nitrogen balances as affected by agricultural practices. *Acta Agric. Scand.* 38:3-11.

Bertoldi, de, M., U. Citernesi, and M. Grisell. 1982. Microbial populations in compost process. *Biocycle* 23:26-33.

Besson, J.M., V. Lehmann, M. Roulet, and W. Edelmann. 1982. Vergleich dreier Güllebehandlungsmethoden unter kontrollierten Versuchsbedingungen: Lagerung, Belüftung und Methangärung. *Mitt. Schweiz. Landw.* 30:161-168.

Bhat, K.K.S., T.H. Flowers, and J.R. O'Callaghan. 1980. A model for the simulation of the fate of nitrogen in farm wastes on land application. *J. Agric. Sci.* 94:183-193.

Bitzer, C.C. and J.T. Sims. 1988. Estimating the availability of nitrogen in poultry manure through laboratory and field studies. *J. Environ. Qual.* 17:47-54.

Bjarnason, S. 1988. Turnover of organic nitrogen in agricultural soils and the effects of management practices on soil fertility. *Dissertation. Swedish University of Agricultural Sciences.* Department of Soil Sciences. Uppsala, Sweden.

Bless, H.G., R. Beinhauer, and B. Sattelmacher. 1991. Ammonia emissions from slurry applied to wheat stubble and rape in North Germany. *J. Agric. Sci.* 117:225-231.

Bode, de, M.J.C. 1990. Vergleich der Ammoniakemissionen aus verschiedenen Flüssigmistlagersystemen. *Ammoniak in der Umwelt.* Gemeinsames Symposium, 10-12 Oktober 1990, Braunschweig-Völkenrode. KTBL-Schriften-Vertrieb im Landwirtschaftsverlag GmbH, Münster, Germany. p34.1- 34.13.

Bouldin, D.R., S.D. Klausner, and W.S. Reid. 1984. Use of nitrogen from manure. p. 221-245. In: R.D. Hauck (ed.), *Nitrogen in crop production.* Am. Soc. Agron., Madison, Wisconsin.

Boxberger, J. and A. Gronauer. 1990. $NH_3$- Emissionen während der Flüssigmistausbringung. *Ammoniak in der Umwelt.* Gemeinsames Symposium, 10-12 Oktober 1990, Braunschweig-Völkenrode. KTBL-Schriften-Vertrieb im Landwirtschaftsverlag GmbH, Münster, Germany. p42.1-42.5.

Boyle, M. and E.A. Paul. 1989. Carbon and nitrogen mineralization kinetics in soil previously amended with sewage sludge. *Soil Sci. Soc. Am. J.* 53:99-103.

Bremner, J.M. 1955. Nitrogen transformations during the biological decomposition of straw composted with inorganic nitrogen. *J. Agric. Sci.* 45:469-475.

Brink, N. and R. Jernlås. 1982. Leaching after spreading of liquid manure in autumn and spring. Swedish University of Agricultural Sciences, Division of Water Management. *Ekohydrologi* Report 12. Uppsala, Sweden. p. 3-14. (in Swedish).

Brink, N., A.S. Gustavsson, and B. Ulén. 1983. Surface transport of plant nutrients from manured fields. Swedish University of Agricultural Sciences, Division of Water Management. *Ekohydrologi* Report 13. Uppsala, Sweden. p. 3-14. (in Swedish).

Brunke, R., P. Alvo, P. Schuepp, and R. Gordon. 1988. Effect of meteorological parmeters on ammonia loss from manure in the field. *J. Environ. Qual.* 17:431-436.

Bucher, R. 1943. Stallmistbereitung und zusätzliche Herstellung von Strohkompost bei verschiedenen starker Stroheinstreu. *Bodenk. Pflanzenernähr.* 31:63-85.

Buckhouse, J.C. and G.F. Gifford. 1976. Water quality implications of cattle grazing on a semi-arid watershed in southeastern Utah. *J. Range Manage.* 29:109-113.

Buijsman, E., H.F.M. Maas, and W.A.H. Asman. 1987. Anthropogenic $NH_3$ emissions in Europe. *Atmos. Environ.* 21:1009-1022.

Burford, J.R. 1976. Effect of the application of cow slurry to grassland on the composition of the soil atmosphere. *J. Sci. Food Agric.* 27:115-126.

Burford, J.R. and J.M. Bremner. 1975. Relationship between the denitrifcation capacities of soils and total, water-soluble, and ready decomposable organic matter. *Soil Biol. Biochem.* 7:389-394.

Burri, R., E. Herfeldt, and A. Stutzer. 1895. Bakteriologisch-chemische Forschungen über die Ursachen der Stickstoffverluste in faulenden organischen Stoffen, insbesondere im Stallmist und in der Jauche. *J. Landwirtsch.* 43:1-11.

Bågstam, G. 1979. Population changes in microorganisms during composting of spruce bark: II. Mesophilic and thermophilic microorganisms during controlled composting. *Eur. J. Appl. Microbiol. Biotechnol.* 6:279-288.

Campbell, D.J. and P.H.T. Beckett. 1988. The soil solution in a soil treated with digested sewage sludge. *J. Soil Sci.* 39:283-298.

Carlile, F.S. 1984. Ammonia in poultry houses: a literature review. *World's Poultry Sci. J.* 40:99-113.

Castellanos, J.Z. and P.F. Pratt. 1981. Mineralization of manure nitrogen - correlation with laboratory indexes. *Soil Sci. Soc. Am. Proc.* 45:345-357.

CEC (Commission of the European Communities). 1980. Council directive of 15 July 1980 relating to the quality of water intended for human consumption. *Official Journal of the European Communities*, No L 229/11 (80/778/EEC).

CEC (Commission of the European Communities). 1986. Council directive of 12 June 1986 on the protection of the environment, and in particular of the soil, when sewage sludge is used in agriculture. *Officical Journal of the European Communities*, No. L 181/6 (86/278/EEC).

Chae, Y.M. and M.A. Tabatabai. 1986. Mineralization of nitrogen in soils amended with organic wastes. *J. Environ. Qual.* 15:193-198.

Chang, A.C., D.C. Adriano, and P.F. Pratt. 1973. Waste accumulation on a selected dairy corral and its effect on the nitrate and salt of the underlying soil strata. *J. Environ. Qual.* 2:233-237.

Chang, A.C., W.R. Olmstread, J.B. Johanson, and G. Yomashita. 1974. The sealing mechanism of wastewater ponds. *J. Water Pollut. Control. Fed.* 46:1715-1721.

Chescheir, G.M., P.W. Westerman, and L.M. Safley, Jr. 1986. Laboratory methods for estimating available nitrogen in manures and sludges. *Agric. Wastes* 18:175-195.

Christensen, S. 1983. Nitrogen oxide emissions from a soil under permanent grass: Seasonal and diurnal fluctuations as influenced by manuring and fertilization. *Soil Biol. Biochem.* 15:531-536.

Christensen, S. 1985. Denitrification in an acid soil: Effects of slurry and potassium nitrate on the evolution of nitrous oxide and on nitrate-reducing bacteria. *Soil Biol. Biochem.* 17:757-764.

Converse, J.C., G.D. Bubenzer, and W.H. Paulson. 1976. Nutrient losses in surface runoff from winter spread manure. *Trans. ASAE* 19:517-523.

Cook, B.D. and D.L. Allan. 1992. Dissolved organic carbon in old field soils: Total amounts as a measure of available resources for soil mineralization. *Soil Biol. Biochem.* 24:585-594.

Cooper, P. and I.S. Cornforth. 1978. Volatile fatty acids in stored animal slurry. *J. Sci. Food Agric.* 29:19-27.

Culley, J.L.B. and P.A. Phillips. 1982. Bacteriological quality of surface and subsurface runoff from manured sandy clay loam soil. *J. Environ. Qual.* 11:155-158.

Davis, R.E. 1927. The nitrogenous constituents of hens' urine. *J. Biol. Chem.* 74:509-513.

Davis, S., W. Fairbank, and H. Weisheit. 1973. Dairy waste ponds effectively self-sealing. *Trans. ASEA* 16:69-71.

Dewes, T., L. Schmitt, U. Valentin, and E. Ahrens. 1990. Nitrogen losses during the storage of liquid livestock manures. *Biol. Wastes* 31:241-250.

Diaz-Fierros, F., M.C. Villar, F. Gil, M.C. Leiros, M. Carballas, T. Carballas, and A. Cabaneiro. 1987. Laboratory study of the availability of nutrients in physical fractions of cattle slurry. *J. Agric. Sci.* 108:353-359.

Diaz-Fierros, F., M. C. Villar, F. Gil, M. Carballas, M. C. Leiros, T. Carballas, and A. Cabaneiro. 1988. Effect of cattle slurry fractions on nitrogen mineralization in soil. *J. Agric. Sci.* 110:491-497.

Dindal, G.L. 1978. Soil organisms and stabilizing wastes. *Compost Sci.* 19:8-11.

Doak, B.W. 1952. Some chemical changes in the nitrogeneous constituents of urine when voided on pasture. *J. Agric. Sci.* 42:162-171.

Donovan, W.C. and T.J. Logan. 1983. Factors affecting ammonia volatilization from sewage sludge applied to soil in a laboratory study. *J. Environ. Qual.* 12:584-590.

Donsel, van, D.J., E.E. Geldreich, and N.A. Clarke. 1967. Seasonal variations in survival of indicator bacteria in soil and their contribution to storm-water pollution. *Appl. Microbiol.* 15:1362-1370.

Doran, J.W. and D.M. Linn. 1979. Bacteriological quality of runoff water from pastureland. *Appl. Environ. Microbiol.* 37:985-991.

Dunigan, E.P. and R.P. Dick. 1980. Nutrient and coliform losses in runoff from fertilized and sewage sludge-treated soil. *J. Environ. Qual.* 9:243-250.

Durand, G. 1961. Degradation of the purine and pyrimidine bases in the soil: aerobic degradation of uric acid. *C.R. Acad. Sci.* 252:1687-1689.

Duthion, C. 1981. Nitrogen leaching after spreading pig manure. p. 274-283. In: J.C. Brogan (ed.), *Nitrogen losses and surface run-off from landspreading of manures*. Developments in Plant and Soil Sciences, Vol 2. Martinus Nijhoff/ Dr W. Junk Publishers. The Hague, The Netherlands.

Ekman, P., H. Emanuelson, and A. Fransson. 1949. Investigations concerning the digestibility of proteins in poultry. *Ann. Royal Agric. College Swed.* 16: 749-777.

Emmanuel, B. and B.R. Howard. 1978. Endogenous uric acid and urea metabolism in the chicken. *Brit. Poultry Sci.* 19:295-301.

English, C.J., J.R. Miner, and J.K. Koelliker. 1980. Volatile ammonia losses from surface-applied sludge. *J. Water Pollut. Control./Fed.* 52:2340-2350.

EPA (United States Environmental Protection Agency). 1989. Standards for the disposal of sewage-sludge: Part 2 - Proposed rules. *40 CFR parts 257 and 503* (Including Technical support document). U.S. Environmental Protection Agency, Washington DC.

Epstein, E., D.B. Keane, J.J. Meisinger, and J.O. Legg. 1978. Mineralization of nitrogen from sewage sludge and sludge compost. *J. Environ. Qual.* 7:217-221.

Evans, M.R., M.P.W. Smith, E.A. Deans, I.F. Svoboda, and F.E. Thacker. 1986. Nitrogen and aerobic treatment of slurry. *Agric. Wastes* 15:205-213.

Evans, R.O., P.W. Westerman, and M.R. Overcash. 1984. Subsurface drainage water quality from land application of swine lagoon effluent. *Trans. ASAE* 27:473-480.

Fassen van, H.G. and H. van Dijk. 1979. Nitrogen conversions during the composting of manure straw mixtures. p. 113-120. In: E. Grossbard (ed.), *Straw decay and its effect on disposal and utilization.* John Wiley and Sons, New York, USA.

Fassen van, H.G. and H. van Dijk. 1987. Manure as a source of nitrogen and phosphorus in soils. p. 28-45. In: H.G. van der Meer et al. (eds.), *Animal manure on grassland and fodder crops. Fertilizer or waste?* Martinus Nijhoff Publishers, Dordrecht, The Netherlands.

Fenn, L.B. and D.E. Kissel. 1973. Ammonia volatilization from surface application of ammonium compounds on calcareous soils. I. General theory. *Soil Sci. Soc. Am. Proc.* 37:855-859.

Fenn, L.B., R.M. Taylor, and Matocha, J.E. 1981. Ammonia losses from surface applied nitrogen fertilizer as controlled by soluble calcium and magnesium: General theory. *Soil Sci. Soc. Am. J.* 45:777-781.

Fine, P., U. Mingelgrin, and A. Feigin. 1989. Incubation studies of the fate of organic nitrogen in soil amended with activated sludge. *Soil Sci. Soc. Am. J.* 53:444-450.

Flachowsky, G. and A. Hennig. 1990. Composition and digestibility of untreated and chemically treated animal excreta for ruminants - a review. *Biol. Wastes* 31:17-36.

Fleige, H., M. Renger, O. Strebel, and W. Müller. 1980. Nitrogen leaching and groundwater pollution by sprinkling irrigation of sewage sludge on arable land. *Z. Pflanzenernaehr. Bodenk.* 143:569-580 (in German).

Fleming, G.A. 1986. Soil ingestion by grazing animals; A factor in sludge-treated grassland. p. 43-49. In: R.D. Davis et al. (eds.), *Factors influencing sludge utilisation practices in Europe.* Elsevier Applied Science Publishers, London, England.

Flowers, T.H. and P.W. Arnold. 1983. Immobilization and mineralization of nitrogen in soils incubated with pig slurry or ammonium sulphate. *Soil Biol. Biochem.* 15:329-335.

Fordham, A.W. and U. Schwertmann. 1977a. Composition and reactions of liquid manure (Gülle), with particular reference to phosphate: II. Solid phase components. *J. Environ. Qual.* 6:136-140.

Fordham, A.W. and U. Schwertmann. 1977b. Composition and reactions of liquid manure (Gülle) with particular reference to phosphate: III. pH buffering capacity and organic components. *J. Environ. Qual.* 6:140-144.

Franke, B. 1987. Composting source separated organics. *Biocycle* 28:40-42.

Franz, H. and G. Repp. 1949. Untersuchungen über die Stallmistrotte im Stapel und im Boden. *Bodenkultur* 3:465-486.

Frost, J.P., R.J. Stevens, and R.J. Laughlin. 1990. Effect of separation and acidification of cattle slurry on ammonia volatilization and on the efficiency of slurry nitrogen for herbage production. *J. Agric. Sci.* 115:49-56.

Gale, P.M. and J.T. Gilmour. 1986. Carbon and nitrogen mineralization kinetics for poultry litter. *J. Environ. Qual.* 15:423-426.

Garau, M.A., M.T. Felipo, and M.C. Ruiz de Villa. 1986. Nitrogen mineralization of sewage sludges in soils. *J. Environ. Qual.* 15:225-228.

Garwood, E.A. and J.C. Ryden. 1986. Nitrate loss through leaching and surface runoff from grassland: effects of water supply, soil type and management. p. 99-113. In: H.G. van der Meer, J.C. Ryden and G.C. Ennik (eds.), *Nitrogen fluxes in intensive grassland systems.* Developments in Plant and Soil Sciences, Vol. 23. Martinus Nijhoff Publishers, Dordrecht, The Netherlands.

Georgacakis, D., D.M. Sievert, and E.L. Iannotti. 1982. Buffer stability in manure digesters. *Agric. Wastes* 4:427-441.

Germon, J.C., J.J. Giraud, R. Chaussod, and C. Duthion. 1979. Nitrogen mineralization and nitrification of pig slurry added to soil in laboratory conditions. p. 170-184. In: J.K.R. Gasser (ed.), *Modelling nitrogen from farm wastes.* Applied Science Publishers, London, England.

Gerritse, R.G. 1981. Mobility of phosphorus from pig slurry in soils. p. 347-369. In: T.W.G. Hucker and G. Catroux (eds.), *Phosphorus in sewage sludge and animal waste slurries.* Reidel, Dordrecht, The Netherlands.

Gerritse, R.G. and I. Zugec. 1977. The phosphorus cycle in pig slurry measured from $^{32}PO_4$ distribution rates. *J. Agric. Sci.* 88:101-109.

Giddens, J. and A.P. Barnett. 1980. Soil loss and microbial quality of runoff from land treated with poultry litter. *J. Environ. Qual.* 9:518-520.

Gilmour, J.T., M.D. Clark, and G.C. Sigua. 1985. Estimating net nitrogen mineralization from carbon dioxide evolution. *Soil Sci. Soc. Am. J.* 49:1398-1402.

Glathe, H. and O. von Metzen. 1937. Vergleichende Stalldüngerlagerungsversuche. *Bodenk. Pflanzenernähr.* 5:192-208.

Glathe, H. and W. Seidel. 1937. Untersuchungen über die Aufbewahrung des Stalldüngers unter streng anaeroben Bedingungen. *Bodenk. Planzenernähr.* 5:118-128.

Godden, B., M. Penninckx, A. Piérard, and R. Lannoye. 1983. Evolution of enzyme activities and microbial populations during composting of cattle manure. *Eur. J. Appl. Microbiol. Biotechnol.* 17:306-310.

Gordon, R., M. Leclerc, P. Schuepp, and R. Brunke. 1988. Field estimates of ammonia volatilization from swine manure by a simple micrometeorological technique. *Can. J. Soil Sci.* 68:369-380.

Guenzi, W.D. and W.F. Beard. 1981. Volatile fatty acids in a redox-controlled cattle manure slurry. *J. Environ. Qual.* 10:479-482.

Guenzi, W.D., W.E. Beard, F.S. Watanabae, S.R. Olsen, and L.K. Porter. 1978. Nitrification and denitrification in cattle manure amended soil. *J. Environ. Qual.* 7:196-202.

Hadas, A., B. Bar-Yosef, S. Davidov, and M. Sofer. 1983. Effect of pelleting, temperature, and soil type on mineral nitrogen release from poultry and dairy manures. *Soil Sci. Soc. Am. J.* 47:1129-1133.

Harding, S.A., C.E. Clapp, and W.E. Larson. 1985. Nitrogen availability and uptake from field soils five years after addition of sewage sludge. *J. Environ. Qual.* 14:95-100.

Harmsen, G.W. and D.A. van Schreven. 1955. Mineralization of organic nitrogen in soil. *Adv. Agron.* 7:299-398.

Healy, W.B. 1968. Ingestion of soil by dairy cows. *N. Z. J. Agric. Res.* 11:487-499.

Hébert, M., A. Karam, and L.E. Parent. 1991. Mineralization of nitrogen and carbon in soils amended with composted manure. *Biol. Agric. Horticul.* 7:349-361.

Heck, A.F. 1931. The availability of the nitrogen in farm manure under field conditions. *Soil Sci.* 31:467-479.

Hensler, R.F., R.J. Olsen, S.A. Witzel, O.J. Altoe, W.H. Paulson, and R.F. Johannes. 1970. Effects of manure handling on crop yields, nutrient recovery and runoff losses. *Trans. ASAE* 13:726-731.

Herbauts, J. 1980. Direct evidencc of water-soluble organic matter leached in brown earths and slightly podzolized soils. *Plant Soil* 54:317-321.

Herbst, F., F. Aziz, and J. Garz. 1987. Stickstoffumsetzungen im Boden nach Zugabe von $^{15}$N-markierter Gülle. *Arch. Acker- Pflanzenbau Bodenkd.* 31:169-175.

Herrmann, V., H. Görlitz, and F. Asmus. 1983. Lysimeteruntersuchungen zur Nährstoffverlagerung nach Gülledüngung in einer Sand-Rosterde - Sickerwasser und Stickstoff. *Arch. Acker- Pflanzenbau Bodenk.* 27:509-515.

Hesse, W. 1938. Untersuchungen an Kalt-, Stapel- und Heissmist unter besonderer Berücksichtigung der chemischen und bakteriologischen Vorgänge während der Lagerung und Wirkung auf die Pflanzenproduktion. *Bodenk. Pflanzenernähr.* 52:303-359.

Hill, K.J. 1970. Developmental and comparative aspects of digestion. p. 409-423. In: M.J. Swenson (ed.), *Duke's physiology of domestic animals*. 8th ed. Comstock Publishing Assoc., London, England.

Hills, D.J. 1976. Infiltration characteristics from anaerobic lagoons. *J. Water Pollut. Control Fed.* 48:695-709.

Hsieh, Y.P., L.A. Douglas, and H.L. Motta. 1981a. Modelling sewage sludge decomposition in soil: I. Organic carbon transformation. *J. Environ. Qual.* 10: 54-59.

Hsieh, Y.P., L.A. Douglas, and H.L. Motta. 1981b. Modelling sewage sludge decomposition in soil: II. Nitrogen transformations. *J. Environ. Qual.* 10: 59-64.

Hoff, J.D., D.W. Nelson, and A.L. Sutton. 1981. Ammonia volatilization from liquid swine manure applied to cropland. *J. Environ. Qual.* 10:90-95.

Horlacher, D. and H. Marschner. 1990. Schätzrahmen zur Beurteilung von Ammoniakverlusten nach Ausbringung von Rinderflüssigmist. *Z. Pflanzenerbähr. Bodenk.* 153:107-115.

Hrubrant, G.R. and R.W. Detroy. 1979. Composition and fermentation of feedlot wastes. p. 411-424. In: M. Moo-Young and G.J. Farquhar (eds.), *Waste treatment and utilization.* Pergamon Press, Oxford, England.

Iversen, K. 1936. Forsög med biologisk-dynamisk gödskning. *Tidsskr. Planteavl* 41:210-222.

Iversen, K. 1957. Stalgödningens opbevaring og udbringning. *Tidsskr. Planteavl* 60:1-19.

Iversen, K. and K. Dorph-Petersen. 1952. Konservering af staldgödning med superfosfat. *Tidsskr. Planteavl* 55:282-302.

Iversen, K. and K. Dorph-Petersen. 1949. Forsög med stalgödningens opbevaring og anvendelse. *Tidsskr. Planteavl* 52:70-108.

Jacobs, L.W., G.A. O'Connor, M.A. Overcash, M.J. Zabik, and P. Rygiewicz. 1987. Effect of trace organics in sewage sludge on soil-plant systems and assessing their risk to humans. p 101-143. In: A.L. Page et al. (eds.), *Land application of sludge.* Lewis Publishers Inc, Chelsea, MI.

Jansson, S.L. 1963. Balance sheet and residual effects of fertilizer nitrogen in a 6-year study with N15. *Soil Sci.* 95:31-37.

Jansson, S.L. 1977. Återförandet av tätorternas växtnäringshaltiga avfall till odlingsmarken. II Teknik och ekologiska synpunkter. *Skogs- o. Lantbr. akad. Tidskr.* 116:65-84.

Japenga, J. and K. Harmsen. 1990. Determination of mass balances and ionic balances in animal manure. *Netherl. J. Agric. Sci.* 38: 353-367.

Jarvis, S.C. and B.F. Pain. 1990. Ammonia volatilization from agricultural land. *Fert. Soc. Proc.* (London) 298:1-35.

Jarvis, S.C., M. Sherwood, and J.H.A.M. Steenvoorden. 1987. Nitrogen losses from animal manures: from grazed pastures and from applied slurry. p. 195-212. In: H.G. van der Meer, R.J. Unwin, T.A. van Dijk, and G.C. Ennik (eds.), *Animal manure on grassland and fodder crops. Fertilizer or waste?* Developments in Plant and Soil Sciences, No 30. Martinus Nijhoff Publishers. Dordrecht, The Netherlands.

Jenkinson, D.S. 1977. The nitrogen economy of the Broadbalk experiments. I. Nitrogen balance in the experiment. *Rothamsted Experimental Station Report for 1976*. Part 2. Rothamsted Exp. Stn., Harpenden, Herts, England. p. 103-110.

Jenkinson, D.S. 1981. The fate of plant and animal residues in soils. p. 505-561. In: D.J. Greenland and M.H.B. Hayes (eds.), *The chemistry of soil processes*. John Wiley & Sons Ltd, New York.

Jenkinson, D.S. and D.S. Powlson. 1976. The effect of biocidal treatments on metabolism in soil. I. Fumigation with chloroform. *Soil Biol. Biochem.* 8:167-177.

Jenkinson, D.S. and A.E. Johnston. 1977. Soil organic matter in the Hoosfield continuous barley experiment. *Rothamsted Experimental Station Report for 1976*. Part 2. Rothamsted Exp. Stn., Harpenden, Herts, England. p. 87-101.

Jensen, S.T. 1928. Investigations in ammonia volatilization in connection with nitrogen losses during spreading of natural manures. *Tidsskr. Planteavl* 34:117-147 and 35:59-80.

Jensen, S.T. 1930. Investigations in ammonia volatilization and nitrogen losses during slurry spreading. *Wissensch. Arch. Landwirt.* 3:161-180.

Johnston, A.E. 1975. Experiments made on Stackyard field, Woburn, 1879-1974. *Rothamsted Experimental Station Report for 1974*. Part 2. Rothamsted Exp. Stn., Harpenden, Herts, England. p. 45-78.

Jones, R.L. and T.D. Hinesly. 1988. Nitrate in waters from sewage-sludge amended lysimeters. *Environ. Pollut.* 51:19-30.

Jordan, C. and R.V. Smith. 1985. Factors affecting leaching of nutrients from an intensively managed grassland in country Antrim, Northern Ireland. *J. Environ. Management* 20:1-15.

Kaila, A. 1950. Use of superphosphate with farm manure. *Finnish State Agric. Res. Board* 134:1-35.

Kemppainen, E. 1987. Ammonia binding capacity of peat, straw, sawdust and cutter shavings. *Ann. Agric. Fenn.* 26:89-94.

Khaleel, R., K.R. Reddy, and M.R. Overcash. 1980. Transport of potential pollutants in runoff water from land areas receiving animal wastes: A review. *Water Res.* 14:421-436.

Kirchmann, H. 1985. Losses, plant uptake and utilisation of manure nitrogen during a production cycle. *Acta Agric. Scand. Supplementum* 24:1-77.

Kirchmann, H. 1989. A 3-year N balance study with aerobic, anaerobic and fresh $^{15}$N-labelled poultry manure. p 113-125. In: J. A. Hansen and K. Henriksen (eds.) *Nitrogen in organic wastes applied to soils*. Academic Press, London, England.

Kirchmann, H. 1990. Nitrogen interactions and crop uptake from fresh and composted $^{15}$N-labelled poultry manure. *J. Soil Sci.* 41:379-385.

Kirchmann, H. 1991. Carbon and nitrogen mineralization of fresh, aerobic and anaerobic animal manures during incubation with soil. *Swed. J. Agric. Res.* 21:165-173.

Kirchmann, H. and E. Witter. 1989. Ammonia volatilization during aerobic and anaerobic manure decomposition. *Plant Soil* 115:35-41.

Kirchmann, H. and A. Lundvall. 1992. Relationship between N immobilization and volatile fatty acids in soil after application of pig and cattle slurries. *Biol. Fert. Soils* (in press).

Kirchmann, H. and E. Witter. 1992. Composition of fresh, aerobic and anaerobic farm animal dung. *Bioresource Technol.* 40:137-142.

Klausner, S.D., P.J. Zwerman, and D.F. Ellis. 1976. Nitrogen and phosphorus losses from winter disposal of dairy manure. *J. Environ. Qual.* 5:47-49.

Knapp, E.B., L.F. Elliott, and G.S. Campbell. 1983. Carbon, nitrogen and microbial biomass interelationships during the decomposition of wheat straw: A mechanistic simulation model. *Soil Biol. Biochem.* 15:455-461.

Krajenbrink, G.J.W., D. Ronen, W. van Duijvenbooden, M. Magaritz, and D. Wever. 1988. Monitoring of recharge water quality under woodland. *J. Hydrol.* 98:83-102.

Krajenbrink, G.J.W., L.J.M. Boumans, and C.R. Meinardi. 1989. Hydrochemical processes in the top layer of groundwater under pasture land. p. 317-333. In: J. A. Hansen and K. Henriksen (eds.) *Nitrogen in organic wastes applied to soils*. Academic Press, London, England.

Krauss, P., R. Blessing, and U. Korherr. 1987. Heavy metals in compost from municipal refuse strategies to reduce their content to acceptable levels. p. 254-265. In: M. de Bertoldi et al. (eds.), *Compost: Production, quality and use*. Elsevier Applied Science, London.

Krogdahl, A. and B. Dahlsgård. 1981. Estimation of nitrogen digestibility in poultry: Content and distribution of major urinary nitrogen compounds in excreta. *Poultry Sci.* 60:2480-2485.

Köhnlein, J. and H. Vetter. 1953. Die Stalldüngerrotte bei steigender Stroheinstreu. *Z. Planzenernähr. Düngung Bodenk.* 63:119-141.

Lantbruksstyrelsen 1990. Lantbruksstyrelsens föreskrifter om begränsning av antalet djur i ett jordbruk. *Lantbruksstyrelsens författningssamling* 1988:44. Jönköping, Sweden.

Lanyon, L.E., L.J. Stearns, H.D. Bartlett, and S.P. Persson. 1985. Nutrient changes during storage of anaerobic digester effluent and fresh dairy cattle manure with phosphoric acid. *Agric. Wastes* 13:79-91.

Lauer, D.A., D.R. Bouldin, and S.D. Klausner. 1976. Ammonia volatilization from dairy manure spread on the soil surface. *J. Environ. Qual.* 5:134-141.

Laura, R.D. and M.A. Idnani. 1972. Mineralization of nitrogen in manures made from spent-slurry. *Soil Biol. Biochem.* 4:239-243.

Lembke, W.D. and M.D. Thorne. 1980. Nitrate leaching and irrigated corn production with organic and inorganic fertilizers on sandy soil. *Trans. ASAE* 23:1153-1156.

Liebhardt, W.C., C. Golt, and J. Tupin. 1979. Nitrate and ammonium concentrations of ground water resulting from poultry manure applications. *J. Environ. Qual.* 8:211-215.

Limmer, A.W. and K.W. Steele. 1983. Effect of cow urine upon denitrification. *Soil Biol. Biochem.* 15:409-412.

Linak, W.P., J.A. McSorley, R.E. Hall, J.V. Ryan, R.K Srivastava, J.O.L. Wendt, and J.B. Marek. 1990. Nitrous oxide emissions from fossil fuel combustion. *J. Geophysic. Res.* 95:7533-7541.

Lindemann, W.C. and M. Cardenas. 1984. Nitrogen mineralization potential and nitrogen transformations of sludge-amended soil. *Soil Sci. Soc. Am. J.* 48:1072-1077.

Lindemann, W.C., G. Connel, and N.S. Urquhart. 1988. Previous sludge addition effects on nitrogen mineralization in freshly amended soil. *Soil Sci. Soc. Am. J.* 52:109-112.

Lindhard, J. 1954. Undersögelser over tabet af ammoniakkvälstof fra gödningspröver udtaget i kostalden. *Tidsskr. Planteavl* 57:108-120.

Lindsay, W.L. and H.F. Stephenson. 1959. Nature of the reactions of monocalcium phosphate monohydrate in soils: I. The solution that reacts with the soil. *Soil Sci. Soc. Am. Proc.* 23:12-18.

Linke, B., P. Wedekind, and H. Koriath. 1987. Zur N-dynamik bei der aerob-biologischen Aufbereitung von Schweinegülle. *Arch. Acker- Pflanzenbau Bodenkd.* 31:677-687.

Lockyer, D.R. and B.F. Pain. 1989. Ammonia emissions from cattle, pig and poultry wastes applied to pasture. *Environ. Pollut.* 56:19-30.

Lockyer, D.R. and D.C. Whitehead. 1990. Volatilization of ammonia from cattle urine applied to grassland. *Soil Biol. Biochem.* 22:1137-1142.

Long, F.L. 1979. Runoff water quality as affected by surface-applied dairy cattle manure. *J. Environ. Qual.* 8:215-218.

Long, F.L., Z.F. Lund, and R.E. Hermanson. 1975. Effect of soil-incorporated dairy cattle manure on runoff water quality and soil properties. *J. Environ. Qual.* 4:163-166.

Loynachan, T.T., W.V. Bartholomew, and A.G. Wollum. 1976. Nitrogen transformations in aerated swine manure slurries. *J. Environ. Qual.* 5:293-297.

Maag, M. 1989. Denitrifcation losses from soil receiving pig slurry or fertilizer. p. 236-246. In: J.A. Hansen and K. Henriksen (eds.), *Nitrogen in organic wastes applied to soils*. Academic Press, London, England.

Magdoff, F.R. and J.F. Amadon. 1980. Nitrogen availability from sewage sludge. *J. Environ. Qual.* 9:451-455.

Maiwald, K. and O. Siegel. 1937. Untersuchungen über Lagerung und Wirkung von Stalldünger. *Bodenk. Pflanzenernähr.* 5:70-104.

Maiwald, K., and G. Steigmiller. 1938. Untersuchungen über Lagerung und Wirkung von Stalldünger. *Bodenk. Pflanzenernähr.* 9:180-217.

Martin-Smith, M. 1963. Uricolytic enzymes in soil. *Nature* 26:361-362.

Mattingly, G.E.G. 1974. The Woburn organic manuring experiments. *Rothamsted Experimental Station Annual Report for 1973*. Part 2. Rothamsted Exp. Stn., Harpenden, Herts, England. p. 98-151.

Mc Grath, D., D.B.R. Poole, G.A. Fleming, and J. Sinnott. 1982. Soil ingestion by grazing sheep. *Ir. J. Agric. Res.* 21:135-145.

Mc Leod, R.V. and R.O. Hegg. 1984. Pasture runoff water quality from application of inorganic and organic nitrogen sources. *J. Environ. Qual.* 13:122-126.

McNabb, F.M.A. and R.A. McNabb. 1975. Proportions of ammonia, urea, urate and total nitrogen in avian urine and quantitative methods for their analysis on a single urine sample. *Poultry Sci.* 54:1498-1505.

Meek, B.D., A.J. MacKenzie, T.J. Donovan, and W.F. Spencer. 1974. The effect of large applications of manure on movement of nitrate and carbon in an irrigated desert soil. *J. Environ. Qual.* 3:253-258.

Meyer, M. and H. Sticher. 1983. Die Bedeutung des Strohgehaltes für die Erhaltung des Stickstoffs während der Kompostering von Rindermist. *Z. Pflanzenernähr. Bodenk.* 146:199-206.

Miller, M.H., J.B. Robinson, and R.W. Gillham. 1985. Self-sealing of earthen liquid manure storage ponds: I. A case study. *J. Environ. Qual.* 14:533-538.

Milne, C.W. 1976. Effect of a livestock wintering operation on a western mountain stream. *Trans. ASAE* 19:749-752.

Mininni, G. and M. Santori. 1987. Problems and perspectives of sludge utilization in agriculture. *Agric. Ecosystems Environ.* 18:291-311.

Mitchell, J.K. and R.W. Gunther. 1976. The effect of manure applications on runoff, erosion and nitrate losses. *Trans. ASAE* 19:1104-1106.

Molen, van der, J., D.W. Bussink, N. Vertregt, H.G. van Fassen, and D.J. den Boer. 1989. Ammonia volatilization from arable and grassland soils. p. 185-201. In: J. A. Hansen and K. Henriksen (eds.), *Nitrogen in organic wastes applied to soils.* Academic Press, London, England.

Molina, J.A.E., C.E. Clapp, M.J. Shaffer, F.W. Chichester and W.E. Larson. 1983. NCSOIL, A model of nitrogen and carbon transformations in soil: Description, calibration, and behavior. *Soil Sci. Soc. Am. J.* 47:85-91.

Montin, T., P.E. Persson, and S. Sundqvist. 1984. Separering, kompostering. Rapport från 19 svenska avfallsverk. *Swed. Nat. Environ. Agency,* PM 1805:1-159. Solna, Sweden.

Mortland, M.M., and A.R. Wolcott. 1965. Sorption of inorganic nitrogen compounds by soil materials. In: W.V. Bartholomew and F.E. Clark (eds.), Soil nitrogen. Am. Soc. Agron., *Agronomy* 10:150-197. Madison, Wisconsin.

Mosier, A.R., C.E. Andre, and F.G. Viets. 1973. Identification of aliphatic amines volatilized from cattle feedyards. *Environ. Sci. Techn.* 7:642-644.

Mosier, A.R., K. Haider, and F.E. Clark. 1972. Water soluble organic substances leachable from feedlot manure. *J. Environ. Qual.* 1:320-323.

Mosier, A.R., G.L. Hutchinson, B.R. Sabey, and J. Baxter. 1982. Nitrous oxide emissions from barley plots treated with ammonium nitrate or sewage sludge. *J. Environ. Qual.* 11:78-81.

Mote, C.R. and C.L. Griffis. 1982. Heat production by composting organic matter. *Agric. Wastes* 4:65-73.

Muck, R.E. and T.S. Steenhuis. 1982. Nitrogen losses from manure storages. *Agric. Wastes* 4:41-54.

Muck, R.E., R.W. Guest, and B.K. Richards. 1984. Effects of manure storage design on nitrogen conservation. *Agric. Wastes* 10:205-220.

Murwira, H.K., H. Kirchmann, and M.J. Swift. 1990. The effect of moisture on the decomposition rate of cattle manure. *Plant Soil* 122:197-199.

Nakasaki, K., M. Sasaki, M. Shoda, and H. Kubota. 1985a. Change in microbial numbers during thermophilic composting of sewage sludge with reference to $CO_2$ evolution rate. *Appl. Environ. Microbiol.* 49:37-41.

Nakasaki, K., M. Sasaki, M. Shoda, and H. Kubota. 1985b. Characteristics of mesophilic bacteria isolated during thermophilic composting of sewage sludge. *Appl. Environ. Microbiol.* 49:42-45.

Nehring, K. and R. Schiemann. 1951. Beiträge zur Kenntnis der Vorgänge bei der Rotte von Stallmist und Komposten sowie zur Kenntnis der Huminsäuren. *Z. Pflanzenernähr. Düngung Bodenk.* 57:97-113.

Oosterom, H.P. 1984. Influence of some factors for sandy soils on nitrate leaching and transport in groundwater (an experiment with deep lysimeters). *Institute for Land and Water Management Research, Internal Report 1490.* ICW, Wageningen, The Netherlands (in Dutch).

Opperman, M.H., M. Wood, and P.J. Harris. 1989. Changes in inorganic N following the application of cattle slurry to soil at two temperatures. *Soil Biol. Biochem.* 21:319-321.

Pain, B.F., V.R. Philips, C.R. Clarkson, and J.V. Klarenbeek. 1989. Loss of nitrogen through ammonia volatilization during and following the application of pig or cattle slurry to grassland. *J. Sci. Food Agric.* 47, 1-12.

Pain, B.F., R.B. Thompson, Y.J. Rees, and J.H. Skinner. 1990a. Reducing gaseous losses of nitrogen from cattle slurry applied to grassland by the use of additives. *J. Sci. Food Agric.* 50:141-153.

Pain, B.F., V.R. Philips, C.R. Clarkson, T.H. Misselbrook, Y.J. Rees, and J.W. Farrent. 1990b. Odour and ammonia emissions following the spreading of aerobically-treated pig slurry on grassland. *Biol. Wastes* 34:149-160.

Parker, C.F. and L.E. Sommers. 1983. Mineralization of nitrogen in sewage sludges. *J. Environ. Qual.* 12:150-156.

Parnas, H. 1975. Model for decomposition of organic material by microorganisms. *Soil Biol. Biochem.* 7:161-169.

Patni, N.K. and P.Y. Jui. 1985. Volatile fatty acids in stored dairy-cattle slurry. *Agric. Wastes* 13:159-178.

Paul, J.W., and E.G. Beauchamp. 1989a. Biochemical changes in soil beneath a dairy cattle slurry layer: The effect of volatile fatty acid oxidation on denitrification and soil pH. p. 261-270. In: J. A. Hansen and K. Henriksen (eds.) *Nitrogen in organic wastes applied to soils*. Academic Press, London, England.

Paul, J.W. and E.G. Beauchamp. 1989b. Effects of carbon constituents in manure on denitrification in soil. *Can. J. Soil Sci.* 69:49-61.

Paul, J.W. and E.G. Beauchamp. 1989c. Relationship between volatile fatty acids, total ammonia, and pH in manure slurries. *Biol. Wastes* 29:313-318.

Paul, J.W., E.G. Beauchamp, and J.T. Trevors. 1989. Acetate, propionate, butyrate, glucose and sucrose as carbon sources for denitrifying bacteria. *Can. J. Microbiol.* 35:754-759.

Peperzak, P., A.G. Caldwell, R.R. Hunziker, and C.A. Black. 1959. Phosphorus fractions in manures. *Soil Sci.* 87:293-302.

Peschke, H. 1982. Wirkungsvergleich organischer Dünger mittels $^{15}N$-Tracer. *Arch. Acker- Pflanzenbau Bodenkd.* 26:207-216.

Polprasert, C. 1989. Organic waste recycling. John Wiley & Sons, Chichester, England. p. 1-357.

Powlson, D.S., G. Pruden, A.E. Johnston, and D.S. Jenkinson. 1986. The nitrogen cycle in the Broadbalk wheat experiment: recovery and losses of $^{15}N$-labelled fertilizer applied in spring and inputs of nitrogen from the atmosphere. *J. Agric. Sci.* 107:591-609.

Powlson, D.S., P.R. Poulton, T.M. Addiscott, and D.S. McCann. 1989. Leaching of nitrate from soils receiving organic or inorganic fertilizers continuously for 135 years. p. 334-345. In: J.A. Hansen and K. Henriksen (eds.), *Nitrogen in organic wastes applied to soils*. Academic Press, London, England.

Pratt, P.F., S. Davis, and R.G. Sharpless. 1976. A four-year trial with animal manures. I. Nitrogen balances and yields. *Hilgardia* 44:99-112.

Rauhe, K. and H. Bornak. 1970. Die Wirkung von $^{15}N$-markiertem Rinderkot mit verschiedenen Zusätzen im Feldversuch unter Berücksichtigung der Reproduktion der organischen Substanz im Boden. *Albrecht-Thaer-Arch.* 14:937-948.

Rauhe, K. and M. Hesse. 1957. Über die Wirkung verschieden gelagerten Stalldüngers auf leichten und schweren Böden. *Z. Acker- Pflanzenbau* 102:283-298.

Rauhe, K. and V. Koepke. 1967. Der Einfluss unterschiedlicher Verfahren der Stalldunglagerung auf die Stickstoff- und Substanzverluste. *Albrecht-Thaer-Arch.* 11:541-548.

Rauhe, K., E. Fichtner, F. Fichtner, E. Klappe, and W. Drauschke. 1973. Quantifizierung der Wirkung organischer und mineralischer Stickstoffdünger auf Pflanze und Boden unter besonderer Berücksichtigung $^{15}$-markierter tierischer Exkremente. *Arch. Acker- Pflanzenbau Bodenkd.* 17:906-916.

Reddy, K.R., R. Khaleel, and M.R. Overcash. 1980. Nitrogen, phosphorus and carbon transformations in a coastal plain soil treated with animal manures. *Agric. Wastes* 2:225-238.

Rice, C.W., P.E. Sierzega, J.M. Tiedje, and L.W. Jacobs. 1988. Stimulated denitrification in the microenvironment of a biodegradable organic waste injected into soil. *Soil Sci. Soc. Am. J.* 52:102-108.

Richards, I.R. and K.M. Wolton. 1976. A note on the properties of urine excreted by grazing cattle. *J. Sci. Food Agric.* 27:426-428.

Robbins, J.W.D., D.H. Howells, and C.J. Kriz. 1972. Stream pollution from animal production units. *J. Water Pollut. Control Fed.* 44:1536-1544.

Rolston, D.E. and H.J. Liss. 1989. Spatial and temporal variability of water-soluble organic carbon in a cropped field. *Hilgardia* 57(3):1-19.

Rolston, D.E., D.L. Hoffman, and D.W. Toy. 1978. Field measurement of denitrification. I. Flux of $N_2$ and $N_2O$. *Soil Sci. Soc. Am. J.* 42:863-869.

Ronen, D., M. Magaritz, E. Almon, and A. Amiel. 1987. Anthropogenic Anoxification ("Eutrophication") of the water table region of a deep phreatic aquifer. *Water Resour. Res.* 23:1544-1560.

Ross, I.J., S. Sizemore, J.P. Bowden, and C. T. Haan. 1979. Quality of runoff from land receiving surface application and injection of liquid dairy manure. *Trans. ASAE* 22:1058-1062.

Rowsell, J.G., M.H. Miller, and P.H. Groenevelt. 1985. Self-sealing of earthen liquid manure storage ponds: II. Rate and mechanism of sealing. *J. Environ. Qual.* 14:539-543.

Russell, E.J. and E.H. Richards. 1917. The changes taking place during the storage of farmyard manure. *J. Agric. Sci.* 8:495-563.

Ryan, J.A., D.R. Keeney, and L.M. Walsh. 1973. Nitrogen transformations and availability of an anaerobically digested sewage sludge in soil. *J. Environ. Qual.* 2:489-492.

Ryden, J.C., P.R. Ball, and E.A. Garwood. 1984. Nitrate leaching from grassland. *Nature* 311:50-53.

Safley, L.M., D.W. Nelson, and P.W. Westerman. 1983. Conserving manurial nitrogen. *Trans. ASAE* 26:1126-1170.

Safley, L.M., P.W. Westerman, and J.C. Barker. 1985. Fresh dairy manure characteristics and barnlot nutrient losses. p. 191-199. In: *Agricultural waste utilization and management*. Proceedings of the fifth international symposium on agricultural wastes. ASAE, Michigan, USA.

Salo, M.L. 1965. Determination of carbohydrate fractions in animal foods and faeces. *Acta Agralia Fenn.* 105:1-102.

Scheffer, F. and H. Zöberlein. 1937. Untersuchungen über die zweckmässigste Gewinnung und Behandlung des Stallmistes auf der Dungstätte. *Bodenk. Pflanzenernähr.* 5:47-69.

Schefferle, H.E. 1965. The decomposition of uric acid in built up poultry litter. *J. Appl. Bact.* 28:412-420.

Schepers, J.S. and D.D. Francis. 1982. Chemical water quality of runoff from grazing land in Nebraska: I.Influence of Grazing Livestock. *J. Environ. Qual.* 11:351-354.

Schuman, G.E. and T.M. McCalla. 1976. Effect of short-chain fatty acids extracted from beef cattle manure on germination and seedling development. *Appl. Environ. Microbiol.* 31:655-660.

Serna, M.D. and F. Pomares. 1992. Nitrogen mineralization of sludge-amended soil. *Bioresource Technol.* 39:285-290.

Sherwood, M. 1980. The effects of landspreading of animal manures on water quality. p. 379-390. In: J.K.R. Gasser (ed.), *Effluents from livestock*. Applied Science Publishers. London, England.

Sherwood, M. 1986. Nitrate leaching following application of slurry and urine to filed plots. p. 150-157. In: A.D. Kofoed, J.H. Williams, and P. L'Hermite (eds.), *Efficient land use of sludge and manure*. Elsevier Applied Science Publishers, London, England.

Sherwood, M. and A. Fanning. 1981. Nutrient content of surface run-off water from land treated with animal wastes. p. 5-17. In: J.C. Brogan (ed.), *Nitrogen losses and surface run-off from landspreading of manures*. Developments in Plant and Soil Sciences, Vol 2. Martinus Nijhoff/ Dr W. Junk Publishers. The Hague, The Netherlands.

Sibbesen, E. 1990. Kvälstoff, fosfor og kalium i foder, animalsk produktion og husdyrgödning i dansk landbrug i 1980-erne. *Tidsskr. Planteavls Specialserie* 2054:1-21.

Siegel, O. 1936. Experimentelle Grundlagen zur zweckmässigsten Stallmistbereitung unter bäuerlichen Verhätnissen. *Z. Pflanzenernähr. Düngung Bodenk.* 43:186-220.

Siegel, O. and L. Meyer. 1938. Untersuchungen über Lagerung und Wirkung von Stalldünger. *Bodenk. Pflanzenernähr.* 7:190-199.

Sjöqvist, T. and E. Wikander-Johansson. 1985. Vad innehåller slammet? *St. Lantbr.-kem. Lab. Medd.* 52:1-39.

Sluijsmans, C.M. and G.J. Kolenbrander. 1976. De stikstofwerking van stalmest op korte en lange termijn. *Stikstof* 7:349-354.

Smith, L.W. 1973. Recycling animal wastes as protein sources. In: *Alternative sources of protein for animal production.* National Academy of Sciences, Washington, D.C., pp. 147-173.

Smith, J.H. and J.R. Peterson. 1982. Recycling of nitrogen through land application of agricultural, food processing, and municipal wastes. p 791-831. In: F.J. Stevenson et al. (eds.), Nitrogen in agricultural soils. Am. Soc. Agron., *Agronomy* 22. Madison, Wisconsin.

Smith, K.A. and R.J. Unwin. 1983. Fertiliser value of organic manures in the UK. *Fert. Soc. Proc.* (London) 221:1-31.

Smith, K.A. and J.R.M. Arah. 1990. Losses of nitrogen by denitrification and emissions of nitrogen oxides from soils. *Fert. Soc. Proc.* (London) 299:1-34.

SNV (Statens Naturvårdsverk). 1989. Slam från kommunala reningsverk - hantering och miljöproblem. *Swed. Nat. Environ. Agency,* Report 3632:1/55. Solna, Sweden.

Sobel, A.T. and R.E. Muck. 1983. Energy in animal manures. *Energy Agric.* 2:161-176.

Sommer, S.G. and J.E. Olesen. 1991. Effect of dry matter content and temperature on ammonia loss from surface-applied cattle slurry. *J. Environ. Qual.* 20:679-683.

Sommer, S.G., J.E. Olesen, and B.T. Christensen. 1991. Effects of temperature, wind speed and air humidity on ammonia volatilization from surface applied cattle slurry. *J. Agric. Sci.* 117:91-100.

Sommer, S.G., B.T. Christensen, N.E. Nielsen and J.K. Schjörring. 1992. Ammonia volatilization during storage of cattle and pig slurry: Effect of surface cover. *J. Environ. Qual.* (in press).

Sommers, L.E. 1977. Chemical composition of sewage sludges and analysis of their potential use as fertilizer. *J. Environ. Qual.* 6:225-232.

Spatz, G., J. Neuendorff, A. Pape, and C. Schröder. 1992. Nitrogen dynamics under excrement patches in pastures. *Z. Pflanzenernaehr. Bodenk.* 155:301-305 (in German).

Spoelstra, S.F. 1979. Volatile fatty acids in anaerobically stored piggery wastes. *Netherl.. J. Agric. Sci.* 27:60-66.

Stark, S.A. and C.E. Clapp. 1980. Residual nitrogen availability from soils treated with sewage sludge in a field experiment. *J. Environ. Qual.* 9:505-512.

Steenhuis, T.S., G.D. Bubenzer, J.C. Converse, and M.F. Walter. 1981. Winter-spread manure nitrogen loss. *Trans. ASAE* 24:436-449.

Steenvoorden, J.H.A.M. 1986. Nutrient leaching losses following application of farm slurry and water quality considerations in the Netherlands. p. 168-176. In: A.D. Kofoed, J.H. Williams, and P. L'Hermite (eds.), *Efficient land use of sludge and manure.* Elsevier Applied Science Publishers. London, England.

Steenvoorden, J.H.A.M., H. Fonck, and H.P. Oosterom. 1986. Losses of nitrogen from intensive grassland systems by leaching and surface runoff. p. 85-97. In: H.G. van der Meer, J.C. Ryden, and G.C. Ennik (eds.), *Nitrogen fluxes in intensive grassland systems.* Developments in Plant and Soil Sciences, Vol. 23. Martinus Nijhoff Publishers, Dordrecht, The Netherlands.

Stephenson, G.R. and L.V. Street. 1978. Bacterial variations in streams from a southwest Idaho rangeland watershed. *J. Environ. Qual.* 7:150-157.

Stevens, R.J. and I.S. Cornforth, I.S. 1974. The effect of pig slurry applied to a soil surface on the composition of the soil atmosphere. *J. Sci. Food Agric.* 25:1263-1272.

Stevens, R.J. and H.J. Logan. 1987. Determination of the volatilization of ammonia from surface-applied cattle slurry by the micrometeorological mass balance method. *J. Agric. Sci.* 109:205-207.

Stevens, R.J., R.J. Laughlin, and J.P. Frost. 1989. Effect of acidification with sulphuric acid on the volatilization of ammonia from cow and pig slurries. *J. Agric. Sci.* 113:389-395.

Stewart, B.A. 1970. Volatilization and nitrification of nitrogen from urine under simulated cattle feedlot conditions. *Environ. Sci. Technol.* 4:579-582.

Terman, G.L. 1979. Volatilization losses of nitrogen as ammonia from surface applied fertilizers, organic amendments, and crop residues. *Adv. Agron.* 31, 189-223.

Terpstra, K., and N. De Hart. 1974. The estimation of urinary nitrogen and faecal nitrogen in poultry excreta. *Z. Tierphysiol., Tierernährg. u. Futtermittelkde.* 32:306-320.

Tester, C.F. and J.F. Parr. 1983. Decomposition of sewage sludge compost in soil: IV. Effect of indigenous salinity. *J. Environ. Qual.* 12:123-126.

Thelin, R. and G.F. Gifford. 1983. Fecal coliform release patterns from fecal material of cattle. *J. Environ. Qual.* 12:57-63.

Thiele, H.J. and J.G. Zeikus. 1988. Interactions between hydrogen- and formate producing bacteria and methanogens during anaerobic digestion. p. 537-595. In: L.E. Erickson and D.Y-C. Funk (eds.) *Handbook on anaerobic fermentations*. Marcel Dekker Inc., New York, USA.

Thomas, R.J., K.A.B. Logan, A.D. Ironside, and J.A. Milne. 1986. Fate of sheep urine-N to an upland grass sward. *Plant Soil* 91:425-427.

Thomas, R.J., K.A.B. Logan, A.D. Ironside, and G.R. Bolton. 1988. Transformations and fate of sheep urine-N applied to an upland U.K. pasture at different times during the growing season. *Plant Soil* 107:173-181.

Thompson, R.B. 1989. Denitrifcation in slurry-treated soil: Occurrence at low temperatures, relationship with soil nitrate and reduction by nitrification inhibitors. *Soil Biol. Biochem.* 21:875-882.

Thompson, R.B., J.C. Ryden, and D.R. Lockyer. 1987. Fate of nitrogen in cattle slurry following surface application or injection to grassland. *J. Soil Sci.* 38:689-700.

Thompson, R.B., B.F. Pain, and D.R. Lockyer. 1990a. Ammonia volatilization from cattle slurry following surface application to grassland. *Plant Soil* 125:109-117.

Thompson, R.B., B.F. Pain, and Y.J. Rees. 1990b. Ammonia volatilization from cattle slurry following surface application to grassland. *Plant Soil* 125:119-128.

Tietjen, C. 1977. Die Zusammensetzung und Eigenschaften tierischer Abfälle. p. 17-27. In: B. Strauch et al. (eds.), *Abfälle aus der Tierhaltung*. Verlag Eugen Ulmer, Stuttgart, Germany.

Tinsley, J. and T.Z. Nowakowsky. 1959. The composition and manurial values of poultry excreta, straw-droppings composts and deep litter. II. Experimental studies on composts. *J. Sci. Food Agric.* 10:150-167.

Torstensson, G., A. Gustafson, B. Lindén, and G. Skyggesson. 1992. Dynamics of mineral nitrogen and leaching of nitrogen in a silty soil fertilized with mineral fertilizer and slurry in the county of Halland. Swedish University of Agricultural Sciences, Division of Water Management. *Ekohydrologi* Report 28. Uppsala, Sweden. p. 1-24 (in Swedish).

Travis, D.O., W.L. Powers, L.S. Murphy, and R.I. Lipper. 1971. Effect of feedlot lagoon water on some physical and chemical properties of soils. *Soil Sci. Soc. Am. Proc.* 35:122-126.

Uhlen, G. 1978. Nutrient leaching and surface runoff in field lysimeters on a cultivated soil. II. Effects of farmyard manure spread on a frozen ground and mixed in the soil on water pollution. *Scientific Reports of the Agricultural University of Norway* 57(28):1-23.

Urban, N.R., S.E. Bayley, and S.J. Eisenreich. 1989. Export of dissolved organic carbon and acidity from peatlands. *Water Resour. Res.* 25:1619-1628.

Veen, van, J.A., J.N. Ladd, and M.J. Frissel. 1984. Modelling C and N turnover through the microbial biomass in soil. *Plant Soil* 76:257-274.

Vetter, H. and G. Steffens. 1981. Leaching of nitrogen after the spreading of slurry. p. 251-269. In: J.C. Brogan (ed.), *Nitrogen losses and surface run-off from landspreading of manures.* Developments in Plant and Soil Sciences, Vol 2. Martinus Nijhoff/ Dr W. Junk Publishers. The Hague, The Netherlands.

Vogtmann, H., H. Schmeisky, F. Fricke, T. Turk, D. Bergmann, and W. Rehm. 1984. Die Aktion "Grüne Mülltonne". *Müll und Abfall* 8:247-248.

Vuori, A.T. and J.M. Nasi. 1977. Fermentation of poultry manure for poultry diets. *Brit. Poultry Sci.* 18:257-264.

Vuuren, van A.M. and J.A.C. Meijs. 1987. Effects of herbage composition and supplement feeding on the excretion of nitrogen in dung and urine by grazing dairy cows. p. 17-26. In: H.G. van der Meer et al. (eds.), *Animal manure on grassland and fodder crops. Fertilizer or waste?* Martinus Nijhoff Publishers, Dordrecht, The Netherlands.

Wedekind, P. and G. Kuehn. 1971. Über Substanz- und Nährstoffverluste bei der Lagerung von Gülle und ihren festen und flüssigen Komponenten. *Arch. Bodenfruchtbark. Pflanzenprod.* 15:333-343.

Wermke, M. 1987. Nitrate in soil water at different depth during the growing season on permanent pasture (Comparison of mineral fertilizer and liquid manure). p. 373-376. In: H.G. van der Meer, R.J. Unwin, T.A. van Dijk, and G.C. Ennik (eds.), *Animal manure on grassland and fodder crops. Fertilizer or waste?* Developments in Plant and Soil Sciences, No 30. Martinus Nijhoff Publishers. Dordrecht, The Netherlands.

Westerman, P.W., M.R. Overcash, R.O. Evans, L.D. King, J.C. Burns, and G.A. Cummings. 1985. Swine lagoon effluent applied to "coastal" bermudagrass: III. Irrigation and rainfall runoff. *J. Environ. Qual.* 14:22-25.

Whitehead, D.C. and N. Raistrick. 1990. Ammonia volatilization from five nitrogen compounds used as fertilizers following surface application to soils. *J. Soil Sci.* 41:387-394.

Whitehead, D.C., D.R. Lockyer, and N. Raistrick. 1989. Volatilization of ammonia from urea applied to soil: Influence of hippuric acid and other constituents of livestock urine. *Soil Biol. Biochem.* 21:803-808.

WHO. 1984. Guidelines for drinking-water quality. Vol 1. Recommendations. WHO, Geneva, Switzerland.

Witter, E. 1991. Use of $CaCl_2$ to decrease ammonia volatilization of fresh and anaerobic chicken slurry to soil. *J. Soil Sci.* 42: 369-380.

Witter, E. and J. Lopez-Real. 1988. Nitrogen losses during composting of sewage sludge and the effectiveness of clay soil, zeolite and compost in adsorbing the volatilized ammonia. *Biol. Wastes* 23:279-294.

Witter, E. and H. Kirchmann. 1989a. Peat, zeolite and basalt as adsorbents of ammoniacal nitrogen during manure decomposition. *Plant Soil* 115:43-52.

Witter, E. and H. Kirchmann. 1989b. Effects of addition of calcium and magnesium salts on ammonia volatilization during manure decomposition. *Plant Soil* 115:53-58.

Xin-Tao, H., S.T. Traina, and T.J. Logan. 1992. Chemcial properties of municipal solid waste composts. *J. Environ. Anal.* 21:318-329.

Yokoyama, K., H. Hai, and H. Tsuchiyama. 1991. Paracoprid dung beetles and gaseous loss of nitrogen from cow dung. *Soil Biol. Biochem.* 23:643-647.

Young, R.A. and C.K. Mutchler. 1976. Pollution potential of manure spread on frozen ground. *J. Environ. Qual.* 5:174-181.

Zucconi, F. and M. de Bertoldi. 1987. Compost specifications for the production and characterization of compost from municipal solid waste. p. 30-50. In: M. de Bertoldi et al. (eds.), *Compost: Production, quality and use*. Elsevier Applied Science, London, England.

# Soil and Water Contamination by Heavy Metals

Bal Ram Singh and Eiliv Steinnes

I. Introduction . . . 233
II. Background Levels of Metals in Soils and Waters . . . 234
A. Heavy Metals in Soils . . . 235
B. Heavy Metals in Waters . . . 237
III. Anthropogenic Sources of Metals in Soils and Waters . . . 238
A. Fertilizers and Amendments . . . 238
B. Sewage Sludge . . . 241
C. Pesticides . . . 243
D. Industry and Mining . . . 244
E. Atmospheric Deposition . . . 247
IV. Soil Processes . . . 249
A. Weathering and Accumulation . . . 249
B. Soil pH and Redox Conditions . . . 249
C. Adsorption and Desorption . . . 251
D. Mobility and Transport . . . 252
E. Microbial Activity . . . 253
V. Heavy Metals in Water . . . 255
A. General Aspects . . . 255
B. Processes in Water . . . 256
C. Runoff Waters, Rivers and Streams . . . 257
D. Lakes . . . 257
E. Groundwater . . . 258
F. Fresh Water Sediments . . . 258
VI. Ecological Implications and Regulatory Control . . . 259
VII. Future Outlook and Research Needs . . . 262
References . . . 263

## I. Introduction

In recent years, greater attention has been focused on heavy metals in soils and waters, partly because of scientific and public awareness of environmental issues, and partly because modern analytical techniques available for determina-

0-87371-980-8/94/$0.00+$.50

tion of very small quantities of these metals. In many parts of the world, particularly in the vicinity of urban and industrial areas, abnormally high concentrations of heavy metals have been reported. Some of these metals such as Cu, Zn, Mn, Fe, and perhaps also Cr are biologically essential and play an important role as cofactor in enzymatic processes.Others such as As, Cd, Pb, and Hg are not known to have any essential functions, but may create health problems when entering the food chain. Therefore, increased attention has been paid in recent decades to the monitoring and management of these non-essential but potentially toxic metals to prohibit their entry into the food chain. A number of cases of health problems related to environmental Cd and Hg poisoning, and elevated levels of Pb in the blood of infants have been reported. Of the various metals known mainly Cd, Cr, Co, Cu, Fe, Hg, Mn, Mo, Ni, Pb, Sn, and Zn are used in industry and out of them Cd, Cu, Hg, Ni, Pb, and Zn are considered most dangerous. In this paper attention will be paid to these heavy metals. In addition As and Cr, also sometimes associated with health problems, have been included.

Heavy metals initially present or added as pollutants are distributed in solid and liquid phases, and the partitioning of metals between solid and liquid phases is influenced by a number of factors. Complexation of heavy metals with various organic and inorganic ligands, oxidation and reduction reactions, ion exchange-adsorption reactions, and dissolution-precipitation of solid phases along with mass transfer continually dictate the quantities and concentrations of metals in each phase. These complex interactions are the main controlling factors for mobility and bioavailability of heavy metals in soils.

Soil and water systems are interrelated through a complex web of interactions and any change in one system will cause varying degree of changes in the other. In several instances the metal enrichment exceeds that of the natural environments many times: the levels of Hg, Cd, and Zn in organisms are 10-100 times greater in polluted inland and coastal waters than those in less contaminated areas (Bryan, 1976). A similar increase above "normal" levels of these and other metals was established in water samples from polluted rivers such as the lower Rhine River (Förstner, 1981).

In this paper attempts are made to summarize information on background levels of metals in soils and waters, anthropogenic sources of metals to soil and water systems, important soil processes which dictate the mobility and bioavailability of metals, and contamination of different water systems with heavy metals. It is not the intention of this paper to provide a complete review of the literature but rather to focus on some important aspects of this subject from recent literature.

## II. Background Levels of Metals in Soils and Waters

The heavy metals of concern in terms of environmental pollution are present in rocks, soils, waters, and air in small amounts originated from natural geological materials which in turn may directly or indirectly affect the chemical composition of potable water supplies, foodstuff and animal feed, airborne particulates and dust. It is, therefore, of utmost importance to know the average background amounts of each metal in the natural uncontaminated geological materials, soils, and waters before assessing the contribution of anthropogenic sources to environmental contamination.

### A. Heavy Metals in Soils

The heavy metals in soils are derived either from the weathering of the parent material or from numerous external contaminating sources. For some metals, such as lead, contamination from other sources often far exceeds the contribution from natural sources, whereas for the less used metals this may not be the case. The metal content of soils is the result of soil-forming factors acting through time. Among these, the main factor that dictates the metal content of a soil is the composition of parent material. A close relationship between the metal content of parent materials and soils that develop during in situ weathering has been observed in a number of studies. The mean concentration of some of the heavy metals in major rock types is given in Table 1. It can be seen from this table that the heavy metal concentrations in major rock type is generally low. In some sedimentary rocks, however, several heavy metals may reach much higher levels, such as in the Norwegian shale included in Table 1. The level of heavy metals in the Norwegian shale is generally higher than in black shales (Spencer, 1966; Rose et al. 1979). The high levels of metals in shales have consequences for the soil developed on these rocks. A typical example of this is shown in Table 2 where soils developed on shales contained many times higher levels of metals as compared to those developed on other materials. The table also shows that not only the total concentration of metals in shale was higher but also the $NH_4$Ac-EDTA extractable fractions of metals were much higher than in other soils. These higher levels may have practical and ecological consequences in terms of their transfer to foodstuff and fodder as metals extracted by $NH_4$Ac-EDTA are considered as the easily soluble and plant available fraction by many investigators (Singh and Narwal, 1984; Ervio et al. 1990; Jeng and Bergseth, 1992; He and Singh, 1993a).

Although the true background concentrations of metals in soils may not be easy to obtain, the concentrations found in some remote areas or in areas with minimum of anthropogenic effects may be taken as background levels for practical purposes. Probable background and typical normal ranges of heavy metals in soils are presented in Table 3. For a number of metals e.g., As and Hg, the differences between background levels and normal range are not large but for others such as Cd, Cu, Pb, and Zn these differences are greater.

**Table 1.** Mean concentrations of heavy metals in rocks

| | -----Igneous rocks----- | | ----------Sedimentary rocks------------ | | |
|---|---|---|---|---|---|
| Element | Basalt | Granite | Limestone | Sandstone | Nor. shale[a] |
| | mg $kg^{-1}$ | | | | |
| As | 1.5 | 1.5 | 1.0 | 1.0 | - |
| Cd | 0.13 | 0.09 | 0.03 | 0.05 | 5.7 |
| Cr | 200 | 4 | 11 | 35 | - |
| Cu | 90 | 13 | 5.5 | 30 | 118 |
| Hg | 0.01 | 0.08 | 0.16 | 0.29 | - |
| Ni | 150 | 0.5 | 7 | 9 | 223 |
| Pb | 3 | 24 | 5.7 | 10 | 84 |
| Zn | 100 | 52 | 20 | 30 | 178 |

[a] Norwegian shale (Average values of four shale samples of varying composition).
From Krauskopf, 1967; Bowen, 1979; Thornton, 1981; Fergusson, 1990; Alloway, 1990; Jeng, 1992.

**Table 2.** Heavy metal concentrations in soils developed on Norwegian alum shales as compared to those developed on other geological material

| | Conc. aqua regia extract | | $NH_4OAc$-EDTA extract | |
|---|---|---|---|---|
| Metals | Mean | s.d. | Mean | s.d. |
| | mg $kg^{-1}$ | | | |
| | (Soils developed on shales) | | | |
| Cd (n = 4) | 2.8 | 0.3 | 2.0 | 0.1 |
| Cu | 97.0 | 16.8 | 16.3 | 1.9 |
| Ni | 114.8 | 22.8 | 27.0 | 3.2 |
| Pb | 29.9 | 1.6 | 4.3 | 0.9 |
| Zn | 217.3 | 0.9 | 33.5 | 10.6 |
| | (Soils developed on other geological material) | | | |
| Cd (n=10) | 0.21 | 0.05 | 0.12 | 0.03 |
| Cu | 16.6 | 5.7 | 4.1 | 2.0 |
| Ni | 20.1 | 10.3 | 1.4 | 1.0 |
| Pb | 20.2 | 4.4 | 4.8 | 1.8 |
| Zn | 80.7 | 33.1 | 4.7 | 4.0 |

(Jeng and Bergseth, 1992.)

**Table 3.** Probable background levels and typical concentrations of some heavy metals in soils

| Element | Background concentration | Typical normal range |
|---|---|---|
| | mg $kg^{-1}$ | |
| As | 0.1-40 | 0.1-50 |
| Cd | 0.01-0.2 | 0.01-2.4 |
| Cr | 80-200 | 5-1500 |
| Cu | 6-60 | 2-250 |
| Hg | 0.06 | 0.01-0.3 |
| Ni | 1-100 | 2-1000 |
| Pb | 12-20 | 2-300 |
| Zn | 17-125 | 10-300 |

(Bowen, 1979; Balsberg-Påhlsson et al., 1982; Kabata-Pendias and Pendias, 1984; Fergusson, 1990; and Alloway, 1990.)

## B. Heavy Metals in Waters

The water present in streams, rivers and lakes constitutes only about 0.002% of the total amount of water present on the surface of the earth (Garrels and Mackenzie, 1971). Still this fraction of natural waters is by far the most important one for mankind. The amount of water discharged annually by rivers to the world oceans is of the same order as the total amount of water present in rivers and lakes. The amount of fresh water present as groundwater is much larger, possibly about 30-fold the surface fresh water (Lvovitch, 1970), but is not so readily accessible to man and other biota.

Two physical factors, the residence time and the pathways along which the water moves through the system are particularly important with respect to the chemical composition of natural waters (Bricker, 1987). The larger the residence time, the better opportunity for reaction between water and materials in contact with it. The pathways determine which materials the water contacts during its passage through the system. In general water following shallow pathways gets in contact with more weathered and consequently less reactive material then along deeper pathways.

The two factors dominating the chemical composition of natural fresh surface waters are the atmosphere and the mineral material in contact with the water. Trace elements, including the heavy metals discussed in this paper, are also derived from these sources. The biosphere is also important in influencing the chemical composition of surface waters, both by direct action of biota and through the dead organic material produced from them.

Natural fresh waters show a very wide range in their concentrations of heavy metals, as indicated in Table 4, depending on variations in climate, geology, and anthropogenic activities (Bowen, 1979). The median values however are generally well below the limits considered safe for human drinking water

**Table 4.** Heavy metal compostition of drinking water

| Metal | Median | Range |
|---|---|---|
| | ------$\mu g\ L^{-1}$------ | |
| As | 0.5 | 0.2-230 |
| Cd | 0.1 | 0.01-3 |
| Cr | 1 | 0.1-6 |
| Cu | 3 | 0.2-30 |
| Hg | 0.1 | 0.0001-2.8 |
| Ni | 0.5 | 0.02-27 |
| Pb | 3 | 0.06-120 |
| Zn | 15 | 0.02-100 |

(Bowen, 1979.)

consumption. In the case of Hg, recent data (Lindqvist et al., 1991) indicate that the median value in Table 4 may be too high, probably due to improper sample handling in some of the earlier investigations.

## III. Anthropogenic Sources of Metals in Soils and Waters

Anthropogenic metal inputs are a major source of metal contamination of soils and waters. It is estimated that the anthropogenic emissions of As, Cd, Cu, Ni, and Zn exceed the inputs of these metals from natural sources by two-fold or more; in the case of Pb, the ratio of anthropogenic to natural emission rate is about 17 (Nriagu and Pacyna,1988). The sources of anthropogenic inputs of heavy metals to soils are either primary sources such as fertilizers and amendments, sewage sludge, and pesticides or secondary sources where metals are added to the soil as a consequence of a nearby activities such as mining and industry or aerosol deposition form more distant sources. Figure 1 shows schematically different types of anthropogenic metal input to soils and waters.

### A. Fertilizers and Amendments

In spite of the vital importance of fertilizers and amendments for increased crop production, when used in large amounts they can cause contamination of soils and waters because they contain some undesired impurities such as heavy metals. It is known that fertilizers and amendments contain many heavy metals in significant concentrations (Swaine, 1962; Singh, 1991). The concentration of metals in fertilizers is dependent on the concentrations of these metals in rock phosphate and also on the gain or losses that occur during the manufacturing process. Table 5 shows the mean concentrations of heavy metals in some important phosphate rock sources. The concentrations of Hg and Pb in most of

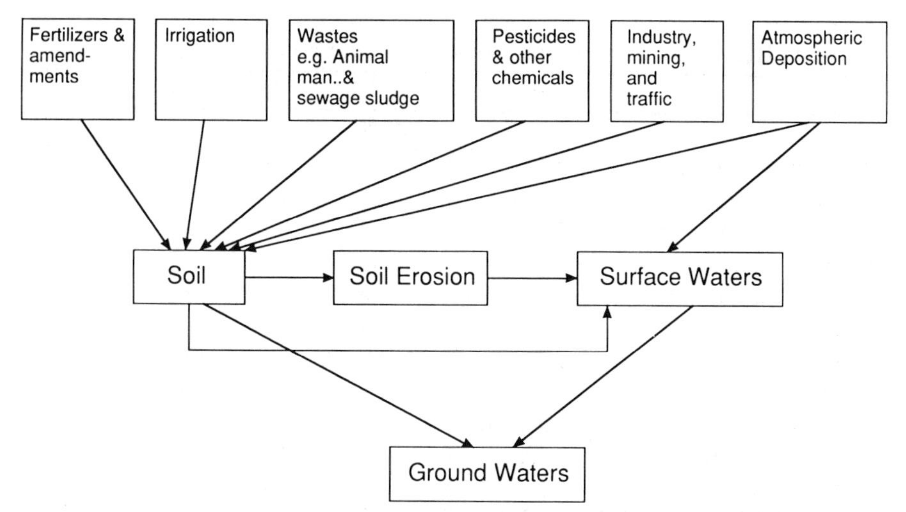


**Figure 1.** Anthropogenic sources of metal contamination of soils and waters.

the rocks are low and hence of minor importance but the concentration of Ni may be of greater significance.

Ranges of heavy metal concentrations in N, P, and NPK fertilizers and in liming materials are presented in Table 6. In some of the fertilizers, the maximum concentrations of As, Cd, Cu, and Zn clearly exceed the background levels of these metals in soils. It may therefore be expected that use of such fertilizers can result in the accumulation of these metals in cultivated soils. Cadmium in particular is singled out as one of the more dangerous contaminants because of the possibility of its transfer from the soil to the human food chain, and its association with various health problems. Commercial fertilizers together with atmospheric deposition are major sources for Cd input to cultivated soils and hence Cd contamination of soils and plants has received increased attention by many investigators in recent decades. Five- to twelve-fold increase in Cd content in the upper 10 cm of soil as a result of phosphate fertilization was observed by Williams and David (1973, 1976). Mulla et al. (1980) found a fourteen-fold increase in Cd content in surface soils after continuous fertilization with high rate of triple superphosphate (1975 kg P $ha^{-1}$ $y^{-1}$) for a period of 36 years. In a comprehensive study on Cd levels in soils and plants after long-term use of commercial fertilizers in Norway, it was found that the soils from long-term cultivated fields (> 40 years) contained significantly higher total Cd (Bærug and Singh, 1990) as well as extracatble Cd (He and Singh, 1993a and b) than those from newly cultivated fields (< 4 years).

A better insight was obtained by comparing the extractable Cd between the two groups of soils according to field types, grown with different plant species in different regions (Figure 2). Highly significant differences in the extractable Cd between the newly and old cultivated soils were found for soils from timothy grass fields in central Norway and those of mixed grass fields in southeastern

**Table 5.** Heavy metal contents in some phosphate rocks

| | Cd | Hg | Pb | Ni | Zn | As |
|---|---|---|---|---|---|---|
| | ---------------mg kg$^{-1}$-------------- | | | | | |
| Khouribga (Morocco) | 18 | 0.04 | 2 | 30 | 270 | 10 |
| Pebble (Florida) | 5 | 0.09 | 12 | 13 | 80 | 5 |
| West reg. rocks (USA) | 100 | - | 12 | 85 | 870 | 92 |
| Palabora rock (S.Afr.) | 0.15 | - | 35 | 35 | 6 | 6 |
| Gafsa (Tunisia) | 34 | 0.03 | 2 | 16 | 290 | 2.1 |
| Taiba (Senegal) | 71 | 0.33 | 4 | 53 | 500 | 2.3 |
| Israel rock | 20 | - | - | 40 | 450 | 6 |
| Jordian rock | 6 | - | 4 | 17 | 250 | 7 |
| Kola (Russia) | 0.2 | 0.01 | 4 | 0.5 | 20 | <1 |
| Gränges (Sweden) | 1.1 | 0.04 | 3 | 15 | 20 | 235 |
| Lkab (Sweden) | 0.3 | 0.15 | 6 | 20 | - | 220 |

(Gunnarsson, 1983; Konghaug et al., 1992; Schultz et al., 1992.)

**Table 6.** Range of concentration of heavy metals in some fertilizers and lime materials

| Element | N[a] | P[a] | NPK[b] | Lime[a] |
|---|---|---|---|---|
| | -----------------------mg kg$^{-1}$----------------------- | | | |
| As | 2.2-120 | 2-1200 | - | 0.1-24 |
| Cd | 0.05-8.5 | 0.1-170 | 0.1-10 | 0.04-0.1 |
| Cr | 0.3-2.9 | 66-245 | 20-72 | 10-15 |
| Cu | <1-15 | 1-300 | 4-38 | 2-125 |
| Hg | 0.3-2.9 | 0.01-1.2 | 0.01-0.1 | 0.05 |
| Ni | 7-34 | 7-38 | 9-20 | 10-20 |
| Pb | 2-27 | 7-225 | 10-130 | 20-1250 |
| Zn | 1-42 | 50-1450 | 22-350 | 10-450 |

([a] Kabata-Pendias and Pendias, 1984; [b] Gunnarsson, 1983.)

Norway. It was, therefore, concluded that fertilization with Cd-containing fertilizer or with farmyard manure seems responsible for the increased Cd concentration in the old cultivated soils because the paired samples were collected from an area with similar parent material or atmospheric deposition and the sampled fields did not have a history of sludge application.

A summary of results from other parts of the world presented in Table 7 shows the same trend as reported from Norway. In general, Cd accumulation in soils as a result of commercial fertilizer or farmyard manure application was found in most of the cases. Although the concentrations of metals such as Cu, Pb, and Zn in some liming materials can exceed their background levels in

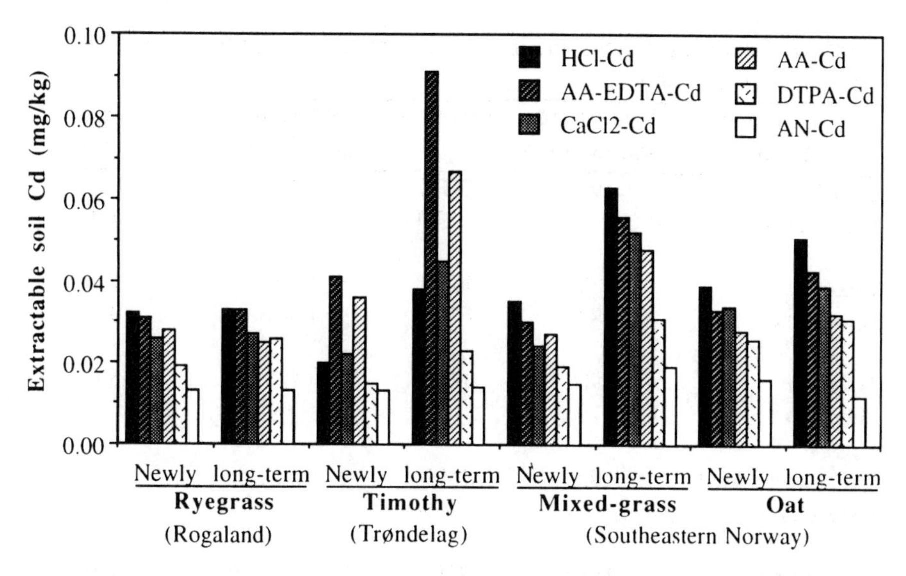


**Figure 2.** Extractable Cd in the newly and long-term cultivated soils from the fields grown with different crops (AA, DTPA, AA-IDTA, An, Hcl, and $CaCl_2$ stand for $NH_4OAc$, DTPA, $NH_4OAc$-EDTA, $NH_4NO_3$, HCl, and $CaCl_2$-extractable Cd, respectively). (He and Singh, 1993a.)

soils, cases of soil contamination with this source has not been reported for these elements.

## B. Sewage Sludge

Metal-containing sewage sludge when applied to agricultural soils can cause contamination of soils, crops, and waters because soil works as a dynamic entity through which metals pass during recycling from the waste to plants, water or air. Sewage sludge due to its high organic matter, N, and P contents is considered suitable for land disposal. It is estimated that about 42% of the 5.6 million tonnes/year of sludge produced in the USA is applied to agricultural land and similarly about 40% of the 5.9 million tonnes/year produced in western Europe is also applied to farmlands (Brown and Jacobson, 1987). Unfortunately, sludges contain a considerable amount of heavy metals which may persist in surface soils long after their application.

Some typical concentration ranges of heavy metals in sludges from several industrialized countries are presented in Table 8. There is a fairly large variation in metal concentrations in sludges from different countries which may reflect differences in the nature of the industrial inputs to waste water systems and the degree of stringency in regulations enforced to control effluents from the industry. The concentrations of most of the metals are generally higher in

**Table 7.** Cadmium enrichment in soils as a result of P-fertilization in some long-term experiments

| Place and country | Year of start and duration | Fertilizer | P-amount kg ha$^{-1}$ year$^{-1}$ | Cd-enrich g ha$^{-1}$ year$^{-1}$ |
|---|---|---|---|---|
| Halle[a] (E. Germany) | 1978 (103) | NPK<br>FYM | 24<br>11 | 0.3<br>4.4 |
| Askov[a] (Denmark) | 1894 (80) | NPK<br>FYM | 17-36<br>17-36 | 1.0<br>0.5 |
| Grignon[a] (France) | 1902 (79) | NPK | 40 | 5.0 |
| Crossenzersdorf[a] (Austria) | 1906 (75) | NPK<br>FYM | 44 | 1.9<br>0.9 |
| Rothamsted-Broadback (UK)[b] | 1881 (102) | NIC<br>NPK | -<br>35 | 2.9<br>2.6 |
| Morrow Plots[c] Illinois (USA) | 1976 (102) | TSP<br>FYM | 20<br>22 | -0.5<br>0.5 |

([a]Dam Kofoed and Søndergård-Klausen, 1983; [b]Jones and Johnston, 1989; [c]Mortvedt, 1987.)

**Table 8.** Concentrations of heavy metals in sewage sludge

| Country | Cd | Cr | Cu | Hg | Ni | Pb | Zn |
|---|---|---|---|---|---|---|---|
| | mg kg$^{-1}$ | | | | | | |
| U.S.A. (200 tmt. plants) | 3-3,410 | 10-99,000 | 84-10,400 | 0.5-10,600 | 2-3,520 | 13-19,730 | 101-27,800 |
| U.K. (42 tmt. plants) | 60-1,500 | 40-8,600 | 200-8,000 | - | 20-5,300 | 120-3,000 | 700-49,000 |
| Norway (38 tmt. plants) | <1-14 | 8-3,700 | 57-1,650 | 0.6-290 | 5-272 | 5-432 | 57-1,567 |
| Denmark (22 tmt. plants) | 5-58 | 37-3,675 | 106-2,264 | 2.7-38 | 17-327 | 188-3,898 | 1218-17,414 |
| W. Germany (500 samples) | 4-193 | 1-7,825 | 17-4,864 | 4-24 | 1-1,930 | 1-1,930 | 25-7,588 |

(Sommers et al., 1976; Berrow and Webber, 1972; Hucker, 1980; Hallberg and Vigerust, 1981; Pauly and Simonsen, 1974; C.E.C., 1982.)

**Table 9.** Concentrations of Cd and Zn in sludge-amended soils using various extractions

| Treatment | Soil pH | Extractant | Cd | Zn |
|---|---|---|---|---|
| | | | -------------mg $kg^{-1}$------------ | |
| Control | 5.35 | $Ca(NO_3)_2$ | 0.004 ± 0.0008 | 0.21 ± 0.05 |
| | | DTPA | 0.099 ± 0.024 | 0.13 ± 0.20 |
| | | Total[a] | 0.160 ± 0.03 | 5.4 ± 4.8 |
| Sludge | 5.77 | $Ca(NO_3)_2$ | 0.13 ± 0.19 | 2.3 ± 0.14 |
| | | DTPA | 2.4 ± 0.29 | 3.0 ± 0.18 |
| | | Total | 5.0 ± 0.33 | 153 ± 6.8 |

[a]Total metal concentration is by aqua regia digestion.
(Bell et al., 1991.)

sludges from the United States than in those from the European countries, especially from Scandinavia.

It is widely reported that the treatment of soils with sludge often results in increased concentrations of metals in the food chain, particularly of Cd, Cu, Ni, and Zn which are considered to be phytotoxic, . However, the extent of contamination of soils and waters and eventually food chains depends upon the metal concentrations in the sludge applied, soil type, topography, climate, vegetation, and soil pH.

Numerous studies have shown large increases in metal concentrations in soils as a result of sludge addition e.g., Kelling et al. (1977); Sposito et al. (1982); Bell et al. (1991). Bell et al. (1991) showed many fold increase in available content of metals in sludge-amended soils from a long-term experiment (Table 9). Kelling et al. (1977) also found that application of sludge significantly increased the concentrations of DTPA-extractable Cu, Cd, N, and Zn in the soil.

## C. Pesticides

Inorganic pesticides have long been used to control pests in fruit orchards and vegetable crops. Although their use in many countries, especially in developed ones, has been supplemented by synthetic organic pesticides, many cases of metal contamination of soils and waters have been reported because of long persistence of metal-containing pesticides. Frank et al. (1976) reported levels as high as 890 mg $kg^{-1}$ Pb and 126 mg $kg^{-1}$ As in the surface soils of apple (*Malus pumila*) orchards in Ontario, Canada (Figure 3) while Tiller and Merry (1982) found 1000 mg $kg^{-1}$ Cu in the surface soils of orchards where Bordeaux spray (Cu based) were used over long period. These concentrations are orders of magnitude higher than their background levels in soils.

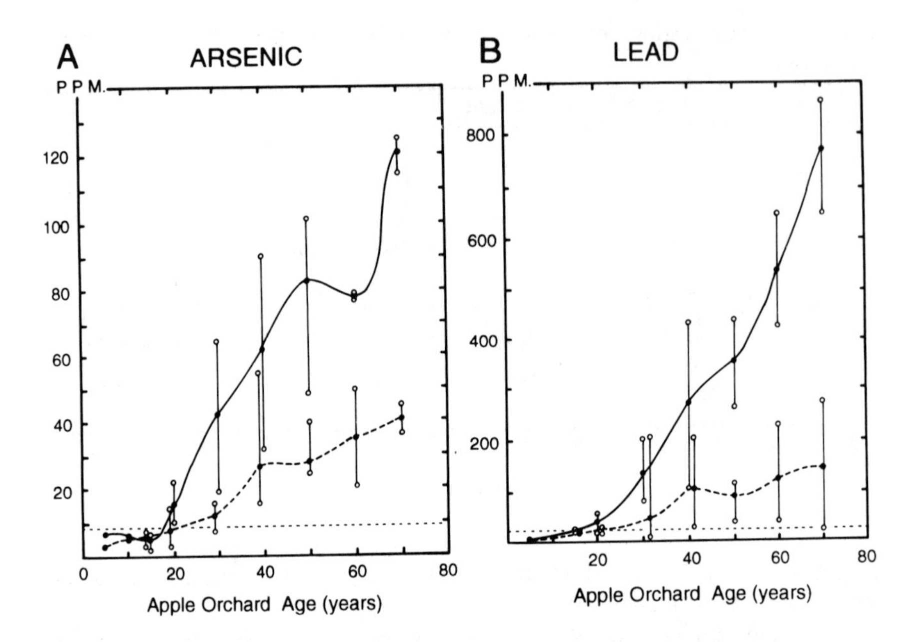


**Figure 3.** The accumulation of As(A) and Pb(B) in the surface soil (0-15 cm) and subsurface soil (15-30 cm) in 31 apple orchards ranging in age from 5 to 70 yr and to which lead arsenate has been applied at various annual rates over the life of the orchard (———— surface soil, ------- subsurface soil, ……. background soil contents). (Frank et al., 1976.)

Mercurial fungicides contribute to the environmental Hg burden because they are persistent and slowly degrading products. The use of organic mercurials as seedcoat dressing or to control turf fungal disease has been shown to raise the levels of Hg in soils. Gowen et al. (1976) stated that up to 10 g Hg $ha^{-1}$ could be added to soils through seed dressing. Maclean et al. (1973) reported Hg concentrations ranging from 24 to 120 mg $kg^{-1}$ in the surface (0-5 cm) soil of a golf course.

## D. Industry and Mining

It is estimated that for heavy metals such as Cd, Cr, Cu, Hg, Pb, and Zn, the rate of mobilization associated with industrial activities exceeds by a factor of ten or more the rate of mobilization from natural cycling. This is caused either by metal contaminated waste from mining or industrial activities such as chemical manufacturing, oil refining, metal processing and plating, and fertilizer production.

**Table 10.** Heavy metal concentrations in surface soils in the vicinity of metal smelters

| Location | Cd | Cu | Hg | Ni | Pb | Zn |
|---|---|---|---|---|---|---|
| | | | mg $kg^{-1}$ | | | |
| Odda, Norway | 114 | 1250 | 4.2 | 34 | 1200 | 21660 |
| Avonmouth, U.K. | 32 | - | - | - | 600 | 5000 |
| Sudbury, Canada | - | 2071 | - | 3309 | 84 | 75 |
| Kellog, U.S.A. | 140 | - | - | - | 7900 | 29000 |
| Glogow, Poland | 1 | 855 | - | - | 390 | 90 |

(Låg, 1972; Burkitt et al., 1972; Hutchinson and Whitby, 1973; Ragaini et al., 1977; Kabata-Pendias and Gondek, 1978.)

Impact of mining and smelting activities on metal contamination of soils is extensively documented (Lagerwerff et al. 1972; Buchauer, 1973; Hutchinson and Whitby, 1973; Singh and Låg, 1976; Ragaini et al. 1977; Låg, 1978; Elsokkary and Låg 1978; Blom 1986). Surface soils near a Zn smelter are known to contain 50,000-80,000 mg $kg^{-1}$ Zn, 900-1,500 mg $kg^{-1}$ Cd, 600-1,200 mg $kg^{-1}$ Cu, and 200-1,100 mg $kg^{-1}$ Pb (Buchauer,1973). Near a Zn smelter in western part of Norway, Låg (1972) found very high concentrations of heavy metals (Table 10) and similar concentrations have also been reported by others (Burkitt et al. 1972; Hutchinson and Whitby, 1973; Ragaini et al. 1977; Kabata-Pendias and Gondek, 1978).

The concentrations in Table 10 reflect upon the type of smelting activities. For example at Odda, Norway, the very high concentration of Zn resulted from Zn smelting activities while at Sudbury, Canada, Cu-Ni smelting plants are the source of very high concentrations of Cu and Ni.

At an old industrial site at Modum, Norway where large quantities of cobalt ore containing As were mined in the last century, Låg (1978) found still very high levels of As in soils (72-230 mg $kg^{-1}$ As) and vegetation (0.3-2.3 mg $kg^{-1}$ DM) around this abandoned site. These levels of As in both soils and vegetation are much higher than normally found in Norway (0.08-1.7 mg $kg^{-1}$) (Låg and Steinnes, 1978).

Not only are the total concentrations of these metals higher but also the concentrations of extractable metals were found to be much higher near smelters. Singh and Låg (1976) found that the surface soils from Odda contained as much 710 mg $kg^{-1}$ 0.2 M HCl extractable Zn. Elsokkary and Låg (1978) fractionated metals from polluted and unpolluted sites and found that on average about 17, 1.9, 12.5, and 2.5% of the total Cd, Pb, Zn, and Cu were extractable by 0.05 M $CaCl_2$. Different fractions of these metals are shown in Table 11. Zinc showed a pattern similar to Cd and Cu similar to Pb. These results point to a relatively low solubility and bioavailablity of Pb and Cu while higher fractions of the total Cd and Zn may be taken up by the vegetation in toxic amounts. High concentrations of several metals that were extracted with acetic

**Table 11.** Fractions of metals in polluted and unpolluted soils near Odda, Norway

| Metal | Fraction | Conc. Range mg kg$^{-1}$ | Average |
|---|---|---|---|
| Cd | CA | 0.01-3.06 | 0.54 |
| | AAC | 0.02-4.28 | 0.68 |
| | PYR | 0.10-1.66 | 0.47 |
| | OX | 0.03-1.40 | 0.39 |
| | RES | 0.10-1.15 | 0.28 |
| | Total | 0.46-12.00 | 2.48 |
| Pb | CA | 0.25-2.80 | 0.95 |
| | AAC | 0.30-10.45 | 2.15 |
| | PYR | 5.00-84.00 | 21.67 |
| | OX | 5.50-71.50 | 21.08 |
| | RES | 2.50-40.20 | 7.85 |
| | Total | 17.00-204.5 | 59.70 |

CA: 0.05 M $CaCl_2$, AAC: 2.5% acetic acid, PYR: 0.1 M sodium pyrophosphate, OX: 0.1 M oxalate acid plus 0.175 M ammonium oxalate, and RES: $HNO_3$ plus $HClO_4$. The total metal content was also determined in a $HNO_3$ plus $HClO_4$ mixture.
(From Elsokkary and Låg, 1978)

acid or EDTA have also been reported (Little and Martin, 1972; Elsokkary and Låg, 1978).

Enhanced concentrations of metals in plants grown in soils around industrial sites have also been reported (Hutchinson and Whitby, 1973; Singh and Steinnes, 1976; Låg and Elsokkary, 1978; Cox and Hutchinson, 1980).

Metal concentration in soils decline rapidly with increasing distance from the pollution source and the rate of decline depends upon predominant wind direction, topography, vegetation, and precipitation (Muller and Kramer, 1977; Freedman and Hutchinson, 1980; Blom, 1986). Blom (1986) reported a very drastic decrease in metal concentration away from an electro-metallic industry in southeastern Norway (Table 12). Highly significant correlations between the concentration of Cd, Cu, Pb, and Zn and the distance from the industry were found in these investigations. Similarly Semu et al. (1986) found that soils in the immediate vicinity of a mercury battery factory in Tanzania contained as much 472 mg kg$^{-1}$ Hg but downwind the concentration decreased with increasing distance from the factory resulting in a soil concentration of only 1.0 mg kg$^{-1}$ Hg about 2 km away.

**Table 12.** Concentration of metals in humus samples from uncultivated fields with increasing distance from an electro-metallic industry

| Distance | Cd | Cr | Cu | Hg | Pb | Zn | Ni |
|---|---|---|---|---|---|---|---|
| (m) | | | | mg $kg^{-1}$ | | | |
| 86 | 57.8 | 35 | 749 | - | 3324 | 18362 | 31 |
| 175 | 22.8 | 34 | 165 | 0.44 | 424 | 7783 | 41 |
| 310 | 11.9 | 18 | 214 | - | 719 | 1418 | 18 |
| 732 | 4.1 | 43 | 100 | - | 255 | 1995 | 36 |
| 1538 | 0.9 | 26 | 86 | 0.68 | 117 | 154 | 23 |

(From Blom, 1986)

### E. Atmospheric Deposition

From the examples presented in Section III, D it is obvious that very severe soil contamination with heavy metals can occur in the vicinity of an industrial source of air pollution. It has become increasingly evident however that even in areas very far from the source region heavy metals may be supplied to surface soils in significant amounts, due to the now very well documented long range atmospheric transport of pollutants. The extent of this diffuse soil contamination may be illustrated by taking the situation in Norway as an example.

The southernmost part of Norway is considerably affected by air pollutants supplied to this area from other parts of Europe. The deposition rates of volatile trace elements such as Pb, Zn, Cd, As, and Sb are about ten-fold higher in this area than in some more northerly regions of the country (Steinnes, 1980). Studies of air particulate samples at Birkenes, S. Norway, in conjunction with air trajectory data show that the supply of the above elements occur with winds from the sector SE-SW to at least 80% (Pacyna et al., 1984, Amundsen et al., 1992). Furthermore, precipitation events occur mainly with the same wind directions. The atmospheric supply of these metals, along with sulfur and nitrogen oxides, is therefore considerable in spite of the fact that the main source regions are located at distances of 1000 km or more.

In order to discuss the extent of soil contamination from this diffuse atmospheric deposition of metals from distant sources, it is appropriate to consider agricultural soils and "undisturbed" natural soils separately. We shall use as an example the situation in southernmost Norway. In Table 13 are listed atmospheric deposition rates for the heavy metals in question. Listed are also typical concentration levels of the metals in arable soils (Bowen, 1979; Page and Steinnes, 1990) and annual increase in heavy metal concentrations in the upper 0.15 m of the soil, assuming a bulk density for soil of $1.5x10^3$ kg $m^{-3}$ and complete mixing within the plough layer. The deposition values are from Steinnes (1984) and reflects the situation around 1980. These values are probably good estimates of the average metal deposition levels during the last 20 years.

**Table 13.** Distribution from atmospheric deposition of pollutants to the total concentration of heavy metals in the upper 0.15 m of cultivated soil in southernmost Norway

| Metal | Atmospheric deposition[a] | Typical soil concentration | Annual increase rate |
|---|---|---|---|
| | $mg\ m^{-2}y^{-1}$ | $mg\ kg^{-1}$ | % |
| V | 2 | 90 | $1 \times 10^{-2}$ |
| Cr | 0.5 | 70 | $3 \times 10^{-2}$ |
| Ni | 0.6 | 50 | $1 \times 10^{-2}$ |
| Cu | 2 | 30 | $3 \times 10^{-2}$ |
| Zn | 15 | 5 | 0.13 |
| As | 0.8 | 5 | $7 \times 10^{-2}$ |
| Se | 0.4 | 0.4 | 0.5 |
| Cd | 0.3 | 0.35 | 0.4 |
| Sb | 0.3 | 1 | 0.13 |
| Hg | 0.03[b] | 0.05 | 0.3 |
| Pb | 10 | 15 | 0.3 |

([a]Steinnes, 1984; [b]Steinnes and Andersson, 1991)

It is clear from Table 13 that annual increase rates of heavy metals to surface layers of agricultural soils in this region is less than 1% even for the metals most readily released in volatile forms by high-temperature anthropogenic processes. Still the increment may be significant over a long period of time if the air pollution situation persists. Moreover it is important to consider the fact that normally only a small fraction of indigenous metals in soils are present in plant available form, while probably a major part of the airborne material adds to the bioavailable pool of the metals.

When it comes to undisturbed natural soils, the conditions may be quite different from that of agricultural soils where the supplied heavy metals will be mixed within the surface layer. In the natural soil, the deposited metals in most cases tend to concentrate in the uppermost few cm of the soil. Moreover, the surface layer of natural soils may often have a high content of organic matter, in which case it may serve as a very efficient trap for some metals. A nation-wide survey of natural surface soils in Norway revealed that the southern soils had many-fold higher contents of Pb, Cd, As, and Sb in the humus layer than soils from central and northern Norway. Also for Zn and Se a distinct N-S gradient was evident, while for Cu there were no strong regional differences (Allen and Steinnes, 1979). The correspondence with the geographical atmospheric deposition patterns for these elements was remarkable. In a similar, more recent survey, samples at different depth in the soil profile were collected (Bølviken and Steinnes, 1987). It appears that the $HNO_3$-extractable Pb in the humus layer (3-5 cm depth) is consistently more than 10 times higher in southernmost Norway than in areas far north, whereas no N-S gradient is evident for Pb in the C-horizon (60 cm depth). This strongly indicates that what is observed is a large-scale contamination from long-range atmospheric transport, probably over a long period of time. Similar but somewhat less pronounced trends were observed for Cd, Zn, and V.

All in all the contribution from long-range atmospheric transport to the heavy metal content of surface soils may in some cases be considerable, and should be given greater attention than has been the case in the past.

## IV. Soil Processes

### A. Weathering and Accumulation

Weathering, a process of physical disintegration and chemical decomposition of minerals, helps in passing the metals from the parent materials to solutions and suspensions. Translocation and accumulation of released metals by soil constituents, such as clays, hydrous oxides, and organic matter are controlled by many soil forming factors which affect the intensity of the pedogenic process. The important chemical weathering processes are dissolution, hydration, hydrolysis, oxidation, reduction, and carbonation (Kabata-Pendias and Pendias, 1984). Some of the mineral materials in rocks weather more rapidly than others and as a result the mineral in soils have either remained behind from weathering, or have been transported. Jeng (1992) studied the effects of different treatments on the release of Fe, Cu, Zn, Ni, Cd, Pb, and Mn in a laboratory experiment and reported that, with time, large amounts of heavy metals were released from alum shale. One particular sample, with a higher buffering capacity against acid due to its calcareous nature, was found to be slow in heavy metal release (Table 14). Most of the heavy metals in the acid samples were affected in a similar way i.e. a decrease in pH accompanied by an increase in heavy metal concentration into solution. Liming was effective in rendering the heavy metals immobile. Cadmium was found to be a relatively mobile element, being released easily under acidic conditions and remaining in solution even under weakly acidic to neutral conditions.

### B. Soil pH and Redox Conditions

The status of several metal ions is affected by the redox conditions, particularly for Fe and Mn, but also for Cr, Cu, As, Hg, and Pb. The redox potential can change directly the oxidation state of a heavy metal. Indirectly the chemical form of a metal ion can be changed through a change in the oxidation state of a ligand atom such as C, N, O, and S.

Redox reaction in soils are generally slow but are catalyzed by soil microorganisms which are able to live over the full range of pH and pE conditions normally found in soils (pE + 12.7 to -6.0 and pH 3-10) (Sposito and Page,

**Table 14.** Heavy metal release (ratio of water soluble to aqua regia extractable content of unweathered material expressed in percent) from two Norwegian alum shales during a 100 day incubation period in the laboratory at 20 °C and relative humidity of 60%

| pH | days | Cu | Zn | Ni | Pb | Cd |
|---|---|---|---|---|---|---|
| | | % of total in solution after incubation (days) | | | | |
| Brum | | | | | | |
| 4.53 | 0 | 0.00 | 15.68 | 2.59 | 1.45 | 6.31 |
| 3.54 | 10 | 0.00 | 23.62 | 3.55 | 4.21 | 10.02 |
| 3.10 | 50 | 4.01 | 76.43 | 5.88 | 1.26 | 24.65 |
| 2.74 | 100 | 12.40 | 96.26 | 8.10 | 0.83 | 21.84 |
| Munch | | | | | | |
| 8.37 | 0 | 0.00 | 0.22 | 0.06 | 0.00 | 0.00 |
| 7.90 | 10 | 0.00 | 0.73 | 0.00 | 0.00 | 0.00 |
| 7.70 | 50 | 0.00 | 1.89 | 0.00 | 0.00 | 0.00 |
| 7.62 | 100 | 0.00 | 1.09 | 0.21 | 1.35 | 1.67 |

(From Jeng, 1992.)

1985). If the oxygen in a zone of soil becomes exhausted, microorganisms with anaerobic respiration predominate and susceptible elements such as Mn, Cr, Hg, Fe, and Cu are gradually reduced (Sposito and Page, 1985; Rowell, 1981). In general, reducing conditions cause a pH increase, and oxidation brings about a decrease. Oxidation of pyrite in a soil parent material can cause a marked drop in pH. However, there are several mechanisms in soils which serve to buffer pH to varying extents, including hydroxyaluminium ions, $CO_2$, carbonates; and cation exchange reactions (Bache,1979). The effect of pE and pH conditions can best be seen on Fe and Mn forms. The oxides of both Fe and Mn can be dissolved by either decreasing pH or Eh, but Mn oxides are more easily dissolved than Fe oxides. With increasing Eh and pH, Fe oxides precipitate before those of Mn. Small changes in Eh or pH can give rise to either extensive dissolution or precipitation of Fe oxides.

Under the indirect effects of reducing conditions, sulfate ions are reduced to sulfide and this can lead to the precipitation of metal sulfides such as $FeS_2$, HgS, CdS, CuS, MnS, and ZnS (Sposito and Page, 1985). When reducing conditions cause the dissolution of hydrous Mn and Fe oxides, their co-precipitated metals are released into the soil solution (Sposito, 1983). Organic matter, Fe(II) minerals, and other reducing agents in soil are known to reduce Cr(VI) to Cr(III) under acidic conditions and heat.

### C. Adsorption and Desorption

Adsorption of metals from the aqueous phase to the solid phase and their subsequent desorption are the most important chemical processes affecting the mobility and bioavailabilty of metals in soils. Metals in soils can be adsorbed by ion exchange, precipitation or co-precipitation, and chelation but often it is difficult to identify one precise mechanism responsible for the adsorption of metals in any particular soil. The process of adsorption/desorption is controlled by pH, redox potential, ionic strength, competing ions, and the soil constituents (organic and inorganic) but the relative importance of these factors is different for different metals.

The pH at which metal cations are removed as insoluble hydroxides depends upon the type of soil constituent present. Farrah and Pickering (1977) showed that in the presence of a clay suspension, total removal of metal ions (Cu, Cd, Pb, and Zn) as insoluble hydroxy species occurred at a slightly lower pH than in the absence of clay (Figure 4). Tills and Alloway (1983) found that at pH below 5.0, 80-90% of Cd was as $Cd^{2+}$, while neutral species became important only in the neutral to alkaline range. Most of the metals are precipitated or coprecipitated in the presence of carbonates and are rendered immobile. McBride (1980) showed that Cd was effectively coprecipitated on calcite surfaces and similar effect was found for Zn by Udo et al. (1970).

The presence of competing ions plays an important role in controlling the sorption of metals. Garcia-Miragaya and Page (1977) found that at pH 4.6 $Al^{3+}$ was more effective in preventing Cd sorption than $Ca^{2+}$, $K^{+}$, and $Na^{+}$. One may therefore expect that in the presence of $Al^{3+}$ or positively charged Al-polymers, more Cd would be present in solution. Reduction in Cd sorption by 30% was reported by Christensen (1984a) when the Ca concentration was increased by ten-fold, and he attributed this to competition by Ca.

The effects of organic ligands on heavy metal solubility have been investigated by Farrah and Pickering (1977), Garcia-Miragaya and Page (1976), Christensen (1984b), and Brummer and Herms (1983). Farrah and Pickering (1977) found the Pb concentration in solution to be reduced by increasing the concentration of $SO_4$ or $PO_4$. Organic ligands such as EDTA was found effective in desorbing Cd form adsorbed surfaces (Christensen, 1984b). The metals are immobilized by soil organic matter in the order Cu > Cd > Zn > Pb and the immobilization effect depends on soil reaction and the kind of mineral components present in soils (Brummer and Herms, 1983). They found that at acid soil reaction, the concentration of Cd and Zn in the equilibrium solution of soils were lower with higher organic matter content compared to soils with lower organic matter content. But at pH 6 to 8 higher concentrations of these metals were measured for soils with higher organic matter content. Therefore they concluded that organic matter of the soils immobilizes heavy metals at strongly acid conditions and mobilizes metals at weakly acid to alkaline reaction by forming insoluble or soluble organic metal complexes, respectively. Semu et al. (1987) found severe reduction (up to 95%) in Hg adsorption in some soils

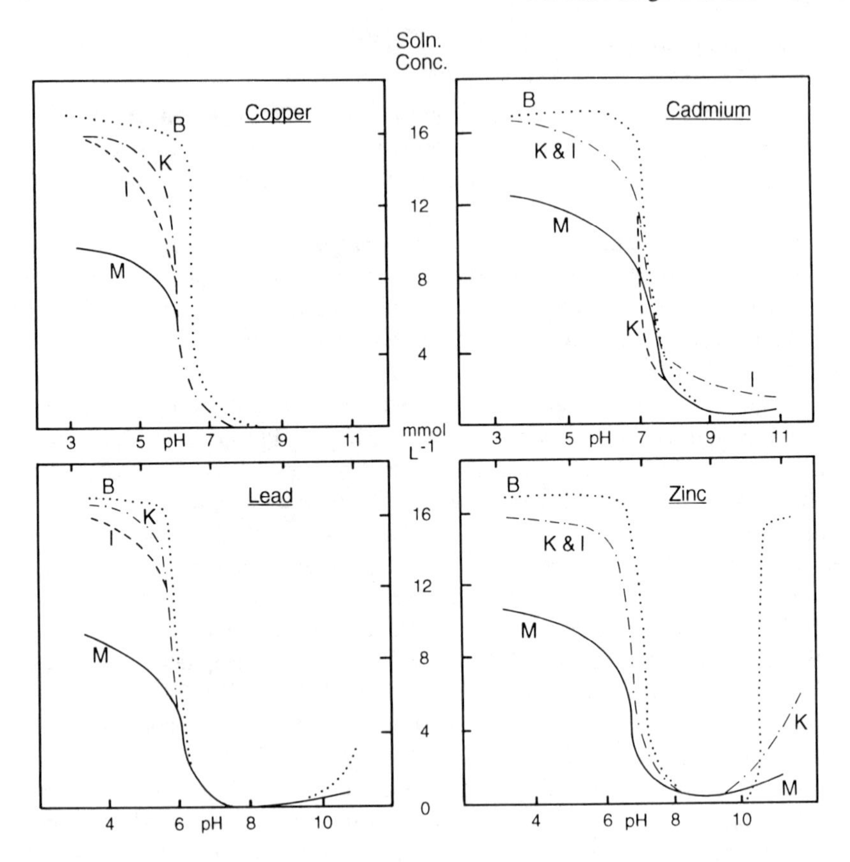


**Figure 4.** Effect of pH on the concentration of heavy metal ions in solution, in the presence of clay suspensions (B, clay absent; K, kaolinite, 0.05% w/v; I, illite, 0.025% w/v; M, montmorillonite, 0.02 % w/v). (Farrah and Pickering, 1977.)

from Tanzania as a result of organic matter removal. Significant reduction in plant availability of Cd with increasing rates of organic matter addition was reported by several investigators (Eriksson, 1990; He and Singh, 1993c).

Although reversibility of adsorbed metals on soil colloids in general is rather weak, Christensen (1984b) found full reversibility of Cd sorption in $10^{-3}$ M Ca $Cl_2$ at pH 6 in a loamy sand but the reversibility was only partial in a sandy loam soil. However, no irreversibility was observed at higher concentrations of Ca $Cl_2$ ($10^{-2}$).

## D. Mobility and Transport

Under cool and humid climatic conditions precipitation water percolates through soils and may transport soluble heavy metals with the soil solution. In warm dry climate, and also to some extent in humid hot climate, upward translocation of these metals in the soil profile is the most common movement. The transport of metals within the soil or even to the groundwater depends on the metal concentration of the solution phase. However, the processes which determine the concentrations of heavy metals in the soil solution are not fully understood. Changes in soil water content, departures from equilibrium or steady state to non-equilibrium conditions and changes in the activity of microorganisms which influence redox conditions, contents of soluble chelating agents, and the composition of soil atmosphere require simultaneous consideration. These processes affect chemical reactions with heavy metals such as precipitation-dissolution, adsorption-desorption, and complex and ion-pair formation and thus influence the distribution of the various metal species in the solid phase of the soil and in the soil solution. Theoretical reviews of mechanisms involved in the transport and accumulation of soluble soil components are given by Bolt and Bruggenwert (1976) and Lindsay (1979).

Depletion of heavy metals in soil occurs due to their downward movement with percolating water through the profile of freely drained acid soils and also due to their uptake by plants. The relative mobility of the different metals can roughly be expressed by the ratio of dissolved to bound amount of each metal in relation to pH. According to this interpretation and in close agreement with the concept of metal ion hydrolysis, the mobility of heavy metals in acid soils decreases in the order Cd > Ni > Zn > Mn > Cu > Pb > Hg (Herms, 1982).

Organic compounds naturally occurring in soils are effective agents for chelating heavy metals and the solubility of metal chelates depends on both the binding strength and the mobility of the chelates thus formed. Strong binding of metals with a low molecular organic substance will appreciably increase its mobility in soil. In spite of a high mobilization of heavy metals, the forest floor is also well known as an important sink of heavy metals. Dumontet et al. (1990) found that the amounts of heavy metals (Cu, Zn, Ni, Cd, and Pb) moving and accumulating in the anaerobic zone of the peat profile near a smelter in the Noranda region of Quebec were limited. The distributions and the enrichment ratios in the profile showed that Cu, Zn, and Cd would have relatively higher mobility than Pb.

### E. Microbial Activity

Heavy metals in soils exert deleterious effects on the metabolic function of soil microorganisms, for example. both the decomposition of organic matter and microbial activities were inhibited at soil sites polluted with heavy metals (Fritze et al., 1989). The addition of heavy metals such as Cd, Cr, Cu, Ni, Pb, and Zn to soil significantly inhibited soil respiration (Doelman and Haanstra, 1979; Chang and Broadbent, 1981). Chang and Broadbent (1982) reported that Pb did not inhibit N mineralization and nitrification; the order of decreasing inhibition

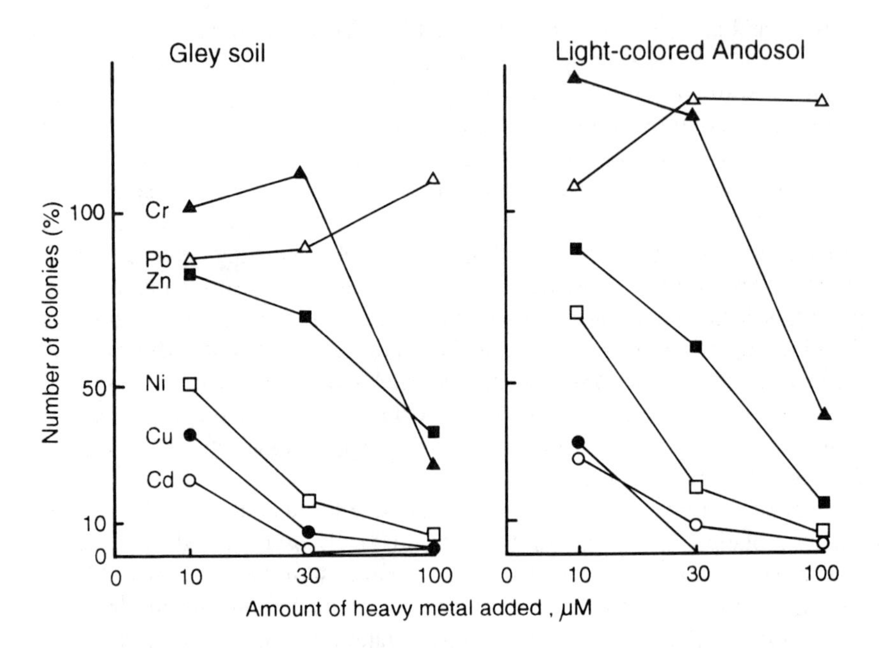


**Figure 5.** Effect of heavy metals on colony formation by bacteria from Gley soil and light colored Andosol, expressed as percentage of the value without heavy metals. (Hattori, 1992.)

was Cr > Cd > Cu > Zn> Mn > Pb. Doelman (1985) showed that the effect of Hg was the most significant and the effect of Pb was the least significant among the effects of Cd, Cr, Cu, Hg, Ni, Pb, and Zn on soil microbial activities. McNeilly et al. (1984) studied a Pb-Zn-contaminated pasture and concluded that plant productivity was less sensitive to metal contamination than the decomposition processes which follow its death. The sensitivities to metals could be summarized as productivity < litter accumulation < soil organic matter breakdown < soil humus decomposition.

Hattori (1992) found that the number of bacterial colonies formed from the suspension of a gley soil or light-colored andosol decreased markedly (Figure 5) by the addition of 10 $\mu$mol $L^{-1}$ of Cd, Cu, or Ni . The addition of Pb did not affect adversely the colony formation. The toxicity of heavy metals to colony formation by soil bacteria was high for Cd and Cu, which inhibited $CO_2$ evolution substantially and furthermore the inhibition of $CO_2$ evolution by the heavy metals depended mainly on the degree of toxicity of the metals to soil bacterial growth and the amount of water-soluble heavy metals in soil.

Doelman (1985) summarized the effects of Cd, Cr, Cu, Hg, Ni, Pb, and Zn on soil respiration, N mineralization and nitrification processes (Figure 6). In most cases the lowest concentrations of metals causing measurable effects were found in sandy soil and the highest in clay and organic soils. This involves the

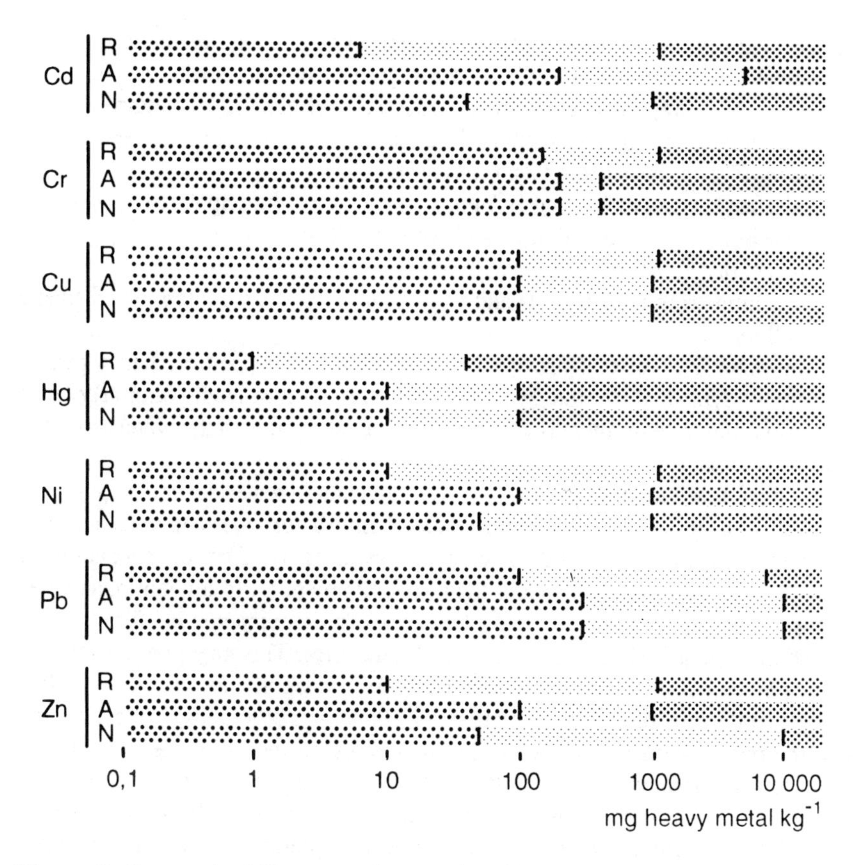


**Figure 6.** Summarized literature data on the effects of Cd, Cr, Cu, Hg, Ni, Pb, and Zn on soil respiration (R), nitrogen mineralization (A), and nitrification (N) (left portion of bar represents the heavy metal concentration range where inhibition was never recorded; middle portion where inhibition was sometimes recorded, and right portion where inhibition was always recorded). (Doelman, 1985.)

consideration of the buffering capacity e.g., of soils to metals and consequently the availability of heavy metals.

# V. Heavy Metals in Water

## A. General Aspects

Although all naturally occurring elements are supplied to waters by purely natural processes, their concentration levels are often significantly enhanced by human activities. This may lead to situations that are unacceptable from an ecological point of view as well as with regard to human consumption.

Regarding heavy metals, various processes influenced by anthropogenic activities may contribute to increased concentrations in natural waters:

- Runoff from agricultural and urban areas.
- Discharge from a) Mining; b) Factories; and c) Municipal sewer systems.
- Leaching from waste dumps and former industrial sites.
- Atmospheric deposition.
- Leaching from natural soils and vegetation previously contaminated by air pollution.

If the water is acidified from discharges or from acid deposition, heavy metals previously deposited to the sediment may be re-dissolved. Similarly acid deposition may release increased amounts of heavy metals to the water from the catchment area.

The transport and bioavailability of heavy metals in water depend strongly on their distribution between different physico-chemical forms, i.e. their speciation. Parameters such as pH and humus content may strongly affect the speciation of water-borne heavy metals, and thus their toxicity potential. In conventional water analysis particles are often removed prior to measurements by filtering the sample through a 0.45 mm pore size membrane filter. This may give misleading results, since colloidal particles often containing a significant fraction of some heavy metals, are likely to pass the filter (Beneš and Steinnes, 1975). In this way the potential hazard to biota may be overestimated. On the other hand, some aquatic organisms are able to extract chemical species from particulate material. Physical separation methods suitable to separate truly soluble species from colloids include dialysis in situ (Beneš and Steinnes, 1974) and hollow-fiber ultrafiltration (Salbu et al., 1984). Voltametric methods appear to be useful to determine the bioavailable fractions of heavy metals in salt water, but their use in fresh water requires addition of an electrolyte (Florence, 1982).

## B. Processes in Water

Trace elements in natural waters participate in many processes which change their physico-chemical forms (speciation) or distribution in space (migration) and affect their uptake by organisms. The metals considered in this paper are normally observed in natural waters at concentrations lower than $10^{-5}$ M.

At such low levels chemical substances are affected by processes which may play only a minor role at macro concentration levels. These processes, often named trace chemistry processes (Beneš and Major, 1980) are the following: oxidation/reduction, association/dissociation in solution, adsorption/desorption, precipitation/dissolution, and aggregation/disaggregation. The following examples illustrate differences in behavior of substances at trace concentration levels and at higher concentrations:

1. At low concentrations substances tend to adsorb on the surface of solid particles of low adsorption capacity or on compact solid phases of similar capacity in contact with the solution. These effects are negligible at higher concentration levels.
2. Precipitation of a substance present in very low concentration, if occurring at all, cannot lead to the formation of particles larger than colloidal size. Such particles often show an unpredictable behavior.
3. The behavior of a substance at low concentration can be affected by its interaction with other substances present in the water at low concentrations, often as impurities.

The behavior of heavy metals in natural waters may therefore often be difficult to predict from a purely thermodynamic point of view using simple models.

## C. Runoff Waters, Rivers and Streams

Surface runoff, in particular from urban areas, can supply considerable amounts of heavy metals to lakes and rivers both in dissolved form and attached to particulates. Metal-containing water first penetrating soil layers before reaching the lake or river is much less likely to constitute a substantial source of supply, since most of the metals would tend to be rather strongly retained in the soil.

The residence time of water in rivers and streams is typically of the order of days to a few weeks. Therefore pollutants reaching the river in dissolved or colloidal form may travel a considerable length in the river before being removed to any great extent. This means that release of high amounts of a metal at one point may affect the water quality and consequently the river ecosystem at considerable length downstream the point where the discharge occurred.

A particularly unfortunate situation may occur where drainage water from existing or previous mining activities is reaching the river. Acid produced from sulfate oxidation along with heavy metals is not only toxic to aquatic organisms in itself, but also renders a higher fraction of metals soluble and thus more likely bioavailable. An ecological disaster in a river or stream associated with mining activity can therefore extend to considerable distance downstream the point of discharge.

## D. Lakes

As compared to rivers, the residence time of water in lakes is long; of the order of months - years depending on size. Heavy metals supplied to the lake are therefore quite likely to be removed from the water column by sedimentation. A water contamination problem is then transformed to one of sediment contamination. In order to illustrate different patterns of heavy metal contamination in lakes, we shall consider two extreme cases:

In small lakes of southern Fennoscandia, the water (Steinnes et al., 1989) and sediment (Johansson, 1988) is being contaminated with heavy metals such as Zn, Pb, and Cd from long range atmospheric transport, either by direct deposition on the water surface or via the catchment. Many of these lakes are also undergoing water acidification because the initially limited alkalinity has been used up by acid deposition. In these cases heavy metals stored in the sediments are being re-dissolved from the surface layer (Johansson, 1988).

To the other extreme we consider huge fresh-water lakes such as the North-American Great lakes and lake Baikal in Siberia. These lakes are generally well buffered, and metals entering the lakes through tributaries are likely to be removed from the water column predominantly in areas fairly close to the shore. In the more pelagic part of the lakes, atmospheric deposition may be a very significant source for some metals (Dolska and Sievering, 1979). Thus air pollutants can play an important role for heavy metal contamination both in very small and very large lakes.

## E. Groundwater

Groundwaters are rather insignificant from a purely ecological point of view, but are becoming increasingly important all over the world for drinking water and irrigation purposes. Since groundwater reservoirs are generally protected by overlying soils, they are not likely to be severely affected by human activities, and thus heavy metal contents are likely to be controlled by natural processes. For this reason, and since they are often more alkaline than surface waters, their heavy metal contents are generally quite low compared to legal standards. Exceptions may occur for elements such as arsenic and molybdenum which tend to become more soluble with higher pH. The rapidly increasing pollution of groundwaters with nitrate experienced in some countries, however, indicates that these valuable resources may eventually be contaminated with less mobile constituents such as heavy metals. Another incipient problem in some regions is acidification of shallow ground waters from surface acid deposition. In southern Sweden this appears to be a significant problem (Hultberg and Johansson, 1981), which is likely to be accompanied with increased solubility of heavy metals in the subsoil.

## F. Fresh Water Sediments

Fresh water sediment is the general sink for heavy metals in lakes, and is also important for removal of metals from the water column in rivers. In industrialized regions of the world and areas affected by airborne transport of pollutants from these areas, the general fresh water contamination has resulted in substantial enrichment of the upper sediment layers with heavy metals. This removal by sedimentation improves the lake or river from a drinking water point of view and normally protects water-living organisms, but may constitute a serious problem for species living in and on the bottom sediment. For this reason it is also important to watch critically metal levels in fresh-water sediments.

The behavior of mercury in sediments is different from that of other heavy metals discussed in this paper. Jensen and Jernelöv (1969) discovered that Hg in the sediment is being converted by microorganisms to monomethyl mercury, which can enter the food chains. Lakes appreciably contaminated with mercury frequently have fish with Hg contents above what is acceptable for human consumption (Lindqvist et al., 1991). In recently established artificial water reservoirs it has been noticed that fish contain higher amounts of mercury. The reason for this is apparently that organic top soils, which normally act as a preliminary sink for atmospheric mercury, turn into sediments after flooding and thereby support microorganisms methylating some of the accumulated mercury.

## VI. Ecological Implications and Regulatory Control

Ecological consequences of metal contamination of soils and waters can be assessed by its impact on soil and aquatic organisms and plants. If metals enter the food chain they may also create human health problems. Studies of soil contamination from different sources generally include measurements of metal concentration in plants growing in contaminated soils or in soil organisms and similarly studies of water contamination deal with bioavailability and toxicity of metals to aquatic organisms. A number of such examples on ecological impacts of metal contamination of terrestrial and aquatic environments have been cited in the preceding text (III) under anthropogenic sources of contamination. However, a few examples of worst cases of naturally heavy metal poisoned areas or anthropogenic pollution or large-scale environmental pollution are also reported here to illustrate the region specific severity of the problem. Låg and Bølviken (1974) reported occurrence of naturally lead-poisoned soil and vegetation in five different areas of Norway where galena occur in the bedrock. Soil samples from lead affected patches contained as much as 2.5% Pb and up to 400 mg $kg^{-1}$ in vegetative dry matter, corresponding approximately to 400 and 70 times, respectively, the contents found at "background stations". They concluded that natural heavy metal poisoning of soil and vegetation is more common than earlier expected and that certain features of the natural geochemical environment may be noxious to animals. Stone and Timmer (1975) reported a Cu concentration as high as 10% in surface peat that was filtering Cu-rich spring water emerging into a marsh in New Brunswick. Forgeron (1971) described surface soil with up to 3% Pb+ Zn at site on Baffin Island, Canada. Warren et al. (1966) reported a Hg concentration of 1-10 mg $kg^{-1}$ in soil overlying a cinnabar (HgS) deposit in British Columbia. A case of large-scale environmental pollution by Cd, Pb, and Zn in a populated area occurred in the village of Shipham in Somerset, UK, where Zn was mined during the eighteenth and nineteenth centuries. Geochemical and soil surveys revealed that the range and median values (mg $kg^{-1}$), respectively, for 329 soil samples from Shipham were: Cd 2-360 (91), Zn 250-37,200 (7,600), and Pb 108-6,450 (2340) (Sims and Morgan, 1988). Some of the most contaminated vegetables from this village contained 15 to 60 times higher Cd levels than those grown in ordinary soils. Health studies on 500 people revealed small but significant differences in some biochemical parameters, but there was evidence of adverse health effects (Morgan and Sims, 1988). Even in some non-industrialized countries such as in Africa elevated levels of toxic metals in blood or other human organs have been reported (Nriagu, 1992). For example, the blood lead concentration of urban and rural residents of Egypt have been reported to be 170-360 and 140-250 $\mu g$ $L^{-1}$, respectively (Kamal et al., 1991). These values are higher than those being observed in the population of most of the developed countries (WHO, 1987). Ahmed and Almubarak (1990) found significantly higher concentrations of Pb in the hair of adult males in Khrtoum, Sudan (mean = 14 $\mu g$ $g^{-1}$) compared to those of West Germans (7.6 $\mu g$ $g^{-1}$). Nriagu (1992) has given an excellent review on toxic metal pollution in Africa.

Increased attention has been paid to ecological consequences of metal contamination of soils and waters in recent years and therefore, in many countries, guidelines to protect the human population from the adverse effects of contamination of soils and waters have been established. In the case of soils this applies especially to the disposal of sewage sludge and the use of chemicals for plant protection. Similar guidelines have also been established for the use of Cd containing commercial fertilizers in some European countries. The regulatory control measures to reduce unnecessary industrial use of metals and strict control measures for industrial emissions of metals to atmosphere or waterways have given positive results in many areas and the atmospheric deposition of metals have declined in recent years in many areas.

It may be difficult to establish the precise critical values of heavy metals in soils because their transfer from soil to waters or plants is governed by a complex web of interactions and edaphic factors. However, some critical levels of metals in soils have been suggested in literature beyond which harmful effects of metals may be expected (Table 15). The maximum allowable concentrations of metals in sludges for different countries are presented in Table 16. The table shows that the Norwegian standards for metal concentrations in sewage sludge are more stringent than EEC and U.S. The rate of sludge application in Norway should not exceed 20 tons dry matter $ha^{-1}$ in a 10 year period. However, in some cases, such as organic matter poor sand or sandy loam soils, the rate could be doubled after special considerations.

The average maximum concentration of Cd in commercial fertilizers in Finland and Norway is 100 mg $kg^{-1}$ P applied, and in Denmark is 200 mg Cd $kg^{-1}$ P which will be reduced to 150 in 1993 and to 110 in 1995. This concentra-

**Table 15.** Total concentration of metals in soils considered as critical for plant growth and animal and human consumption

| Metal | $mg\ kg^{-1}$ |
|---|---|
| As | 20 |
| Cd | 8.0 |
| Cr | 75 |
| Cu | 100 |
| Hg | 5.0[a] |
| Ni | 100 |
| Pb | 200 |
| Zn | 400 |

(From Linzon, 1978; [a]El-Bassam and Tietjen, 1977.)

**Table 16.** Maximum allowable concentrations of metals in sewage sludges in different countries

| Metal | EEC[a] | USA[b] | Norway[1] |
|---|---|---|---|
| | ----------------------mg $kg^{-1}$ dry matter--------------------- | | |
| Cd | 20-40 | 20 | 4 |
| Cu | 1000-1750 | 1200 | 1500 |
| Cr | - | - | 125 |
| Hg | 16-25 | - | 5 |
| Pb | 750-1200 | 300 | 100 |
| Ni | 300-400 | 500 | 80 |
| Zn | 2500-4000 | 2750 | 700 |

[a]Limits for 1993 according to Haraldsen, 1992; [b]Based upon the tentative revised limits for application of sludge to land at a rate of 1000 Mg $ha^{-1}$, Page, 1992.

tion in Japan is 340 mg Cd $kg^{-1}$ P and for EEC the proposed limit is 200 but is not yet enforced.

Also for waters, regulatory control of heavy metal contents has been introduced in most countries, mainly from a drinking water point of view. These standard limit values, some examples of which are shown in Table 17, vary considerably from country to country and may seem somewhat arbitrary. An assessment and comparison of criteria and legislation forming basis of standards for heavy metals in potable water is presented by Kirk and Lester (1984). The kind of considerations used are generally upper limits for total daily human intake on the basis of classical toxicological evidence, and the significance of drinking water relative to other food and drink with respect to metal exposure.

**Table 17.** Recommended standards for drinking water quality

| Metal | WHO (1984) | USEPA (1986) | Denmark (1983) | Norway (1987) | CEC (1980) |
|---|---|---|---|---|---|
| | $\mu g\ L^{-1}$ | | | | |
| As | 50 | 50 | 50 | 10 | 10 |
| Cd | 5 | 5 | 5 | 1 | 1 |
| Cr | 50 | 50 | 50 | 10 | 50 |
| Cu | 1000 | 1300 | 100 | 100 | 20 |
| Hg | 1 | 3 | 1 | 0.05 | 0.5 |
| Pb | 50 | 20 | 50 | 5 | 50 |
| Zn | 5000 | - | 100 | 300 | 500 |

(From SIFF, 1987.)

**Table 18.** Water quality standards for fresh waters based on ecological criteria

| Metal | USA[a] | USA[a] | Canada[b] | Other[3] |
|---|---|---|---|---|
| | 10 mg $CaCO_c$/L | 50 mg $CaCO_3$/L | 0-10 mg $CaCO_3$/L | 0-10 mg $CaCO_3$/L |
| | $\mu g\ L^{-1}$ | | | |
| As | | 190 | 50 | 8.6 |
| Cd | 0.18 | 0.66 | 0.2 | 0.16 |
| Cr | | 120 | 20 | 2.0 |
| Cu | 1.6 | 6.5 | 2 | 1.7 |
| Hg | | 0.01 | 0.1 | 0.01 |
| Ni | 22.5 | 88 | 25 | 1.4 |
| Pb | 0.16 | 1.3 | 1 | 2.0 |
| Zn | 15.0 | 59 | 30 | 1.6 |

( From [a]EPA, 1986; [b]CCREM, 1987; [c]Van de Meent et al., 1990.)

More recently work has been conducted to formulate water quality criteria for heavy metals and other micropollutants based on their potential ecological hazard. Some recommended upper limits for heavy metals on this basis are listed in Table 18. The limits are in some cases specified relative to water hardness. In general the upper limit values based on ecological criteria are distinctly lower than those formulated on the basis of human health considerations. Ecologically based criteria are now also being developed for heavy metals in sediments.

## VII. Future Outlook and Research Needs

Concerns associated with a slow but steady build up of heavy metals in our soil and water resources due to industrial, mining, and agricultural activities demand greater attention both from legislators and scientists. Soil scientists should continue to play a major role in further development and refinement of guidelines for disposal of toxic substances on agricultural lands, where as the legislators should ensure that the guidelines are followed

On the part of soil scientists an understanding of regional pollution problems, background levels of soils, soil chemical processes controlling transport and mobility of metals, various sources of metal inputs to soil-plant-animal system, soil factors affecting the phytoavailability of metals is required. In case of metals where contribution of soil to the total metal burden of plants is smaller than atmosphere, knowledge on the relative uptake of metals from these sources may be quite essential. Long term perspectives of heavy metals accumulation in soils is of utmost importance both in terms of their uptake by plants and their transfer to water bodies because soil factors and management practices are known to have a key influence on the amounts of metals entering the food chain. Climatic factors, precipitation, and temperature, influence the mobility, deposition, and plant uptake of metals and their role on metal management in soil-plant and soil-water systems need to be understood. Chemical monitoring of metal status of soils, waters, and food products is an important tool to provide a warning for metal transfer among various systems.

Concerning heavy metals in fresh waters, much research also needs to be done. The development of modern analytical techniques has facilitated accurate determination of heavy metals at the levels at which they exist even in unaffected water resources. Still, however, much needs to be done in terms of sampling and pre-analysis handling of water samples in order to avoid errors e.g., due to contamination and loss of analyte, cf. recent experience on mercury (Lindqvist et al., 1991). Most studies of heavy metals in water are still based on total or 0.45 $\mu$m filterable concentration. Since the mobility and bioavailability of heavy metals in aquatic systems depend strongly on their physico-chemical forms, it is important that speciation work is introduced to a larger extent than presently. Further work on the ecological effects of heavy metals in fresh-water ecosystems should be conducted, and the results from this work should form basis for improved water quality criteria. The impact of heavy metals in water in relation to human health also deserves further attention, e.g., in connection with acidification problems which also affect heavy metals.

## References

Ahmed, A.F.M. and A.H. Almubarak. 1990. Lead and Cd in human hair: A comparison among four countries. *Bull. Environ. Contam. Toxolol.* 45:139-148

Allen, R.O. and E. Steinnes. 1979. Contribution from long-range transport to the heavy metal pollution of surface soil. p. 271-274. In *Heavy Metals in the Environment*, CEP Consultants, Edinburgh, Scotland.

Alloway, B.J. 1990. *Heavy Metals in Soils*. Blackie, London.

Amundsen, C.E., J.E. Hanssen, S. Semb, and E. Steinnes. 1992. Long-range atmospheric transport of trace elements to southern Norway. *Atmos. Environ.* 26A: 1309-1324.

Bache, B.W. 1979. p. 487-492. In Fairbridge, R.W. and C.W.Finkl (eds) *The Encyclopedia of Soil Science*, Hutchinsen and Ross, Stroudsburg.

Balsberg-Påhlsson, A.M., G. Lithner, and G. Tyler. 1982. Krom i miljøn, Statens Naturvårdsverk Rapport SNV pm 1570, Solna, Sweden (in Swedish).

Bell, P.F., B.R. James, and R.L. Chaney. 1991. Heavy metal extractability in long-term sewage sludge and metal salt-amended soils. *J. Environ. Qual.* 20:481-486.

Beneš, P. and U. Major. 1980. *Trace Chemistry of Aqueous Solutions.* Academia, Praha, 252 pp.

Beneš, P. and E. Steinnes. 1974. In situ dialysis for the determination of the state of trace elements in natural waters. *Water Res.* 8: 947-953.

Beneš, P. and E. Steinnes. 1975. Migration forms of trace elements in natural fresh waters and the effect of the water storage. *Water Res.* 9: 741-749.

Berrow, M.L. and J. Webber. 1972. Trace element in sewage sludge. *J. Sci. Food Agric.* 23:93-100.

Blom, H. A. 1986. Heavy metal contamination of soils around the cities of Østfold county, Norway (In Norwegian). Ph.D. Thesis, Norway Agric. Univ.

Bolt, G.H. and M.G.M. Bruggenwert. 1976. *Soil Chemistry. A. Basic Elements* Elsevier, Amsterdam. 281 pp.

Bowen, H.J.M. 1979. *Environmental Chemistry of the Elements.* Academic Press, Troy, MO.

Bricker, O.P. 1987. Catchment flow systems. In: Luman, A. and M. Meybek (eds.) *Physical and Chemical Weathering in Geochemical Cycles.* NATO ASI Series 251: 33-60.

Brown, L.R. and J. Jacobson. 1987. Assessing the future of urbanization. p. 38-56. In State of the World 1987. Worldwatch Institute. Norton. New York.

Brummer, G. and U. Herms. 1983. Influence of soil reaction and organic matter on the solubility of heavy metals in soils. p. 233-243. In Ulrich, B. and J.Pankrath (eds.) *Effects of Accumulation of Air Pollutants in Forest Ecosystems.* Reidel Pub. Comp.

Bryan, G.W. 1976. In Johanston, R. (ed) *Marine Pollution.* Academic Press, London, 185 pp.

Buchauer, M.J. 1973. Contamination of soil and vegetation near a zinc smelter by Zn, Cd, Cu, and Pb. *Environ Sci.Technol.* 7:131-135.

Burkitt, A., P. Lester, and G. Nickeless. 1972. Distribution of heavy metals in the vicinity of an industrial complex. *Nature* 238:327-328.

Bærug, R and B.R. Singh. 1990. Cadmium levels in soils and crops after long-term use of commercial fertilizers. *Norw. J. Agric. Sci.* 4:251-260.

Bølviken, B. and E. Steinnes. 1987. Heavy metal contamination of natural surface soils from long-range atmospheric transport. Further evidence from analysis of different soil horizons. p. 291-293. In Heavy Metals in the Environment, Vol. I. CEP Consultants, Edinburgh, Scotland.

CCREM. 1987. Canadian Water Quality Guidelines. March 1987. Canadian Council of Resource and Environment Ministers. Environment Canada, Ottawa.

C.E.C. 1982. Proposal for a Council Directive on the use of sewage sludge in agriculture. Comm-Rur. Communities Eur. Rep. C 264:3-8.

Chang, F.H. and F.E. Broadbent. 1981. Influence of trace metals on carbon dioxide evolution from a Yolo soil. *Soil Sci.* 132:416-421.

Chang, F.H. and F.E.Broadbent. 1982. Influence of trace metals on some soil nitrogen mineralization. *J.Environ.Qual.* 11:115-119.

Christensen, T.H. 1984a. Cadmium soil sorption at low concentrations. I. Effect of time, Cd load, pH, and Calcium. *Water, Air, and Soil Pollut.* 21:105-114.

Christensen, T.H. 1984b. Cadmium soil sorption at low concentrations. II. Reversibility, effect of change in solute composition, and effect of aging. *Water, Air, and Soil Pollut.* 21:115-125.

Cox, R.M. and T.C. Hutchinson.1980. Multiple metal tolerances in the grass *Deschampsia caespitosa. Nature* 279:231-233.

Dam Kofoed, A. and P. Søndergård-Klausen. 1983. Effect of fertilization on Cd content of soil and plants (in Danish). *Tidsskr. Planteavl.* 87:23-32.

Doelman, P. 1985. Reisitance of soil microbial communities to heavy metals. p. 369-384. In V. Jensen, A. Kjoller, and L.H. Sorensen (eds.) *Micorobial Communities.* Elsevier Applied Science Publishers, London.

Doelman, P. and I. Haanstra. 1979. Effect of lead on soil respiration and dehydrogenase activity. *Soil Biol. Biochem.* 11:475-479

Dolska, P. A. and H. Sievering. 1979. Trace elements loading of southern Lake Michigan by dry deposition of atmospheric aerosol. *Water, Air, Soil Pollut.* 12: 485-502.

Dumontet,S.,M.Levesque,and S.P.Mathur.1990. Limited downward migration of pollutant metals(Cu,Zn,Ni,and Pb) in acidic virgin peat soil near a smelter. Water,Air,and Soil Pollut. 49:329-342.

El-Bassam, N. and C. Tietjen. 1977. Municipal sludge as organic fertilizer with special reference to the heavy metals constituents. In: *Soil Organic Matter Studies.* Vol. 2, IAEA, Vienna. 253 pp.

Elsokkary, I.H. and J.Låg. 1978. Distribution of different fractions of Cd, Pb, Zn, and Cu in industrially polluted and non-polluted soils of Odda region, Norway. *Acta. Agric. Scand.* 23:362-368.

EPA. 1986. Quality criteria for water, 196. EPA No. 440/5-86-001. U.S. Environmental Protection Agency.

Eriksson, J.E. 1990. Factors influencing Cd levels in soils and in grain of oats and winter wheat: A field study on Swedish soils. *Reports and Dissertations* 4. Uppsala, Sweden

Eriksson, J.E. 1989. The influence of pH, soil type, and time on adsorption and uptake by plants of Cd added to the soil. *Water, Air, and Soil Pollut.* 48:317 335.

Ervio, R.R., Makela-Kurtto, and J. Sippola. 1990. Chemical characteristics of Finish agricultural soils in 1974 and in 1987. In Kauppiet el. (eds) Acidification in Finland. Springer-Verlag. Berlin, Heidelberg.

Farrah, H. and W.F. Pickering. 1977. Influence of clay-solute interaction on aqueous heavy metal ion levels. *Water, Air, and Soil Pollut.* 8:189-197.

Fergusson, J.E. 1990. *The Heavy Elements: Chemistry, Environmental Impact and Health Effects.* Pregamon Press, Oxford.

Florence, T.M. 1982. The speciation of trace elements in waters. *Talanta* 29: 345-364.

Forgeron, F.D. 1971. Soil geochemistry in the Canadian Shield. *CIM Bull.* 64:37-42.

Förstner, U. 1981. In Wolf, K.H. (ed.) *Handbook of Strat-Bound and Stratiform Ore Deposite* Vol. 9. Elsevier, Amsterdam. 271 pp.

Frank, R., H.E. Braun, K. Ishida, and P. Suda. 1976. Persistent organic and inorganic pesticide residues in orchards soils and vineyards of southern Ontario. *Can. J. Soil Sci.* 56:463-484.

Freedman, B. and T.C. Hutchinson. 1980. Pollutant inputs from the atmosphere and accumulation in soils and vegetation near a nickel-copper smelter at Sudbury, Ontario, Canada. *Can. J. Bot.* 58:108-132.

Fritze, H., S. Nini, K. Mikkola, and A. Makinen. 1989. Soil microbial effects of a Cu-Zn smelter in southwestern Finland. *Biol. Fertil. Soils.* 8:87-94.

Garcia-Miragaya, J. and A.L. Page. 1976. Influence of ionic strength and inorganic complex formation on the sorption of trace amounts of Cd by montmorillonite. *Soil Sci. Soc. Am. J.* 40:658-663.

Garcia-Miragaya, J. and A.L. Page. 1977. Influence of exchangeable cation on the sorption of trace amounts of Cd by montmorillonite. *Soil Sci. Soc. Am. J.* 41:718-721.

Garrels, R.M. and F.T. Mackenzie. 1971. *Evolution of Sedimentary Rocks; w.w.* Norton, New York.

Gowen, J.A., G.B. Wiersma, and H.Tai. 1976. Mercury and 2,4-D levels in wheat and soils from sixteen states. *Pestic. Monit. J.* 10:111-113.

Gunnarsson O. 1983. Heavy metals in fertilizers: Do they cause environmental and health problems. *Fertilizers and Agriculture* 85:27-42.

Hallberg, P.A. and E. Vigerust. 1981. Slamdisponering 3. Tungmetaller i kloakslam (in Norwegian). *Utvalg for Fast Avfall NTNF*, Norway.

Haraldsen, S. 1992. State Control Board's strategy to reduce toxic elements from sludges. Lecture to the Society of Norwegian Ingineers, 15 Oct. 1992. Norwegian State Control Board, Oslo.

Hattori, H.1992. Influence of heavy metals on microbial activities. *Soil Sci. Plant Nutr.* 38:93-100.

He, Q.B. and B.R. Singh 1993a. Plant availability of Cd in soils: I. Extractability of Cd in newly- and long-term cultivated soils. Acta Agric. Scand. B. *Soil and Plant Sci.* (in press).

He, Q.B. and B.R. Singh 1993b. Plant availability of Cd in soils: II. Factors affecting the extractability and plant uptake of Cd in cultivated soils. *Acta Agric. Scand. B. Soil and Plant Sci.* (in press).

He, Q.B. and B.R. Singh 1993c. Cadmium distribution and extractability in soils and its uptake by plants as affected by organic matter and soil type. *J. Soil Sci.* (in press).

Herms, U. 1982. Investigation on heavy metal soulubility in contaminated soils and composted plant residue in relation to soil reaction, redox conditions, and nutrient status (in German). Ph.D. Dissertation, Kiel, Germany. 269 pp.

Hucker, G. 1980. Cadmium in municipal sewage sludge. E F Cost-Project. WP 5 Mimograph.

Hultberg, H. and S. Johansson. 1981. Acid groundwater. *Nordic Hydrology* 12:51-64.

Hutchinson,T.C. and L.M. Whitby. 1973. A study of airborne contamination of vegetation and soils by heavy metals form the Sudbury, Ontario, copper-nickel smelters. In D.D. Hemphill (ed.) *Trace Substances Environ. Health* Vol. VII. Univ. of Missouri,Columbia.

Jeng, A.S. 1992. Weathering of some Norwegian alum shales. II. Laboratory simulations to study the influence of aging,acidification and liming on heavy metal release. *Acta Agric. Scand. B. Soil and Plant Sci.* 42:76-87.

Jeng, A.S. and H. Bergseth. 1992. Chemical and mineralogical properties of Norwegian alum shale soils with special emphasis on heavy metal content and availability. *Acta Agric. Scand. B. Soil and Plant Sci.* 42:88-93.

Jensen, S. and A. Jernelöv. 1969. Biological methylation ofmercury in aquatic organisms. *Nature* 223: 753-754.

Johansson, K. 1988. Heavy metals in Swedish forest lakes - factors influencing the distribution in sediments. Dissertation, Swedish University of Agric., Uppsala, Sweden.

Jones, K.C. and A.E. Johnston 1989. Cadmium in cereals grain and herbage from long-term experimental plots at Rothamsted, U.K. *Environmental Pollut.* 57:199-216.

Kabata-Pendias, A. and H. Pendias. 1984. *Trace Elements in Soils and Plants.* CRC Press, Boca Raton,Fla.

Kabata-Pendias, A. and Gondek. 1978. Bioavailabilty of heavy metals in the vicinity of a copper smelter. p. 523-531. In D.D. Hemphill (ed.) *Trace Substances in Environmental Health XII.* University of Missouri, Columbia.

Kamal, A.A.M., S.E.Eldamaty, and R. Farris. 1991. Blood lead level of Cairo traffic policemen. *Sci. Total Environ.* 105:165-170.

Kelling, K.A., D.R. Keeney, L.M. Walsh, and J.A. Ryan. 1977. A field study of the agricultural use of sewage sludge: III. Effect on uptake and extract-abilty of sludge-borne metals. *J. Environ. Qual.* 6:352-358.

Kirk, P.W.W. and J.N. Lester. 1981. Significance and behaviour of heavy metals in waste water treatment processes. IV. Water quality standards and criteria. *Sci. Total Environ.* 40:1-44.

Konghaug, G., O.C. Bøckman, O. Kårstad, and H. Morka. 1992. Input of trace elements to soil and plants. Paper presented at the International Symposium Chemical Climatology and Geomedical Problems. *Norwegian Academy of Sciences and Letters*, Oslo, Norway.

Krauskopf, K.B. 1967. *Introduction to Geochemistry*. McGraw-Hill. New York.

Lagerwerff, J.V., D.L.Brown, and G.T. Biorsdorf. 1972. p. 523-531. In D.D. Hemphill (ed.) *Trace Substances in Environmental Health XII*. University of Missouri, Columbia.

Lindqvist, O., K. Johansson, M. Aastrup, A. Andersson, L. Bringmark, G. Hovsenius, L. Håkansson, Å. Iverfeldt, M. Meili and B. Timm. 1991. Mercury in the Swedish Environment. *Water, Air, Soil Pollut*. 55: 1-261.

Lindsay, W.L. 1979. *Chemical Equilibrium in Soils*. Wiley-Intersience, New York 499 pp..

Linzon, S.N. 1978. Phytotoxicology excessive levels for contaminants in soil and vegetation. Report of Ministry of the Environment. Ontario, Canada.

Little, P. and M.H. Martin. 1972. A survey of zinc, lead, and cadmium in soil and natural vegetation around a smelting complex. *Environ. Pollut*. 3:241-254.

Lvovitch, M.I. 1970. World water balance: general report. In: Proc. Symp. World Water Balance. *Inter. Assoc. Sci. Hydrol.* 2: 401-405.

Låg, J.1972. Norwegian soil research in relation heavy metal pollution (in Norwegian). Symposium om Tungmetallforurensninger. NAVF, NLVF, and NTNF, p 52-58.

Låg,J. 1978. Arsenic pollution of soils at old industrial sites. *Acta Agric. Scand.* 28:97-100.

Låg J. and B.Bølviken. 1974. Some naturally heavy metal poisoned areas of interest in prospecting,soil chemistry and geomedicine. *NGU* 304:73-96.

Låg, J. and Elsokkary I.H. 1978. A comparison of chemical methods for estimating Cd, Pb, and Zn availability to six food crops grown in industrially polluted soils at Odda, Norway. *Acta Agric. Scand*. 28:76-80.

Låg, J. and E. Steinnes. 1978. Regional distribution of selenium and arsenic in humus layers of Norwegian forest soils. *Acta. Agric. Scand*. 20:3-14.

Maclean, A.J., B. Store, and W.B. Cordukes. 1973. Amounts of mercury in soil of some golf course sites. *Can. J. Soil Sci*. 53:130-132.

McBride, M.B. 1980. Chemisorption of $Cd^{+2}$ on calcite surfaces. *Soil Sci. Soc. Am. J.* 44:26-28.

McNeilly, T., S.T. William, and P.J. Christian. 1984. Lead and zinc in a contaminted pasture at Minera, North Wales, and their impact on productivity and organic matter breakdown. *Sci. Total Environ*.38:183-198.

Morgan, H. and D.L. Sims. 1988. *Sci.Total Environ*. 75:135-143.

Mortvedt, J.J. 1987. Cadmium levels in soils and plants from some long-term soil fertility experiments in the USA. *J.Envirn. Qual*. 16:137-142.

Mulla, D.J., A.L. Page and T.J.Ganje. 1980. Cd accumulation and bioavailability in soils from long-term fertilization. *J.Envron.Qual*. 9:408-412.

Muller, E.F. and J.R. Kramer. 1977. Precipitation scavenging in central and northern Ontario. *ERDA Symp. Ser.* 41:590-601.

Nriagu, J.O. 1992. Toxic metal pollution in Africa. *Sci. Total Environ.* 121:1-37.

Nriagu, J.O. and J.M. Pacyna. 1988. Quantitative assessment of worldwide contamination of air, water and soils by trace metals. *Nature* Vol. 333 No. 6169:134-139.

Pacyna, J.M., A. Semb, and J.E. Hanssen, 1984. Emission and long-range transport of trace elements in Europe. *Tellus* 36B: 163-178.

Page, A.L. 1992. Scientific basis for the development of standards for land application of municipal sewage sludge in the United States of America. Guest lecture to the Agric. Univ. of Norway. 18 Sept. 1992.

Page, A.L. and E. Steinnes. 1990. Atmospheric deposition as a source of trace elements in soils. *Palaeogeogr., Palaoclimat., Palaeoecol.* (Global and Planetary Change Section) 82: 141-148.

Pauly, H. and A. Simonsen. 1974. 22 Danish sewage sludge plants. Minralogisk Institutt, Mimiograph.

Ragaini, R.C., H.R.Ralston, and N. Roberts. 1977. Environmental trace metal contamination in Kellogg, Idaho, near a lead smelting complex. *Environ. Sci. Technol.* 11:773- 781.

Rose, A.W., H.E. Hawkes, and J.S. Webb. 1979. *Geochemistry in Mineral Exploration*, 2nd ed. Academic Press, London.

Rowell, D.L. 1981. p. 401-462. In: Greenland, D.J. and M.H.B.Hayes (eds.) *Chemistry of Soil Processes.* John Wiley, Chichester.

Salbu, B., E. Steinnes, and H. Bjørnstad. 1984. Use of different physical separation techniques for trace element speciation studies in natural waters. In: Eriksson, E. (ed.) Hydrochemical Balances of Freshwater Systems. *IAAS-AIHS Publ.* No. 150: 203-213.

Schultz, J.J., I. Gregory, and O.P.Engelstad. 1992. Phosphate fertilizers and the environment. A discussion paper. IFDC, Muscle Shoals, Alabama, USA.

Semu, E., B.R. Singh, and A.R. Selmer-Olsen. 1986. Mercury pollution of effluent, air, soil near a battery factory in Tanzania. *Water, Air, and Soil Pollut.* 27:141-146.

Semu, E., B.R. Singh, and A.R. Selmer-Olsen. 1987. Adsorption of mercury compounds by tropical soils. II. Effect of soil:solution ratio, ionic strength, pH, and organic matter. *Water, Air, and Soil Pollut.*32:1-10.

SIFF. 1987. *Kvalitetsnormer for drikkevann.* Statens institutt for Folkehelse (National Institute for Public Health), Oslo, Norway. ISBN 82-7364-013-2. 72 pp.

Sims, D.L. and H. Morgan. 1988. *Sci. Total Environ.* 75:1-10

Singh, B.R. 1991. Unwanted components of commercial fertilizers and their agricultural effects. Proc. No. 312. The Fertilizer Society, Peterborough, UK.

Singh, B.R. and R.P. Narwal. 1984. Plant availability of heavy metals in a sludge-treated soil: II. Metal extractability compared with plant metal uptake. *J. Environ. Qual.* 13:344-349.

Singh, B.R. and Steinnes. 1976. Uptake of trace elements by barley in zinc polluted soils: 2. Lead, cadmium, mercury, selenium, arsenic, chromium, and vanadium in barley. *Soil Sci.* 121:38-43.

Singh, B.R. and J. Låg. 1976. Uptake of trace elements by barley in zinc polluted soils: 1. Availability of zinc and the effects of excessive zinc on the growth and chemical composition of barley. *Soil Sci.* 121:32-37.

Sommers, L.E., D.W. Nelsen, and K.J. Yost. 1976. Variable nature of chemical composition of sewage sludges. *J. Environ. Qual.* 3:303-306.

Spencer, D. 1966. Factors affecting element distribution in a silurian gratolite band. Chem.Geol. 1:221-249.

Sposito, G., L.J. Lund, and A.C. Chang. 1982. Trace metal chemistry in arid - zone soils amended with sewage sludge: I. Fractionation of Ni, Cu, Zn. Cd, and Pb in solid phases. *Soil Sci. Soc. Am. J.* 46:260-264.

Sposito, G. 1983. p. 123-170. In I. Thornton (ed.) *Applied Environmental Chemistry*, Academic Press, London.

Sposito, G. and A.L. Page. 1985. p. 287-332. In Sigel, H. (ed.) *Metal Ions in Biological Systems Vol 18, Ciculation of Metals in the Environment*, Marcel Dekker, New York.

Steinnes, E. 1980. Atmospheric deposition of heavy metals in Norway studied by analysis of moss samples using neutron activation analysis and atomic absorption spectrometry. *J. Radional. Chem.* 58: 387-391.

Steinnes, E. 1984. Contribution from long range atmospheric transport to the deposition of trace metals in Southern Scandinavia. Report NILU OR 29/84. Norwegian Institute for Air Research, Lillestrøm, Norway. 40 pp.

Steinnes, E. and E. Andersson, 1991. Atmospheric deposition of mercury in Norway: Temporal and spatial trends. *Water, Air, Soil Pollut.* 56: 391-404.

Steinnes, E., H. Hovind, and A. Henriksen. 1989. Heavy metals in Norwegian surface waters, with emphasis on acidification and atmospheric deposition. p. 36-39. In J.P. Vernet (ed.) *Heavy Metals in the Environment.* September, 1989. Geneva.

Stone, E.L. and V.R.Timmer. 1975. On the Cu content of some northern conifers. *Can. J. Bot.* 53:1453-1456.

Swaine, D.J. 1962. *The Trace Element Content of Fertilizers.* Commonwealth Agricultural Bureau. Farmham Royal, Bucks,UK.

Thornton, I. 1981. Geochemical aspects of the distribution and forms of heavy metals in soils. p. 1-33. In N.W. Lepp (ed.) Effect of heavy metal pollution on plants. *Metals in the Environment.* Vol.2 Applied Science Publishers.

Tiller, K.G. and R.H. Merry. 1982. Copper pollution of agricultural soils. p. 119-140. In: J.F. Loneragen, A.D.Robson, and R.D. Graham (eds.) Copper in Soils and Plants. Academic Press, London.

Tills, A.R. and B.J.Alloway. 1983. The speciation of Cd and Pb in soil solutions from polluted soils. p. 1211-1214. In: Heavy Metals in the Environment. Int.Conf., Heidelberg, Vol.2.

Udo, E.J., H.L.Bohn, and T.C. Tucker 1970. Zinc adsorption by calcareous soils. *Soil Sci. Soc. Am. Proc.* 34:405-407.

Van de Meent, D., T. Aldenberg, J.H. Canton, C.A.M. van Gestel, and W. Slooff. 1990. Desire for levels. Background study for the policy document *Setting Environmental Quality Standards for Water and Soil.* Rijksinstituut voor Volksgesundheid en Milieuhygiene, Report 670101 002, Amsterdam.

Warren,H.V., R.V.E. Delevault, and J.Barakso. 1966. Some observations on the geochemistry of Hg as applied to prospecting. *Econ. Geol. Ser. Can.* 61:1010-1028.

WHO. 1987. Global Pollution and Health. Global Environmental Monitoring System Report, World Health Organization, Geneva.

Williams, C.H. and D.J. David 1973. The effect of superphosphate on the Cd content of soils and plants. *Aust. J. Soil Res.* 11:43-56.

Williams,C.H. and D.J. David 1976. The accumulation in soil of Cd residues from phosphate fertilizers and their effect on the Cd content of plants. *Soil Sci.* 121:86-93.

# Water Quality Effects of Tropical Deforestation and Farming System on Agricultural Watersheds in Western Nigeria

R. Lal

I. Introduction . . . . . . . . . . . . . . . . . . . . . . . . . . . . . . 273
II. Land Use and Water Quality . . . . . . . . . . . . . . . . . . . . . . 274
A. Soil Processes and TRF Conversion in Relation to Water Quality . . . . . . . . . . . . . . . . . . . . . . . . . . . . . 274
B. Factors Affecting Water Quality Following Forest Conversion . . . . . . . . . . . . . . . . . . . . . . . . . . . . . . 278
III. Management Effects on Water Quality . . . . . . . . . . . . . . . 286
A. Tillage Methods . . . . . . . . . . . . . . . . . . . . . . . . . . . 286
B. Ground Cover . . . . . . . . . . . . . . . . . . . . . . . . . . . . 290
C. Fertility Management . . . . . . . . . . . . . . . . . . . . . . . . 296
D. Weed Management . . . . . . . . . . . . . . . . . . . . . . . . . 297
IV. Toward Enhancing Water Quality from Agricultural Watershed 297
References . . . . . . . . . . . . . . . . . . . . . . . . . . . . . . . . 299

## I. Introduction

Tropical rainforest (TRF), the climax vegetation of the humid tropics, cover about 1.46 billion ha. TRF occupies about 48% of the land area of the humid tropic ecosystem, 30% of the land area within the tropical region, and 86% of the total tropical forest area comprising all vegetation type of the humid tropics (Forest Resource Assessment, 1990). However, the exact area of the TRF is debatable because it depends on what is considered as TRF. Similarly, the exact rate of deforestation of TRF is also debatable because the rate computations depend on what is considered TRF and what is deforestation. Therefore, published estimates of TRF deforestation range from 11 m ha to 20 m ha (World Resources Institute, 1988-89; 1990-91).

Conversion of TRF to other land uses has become a global issue because of its perceived or potential effects on land and environment. Among major impacts of TRF conversion on land include degradation of soil and water quality through

0-87371-980-8/94/$0.00+$.50

accelerated soil erosion, transport of sediments and dissolved organic and inorganic constituents in surface and sub-surface water, depletion of soil organic matter and plant nutrients through leaching and mineralization, and reduction in activity and biodiversity of soil fauna (Lal, 1986, 1992; NRC, 1982, 1993). An important environmental effect of TRF conversion is emission of radiatively-active gases into the atmosphere with possible consequences related to the "Greenhouse Effect" (Houghton, 1990; Schlesinger, 1984).

Effects of TRF conversion to agricultural land uses on water quality are important in terms of eutrophication of natural waters and its impact on human and livestock population. Natural water, streams or ground water, is directly used for human and livestock consumption with little or no treatment in most TRF regions of Asia, Africa and South and Central America. Consequently, impact of TRF conversion on water quality can pose severe health hazards.

## II. Land Use and Water Quality

Impacts of TRF conversion have been evaluated for changes in soil properties (Cunningham, 1963; Nye and Greenland, 1964; Weert, 1974a,b; Seubert et al., 1977; Lal and Cummings, 1979; Hulugalle et al., 1984; Alegre et al., 1986, 1986a,b; Lal, 1987; Ghuman and Lal, 1989 a,b; Ghuman et al. 1991; Ghuman and Lal, 1991a,b; 1992), crop growth (Couper et al., 1981; Lal, 1987; Lal et al., 1986) and on micro-climate (Lal and Cummings, 1979; Ghuman and Lal, 1987). A considerable progress has also been made in identifying and adapting agronomic systems for sustaining agricultural production (Lal, 1987; Lal et al., 1992; Sanchez and Benites, 1987). However, the impact of TRF conversion on water quality have not been extensively documented. The objective of this report is to explain the impact of forest conversion on soil processes that affect quality of surface and sub-surface water. Examples used in the study are based on a large watershed management project conducted in sub-humid region of southwestern Nigeria (Lal, 1992). Soils of this region are Alfisols, and the mean annual rainfall of about 1250 mm is received in a biomodel distribution. The experiment reported herein was conducted over the 10-year period from 1979 to 1988.

Water characteristics affecting water quality from agricultural land are listed in Figure 1. The nature and degree of pollutants transported in surface and sub-surface waters following TRF conversion to agricultural land use depend on several processes (Figure 2) and factors (Figure 3). It is the interaction between processes and factors that affect the water quality.

### A. Soil Processes and TRF Conversion in Relation to Water Quality

Several soil processes are drastically affected by TRF conversion (Figure 1). Principal among these are hydrologic cycle, energy budget, cycles of carbon and

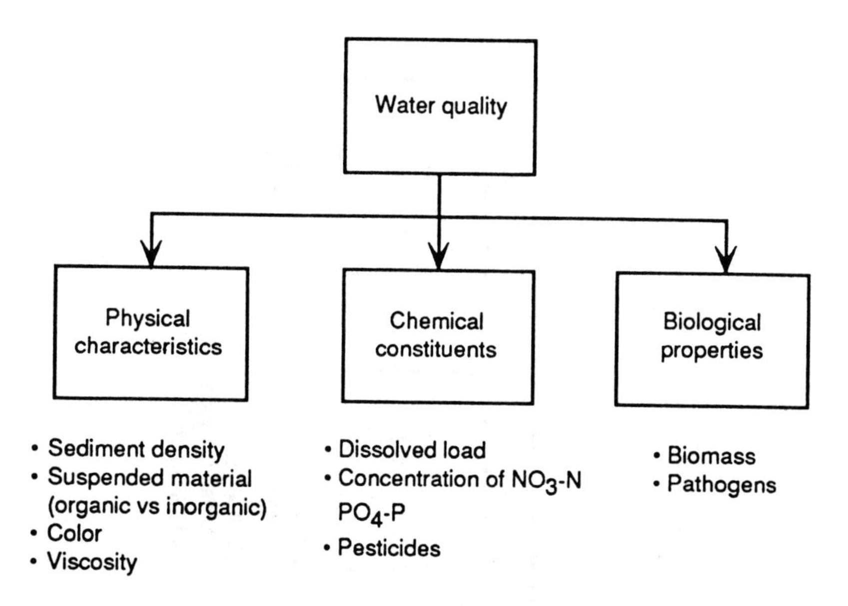


**Figure 1**. Water quality from agricultural land.

principal nutrients, and activity and biodiversity of soil fauna. Hydrologic cycle is a principal process of major relevance to water quality. TRF conversion usually increases surface runoff (Lal, 1981) and interflow (Lawson and Lal, 1981), and decreases soil water storage. Increase in surface runoff and interflow usually increases the total amount of sediment and dissolved elements transported along with surface and sub-surface flow. There may be no direct relationship, however, between the volume of flow and concentration of suspended or dissolved elements. Energy budget and hydrologic cycle are interlinked. For example, the rate and amount of evapotranspiration depend on the energy available at the soil/vegetation surface. An important effect of change in the energy budget following TRF conversion is that on the soil temperature regime. TRF conversion increases the maximum soil temperature by as much as 20°C (Cunningham, 1963; Lal and Cummings, 1979). Biomass burning following TRF conversion, especially that in the windrows, has drastic effects on soil temperature during and after burning (Ghuman and Lal, 1987, 1989a). These conversion-induced changes in soil temperature regime affect mineralization rate of soil organic carbon, flux of radiatively-active gases ($CO_2$, $NO_x$, $H_2O$) into the atmosphere, and the amount of nutrients available for mass flow with surface and sub-surface water flow.

Rate of mineralization of soil organic carbon, and disruption in cycles of C and plant nutrients (N, P, S, K, Ca, Mg, etc.) are major effects of TRF conversion with impact on water quality. Large quantities of nutrients released from biomass burning or mineralization are transported in surface water, or leached with the percolating water and eventually carried into the stream flow.

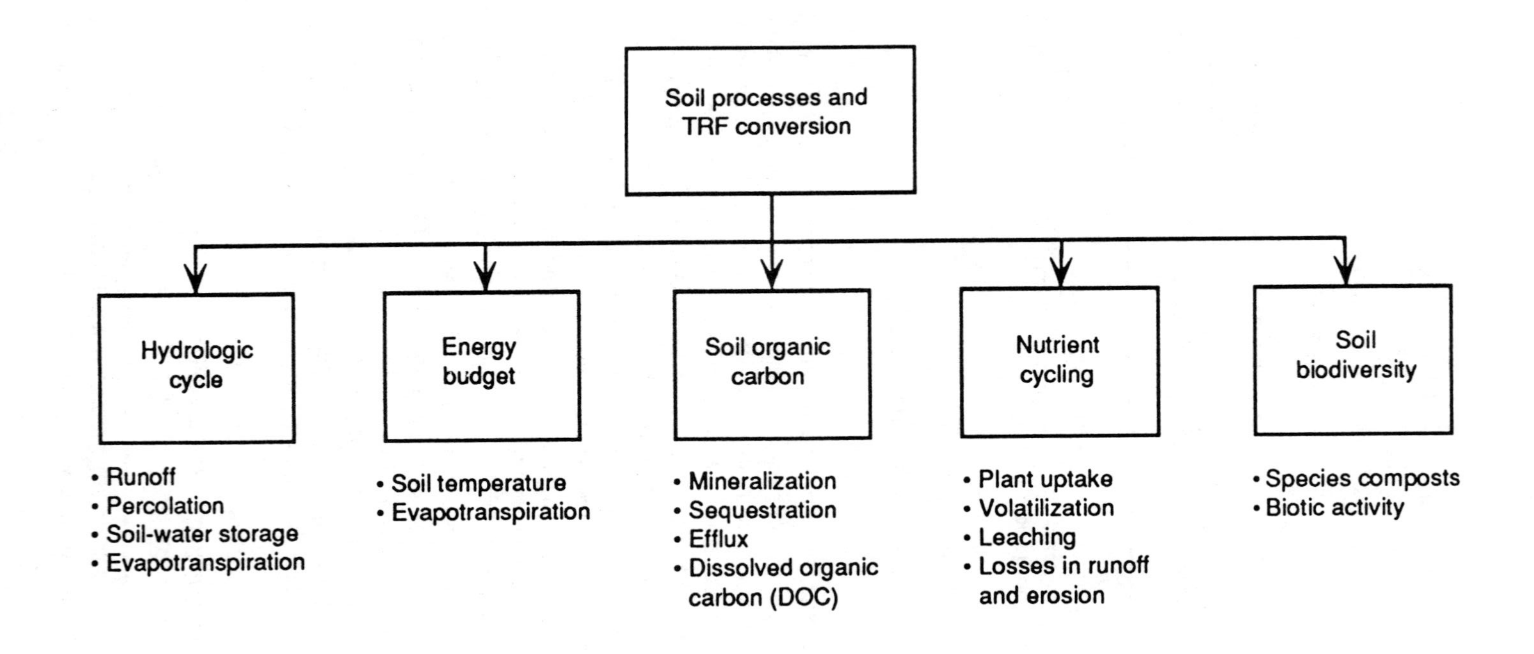


**Figure 2.** Soil processes and water quality in relation to TRF conversion.

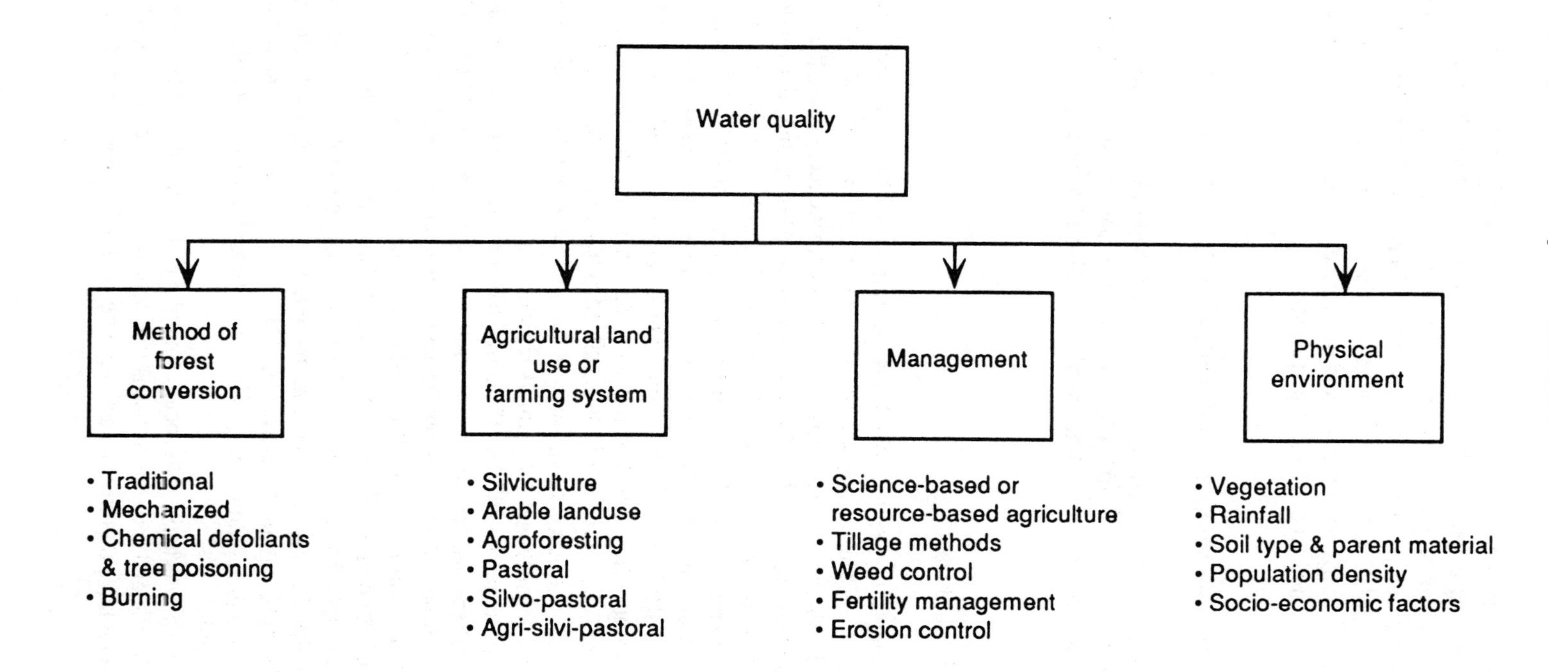


**Figure 3.** Factors affecting water quality following TRF conversion.

Increase in the interflow amount and rate may also increase the quantity and quality of dissolved organic carbon (DOC) transported in the water.

Considering all processes affected by forest conversion, those with major impact on water quality are: (i) runoff and soil erosion, (ii) percolation and leaching, (iii) mineralization of soil organic matter, and (iv) disruption in cycles of principal plant nutrients. Intensity and rate of these processes are affected by several factors related to ecosystem perturbation.

## B. Factors Affecting Water Quality Following Forest Conversion

Principal factors affecting water quality are outlined in Figure 2, and are briefly described below with specific examples of the data from an experiment conducted at the International Institute of Tropical Agriculture in southwestern Nigeria.

### 1. Physical Environment

Parent material, soil type, vegetation, and the rainfall all have important effects on water quality. Climate, vegetation, and soil are all inter-related, and their interactions determine: (i) total nutrient capital within the ecosystem, (ii) amount of nutrient immobilized in the biomass, (iii) total amount and distribution of nutrients within the soil profile, and (iv) mechanisms and rate of nutrient cycling. Apparently, soils derived from basic parent material are relatively fertile and contain higher concentrations of cations than those derived from acidic rocks or sand stone. Soil fertility and intensity and capacity factors of plant-available nutrients, are also related to age or status of weathering -- the younger the soil the more fertile it is, all other factors remaining the same. Furthermore, the climax vegetation, species diversity and the total biomass, depend on soil fertility and the rainfall regime.

It is difficult to generalize the effects of TRF conversion on water quality in relation to inherent soil fertility. In general, however, the amount of nutrients (pollutants) transported in water out of the ecosystems is directly proportional to the initial concentration. The higher the initial concentration in the soil, the higher the potential risks of transport in water following ecosystem perturbations. On the other hand, however, the higher the inherent soil fertility the higher is usually the vegetation growth, thereby immobilizing the nutrients and holding them back within the ecosystem. The time, nature, and magnitude of perturbation also play an important role in release and transport of nutrients out of the ecosystem.

The nature of perturbation has socio-economic and political connotations. The pressure on fragile and ecologically-sensitive ecoregions is directly proportional to the demographic pressure. In fact, the rate of TRF conversion is driven by the demographic pressure and its needs, and is manifested by legislations, development of roads and infra-structure, and local and regional politics.

## 2. Methods of TRF Conversion

Water quality can be significantly influenced by the methods of forest conversion. Commonly prevalent methods are: (i) traditional based on native tools used manually, (ii) semi-mechanized tools such as motorized chain-saw, (iii) motorized equipment using tractor-driven devices, e.g., tree pusher, tree cutter, tree extractor, tree crusher, root extractors, etc., and (iv) chemicals and growth regulators used as defoliants, desiccants, and poisons. The degree of perturbation, and the magnitude of pollutants released in natural waters, depend on the method of TRF conversion. For example, traditional methods based on manual fellings of trees do not pose as much risk of decreasing water quality as motorized methods of forest conversion. Lal (1981) observed that concentration of $PO_4$-P in runoff water was low for plots managed by traditional and manual methods of TRF conversion. With tree pusher and bulldozer clearing, $PO_4$-P concentrations were substantially higher. When biomass was burnt, runoff water had higher concentrations of cations (Ca, Mg, K, Na) in surface runoff than when it was removed to the plot boundaries and/or allowed to decompose in situ. In addition to dissolved load, sediments or suspended load also affect water quality, and methods of TRF conversion have an important effect on sediment concentration in water runoff. Lal (1981) reported significant differences in total soil erosion among different methods of TRF conversion. The data in Table 1 show significant differences in sediment concentrations in water runoff due to methods of forest conversion. These observations were made in the first year beginning soon after forest conversion. In general, the sediment concentration in runoff was in the order of traditional conversion > manual > shear blade > tree pusher/root rake. In other words, the sediment concentration in runoff was directly proportional to the amount of soil disturbance. The more the soil was exposed due to soil disturbance, the more was the sediment transport in water runoff.

## 3. Farming Systems Effects on Quality of Surface Runoff

Land use and farming systems play an important role in water quality. In general, water quality is high in undisturbed or natural ecosystems. Conversion of TRF to agricultural land use reduces water quality due to transport of sediments, plant nutrients, and of agricultural chemicals. Water quality depends as much upon the management as on the farming system. High water quality is expected from good agricultural management, even when high inputs are used.

The data in Tables 2 and 3 show the impact of farming systems on nutrient loss in water runoff. These observations were made for the 1985 rainy seasons.

**Table 1.** Effects of methods of deforestation on sediment density in water runoff during the first season 1979, soon after deforestation (Forest was removed in March/April 1979)

| TRF conversion method | Plot | 25 May | 15 June | 30 June | 23 July | 30 July | 28 Oct. |
|---|---|---|---|---|---|---|---|
| | | g liter$^{-1}$ | | | | | |
| Traditional | 7 | 0.76 | 2.7 | 2.20 | 5.7 | 11.1 | 1.3 |
| | 13 | 0.00 | 2.8 | 0.0 | 5.1 | 6.1 | 0.0 |
| | X | | 2.75∓ | | | | |
| Manual | 1 | 1.75 | 4.00 | 1.50 | 19.5 | 1.6 | 3.2 |
| | 3 | 0.34 | 3.60 | 0.80 | 2.4 | 1.2 | 0.0 |
| | 10 | 2.29 | 2.10 | 0.80 | 14.7 | 3.8 | 0.0 |
| | 12 | 1.34 | 3.30 | 0.70 | 6.5 | 1.9 | 0.0 |
| | 14 | 5.35 | 4.20 | 4.50 | 2.2 | 1.6 | 0.0 |
| | X | | 4.3∓ | | | | |
| Shear blade | 4 | 0.24 | 11.30 | 0.30 | 3.70 | 5.5 | 6.4 |
| | 9 | 2.37 | 5.20 | 5.30 | 4.70 | 4.1 | 2.0 |
| | X | | 8.25 | | | | |
| Tree pusher/ root rake | 2 | 0.84 | 17.80 | 2.90 | 1.00 | 0.6 | 3.2 |
| | 5 | 4.78 | 2.60 | 1.00 | 19.90 | 18.4 | 13.0 |
| | 6 | 0.38 | 9.80 | 0.50 | 15.20 | 8.4 | 3.3 |
| | 8 | 3.80 | 11.30 | 9.10 | 20.10 | 32.3 | 5.2 |
| | X | | 10.40 | | | | |

It is apparent from the data that runoff generated from land used for intensive farming systems, based on growing two crops per year with chemical-intensive inputs, carried more nutrients (pollutants) than land growing cover crops or pastures. Runoff from natural fallow treatments and forested control lost the least amounts of plant nutrients in runoff. Within an intensive farming system, however, quality of water runoff can be regulated by judicious management of soil surface, crop residue, and fertilizers and amendments. The importance of management on water quality will be discussed later.

The total nutrient loss in water runoff depends on nutrient concentration and sediment load, both of which are influenced by the farming system and management. An example of the farming systems effect on nutrient concentration in water runoff is shown by the data in Table 4. High concentrations of $NO_3$-N in water runoff were observed from plots that received chemical fertilizers, and management systems accelerated runoff and sediment transport. Nonetheless, high concentrations of $NO_3$-N were also observed in runoff from

**Table 2.** Farming systems effects on nutrient loss in water runoff during the first season 1985 at IITA, Ibadan, Nigeria

| Treatment | Plot | Ca | Mg | K | Na | $PO_4$-P | $NO_3$-N |
|---|---|---|---|---|---|---|---|
| | | Nutrient loss (kg $ha^{-1}$) | | | | | |
| Alley cropping | 1 | 0.61 | 0.16 | 0.99 | 0.04 | 0.04 | 0.75 |
| | 5 | 0.64 | 0.34 | 3.53 | 0.68 | 0.12 | 0.18 |
| | X | 0.63 | 0.25 | 2.26 | 0.36 | 0.08 | 0.46 |
| Maize-cowpea with *Macuna* fallow | 4 | 5.61 | 1.7 | 11.35 | 3.65 | 1.05 | 2.55 |
| Ley farming | 9 | 0.17 | 0.07 | 1.04 | 0.11 | 0.23 | 0.05 |
| | 10 | 0.02 | 0.00 | 0.01 | 0.01 | 0.00 | 0.02 |
| | X | 0.10 | 0.04 | 0.52 | 0.06 | 0.12 | 0.04 |
| Maize-cowpea | 2 | 2.94 | 1.05 | 2.63 | 0.57 | 0.13 | 2.70 |
| | 3 | 3.06 | 1.11 | 6.49 | 2.90 | 0.22 | 1.37 |
| | X | 3.00 | 1.08 | 4.56 | 1.73 | 0.18 | 2.04 |
| Natural fallow | 14 | 0.98 | 0.18 | 0.62 | 0.18 | 0.18 | 2.04 |

**Table 3.** Farming systems effects on nutrient loss in water runoff during the second season 1986 at IITA, Ibadan, Nigeria

| Treatment | Plot | Ca | Mg | K | Na | $PO_4$-P | $NO_3$-N |
|---|---|---|---|---|---|---|---|
| | | Nutrient loss (kg $ha^{-1}$) | | | | | |
| Alley cropping | 1 | 0.16 | 0.04 | 0.21 | 0.02 | 0.01 | 0.03 |
| | 5 | 1.94 | 0.53 | 2.45 | 0.52 | 0.12 | 0.18 |
| | X | 1.05 | 0.29 | 1.33 | 0.27 | 0.07 | 0.11 |
| Maize-cowpea with *Macuna* fallow | 4 | 2.75 | 0.48 | 6.77 | 0.46 | 0.11 | 0.56 |
| Ley farming | 9 | 2.60 | 0.55 | 2.09 | 0.71 | 0.08 | 0.10 |
| | 10 | 0.13 | 0.03 | 0.01 | 0.02 | 0.00 | 0.07 |
| | X | 1.36 | 0.29 | 1.05 | 0.37 | 0.04 | 0.09 |
| Maize-cowpea | 3 | 2.60 | 0.44 | 0.69 | 0.69 | 0.03 | 0.24 |
| Natural fallow | 14 | 0.29 | 0.04 | 0.01 | 0.01 | 0.01 | 0.03 |

**Table 4.** Farming systems effects on nutrient concentration in water runoff for rainstorms received on various dates

| Farming systems | Plot | ---23 May 1983--- | | ---10 April 1984--- | | --29 October 1985-- | | ---26 June 1987--- | |
|---|---|---|---|---|---|---|---|---|---|
| | | $PO_4$-P | $NO_3$-N | $PO_4$-P | $NO_3$-N | $PO_4$-P | $NO_3$-N | $PO_4$-P | $NO_3$-N |
| | | Concentration (mg $kg^{-1}$) | | | | | | | |
| Alley cropping | 1 | 0.17 | 6.50 | 1.44 | 28.50 | 0.05 | 10.64 | 0.60 | 13.3 |
| | 5 | 0.90 | 7.60 | - | - | 0.15 | 0.50 | 0.01 | 0.50 |
| *Mucuna* fallow | 4 | 0.42 | 7.12 | - | - | 0.50 | 2.95 | 0.20 | 0.83 |
| Ley farming | 9 | 1.74 | 9.76 | - | - | 0.30 | 0.60 | 0.70 | 0.67 |
| | 10 | 0.82 | 6.00 | - | - | - | - | 0.10 | 0.33 |
| Maize-cowpea | 2 | 0.58 | 11.40 | 0.22 | 13.50 | - | - | - | - |
| | 3 | 0.50 | 10.00 | - | - | - | - | - | - |
| Natural fallow | 14 | 0.26 | 1.48 | - | - | 0.15 | *15.20 | 0.04 | **13.51 |

**Table 5.** Farming systems effects on sediment concentrations in runoff from a watershed planted to cowpea

| Treatment/crop | Plot | Date of observation | Sediment concentration |
|---|---|---|---|
| A. Crop | | | ($g\ liter^{-1}$) |
| Maize | 3 | 15 June, 1979 | 3.60 |
| Cassava | 3 | 29 May, 1980 | 0.54 |
| Cowpea | 3 | 29 Sept., 1981 | 0.12 |
| *Mucuna* | 3 | 23 Sept., 1983 | 0.00 |
| B. Farming system | | | |
| Alley cropping | 5 | 10 June, 1986 | 0.12 |
| *Mucuna* fallow | 3 | 10 June, 1986 | 0.38 |
| Maize-cowpea | 4 | 10 June, 1986 | 3.48 |
| Ley farming | 10 | 10 June, 1986 | 0.12 |
| Natural fallow | 7 | 10 June, 1986 | 0.04 |

natural fallow. Nitrate release from natural mineralization process within forest ecosystem and natural fallow can also be high.

Crops and farming systems effects on sediment concentration in water runoff are shown in the data in Table 5. All other factors remaining the same, sediment concentration is more in open-row crops than in those with close canopy. The sediment concentration was generally in the order of maize (*Zea mays*) > cassava (*Manihot esculent)* > cowpea (*Vigna unjuiculete*) > *Mucuna.* Similarly, soil erosion and sediment transport is often more from farming systems characterized by open space and exposed soil surface, e.g., arable land use. The data in Table 5 show that sediment concentration in surface runoff was in the order arable land use > cover crop > alley cropping = ley farming > natural fallow. These general trends are also verified by the data from other regions with similar soil properties and climatic environments.

## 4. Farming Systems Effects on Quality of Seepage Water

The effects of farming systems on ground water quality are related to the type and amount of nutrient input, water and nutrient requirement of the farming system (input - output), and the climatic factors, e.g., rainfall, evapotranspiration, and water balance. There are few, if any, data relating land use and farming systems to quality of seepage water for tropical ecosystem. Factors affecting sediment density in seepage water are outlined in Figure 4. The data in Table 6 show the importance of land use and farming system on chemical composition of seepage water. The analyses made on May 31, 1982 indicated high $NO_3$-N concentration in treatments cultivated to maize. Even fallowing with

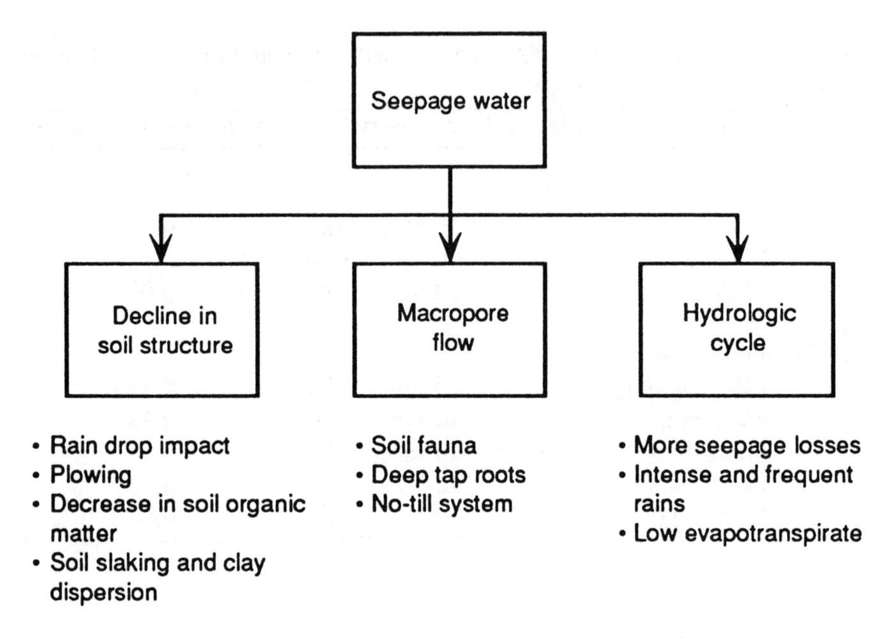


**Figure 4.** Factors affecting sediment density in seepage water.

*Mucuna* and forested control, both treatments receiving no fertilizer, had relatively high concentrations of $NO_3$-N. Expectedly, P concentration in seepage water was low in all treatments. However, the leaching losses of cations were extremely high especially those of Ca and Na. High losses of bases may be one of the reasons for soil acidification. In fact, soil pH in these plots declined by about 1 unit over a 3-year period (Lal, 1992).

The data of chemical analyses for 16 July, 1983 is also interesting (Table 6). Even the pasture and forested control treatments that received no chemical fertilizers, had high concentrations of $NO_3$-N and $PO_4$-P. Losses of bases were also high. These data may have tremendous implications toward developing a strategy for natural resource management in relation to water quality. It is apparent that addition of chemical fertilizers is not the only reason of high concentrations of nitrates in ground water or in spring water draining into the streams. High concentrations of nitrates into ground water or stream flow can also be added by mineralization of nitrogen from forested fallow or from heavily grazed pastures.

Some experiments conducted in West Africa and elsewhere in the tropics have shown that seepage water can also transport dispersed clay as suspended material. In fact, the color of percolating water from subsoil rich in kaolinitic clay is often indicative of the clay transported in the seepage water. The data in Table 7 from lysimetric studies conducted at IITA indicate substantial amount of suspended load transported in the seepage water. The sediment density was especially high for the analyses conducted on May 1, and July 12, 1986.

**Table 6.** Land use and cropping systems effects on chemical composition of seepage water

| Treatment | Plot | Date | ------Chemical composition (mg $kg^{-1}$)------ | | | | | |
|---|---|---|---|---|---|---|---|---|
| | | | Ca | Mg | K | Na | $PO_4$-P | $NO_3$-N |
| Tree pusher (Maize) | | 31 May 1982 | 12.90 | 1.98 | 1.60 | 62.50 | 0.08 | 39.80 |
| Shear blade (*Mucuna)* | | 31 May 1982 | 2.90 | 0.46 | 1.00 | 42.50 | 0.08 | 7.50 |
| Shear blade (Maize) | | 31 May 1982 | 3.50 | 1.46 | 7.00 | 137.50 | 0.03 | 22.80 |
| Forested control | | 31 May 1982 | 4.70 | 1.82 | 1.50 | 60.00 | 0.03 | 9.00 |
| Tree pusher (*Mucuna)* | | 16 July 1983 | 3.10 | 1.50 | 1.50 | 40.00 | 2.00 | 9.90 |
| Shear blade (Maize) | | 11 July 1983 | 9.80 | 2.60 | 7.60 | 47.50 | 2.00 | 8.72 |
| Tree pusher (Pasture) | | 16 July 1983 | 2.10 | 0.60 | 2.30 | 45.00 | 2.00 | 11.70 |
| Forested control | | 16 July 1983 | 2.90 | 2.81 | 3.00 | 45.00 | 2.20 | 18.00 |

(Lysimetric data from IITA, Ibadan, Nigeria.)

High sediment density in percolating water is indicative of several factors outlined in Figure 3.

Decline in soil structure is a principal factor responsible for high sediment density in percolating water. Apparently, use of plow-based system to grow open-row erosion-promoting crops (e.g., maize) would increase clay dispersion and slaking. The data in Table 7 substantiates this hypothesis. The highest sediment density was observed in percolating water from lysimeters cultivated to maize. The macropore flow, with high and turbulent flow, can be another important factor. This factor is especially important in less disturbed or undisturbed soil profile, e.g., forested control, *Mucuna* fallow, or no-till system. Furthermore, a farming system that increases deep seepage or interflow component of the hydrologic cycle is likely to enhance the sediment density in percolating water. Replacement of deep-rooted with shallow-rooted crops, or frequent and intense rains would increase seepage flow and enhance sediment transport in percolating water.

**Table 7.** Farming systems effects on sediment concentrations in seepage water

| Treatment | Plot | Date | Sediment concentration (g liter$^{-1}$) |
|---|---|---|---|
| Maize-cowpea | 2 | 10 July 1982 | 0.10 |
| *Mucuna* fallow | 4 | 10 July 1982 | 0.10 |
| Maize-cowpea | 8 | 10 July 1982 | 0.06 |
| Maize-cowpea | 11 | 10 July 1982 | 0.22 |
| Maize-cowpea | 12 | 10 July 1982 | 0.06 |
| Forested control | 15 | 1 May 1986 | 0.08 |
| Maize-cowpea | 2 | 1 May 1986 | 2.84 |
| Maize-cowpea | 4 | 1 May 1986 | 3.02 |
| Maize-cowpea | 8 | 1 May 1986 | 0.04 |
| Pasture | 11 | 1 May 1986 | 0.12 |
| Pasture | 12 | 1 May 1986 | 0.38 |
| Forested control | 15 | 1 May 1986 | 0.64 |
| Maize-cowpea | 2 | 12 July 1986 | 6.12 |
| Maize-cowpea | 4 | 12 July 1986 | 3.40 |
| Maize-cowpea | 8 | 12 July 1986 | 0.12 |
| Pasture | 11 | 12 July 1986 | 0.10 |
| Pasture | 12 | 12 July 1986 | 0.18 |
| Forested control | 15 | 12 July 1986 | 0.48 |

(Lysimetric data from IITA, Ibadan, Nigeria.)

## III. Management Effects on Water Quality

Agronomic management is an important factor affecting water quality from agricultural land. Principal management factors associated with water quality are outlined in Figure 5. Similar to soil erosion management, agronomic practices that affect water quality include: methods of seedbed preparation, ground cover, crop residue management, soil fertility management, crop rotations and sequences, and weed control. In general, all practices that affect soil erosion would also influence water quality. Using "good farming" practices and applying principles of "land stewardship" are also good for environmental quality in general and water quality in particular. Some specific examples of water quality in relation to agronomic management are described below.

### A. Tillage Methods

Method of seedbed preparation affect water quality both directly and indirectly (Figure 6). Directly, tillage methods affect total porosity and pore-size distribution. Macropore flow, infiltration, aeration and other processes are all influenced

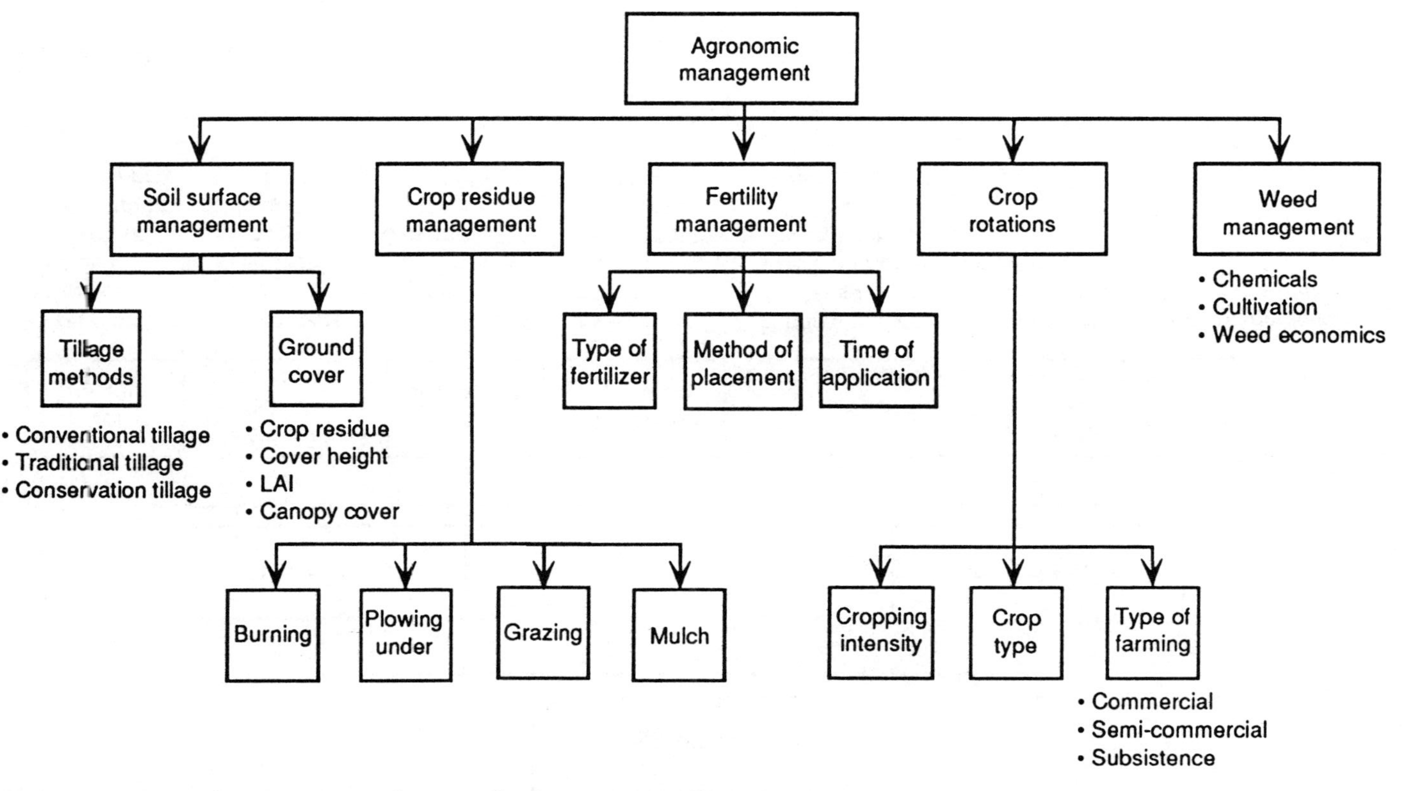


**Figure 5.** Agronomic management factors affecting water quality.

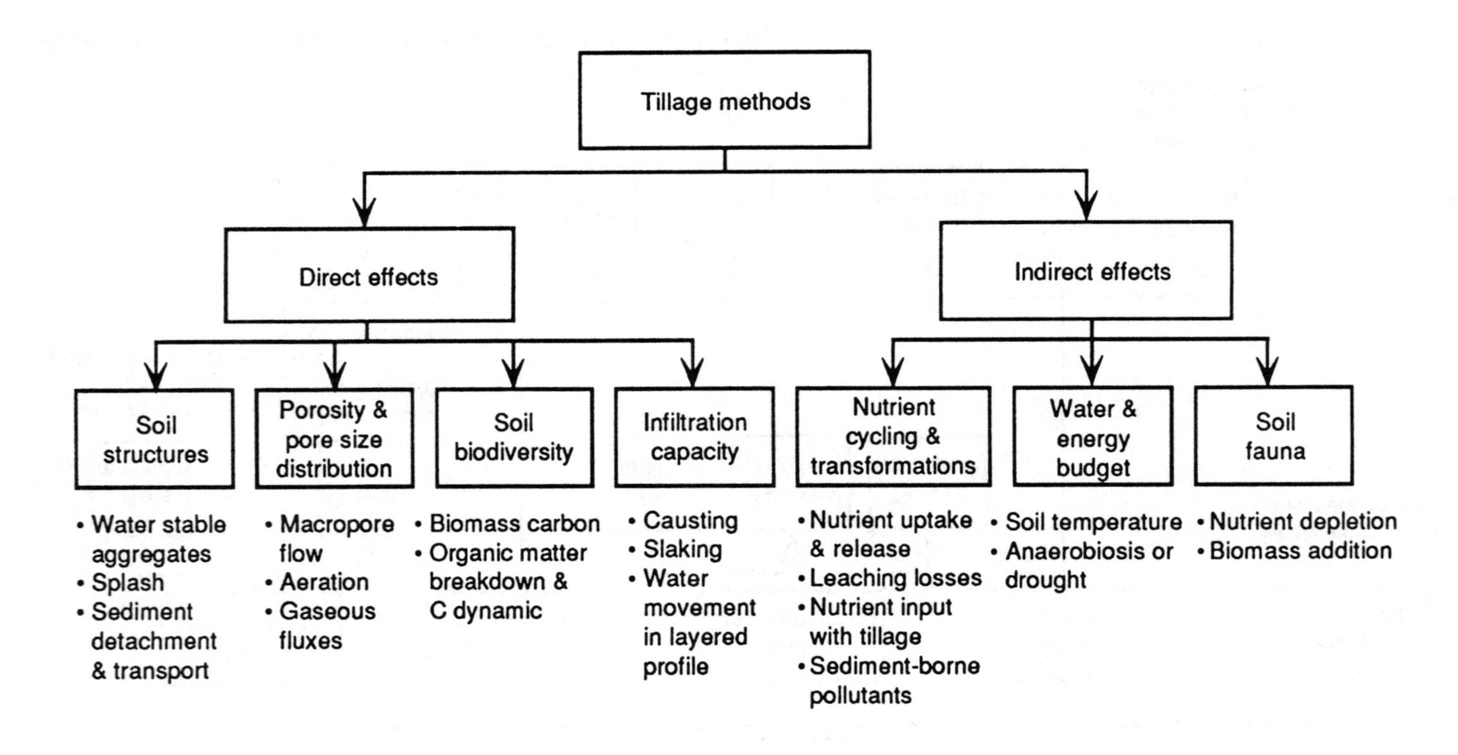


**Figure 6.** Tillage effects on water quality.

**Table 8.** Tillage method effects on sediment load in water runoff from the watershed management experiment at IITA

| Land clearing method | Tillage method | Plot | 25 May 1978 | 15 Aug. 1980 | 13 Oct. 1981 |
|---|---|---|---|---|---|
| Manual | Plow-till | 1 | 1.75 | 16.20 | 0.36 |
| Manual | No-till | 3 | 0.34 | 2.27 | 0.06 |
| Tree pusher/root rake | Plow-till | 5 | 4.78 | 32.00 | 0.26 |
| Tree pusher/root rake | No-till | 6 | 0.38 | 1.66 | 0.07 |
| Shear blade | No-till | 4 | 0.24 | 8.69 | 0.20 |

by tillage methods, which in turn, impact water quality. Tillage methods also influence soil biodiversity, especially activity and species diversity of soil fauna. Of significance to water quality is the activity of soil macrofauna, e.g., earthworms and termites. Earthworm activity influences the total amount and rate of water flowing through the soil profile. Termite activity has a similar input. Relatively, earthworms have a major impact on soil properties in the humid and sub-humid tropics whereas termites have significant impact in semi-arid regions.

Important among indirect effects of tillage methods on water quality are nutrient cycling and transformations, leaching losses, sediment-borne transport of nutrients, and nutrient inputs with tillage, e.g., application of fertilizers, organic and inorganic amendments, and pesticides which can eventually be carried into natural waters. Tillage affects nutrient transformation through its impact on soil temperature regime, which regulates the rate of bio-chemical processes. Nutrient depletion is also affected by the degree and nature of weed infestation.

Tillage affects water quality by influencing the amount and type of sediment and sediment-borne pollutants. In general, sediment availability (from detachment and splash) is more in plow-till than no-till or conservation tillage methods. The data in Table 8 show that regardless of the method of TRF conversions, sediment density in plow-till treatment was several times more than in no-till method. In addition to sediment, tillage methods affect water quality by influencing the nutrient and chemical transport in dissolved and suspended load. Tillage also affects the amount of N and other nutrients mineralized from soil organic matter. The data in Table 9 is an example of the tillage effects on chemicals transported in water runoff from agricultural watersheds. The data does not show any definite trend, except with regard to the concentration of Ca which was more in water runoff from no-till than plowed treatments. For addition to tillage, the nutrient concentration in water runoff depends on many other factors, e.g., total amount and rate of flow, type and method of fertilizer application along with time and rate of its application, type and stage of crop growth in relation to the timing of rainfall event. Most of these factors are discussed in detail by Logan et al (1987).

**Table 9.** Tillage method effects on nutrient concentration in water runoff from the watershed management experiment at IITA during July 1979

| Land clearing | Tillage method | Ca | Mg | K | Na | $NO_3$-N |
|---|---|---|---|---|---|---|
| | | mg liter$^{-1}$ | | | | |
| Manual | Plow-till | 9.9 | 2.1 | 13.4 | 18.8 | 1.19 |
| Manual | No-till | 13.1 | 2.6 | 13.2 | 14.4 | 0.55 |
| Tree pusher/root rake | Plow-till | 7.9 | 1.6 | 12.7 | 16.7 | 2.17 |
| Tree pusher/root rake | No-till | 10.3 | 2.6 | 14.1 | 11.5 | 1.13 |
| Shear blade | No-till | 8.2 | 2.3 | 12.7 | 13.8 | 1.07 |

### B. Ground Cover

The amount and properties of sediment released depend on the ground cover. Characteristics of ground cover related to water quality are outlined in Figure 7. Important among these are nature of the cover (e.g., dead or live), chemical characteristics and the stage of decomposition, etc. In situ residue and its management in relation to water quality is closely associated with tillage methods and soil surface management. In this regard, the amount and type of crop residue are important considerations. While the amount of crop residue affects the percent soil surface exposed to the raindrop impact, the date of decomposition depends on the C:N ratio and chemical composition of the residue. Leguminous residues with low C:N ratio decompose more rapidly than cereal residues with high C:N ratio.

Cover crops, pastures, live hedges, and agro-forestry systems also affect ground cover and water quality by regulating transport of sediments and sediment-borne pollutants. The data in Table 10 is an example of the effect of cover crops vs. grain crop on sediment density in water runoff. With the exception of 3 low-intensity or non-erosive rainfall events, sediment density in water runoff from maize was 1.7 to 23.5 times more than that from *Mucuna*. Maize is an open-canopy, soil degrading, and erosion-promoting crop. It does not protect the soil against raindrop impact. In contrast, *Mucuna* provides an excellent cover close to the ground surface. Raindrop impact is virtually prevented by protective ground cover provided by quick and low-growing *Mucuna*.

Sediment density is also reduced by properly-managed and lightly-grazed pastures. Improper management (e.g., high stocking rate, uncontrolled grazing, low soil fertility, etc.). The data in Table 11 show the impact of maize vs. grazed pasture (*Panicum* plus centro) on sediment density in water runoff. Although not as effective as the *Mucuna* cover, sediment density from maize was 1.04 to 4.2 times more than that from maize.

Vegetative ledges including alley cropping systems are increasingly being used for erosion control and ecological management of soil and water resources. Properly managed, agro-forestry systems can effectively reduce water runoff and

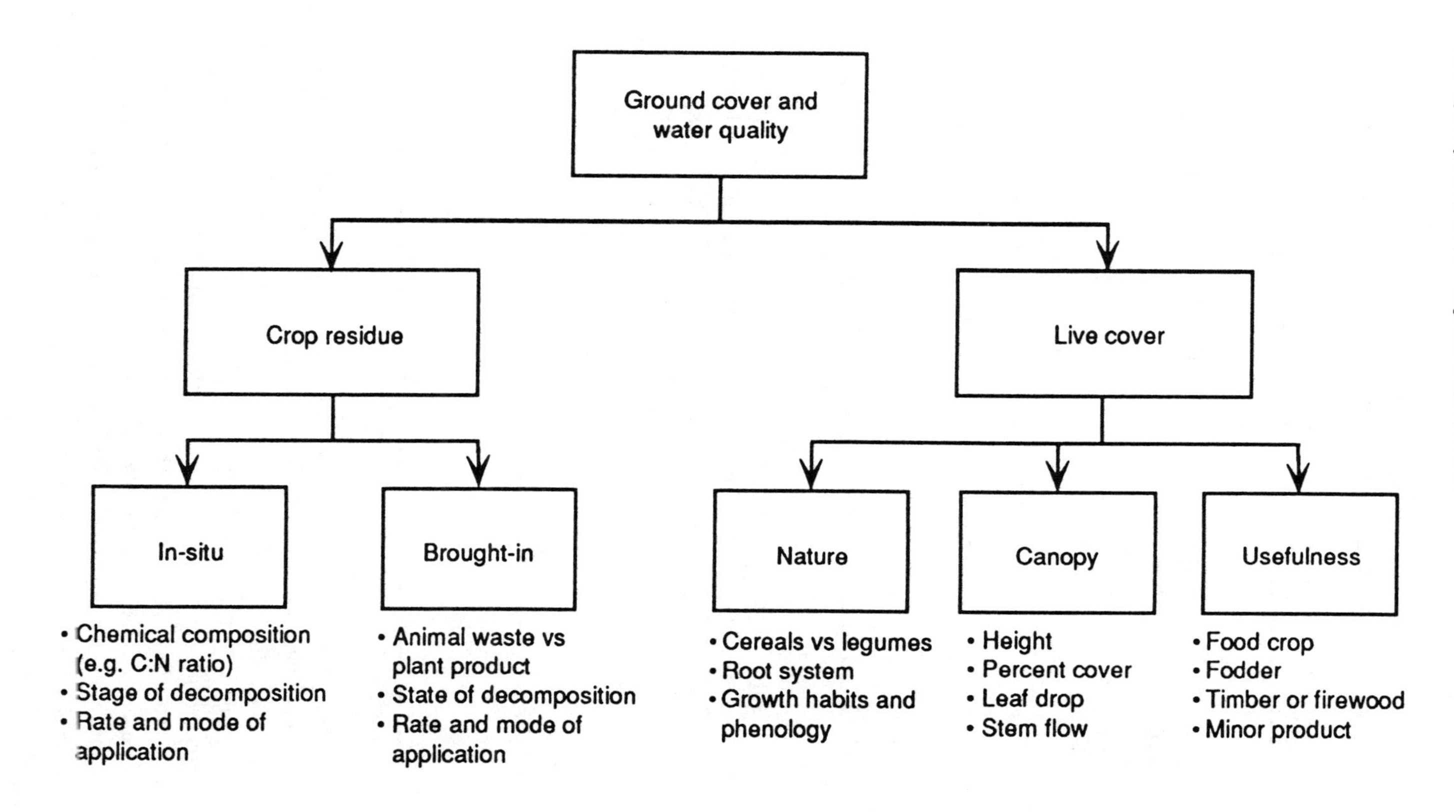


**Figure 7.** Ground cover characteristics in relation to water quality.

**Table 10.** Cover crop effect on sediment density in water runoff in 1982 from plots previously managed by tree pusher/root rake and plowed systems, 1979-1981

| Date of event | Maize (Plot 2) | *Mucuna* (Plot 5) | Maize/*Mucuna* |
|---|---|---|---|
| | ----------------g liter$^{-1}$---------------- | | |
| First Season | | | |
| 27 March | 0.90 | 0.40 | 2.25 |
| 10 April | 1.26 | 0.16 | 7.87 |
| 23 May | 1.42 | 0.28 | 5.00 |
| 30 May | 0.04 | 0.68 | 0.06 |
| 1 June | 0.08 | 0.16 | 0.50 |
| 20 June | 0.04 | 0.08 | 0.50 |
| 29 June | 0.14 | 0.04 | 3.50 |
| 2 July | 0.10 | 0.06 | 1.67 |
| Second Season | | | |
| 28 August | 1.88 | 0.08 | 23.5 |

**Table 11.** Pasture effect on sediment density in water runoff in 1983 from plots previously (1979) managed by tree pusher/root rake and no-till system

| Date of event | Maize (Plot 6) | *Mucuna* (Plot 8) | Maize/Pasture |
|---|---|---|---|
| | ----------------g liter$^{-1}$---------------- | | |
| 3 May | 4.40 | 4.22 | 1.04 |
| 10 May | 4.26 | 2.40 | 1.78 |
| 12 May | 3.32 | 1.20 | 2.77 |
| 15 May | 13.76 | 3.28 | 4.20 |
| 23 May | 4.60 | 3.48 | 1.32 |
| 28 May | 3.06 | 2.12 | 1.44 |
| 16 July | 1.98 | 1.44 | 1.38 |

soil erosion (Lal, 1989). The effectiveness of alley cropping system on sediment density and nutrient concentrations in water runoff are shown by the data in Table 12 through 15. The data in Tables 12 and 13 show that *Leucaena* hedges at 2 m intervals were extremely effective in reducing sediment load compared with *Gliricidia* hedges at 2 or 4 m or *Leucaena* hedges at 2 m intervals. In addition to reducing sediment load, however, hedgerow spacing and type of species used are also decided by several other considerations. These include multi-purpose uses, slope gradient, erosion hazard, etc. Considering all factors involved, 4-m or wider spacing may be more practical and agronomically feasible than 2-m or narrower spacing.

**Table 12.** Alley cropping effects on sediment load in water runoff for maize (1st season) on alley cropping experiment in Block D, IITA, 1984

| Treatment | 29 April | 7 May | 10 May | 18 May | 26 May | 30 May | 7 June | 11 June | 20 June | 26 June | 8 July |
|---|---|---|---|---|---|---|---|---|---|---|---|
| | Sediment concentration (g liter$^{-1}$) | | | | | | | | | | |
| *Leucaena* - 4 m | 2.52 | 2.36 | 3.70 | 1.26 | 1.28 | 4.14 | 8.50 | 1.12 | 0.38 | 0.78 | 0.50 |
| *Gliricidia* - 4 m | 0.24 | 0.32 | 0.94 | 0.32 | 0.38 | 10.24 | 10.20 | 2.34 | 1.12 | 0.62 | 2.84 |
| No-till | 0.42 | 0.27 | 0.32 | 0.24 | 0.42 | 1.74 | 0.62 | 0.32 | 0.10 | 0.14 | 0.22 |
| *Leucaena* - 2 m | 0.12 | 1.78 | 1.98 | 0.54 | 2.90 | 2.26 | 3.02 | 1.84 | 0.78 | 0.98 | 2.42 |
| *Gliricidia* - 2 m | 0.22 | 3.96 | 1.54 | 1.76 | 2.64 | 23.72 | 7.56 | 7.62 | 2.80 | 4.12 | 9.36 |
| Plowed | 0.84 | 4.49 | 9.98 | 4.08 | 2.73 | 9.26 | 13.14 | 8.48 | 1.44 | 7.16 | 9.14 |

**Table 13.** Alley cropping effects on sediment concentration in water runoff from cowpea (second season) on alley cropping experiment in Block D, IITA, 1984

| Treatment | 1 August | 2 August | 5 August | 8 August | 12 August | 3 September | 13 September | 22 September |
|---|---|---|---|---|---|---|---|---|
| | Sediment concentration (g liter$^{-1}$) | | | | | | | |
| *Leucaena* - 4 m | 0.94 | 0.62 | 1.32 | 0.90 | 1.62 | 2.16 | 0.74 | 0.40 |
| *Gliricidia* - 4 m | 0.76 | 1.38 | 0.56 | 0.52 | 0.38 | 6.32 | 1.96 | 3.82 |
| No-till | 0.18 | 0.38 | 0.22 | 3.30 | 0.42 | 0.42 | 1.10 | 0.50 |
| *Leucaena* - 2 m | 0.34 | 0.72 | 0.26 | 0.18 | 1.26 | 1.50 | 0.52 | 1.88 |
| *Glircidia* - 2 m | 5.46 | 2.44 | 1.02 | 4.40 | 2.02 | 4.06 | 2.20 | 3.18 |
| Plowed | 5.39 | 1.04 | 0.64 | 0.66 | 1.28 | 0.16 | 0.92 | 0.46 |

**Table 14.** Alley cropping effects on concentration of $NO_3$-N in water runoff from Block D experiment at IITA, 1985

| Treatment | First season (maize) | | | | Second season (cowpea) | | | | |
|---|---|---|---|---|---|---|---|---|---|
| | 26 June | 3 July | 7 July | 13 July | 18 Aug. | 4 Sept. | 10 Sept. | 24 Sept. | 3 Oct. |
| | mg kg$^{-1}$ | | | | | | | | |
| *Leucaena* - 4 m | 11.4 | 9.0 | 2.6 | 0.60 | 5.26 | 0.60 | 3.40 | 3.50 | 1.06 |
| *Gilricidia* - 4 m | 4.2 | 8.7 | 9.8 | 0.44 | 11.40 | 0.60 | 6.50 | 7.63 | 1.20 |
| No-till | 32 | 3.0 | 2.0 | 0.40 | 7.70 | 0.50 | 1.40 | 1.50 | 1.00 |
| *Leucaena* - 2 m | 32 | 2.4 | 2.6 | 0.60 | 4.00 | 2.60 | 6.00 | 1.75 | 1.40 |
| *Gliricidia* - 2 m | 0.8 | 0.6 | 1.2 | 0.20 | 2.10 | 2.20 | 1.20 | 1.00 | 0.96 |
| Plowed | 1.4 | 6.6 | 5.6 | 0.20 | 1.80 | 0.80 | 0.60 | 0.75 | 1.06 |

**Table 15.** Alley cropping effects on concentration of $PO_4$-P in runoff water in 1985 from Block D, IITA

| Treatment | First season (maize) | | | | | Second season (cowpea) | | | | |
|---|---|---|---|---|---|---|---|---|---|---|
| | 26 June | 3 July | 7 July | 5 Aug. | 7 Aug. | 18 Aug. | 4 Sept. | 10 Sept. | 24 Sept. | 3 Oct. |
| | mg kg$^{-1}$ | | | | | | | | | |
| *Leucaena* - 4 m | < 0.01 | 0.10 | 0.05 | 0.01 | 0.10 | 0.23 | 0.24 | 0.01 | 0.10 | 0.10 |
| *Gliricidia* - 4 m | < 0.01 | 0.08 | 0.02 | 0.01 | 0.01 | 0.17 | 0.24 | 0.01 | 0.10 | 0.05 |
| No-till | < 0.01 | 0.15 | 0.10 | 0.01 | 0.01 | 0.15 | 0.12 | 0.01 | 0.10 | 0.05 |
| *Leucaena* - 2 m | < 0.01 | 0.08 | 0.02 | 0.01 | 0.01 | 0.20 | 0.16 | 0.01 | 0.10 | 0.05 |
| *Gliricidia* - 2 m | 0.05 | 0.08 | 0.02 | 0.01 | 0.01 | 0.22 | 0.12 | 0.01 | 1.85 | 0.05 |
| Plowed | 0.05 | 0.08 | < 0.01 | 0.01 | 0.01 | 0.25 | 0.04 | 0.01 | 0.20 | 0.05 |

**Table 16.** Alley cropping effects on physical and chemical quality of water runoff from IITA's hydrological plots 5 and 6 during 1985, the first season

| Date | Sediment density | | $NO_3$-N | | $PO_4$-P | |
|---|---|---|---|---|---|---|
| | Alley cropping | Control | Alley cropping | Control | Alley cropping | Control |
| | ------g liter$^{-1}$----- | | ----------------- | mg kg$^{-1}$ | ---------------- | |
| First season (maize) | | | | | | |
| 27 May | 0.02 | 0.32 | 0.70 | 3.36 | 0.60 | 0.30 |
| 2 June | 0.80 | 7.18 | 1.50 | 3.66 | 0.73 | 0.80 |
| 8 June | 0.74 | 5.28 | 1.47 | 3.48 | 0.25 | 0.50 |
| 12 June | 0.10 | 11.34 | 1.40 | 0.90 | 0.58 | 0.32 |
| 19 June | 0.27 | - | - | - | - | - |
| 26 June | 0.22 | 6.86 | 2.00 | 2.10 | 0.53 | 0.00 |
| 3 July | 0.20 | 19.30 | 0.60 | 0.54 | 0.76 | 0.65 |
| 7 July | 0.22 | 4.18 | 1.30 | 1.70 | 0.13 | 0.84 |
| 13 July | 0.04 | 2.88 | 0.40 | 0.60 | 0.17 | 0.00 |
| 20 July | 0.10 | 5.12 | 1.00 | 0.60 | 0.00 | 0.29 |
| 30 July | 0.04 | 3.04 | 0.80 | 0.36 | 0.01 | 0.01 |
| Second season (cowpea) | | | | | | |
| 5 August | 0.27 | 5.02 | 0.44 | 0.16 | 0.01 | 0.20 |
| 7 August | 0.08 | 1.68 | 0.40 | 0.40 | 0.25 | 0.30 |
| 18 August | 0.30 | 7.66 | 0.76 | 0.90 | 0.35 | 0.35 |
| 4 September | 0.54 | 1.16 | 0.40 | 1.80 | 1.29 | 0.48 |
| 5 September | 0.42 | 1.76 | 0.70 | 0.60 | 0.56 | 0.01 |
| 24 September | 0.36 | 0.64 | 1.00 | 2.00 | 0.18 | 0.10 |
| 3 October | 0.86 | 1.82 | 0.80 | 2.40 | 0.10 | 0.80 |
| 11 October | 0.34 | 1.30 | 0.96 | 1.30 | 0.20 | 0.70 |
| 29 October | 0.18 | 1.54 | 0.50 | 2.04 | 0.15 | 0.50 |
| 8 November | 0.10 | 4.68 | 1.20 | 15.50 | 0.05 | 0.25 |

Effects of species and spacing of hedgerow in alley cropping systems on nutrient concentration in water runoff are shown by the data in Tables 14 and 15. While alley cropping systems reduced the total amount and rate of runoff, nutrient concentration was often more in water runoff from alley cropping than no-till and plow-till systems of crop production.

Effectiveness of alley cropping systems in reducing sediment $NO_3$-N and $PO_4$-P loading in water runoff is also demonstrated by the data in Table 16. Alley cropping, even with a mechanized system of pruning, and spreading the cut biomass within the alley rather than leaving it on the contour next to the hedge, effectively reduced sediment density. Similar to the data in Table 13 and 14, alley cropping also effectively reduced sediment density. Although not consistent, alley cropping also reduced concentrations of $NO_3$-N and of $PO_4$-P

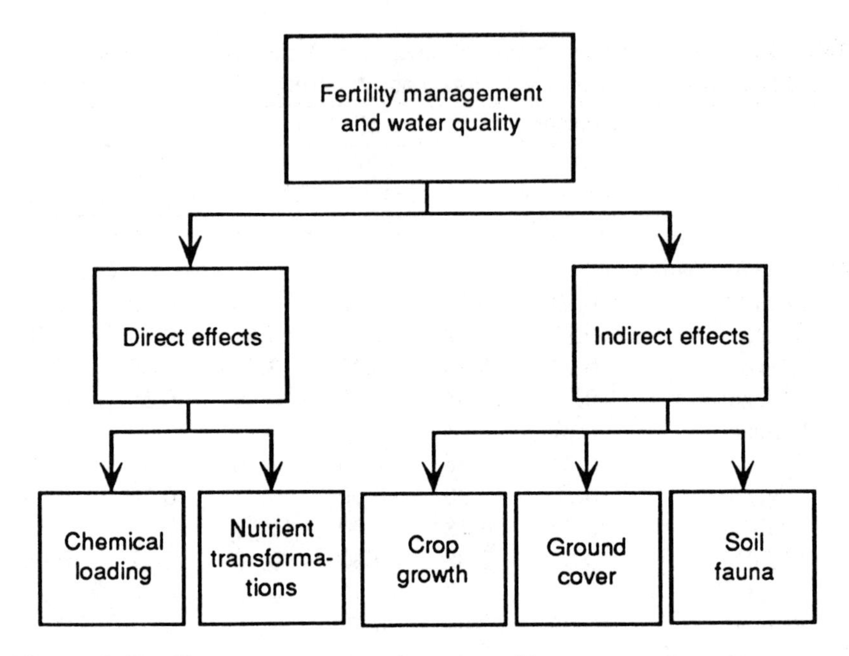


**Figure 8.** Fertility management and water quality.

in water runoff. The highest concentration of $NO_3$-N (15.5 mg $kg^{-1}$) was observed in the control treatment without alley cropping.

### C. Fertility Management

Fertility management affects water quality both directly and indirectly (Figure 8). Directly, addition of organic and inorganic fertilizers and pesticides increases the amount of pollutants available for increasing chemical load in surface and ground waters. Indirectly, fertility affects crop growth and yield. Balanced and correct application of fertilizers and chemicals enhance crop growth, provide ground cover, and facilitate ecologic balance. Good crop growth, involving effective ground cover and prolific root system, protects soil from raindrop impact and reduces rill/inter-rill erosion and sediment load in water runoff. Prolific and actively-growing root system also minimizes soil erosion and reduces chemical loading by minimizing nutrient losses in surface runoff and seepage water.

Method, type and time of application of fertilizer and amendments have significant effect on chemical loading and water quality. Fertilizer application on bare ground without an actively-growth crop cover is likely to increase risks of water pollution. While the startup fertilizer is needed for vigorous growth and good stand, split dose application is ecologically-compatible technique. Although labor-intensive, split dose application of fertilizers is likely to increase fertilizer use efficiency, decrease losses, and reduce risks of water pollution.

### D. Weed Management

The choice of the weed management system is extremely relevant to water quality. Similar to fertilizer use, weed control through application of growth hormones and herbicides is likely to increase chemical loading and risks of polluting surface and ground waters. In contrast, mechanical weed control measures cause soil disturbance, aggravate risks of soil erosion, and increase sediment density in water runoff.

There has been a considerable interest regarding the impact of conservation tillage systems on water quality (Logan et al., 1987). Even with high chemical inputs, water quality risks are not necessarily increased. Macropore flow, direct and rapid as it may be, can have a buffering effect on chemical loading. Most biopores, created by soil fauna or decayed roots are lined by organic compounds that harbor a wide range of soil organisms. Organic compounds and soil organisms lining the macropores act as a sieve in retaining and biodegrading the chemicals passing through them. Most chemicals transported into natural water are generally sediment-ladden. Soil surface management and tillage systems that reduce sediment load also decrease chemical loading.

## IV. Toward Enhancing Water Quality from Agricultural Watershed

The need to increase food production has necessitated intensive land use with chemical-intensive inputs throughout the tropics. Intensive land use also increases risks of polluting natural waters and environment. Because of the high risks involved, judicious and discriminate use of agro-chemicals and amendments is essential.

Strategies for enhancing water quality from agricultural land, and reducing non-point source pollution are outlined in Figure 9. These strategies are broadly classified into two categories: (a) reducing sediment load, and (b) minimizing chemical load. Technological options for achieving these goals are not necessarily mutually exclusive because soil management techniques that decrease sediment load also reduce chemical pollutants. Sediment load in natural waters can be decreased by: (i) conservation tillage and mulch farming, (ii) cover crops, (iii) contour hedges and agro-forestry, and (iv) water management techniques. The latter involves engineering practices of land forming, terraces, waterways and reservoirs, and drop structures. These techniques, properly installed and judiciously implemented, have proven effective in decreasing sediment load from agricultural lands.

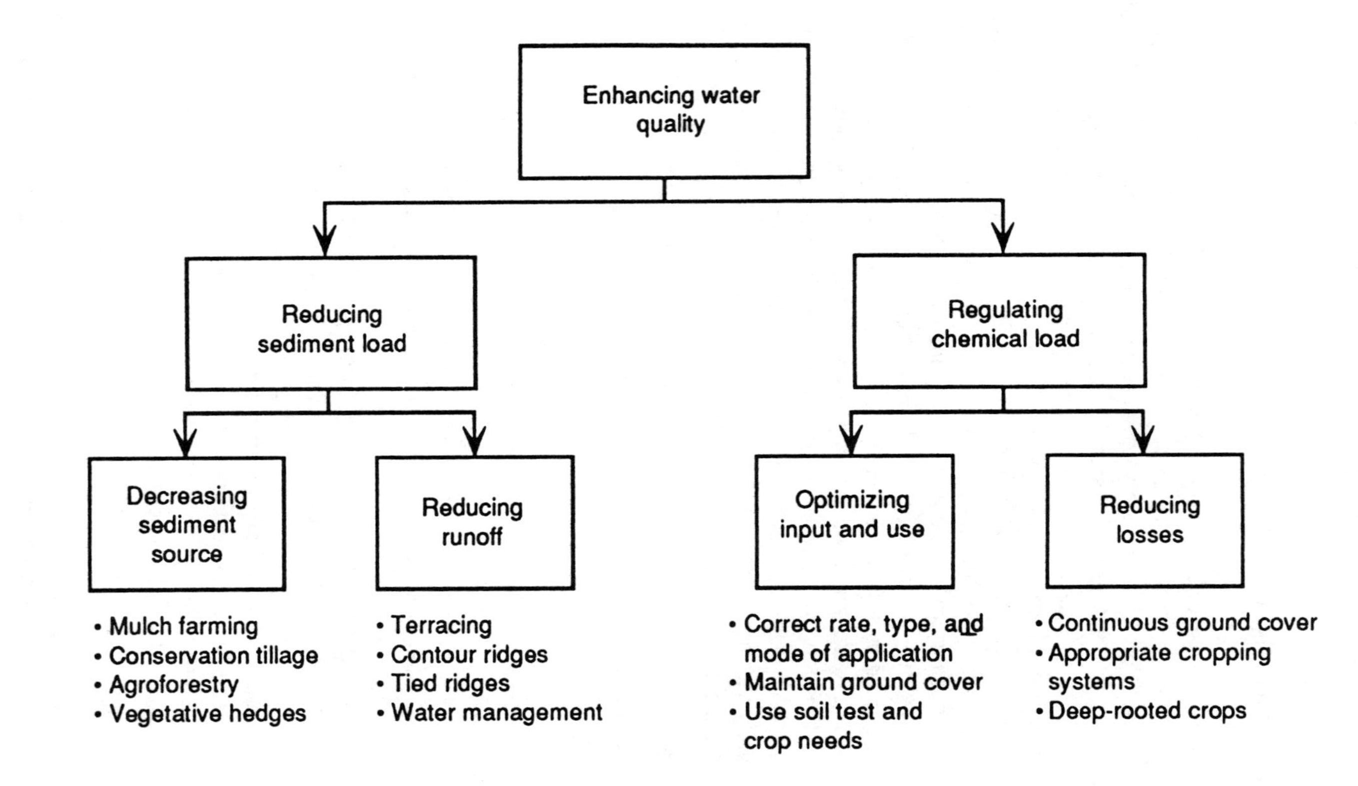


**Figure 9.** Management strategies for enhancing water quality and reducing non-point source pollution.

There are several options for decreasing chemical load. Judicious and discriminate application of fertilizers and pesticides is essential. Excessive use must be avoided. Type and method of application, and formulations used, are critical in chemical loading.

Intensive agriculture can be practiced without necessarily increasing risks of polluting natural waters. The science-based agriculture, properly and judiciously implemented, can enhance and sustain food production and also minimize the risks of polluting water and the environment.

## References

Alegre, J.C., D.K. Cassel, D. Bandy, and P.A. Sanchez. 1986. Effect of land clearing on soil properties of an Ultisol and subsequent crop production in Yurimaguas, Peru. p. 167-180. In: R. Lal et al. (eds.) *Land Clearing and Development in the Tropics.* A. A. Balkema, Rotterdam.

Alegre, J.C., D.K. Cassel, and D.E. Bandy. 1986a. Effects of land clearing and subsequent management on soil physical properties. *Soil Sci. Soc. Am. J.* 50:1379-1384.

Alegre, J.C., D.K. Cassel, and D.E. Bandy. 1986b. Reclamation of an Ultisol damaged by mechanical land clearing. *Soil Sci. Soc. Am. J.* 50:1026-1031.

Couper, D.C., R. Lal, and S. Classen. 1981. Land clearing and development for agricultural purposes in western Nigeria. p. 119-130. In: R. Lal and E. W. Russell (eds.) *Tropical Agricultural Hydrology.* J. Wiley & Sons, Chichester, U. K.

Cunningham, R.K. 1963. The effect of clearing a tropical forest soil. *J. Soil Sci.* 14:334-345.

Forest Resource Assessment. 1990. Project. 1991. Second Interim Report on the State of Tropical Forests. Paper presented at the 10th World Forestry Congress, Paris, France, 17-26 September, 1991.

Ghuman, B.S. and R. Lal. 1987. Effects of partial clearing on micro-climate in a humid tropical forest. *Agric. and Forest Meteorology* 40:17-29.

Ghuman, B.S., and R. Lal. 1989a. Soil temperature effects of biomass burning in windrows after clearing a tropical rainforest. *Field Crops Res.* 22:1-10.

Ghuman, B.S., and R. Lal. 1989b. Biomass burning in windrows after clearing a tropical rainforest: effects on soil properties and crop yields. *Field Crops Res.* 22:247-256.

Ghuman, B.S., R. Lal and W. Shearer. 1991. Land clearing and use in the humid Nigerian tropics. I. Soil Physical Properties. *Soil Sci. Soc. Am. J.* 55:178-183.

Ghuman, B.S. and R. Lal. 1991. Land clearing and use in the humid Nigarian tropics. II. Soil chemical properties. *Soil Sci. Soc. Amer. J.* 55:184-188.

Ghuman, B.S. and R. Lal. 1991a. Land clearing and use in the humid tropics. II. Soil Chemical Properties. *Soil Sci. Soc. J.* 55:184-188.

Ghuman, B.S., and R. Lal. 1992. Effects of soil wetness at the time of land clearing on physical properties and crop response on an Ultisol in southern Nigeria. *Soil & Tillage Res.* 22:1-11.

Houghton, R.A. 1990. The global effects of tropical deforestation. *Env. Sci. Technol.* 24:414-422.

Hulugalle, N.R., R. Lal and C.H.H. ter Kuile. 1984. Soil physical changes and crop root growth following different methods of land clearing in western Nigeria. *Soil Science* 138:172-179.

Lal, R. 1981. Deforestation of tropical rainforest and hydrological problems. p. 131-140. In: R. Lal and E. W. Russell (eds.) *Tropical Agricultural Hydrology*. J. Wiley & Sons, Chichester, U.K.

Lal, R. 1986. Conversion of tropical rainforest agronomic potential and ecologic consequences. *Adv. Agron.* 39:173-264.

Lal, R. 1987. Need for, approaches to, and consequences of, land clearing and development in the tropics. In: R. Lal et al. (eds.) *Tropical Land Clearing for Sustainable Agriculture*. IBSRAM Proc. Series No. 3:15-28, Bangkok, Thailand.

Lal, R. 1989. Agroforestry systems and soil surface management of a tropical Alfisol. Parts I-V: 8:7-29, 97-111, 113-132, 197-215, 239-242.

Lal, R. 1992. *Tropical Agricultural Hydrology and Sustainability of Agricultural Systems*. The Ohio State University, IITA Bulletin, Columbus, Ohio, 303 pp.

Lal, R., and Cummings, D. J. 1979. Clearing a tropical forest. I. Effects on soil and micro-climate. *Field Crops Res.* 22(2):91-107.

Lal, R., P.A. Sanchez, and R.W. Cummings, Jr. (eds.). 1986. Land clearing and development in the tropics. A.A. Balkema, Rotterdam, The Netherlands. 450 pp.

Lal, R., B.S. Ghuman and W. Shearer. 1992. Cropping systems effects of a newly cleared Ultisol in southern Nigeria. *Soil Technology* 5:27-38.

Lawson, T.L. and R. Lal. 1981. Rainfall redistribution, water balance, and microclimate over a cleared watershed. p. 141-152. In: R. Lal and E. W. Russell (eds.) *Tropical Agricultural Hydrology*. J. Wiley & Sons, Chichester, U.K.

Logan, T.L., J.M. Davidson, J.L. Baker and M.R. Overcase. (eds.). 1987. *Effects of Conservation Tillage on Ground Water Quality: Nitrates and Pesticides*. Lewis Publishers, Chelsea, MI. 291 pp.

NRC. 1982. *Ecological aspects of development in the humid tropics*. National Academy of Science, Washington, D.C.

NRC. 1993. *Sustainable agriculture and environment in the humid tropics*. National Academy of Science, Washington, D.C.

Nye, P.H., and D.J. Greenland. 1960. *The soil under shifting cultivation*. Commonwealth Bureau of Soils, Harpenden, England, Tech. Communication No. 51.

Sanchez, P.A., and J.R. Benites. 1987. Low input cropping for acid soils of the humid tropics. *Science* 238:1521-1527.

Schlesinger, W.H. 1984. Soil organic matter: a source of atmospheric $CO_2$. In: G.M. Woodwell (ed) *The Role of Terrestrial Vegetation in the Global Carbon Cycle: Measurements by Remote Sensing*. J. Wiley & Sons, NY.

Seubert, C.E., Sanchez, P.A., and Valverde, C. 1977. Effects of land clearing methods on soil properties on an Ultisol and crop performance in the Amazon jungle of Peru. *Tropical Agriculture (Trinidad)*. 54(4):307-321.

Weert, R. van der. 1974a. The influence of mechanical forest clearing on soil conditions and resulting effects on root growth. *Tropical Agriculture*. 51:325-331.

Weert, R. van der. 1974b. The influence of mechanical forest clearing on soil conditions and resulting on some physical and chemical soil properties. *Surinamise Landbouw* 20(3):2-14.

World Resources Institute. 1988-89. *An assessment of the resource base that supports global economy*. World Resources Institute, Washington, D.C.: 285-293.

World Resources Institute. 1990-91. *A guide to global environment*. World Resources Institute, Washington, D.C.: 101-120.

# Macropore Hydraulics: Taking a Sledgehammer to Classical Theory

E.L. McCoy, C.W. Boast, R.C. Stehouwer, and E.J. Kladivko

I. Introduction . . . . . . . . . . 303
II. Macropore Impact on Water and Chemical Transport . . . . . . . 306
A. Macropore Flow Initiation . . . . . . . . . . 306
B. Funneling of Macropore Flow . . . . . . . . . . 310
C. Factors Affecting Macropore Flow and Short-Circuit Bypass Flow . . . . . . . . . . 311
III. Macropore Impact on Chemical Transport . . . . . . . . . . 313
IV. Macropore Characterization . . . . . . . . . . 315
A. Macropore Numbers, Sizes, and Shapes . . . . . . . . . . 316
B. Macropore Participation in Flow and Continuity . . . . . . . . 317
C. Water Retention and Conductivity . . . . . . . . . . 322
V. Models of Macropore Flow . . . . . . . . . . 326
A. Mechanisms . . . . . . . . . . 329
B. The Water Retention Hypothesis . . . . . . . . . . 330
C. Solution Bypass Models . . . . . . . . . . 330
D. Short-Circuit Bypass Models . . . . . . . . . . 332
E. "Non-Richards" Models . . . . . . . . . . 334
VI. Summary . . . . . . . . . . 336
Acknowledgments . . . . . . . . . . 337
References . . . . . . . . . . 337

## I. Introduction

Macropores and macropore flow have been the subject of tremendous amounts of research in soil science in the past decade. Beven and Germann (1982) pointed out that observations of macropore flow are not new, and they illustrated this with selected excerpts of studies over the past 130 years. However, interest in macropore flow has grown recently with concerns about pollution of ground and surface waters from agricultural chemicals. The arrival of water and chemicals at some point of interest (bottom of soil column, tile drain, groundwater, surface stream) much sooner than predicted by available theories of water

0-87371-980-8/94/$0.00+$.50

and chemical flow, has caused great concern and a reevaluation of classical theory. Thus the observations of rapid flow have stimulated interest in studying the processes controlling flow as well as the physical system (macropores) in which flow occurs, with a further objective of developing appropriate theory of flow in structured soils. This article reviews recent progress in these three areas.

Macropores can be defined as large soil pores that have minimum equivalent cylindrical diameters (ECDs) ranging from 0.075 to 1.0 mm (Luxmoore et al., 1990). Macropores can promote rapid, preferential transport of water and chemicals through soil (Beven and Germann, 1982; White, 1985, Bouma, 1991) not only due to their size but also because they are connected and continuous over sufficient distances to transcend agriculturally and environmentally important soil layers (i.e. the plow depth, the rooting depth, the biologically active zone, etc.). Morphologically, macropores are primarily cylindrical channels formed by soil fauna and plant roots, or planar cracks between soil structural peds (Beven and Germann, 1982; Bouma, 1991).

The element of "surprise" is perhaps a good way to describe many of the consequences of macropore flow in soils. Macropore flow can result in preferential transport of water and chemicals when a macropore and/or the soil surrounding it is not entirely water filled over their complete transport pathway (Beven and Germann, 1982). Presumably, flow in incompletely saturated macropores occurs as free water films along the surfaces of the pores (Beven and Germann, 1981) or as distinct water packets separated by air filled spaces (Phillips et al., 1989; Bouma, 1991). Even more surprising is the observation of vertical movement of free water along continuous macropores through an unsaturated soil matrix (e.g., Bouma et al., 1977; Bouma and Dekker, 1978; Smettem and Trudgill, 1983; Germann et al., 1984; Booltink and Bouma, 1991; Kosmas et al., 1991). This process, called short-circuit bypass flow (Bouma and Dekker, 1978), presents a puzzling riddle for understanding water and chemical transport in aggregated soils containing macropores. Figure 1 illustrates short-circuit bypass flow in a capillary bundle conceptualization of soil pores. Bypass flow invalidates one of the basic tenants of soil water theory (Towner, 1989; Moran and McBratney, 1992) wherein in any representative elemental volume (REV) of soil there exists a pore size such that all pores smaller contain water and all pores larger are air filled (at least for a monotonic wetting process such as water infiltration into soil). This tenant is operationally defined by the soil water retention curve. Given this paradox, it is clear that the fundamental processes and conditions required for macropore flow initiation are complicated.

Other types of macropore flow do not invalidate the basic concept behind the soil water retention curve but still result in more rapid chemical transport than predicted. In these cases the largest water-filled pores have much faster flow rates than smaller pores, and solution in the macropores "bypasses" solution in micropores, with little or no mixing. This type of macropore flow might be called "weak bypass" or "solution bypass." Overall water flow rates and volumes can be adequately described with Darcian concepts, but chemical transport is not accurately described by the convection-dispersion model, due to

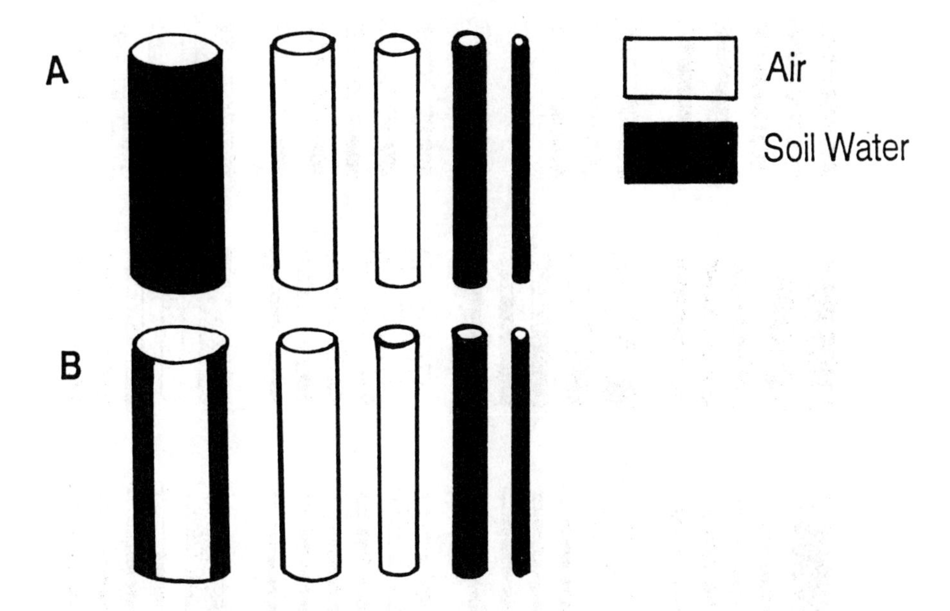


**Figure 1.** Schematic diagram of "by-pass" flow in macropores under two scenarios: A) Macropores and small pores are water filled and flowing while intermediate sized pores are air filled with no water flow. B) Macropores have film flow along pore walls, intermediate sized pores have no water flow and small pores have slow water flow.

non-equilibrium in chemical mixing. Figure 2, adapted from Bouma (1991, Figure 17), illustrates the concepts of solution bypass at several levels.

A number of excellent reviews (Bouma, 1981; Beven and Germann, 1982; White, 1985; Bouma, 1991) and conferences and symposia (Germann, 1988; van Genuchten, et al., 1990; Gish and Shirmohammadi, 1991) have been published on various aspects of macropores and preferential flow. Beven and Germann (1982) focused on water flow in macropores, White (1985) discussed chemical transport as affected by macropores, and Bouma (1981, 1991) linked soil morphology to water flow and chemical transport. In this article we first discuss some of the new concepts that have emerged related to water flow in macropores, and review progress in macropore characterization and its relationship to flow and finally consider models that can be used to describe macropore flow. Several of the questions posed by Beven and Germann (1982): "how do macropores operate hydrologically? How often do they operate? What do macropore structures look like and can they be related to flow properties?", have seen significant study and progress since earlier reviews and will serve as the focus of this article.

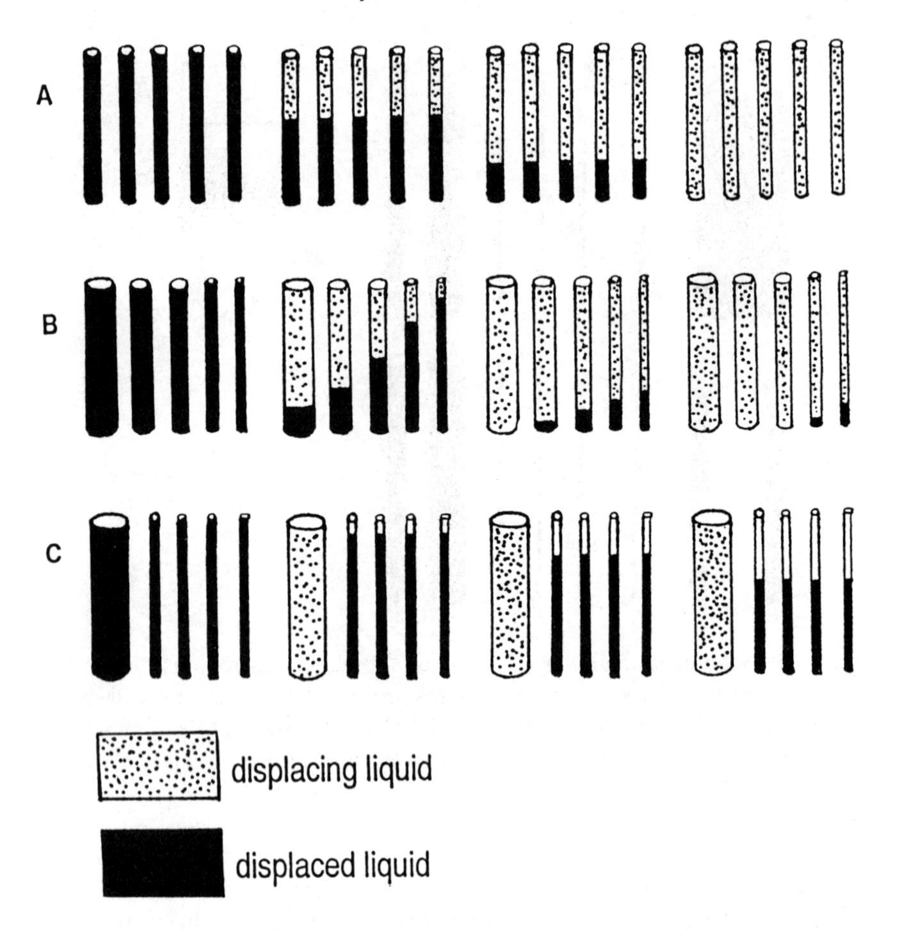


**Figure 2.** Schematic diagram of chemical transport through soil pores under three different scenarios: A) Piston-flow displacement through uniform sized soil pores. B) Dispersion of the convective front due to a relatively small variation in pore sizes. C) "Solution by-pass" flow due to a large variation in pore sizes (macropore and matrix pores). (Adapted from Bouma, 1991.)

## II. Macropore Impact on Water and Chemical Transport

### A. Macropore Flow Initiation

How and when does macropore flow get started in a soil? It has been thought that to initiate macropore flow, the soil must be saturated with water ponded on the entire soil surface (Bouma, 1981; Beven and Germann, 1982). Thus many studies dealing with macropore flow have focused on ponded flow (Bouma and Wösten, 1979; Rao et al., 1980; Seyfried and Rao, 1987; Parlange et al., 1988;

Steenhuis and Muck, 1988; Wilson and Luxmoore, 1988). Studies of macropore flow under non-ponded conditions have tended to interpret initiation of macropore flow as resulting from saturated layers very near the soil surface (Adreini and Steenhuis, 1990), from localized ponded conditions in surface micro-depressions (Shipitalo et al., 1990; Edwards et al., 1992; Trojan and Linden, 1992), or at structural discontinuities such as the interface between tilled and untilled layers (Steenhuis and Muck, 1988; Adreini and Steenhuis, 1990).

There is, however, growing evidence that at least under certain conditions, surface ponding or a distinct subsurface discontinuity may not be a precondition to initiation of macropore flow. Phillips et al. (1989) demonstrated that water in an unsaturated matrix can enter a simulated macropore if the macropore walls are wet. Experiments designed to eliminate ponding by inducing surface runoff still resulted in increased infiltration and deeper penetration of surface applied Br in untilled soils containing earthworms (*Lumbricus rubellus*) vs. no worms (Zachmann and Linden, 1989). Shipitalo et al. (1990) and Edwards et al. (1992) observed ponded water in micro-depressions only during the first few minutes of rainfall. They suggest this surface ponded water may enter macropores, but once pore walls have been wetted macropore flow could continue from subsurface flow into the pores. In certain well-aggregated (or micro-aggregated) tropical soils, macropore flow has been initiated without surface ponding (Russell and Ewel, 1985; Radulovich and Sollins, 1987; Sollins and Radulovich, 1988; Radulovich et al., 1992). In such soils it is proposed that film flow of water can begin on aggregate surfaces as soon as the hydraulic conductivity of the aggregate is exceeded and well before the inter aggregate pores become fully saturated.

For homogeneous soils experiencing steady downward, unsaturated water low, it has been theoretically shown that free water can enter large voids or open cavities located within the macroscopically unsaturated zone (Philip et al., 1989; Knight et al., 1989). In this case, the cavity perimeter behaves as an impermeable boundary of the flow medium, causing the soil matric potential on the upper side of the cavity to increase. For a large enough cavity, the potential reaches zero over a portion of the cavity surface, and this portion becomes a surface for water seepage into the cavity. The ease with which this occurs for a long, horizontal, circular cylindrical cavity of radius $l$ was analyzed by Philip, et al. (1989), and for a spherical cavity of radius $l$ by Knight, et al. (1989). These analyses apply to steady downward flow, spatially uniform far from the cavity, through a homogeneous soil for which the relationship between hydraulic conductivity K and matric potential is given by: $K = K_0 e^{a(\Psi - \Psi_0)}$, where $K_0$ and $\Psi_0$ are the hydraulic conductivity and the matric potential far from the cavity. Seepage into a cylindrical or spherical cavity will occur if $\Psi_0$ is wetter than a critical value:

$$\Psi_0 > -l \frac{\ln(\vartheta_{max})}{2s} = -\ln(\vartheta_{max}) / \alpha$$

where $s = \alpha l/2$, and where $\vartheta_{max}$, which is the ratio of the Kirchhoff potential $\int_{-\infty}^{\Psi} K \, d\Psi$ at the wettest position in the flow (i.e., at the stagnation point above the cavity) to the Kirchhoff potential far from the cavity, is given in Table 1 of each of the two cited articles. Considering a macropore of radius $l$ = 2 mm, and four values of $\alpha$: 1, 0.1, 0.01 and 0.001 $mm^{-1}$ (which could typify soils ranging from extremely coarse to a loam), $s$ ranges from 1 to 0.001. Interpolating in the tables for the smaller $s$ values, $\vartheta_{max}$ ranges from 3.403 to 1.002 for a cylindrical cavity and from 2.133 to 1.001 for a spherical cavity. The four critical far-field matric potential values, $\Psi_0$, are -1.2, -2.1, -2.2 and -2.0 mm for a cylindrical cavity and are -0.76, -1.0, -1.1 and -1.0 mm for a spherical cavity. The first three $\alpha$ values, which represent quite coarse soil, are included in these examples simply to show how little the critical $\Psi_0$ values vary with $\alpha$. For most soils and for cavity sizes in the macropore range, $s$ is quite small, so (using the approximations for very small $s$: $\vartheta_{max} = 1 + 2s$ for a cylindrical cavity and $\vartheta_{max} = 1 + s$ for a spherical cavity) the following generalization can be made: if the steady-flow far-field $\Psi_0$ is wetter than $-l$ for a horizontal cylindrical cavity of radius $l$ or wetter than $-l/2$ for a spherical cavity of radius $l$, then seepage into the cavity will occur. Using strips and disks of zero thickness as approximations to flat-topped cavities, Philip (1989) found a strong dependence of $\vartheta_{max}$ on strip width and disk diameter.

It is important to note that the analyses of Philip et al. (1989), Knight et al. (1989), and Philip (1989), are based on the assumption that incipient seepage requires $\Psi = 0$ at some point on the cavity perimeter. An additional assumption is that the cavity is not water filled but rather contains air, and thereby disrupts the flow field. This second assumption is appropriate for macropores since soil macropores often remain air filled for negative matric potentials. Further, under the conditions of these analyses, a larger cavity can generate $\Psi = 0$ conditions on its perimeter at more negative far-field $\Psi_0$ values because a larger cavity represents a more drastic disruption of the uniform flow field. Knight et al. (1989) argue that the results for cylindrical and spherical cavities indicate that the onset of seepage into a cavity is controlled by the total curvature of the cavity surface, in particular the curvature at the upstream stagnation point. It seems likely that for cavities of more complicated shape than spheres, circular cylinders, strips, and disks, seepage will first occur at the upstream location of the lowest (or concave upward) surface curvature. Thus, seepage into irregular-ly-shaped cavities could occur under much drier macroscopic conditions than those indicated by theories for ideally-shaped cavities.

These analyses are, of course, applicable only to homogeneous soils. Since we are unaware of similar studies on water entry into cavities in heterogeneous soils, we speculate that for a heterogeneous soil in which the flow properties exhibit spatial structure, there would be faster and slower flow paths, and wetter

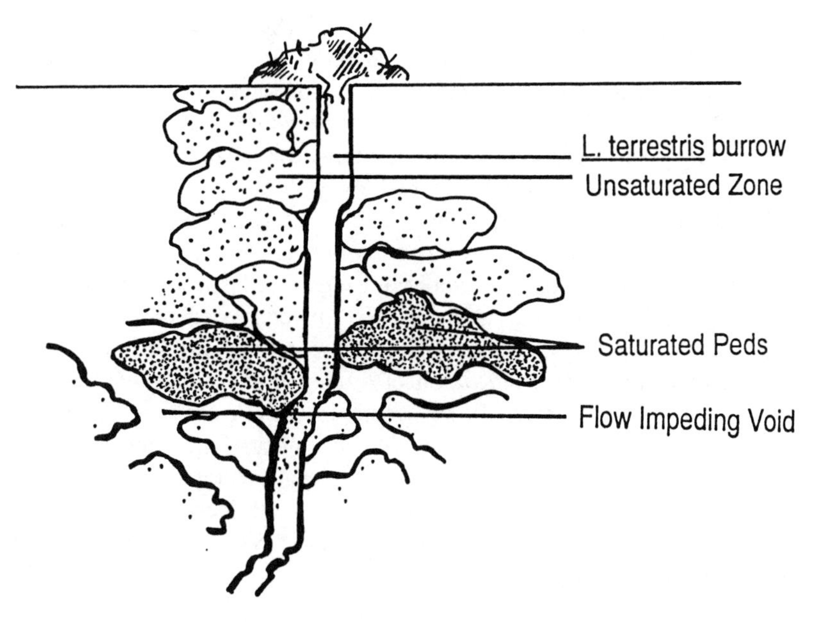


**Figure 3.** Conceptualization of macropore flow initiation due to a localized saturated zone formed above a horizontally oriented planar void.

and drier localized soil zones. If one of these faster flow paths and/or wetter zones happens to intercept a macropore, one could anticipate seepage into the macropore at a time when, overall, the surrounding soil was much drier than the conditions predicted for seepage in a homogeneous soil.

Thus, experimental evidence and theoretical analyses suggest that for soils containing large, essentially vertical macropores such as *Lumbricus terrestris* burrows or planar cracks, and in the absence of textural or morphological differences with depth; water entry into macropores can occur at some location within the soil profile in the absence of surface ponding. The mechanism responsible for this initiation of flow is the disruption of the flow field and momentary delay of the wetting front penetration through soil due to horizontally oriented inter-aggregate voids, such that localized saturated zones are created above these voids (Figure 3). Temporary halting of matrix flow at inter aggregate voids is suggested by the intermittent, step-wise advance of wetting fronts observed in columns of aggregated soil (Collis-George and Lal, 1970). Additionally, horizontally oriented planar cracking voids in clay soils have been shown to halt upward movement of matrix water (Koistra et al., 1987). Development of saturated loci due to matrix flow is explained by the apparent non-equilibrium of matric potentials in soil subjected to transient flow phenomena (Vachaud et al., 1972; Germann et al., 1984). This non-equilibrium is shown to be capable of producing a distinct zone of over saturation in soil during raising of the water table in a sand tank model (Schiegg, 1979 as cited by Germann et al., 1984). We think that horizontal voids in surface soils can have the same effect on the downward movement of matrix water. If the halting is long enough, the ped above the void will become saturated and form a film of free water on its surfaces. This water then can flow laterally until it enters a vertical void and thus initiate macropore flow.

Under the horizontal void hypothesis, the location of water entry into the macropore would occur at a shallower depth for lower antecedent water contents and/or higher rainfall intensities, and at a deeper depth for higher antecedent water contents and/or lower rainfall intensities. Lower antecedent water contents create wider soil cracking and consequently more horizontal planar voids, thus initiating macropore flow at shallower depths. Higher intensity rainfall will also yield free water sooner for shallower depth interception of the macropore. Additionally, transport of surface-applied chemicals (i.e. fertilizers and pesticides) will be greatest with the occurrence of shallow, rapid macropore flow initiation. Attenuation of transported chemicals will increase with deeper, delayed macropore flow initiation, with the most reactive (strongly sorbed) species being most affected.

### B. Funneling of Macropore Flow

The necessary condition for bypass flow is the initiation of flow into macropores, and the funneling of this flow into selected, vertically continuous macropores such that flow volumes and/or velocities are sufficiently large to overcome lateral absorption by the surrounding unsaturated matrix. Once initiated, macropore flow will thus be sustained if lateral flow from several peds converges into relatively few, large, vertically oriented macropores. Such funneling of macropore flow has been suggested by the results of several experiments (Bouma et al., 1977, 1979; Bouma, 1981; Seyfried and Rao, 1987; Edwards et al., 1990; Booltink and Bouma, 1991; Radulovich et al., 1992; Trojan and Linden, 1992), and has been documented by studies in which percolate through undisturbed soil blocks was collected in a grid of discrete sampling cells that allowed quantitative measurement of flow localization (Adreini and Steenhuis, 1990; Shipitalo et al., 1990; Quisenberry and Phillips, 1991; Edwards et al., 1992; Phillips and Quisenberry, 1992; Granovsky et al., 1993). Patterns of localized flow have been repeated in sequential rainfall events (Shipitalo et al., 1990; Granovsky et al., 1993) and have also been associated with earthworm burrows (Edwards et al., 1992; Granovsky et al., 1993).

Funneling of flow has also been observed in field studies (Ehlers, 1975; Edwards et al., 1990) where it has been associated with vertically oriented earthworm burrows. Kung (1990) attributed funneling of flow at the field scale to interbedded textural and structural discontinuities and inclined bedding planes in the profile. He proposed that these features function like the walls of a funnel and concentrate initially unsaturated flow into irregularly spaced columns. He also noted that the resultant shrinkage in conducting area will increase flow velocity.

Kung's work was in the vadose zone of a poorly structured sandy soil. However, similar features may exist in smaller dimensions in surface horizons of untilled, structured soils. Micromorphometric analyses of Ap horizons during 7 years of consecutive no tillage showed that while total macroporosity decreased, remaining macropores tended to become longer, less tortuous, and oriented parallel to the soil surface (Shipitalo and Protz, 1987). There was also much higher bioporosity (primarily earthworm burrows) in the no-tillage pedon. Bathke and Cassel (1991) measured much greater horizontal than vertical hydraulic conductivities in the A horizons of an Ultisol polypedon under permanent pasture. These morphologic and hydraulic characteristics are consistent with the hypothesis that macropore flow can be initiated internal to the soil, in that they would favor funneling of flow along horizontal pathways until a dominant vertical pore (such as an earthworm burrow) is intercepted.

Finally, macropore flow, once established, is maintained and leads to short-circuit bypass flow due to the low lateral absorption of water into the soil matrix. Explanations for this phenomena focus on the high flow velocities and/or large flow volumes in bypass flow conducting macropores. Thus, high flow velocities would suggest that non-Darcian flow in large pores would lead to the observed nonequilibrium conditions (Beven, 1981; Germann et al., 1984; Phillips et al., 1989). Alternatively, macropore flow is maintained by the small contact area of free water with the soil matrix relative to the large volume of water passing through the pore (Hoogmoed and Bouma, 1980), and the occurrence of clay or organic coatings on macropore walls reducing lateral absorption rates (Bouma, 1991; Edwards et al., 1992).

## C. Factors Affecting Macropore Flow and Short Circuit Bypass Flow

Factors reported to affect macropore flow initiation and bypass flow include soil texture, soil morphology, antecedent soil moisture content, and rainfall intensity, (Bouma, 1981, Beven and Germann, 1982; White, 1985; Bouma, 1991). The extent of macropore flow has been shown to be dependent on the degree of inter-connectedness of the macropore system (Dixon and Peterson, 1971; Ritchie et al., 1972; Kissel et al., 1973). Anderson and Bouma (1977a, b) found greater macropore flow in a soil horizon with well developed subangular blocky structure than in a horizon with large prismatic peds, even though the two had identical texture. This was attributed to greater continuity of planar voids between peds with subangular blocky structure. Increased infiltration rates in no-tillage fields have been attributed to the greater number and continuity of macropores in the surface layer of no-tillage than in moldboard plowed soils (Edwards et al., 1988; Dick et al., 1989). High infiltration rates have also often been associated with worm holes (Childs et al., 1957; Ehlers, 1975; Edwards et al., 1979; Beven and Germann, 1982; Germann et al., 1984; Smettem and Collis-George, 1985; Kladivko et al., 1986).

Attempts to relate direct measurements of soil morphology to hydrologic flow (Anderson and Bouma, 1973; Bouma and Denning, 1974; Bouma and Wösten, 1979; Bouma et al., 1979; Hamblin, 1985; Logsdon et al., 1990) have confronted two major difficulties: (1) the strong dependence of flow on pore dimensions (radius or width) (Bouma, 1991), and (2) the transitory and spatially variable nature of the macropore system (Beven and Germann, 1982; Logsdon et al., 1990). The appearance or disappearance of cracking planes as well as their dimensions is strongly related to soil moisture content and illustrates the interdependence of soil moisture, rainfall intensity, and soil structure in determining preferential flow patterns. More detailed discussion of macropore morphology and its relationship to flow are presented later.

Decreasing antecedent moisture content has generally been found to increase short-circuit bypass flow when expressed in terms of the proportion of the soil volume conducting flow and/or times to initial outflow from a given soil volume. Lower antecedent moisture content gave more marked bypass flow in silt loam soil columns subjected to a relatively high rainfall rate of 20 mm $h^{-1}$ (White et al., 1984), and bypass flow from both continuous and discontinuous leaching decreased when undisturbed columns of a structured clay soil had been pre-wetted, increasing water contents by approximately 10% (White et al., 1986). Bypass flow was reduced in the second as compared to the first rain on undisturbed silt loam soil blocks (Shipitalo et al., 1990; Edwards et al., 1992) and was attributed to an approximate 5% increase in soil moisture prior to the second rain. Correspondingly, travel times through soil blocks in both of these studies was increased at the higher antecedent water contents. In field work on a forested soil, Jardine, et al. (1990) found that rains following a long dry period during which the soil dried significantly, caused more rapid channeling of water than did rains on wetter soils.

Alternatively, higher antecedent soil moisture is observed to increase the depth to which macropore flow penetrates soil (Quisenberry and Phillips, 1976) as well as increasing total percolate volume (Shipitalo et al., 1990; Edwards et al., 1992; Granovsky et al., 1993). The increase in total percolate volume observed in the block studies of Shipitalo et al. (1990), Edwards et al. (1992) and Granovsky et al. (1993) was not, however, due to a larger flow volume from the same number of flow paths as under drier antecedent conditions but rather due to an increased number of flow pathways conducting water. This implies that there is reduced bypass flow at higher antecedent water contents even though percolate volume is increased. Thus a distinction must be made between the effects of antecedent moisture on initiation of short-circuit bypass flow and on macropore flow volume. Drier soils appear to favor flow characteristics typical of bypass phenomena presumably due to initiation of macropore flow higher in the profile and more pronounced flow funneling. Yet this flow is then subject to greater matrix absorption as flow proceeds downward, reducing macropore flow volumes. However, matrix absorption in drier soils may also be reduced by organic coatings on certain macropores such as *L. terrestris* burrows (Edwards et al., 1992).

Increasing rainfall intensity appears to increase both initiation of short-circuit bypass flow and macropore flow volume. The mean pore water velocity was increased in four soils ranging in texture from a sandy loam to a silty clay loam with increasing rainfall intensity while not having an appreciable impact on the relatively small soil volume participating in flow (White et al., 1984). Correspondingly, higher rainfall intensity reduced travel times through undisturbed soil blocks and increased percolate volume even though only a few flow paths (in the low antecedent water content soil) produced most of the percolate (Edwards et al., 1992). Also, higher rainfall intensity yielded larger volumes of percolate from soil columns inoculated and incubated with earthworms while neither the number or dimensions of stained burrows, assessed by dissecting the columns, was affected by rainfall intensity (Trojan and Linden, 1992). Field studies have indicated similar effects. Edwards et al. (1989) found the greatest percentage of rainfall flowed in vertically oriented earthworm burrows when storms were high intensity and short duration, while the number of burrows with water flow increased with the amount of rainfall in a storm. Jardine et al. (1990) reported more rapid onset of macropore flow with higher intensity storms.

It is generally presumed that the water application rate to the soil surface must exceed the soil matrix $K_{sat}$ (the intra-ped $K_{sat}$) before macropore flow can be initiated, and that the degree or extent of macropore flow will increase as the water application rate increases above this level (Beven and Germann, 1982; White, 1985; Brusseau and Rao, 1990). We argue, however, that the results of recent studies are consistent with a hypothesis that macropore flow initiation can occur at water application rates below the intra-ped $K_{sat}$, due to the development of localized zones of saturation at some depth below the surface. Further, it is important that the distinction between short-circuit bypass flow initiation and macropore flow volume, as outlined here, be considered in future studies of macropore flow.

## III. Macropore Impact on Chemical Transport

Increased flow of water in macropores does not necessarily result in preferential transport of chemicals. However, the rapid arrival of chemicals at some depth of interest is often the first clue that macropore water flow is occurring. For example, rapid initiation of water flow in subsurface tile drains can often be explained with the concept of a capillary fringe above the water table (Gillham, 1984), whereby small amounts of additional water can initiate drain outflow that is predictable from classical water flow theory. However, the detection of surface-applied chemicals in drain outflow within a few hours or days after the first significant rainfall following chemical application (Bottcher et al., 1981; Richard and Steenhuis, 1988; van Ommen et al., 1989; Kladivko et al., 1991) suggests that at least part of the water arriving at tile drains is "new water" flowing through macropores. Tile drains are now being used as a research tool to study preferential flow of water and chemicals on a field-scale.

The extent to which chemicals are transported in macropore flow will depend on factors such as the chemical reactivity, the location of the chemical, the ratio of macropore to matrix flow, the contact area between bypass flow and matrix water, and the rate of solute diffusion between mobile and immobile water volumes (White, 1985). In most soil systems, actual transport will result from complex interactions among these and other factors.

Solute transport literature generally considers chemicals as either reactive (inorganic cations, phosphate, high $K_d$ organics) or non-reactive (inorganic anions, low $K_d$ organics). The inorganic anions $Cl^-$ and $Br^-$ are highly mobile and have been used extensively as tracers of water movement in soils (White, 1985), although their negative adsorption may exclude them from very small water filled pores (White et al., 1986). Nitrate has also been found to be highly mobile in soils (White, 1985; Kladivko et al., 1991; Kanwar et al., 1985). Conversely, reactive species are generally less mobile (White et al., 1986; Shipitalo et al., 1990; Kladivko et al., 1991; Edwards et al., 1992; Granovsky et al., 1993).

Most reports of the effect of chemical location on transport are interpretations of indirect evidence, rather than actual measurement of spatial distribution. Decreased transport of both reactive and non-reactive chemicals has been observed in the second of two rainfall events (Shipitalo et al., 1990; Edwards et al., 1992; Granovsky et al., 1993), from discontinuous leaching (White et al., 1986), from soil drying (Jardine et al., 1990), and from soil incorporation (Steenhuis and Muck, 1988). These observations have been interpreted to result at least in part from the chemical being moved into the matrix, thereby decreasing contact with subsequent macropore flow.

The ratio of matrix to macropore flow will often be greater in tilled than in un-tilled soil. Leaching of $NO_3^-$ already associated with the soil matrix was greater in tilled than in no-tillage soil (Kanwar et al., 1985), while leaching of surface applied reactive chemicals tended to be greater from untilled soils (Granovsky et al., 1993). Several field studies comparing the effects of tillage on chemical transport have also found greater transport under no-tillage than tilled soil (Hall et al., 1989; Isensee et al., 1990; Sadeghi and Isensee, 1992). Decreased rainfall intensity will also result in a greater portion of matrix flow and has resulted in decreased transport of surface-applied Br (Trojan and Linden, 1992), of Br and atrazine (Boddy and Baker, 1990) and of Br, Sr, and atrazine (Edwards et al., 1992).

Lower antecedent moisture is associated with increased macropore flow and thus more rapid (Jardine et al., 1990) and increased transport of surface-applied solutes (White et al., 1986). Shipitalo et al. (1990), Edwards et al. (1992), and Granovsky et al. (1993) each observed greater transport during the first rain after chemical application than during a second rain when soils were at a higher antecedent moisture. In each of these studies the first rain may have moved solutes into the matrix, decreasing contact with subsequent macropore flow and confounding the effects of antecedent moisture.

Preferential transport of solutes may also be mitigated by the surfaces over which they are flowing. Some macropores such as root channels (Turner and Steele, 1988) and earthworm burrows (Stehouwer et al., 1993) are lined with material that is enriched in organic C. This material sorbs Cd and Mn (Turner and Steele, 1988) and atrazine and metachlor (Stehouwer et al., 1993, 1994) more strongly than the soil matrix. Attenuation of strongly sorbed herbicides was greater during transport through earthworm burrows than through unlined macropores, while for weakly sorbed herbicides there was little attenuation in lined or unlined macropores (Stehouwer et al., 1994).

The general relationships between macropore water flow and preferential transport of chemicals are becoming better understood. Understanding the hydrology of flow in structured soils is an important first step in understanding and predicting chemical transport in those soils, but the processes of sorption, diffusion, degradation, etc., must also be considered. Careful design of preferential flow experiments is essential if the separate but interacting hydrologic and chemical processes are to be described and modeled. The next section focuses on more detailed descriptions of the physical system in which the flow occurs.

## IV. Macropore Characterization

As discussed in the previous sections, macropore flow is affected by many factors, including soil morphology. Since "macropore flow" in structured soils occurs in "macropores" (as opposed to preferential flow due to wetting front instabilities in homogeneous materials, for example), it makes sense to try to describe and predict that flow as a function of macropore characteristics. Macropore morphology can be qualitatively and quantitatively described by soil surveyors, and some of this information is routinely collected and cataloged in data bases that could be subsequently used in water flow models. Although much effort has been expended in trying to relate water flow to soil morphology or structure data, these efforts have probably been more useful for increasing our understanding of underlying processes than as a substitute for direct hydraulic measurements (Bouma, 1992).

Bouma (1981, 1992) has reviewed many of the detailed soil morphology measurements made when trying to predict macropore flow. In this section we review recent approaches to describing macropore structure that may specifically relate to water flow. We also discuss macropore studies of tillage systems as examples, since it is often presumed that no-tillage results in increased macroporosity and subsequently increased macropore flow of both water and chemicals.

## A. Macropore Numbers, Sizes and Shapes

Much of the recent interest in macropores is motivated by the observed dichotomy between the percent pore space occupied by macropores and their contribution to ponded flow and/or saturated hydraulic conductivity ($K_{sat}$). Thus, more than 70% of the measured, saturated hydraulic conductivity of structured soils is generally transmitted by less than 0.1% of the total soil porosity (Watson and Luxmoore, 1986; Wilson and Luxmoore, 1988; Luxmoore et al., 1990; Dunn and Phillips, 1991a). It is important to note, however, that the equivalent diameter of a conducting pore, as calculated in these studies, is controlled primarily by its smallest diameter even though it may represent a fraction of the total pore system length (Dunn and Phillips, 1991b). For macropores formed in saprolite at a depth of 2 m, 93% of the $K_{sat}$ was due to flow in cylindrical channels that occupied less than 2% of the saprolite volume (Vepraskas et al., 1991).

Direct assessments of soil macroporosity are generally conducted by determining the numbers, size distribution, percent of total surface area, and shape of pores along horizontal planes cut through the soil profile. The wide diversity of techniques employed to directly assess macroporosity was recently reviewed by Bouma (1991). Numbers of macropores vary widely depending upon the soil type, management and experimental protocol. Reported values for biopores, generally greater than 1 mm ECD, ranged from 100 to 610 $m^{-2}$ for tilled and 610 to 2200 $m^{-2}$ for no-tillage soils from image analysis of two silt loams (Shipitalo and Protz, 1987). Similarly, biopores in a clay loam subjected to 5 years of tillage treatments ranged from 243 to 1475 $m^{-2}$ for tilled and 666 to 1732 $m^{-2}$ for no-tillage soils with significant differences between tillage treatments above but not below the tillage depth (Gantzer and Blake, 1978). Macropores >0.4 mm diameter ranged from as few as 100 to 3000 $m^{-2}$ from a variety of tillage treatments (Logsdon et al., 1990) to an excess of 14,000 $m^{-2}$ averaged over five depths to 0.3 m in a long-term no-tillage field subjected to frequent manure applications (Edwards et al., 1988). When pores >1.6 mm diameter were considered in 0.05 m depth increments to 0.6 m, no-tillage macropore numbers ranged from 80 to 720 $m^{-2}$ while tilled numbers ranged from 40 to 200 $m^{-2}$, with significantly higher numbers at each depth for the no-tillage soils (Singh et al., 1991a). Further, macropore size distributions are generally log-normal with large numbers of smaller pores and few larger pores (Edwards et al., 1988; Singh et al., 1991a).

Aerial coverage of directly observed macropores >0.4 mm diameter averaged 1.4 % (having a range of 0.4 to 3.8%) over several depths (Edwards et al., 1988) and 0.34% (having a range of 0.03 to 1.66%) over several depths and tillage treatments (Logsdon et al., 1990). Coverage of macropores >1.6 mm diameter as traced on transparent sheets ranged from 3.5% at 0.05 m depth to 0.5% at 0.6 m depth, with no apparent difference due to tillage (Singh et al., 1991a). The tendency for larger macropore areas in the tilled profile and the decrease in macropore area with depth was attributed to wider and deeper cracks as compared with no-tillage. Visible macropores characterized by shape averaged 1.5% for planar and 0.2% for cylindrical pores from surface layers of a calcareous gley soil (Jarvis et al., 1987), and 1.5% for planar and 0.72% for cylindrical pores from an argillic horizon at 0.25 to 0.3 m depth (Lauren et al., 1988). Using the macropore number and aerial porosity data of Edwards et al. (1988) and Logsdon et al. (1990), fractal geometry was proposed as a descriptor of macroporosity based on the fractal Sierpinski carpet (Brakensiek et al., 1992). Yet the observed linearity of water retention in the macropore range of suctions (discussed later) appears to violate the power law function that was used as evidence for the fractal nature of matrix porosity (Tyler and Wheatcraft, 1990).

These studies document the widespread occurrence of visible macropores in a diversity of soils and at a relatively high frequency. Yet their contribution to aerial porosity of the soil is only a few percent. While there is a slight tendency for higher macropore numbers in surface horizons under no-tillage, macropores are also widespread in tilled soils. As will be discussed later, soil management does apparently influence other geometric attributes of macropores, even if management does not particularly impact macropore number and porosity.

Knowing the number, size and areal porosity of macropores may offer little insight into their role in preferential water flow. A lack of significant difference between tillages in macropore number, total perimeter and aerial porosity from detailed image analysis could not be reconciled with the greater degree of preferential flow in no-tillage columns (Singh et al., 1991b) Also, from field studies throughout the period June to October, 1987, never more than 20 of the 50 instrumented *L. terrestris* burrows > 5 mm diameter contributed flow to a depth of 0.3 m (Edwards et al., 1989). The average number of burrows contributing water flow across all storm events was 14.3 throughout this period. Further, while the presence of earthworm macropores was shown to increase water flux in repacked silt loam and clay loam soils as compared to those not containing earthworms, there was no correlation between visible macropore number and measured, zero-tension infiltration rate at the surface or with depth in the soil (Ela et al., 1992). Thus, only a small proportion of the earthworm macropores participated in water flow and contributed to the increased infiltration rates.

### B. Macropore Participation in Flow and Continuity

An alternative to measuring the physical size of individual macropores is to calculate a functional size based on flow characteristics. One such recent approach is to employ macroscopic capillary lengths to calculate characteristic flow-weighted pore sizes for pores participating in flow as soils approach saturation (White and Perroux, 1987; White and Sully, 1987; White and Perroux, 1989; Clothier and Smettem, 1990). The macroscopic capillary length is defined by

$$\lambda_c = [ K_o - K_n ]^{-1} \int_{\Psi_n}^{\Psi_o} K(\Psi)\, d\Psi$$

where K is the hydraulic conductivity as a function of the soil matric potential, $\Psi$, and subscripts o and n denote the supply surface and antecedent conditions (White and Sully, 1987). Since $\lambda_c$ can be thought of as a K-weighted mean soil matric potential, it can be employed to estimate the flow-weighted pore size, $\lambda_m$, using the following modified form of the capillary equation,

$$\lambda_m = \sigma / (\rho g\, \lambda_c)$$

where $\sigma$ is the surface tension, $\rho$ is the water density, and g is the gravitational acceleration. In a strict sense $\lambda_m$ is the flow-weighted hydraulic radius and can thus be used to estimate the equivalent cylindrical pore diameter by multiplication by 4.

Estimates for $\lambda_c$ result from its definition assuming an exponential form of the $K(\Psi)$ relation, or from estimates of sorptivities alone or sorptivities and hydraulic conductivities (White and Sully, 1987). These hydraulic properties were in turn generally estimated from measurements using a disk permeameter (also called a tension infiltrometer) (Perroux and White, 1988) and the water inflow dynamics into soils that were initially air dry and with the permeameters set to various supply pressures near saturation (White and Perroux, 1987; White and Perroux, 1989). Alternatively, disk permeameters of various radii have been employed whereby only steady infiltration rates are measured under various supply pressures (Clothier and Smettem, 1990).

Flow-weighted pore sizes from these studies ranged from 0.37 mm for an undisturbed, silty clay loam to 0.031 mm for a repacked, fine sand (White and Perroux, 1987; White and Sully, 1987). Both of these soils were at air dry antecedent conditions and were provided with a supply pressure of -10 mm of water. This twelve-fold difference, and similar measurements cited in White and Sully (1987), were attributed to biologically formed macropores in the undisturbed soils that yielded pore sizes uncharacteristic of the soil texture. When considering conditions approaching saturation and spanning a much smaller difference between antecedent (-10 mm water) and supply (0 mm water) potentials, flow-weighted pore sizes averaged 0.93 mm for an unstructured sandy loam and 2.2 mm for a fine sandy loam containing numerous surface vented macropores (Clothier and Smettem, 1990).

These studies suggest that calculated $\lambda_m$ values for well graded, simple porous systems provide physically plausible flow-weighted pore sizes (White and Sully, 1987). For undisturbed soils and with supply pressures approaching saturation, however, large $\lambda_m$ values display little or no dependence on soil texture and probably reflect preferential flow pathways dominated by macropores. This approach to estimating macropore sizes is attractive because it measures macropores as they function during flow. There are, however, many computational steps and estimations involved in calculating flow-weighted pore sizes, each of which comes with its own set of assumptions. Obvious extensions to this line of research are to examine the correlation between estimated flow-weighted pore sizes and pore sizes gained from morphological examinations. This approach may help target morphological features subject to routine measurement that contribute to macropore flow. Additionally, flow-weighted pore sizes may be more confidently employed in modeling exercises.

The exact sizes and shapes of macropores may not be as relevant for water flow as is sometimes suggested. As an example, Ahuja et al. (1991) reported that the Root Zone Water Quality Model (RZWQM), which takes macropore flow into account, showed only a small effect of macropore size on the final results. Pore continuity and interconnectedness may be more important than size as even small pores that are continuous through the soil horizon in question can conduct greater water volumes than larger, discontinuous macropores. Dye staining is perhaps the most common method employed to assess macropore continuity and participation in an individual water infiltration event. In pedal clay subsoils ranging from 0.35 to 1.0 m depth, tubular channels and planar voids generally all participated in saturated flow (Bouma et al., 1977). Stained macropores ranged from 59 to 2% of the total number of pores > 0.1 mm diameter, with the higher percent corresponding to soils having a higher $K_{sat}$. Since the size distribution of stained vs. unstained pores was similar, these percentages also apparently applied to macropores ranging in size from 0.3 to 1.0 mm and > 1.0 mm (Bouma et al., 1977). Interestingly, there was no systematic pattern of pore staining percentage with depth through the 0.15 m soil column. Finally, planar voids were intermittently stained along their lengths, suggesting that dyed water flow was controlled by pore necks vertically adjacent to those sampled. Comparison of stained and unstained areas of pores >0.03 mm diameter yielded from 5.0 to 6.4% of the total soil area consisting of channels and planar voids while stained areas ranged from 0.46 to 0.83 percent of the total soil area (Bouma and Wösten, 1979).

While these studies suggest that relatively few macropores and only a fraction of the macropore area participate in saturated flow, the total macropore number and area may have been overestimated since these values were determined from freeze dried and resin-impregnated thin sections. The drying of these pedal, clay soils may have created an increased number and area of planar voids that would not have been considered macropores when the soils were more moist. Subsequently, measurement of dye staining in conjunction with a pore interaction model provided an adequate estimation of $K_{sat}$ in seven pedal clay subsoils that did not contain large, vertically continuous, tubular macropores. In this study, stained voids occupied <1 % of the total soil area (Bouma et al., 1979). In another study, fractal dimensions related to the shape and perimeter of stained macropores observed in thin sections were shown to adequately describe the shape of $Cl^-$ breakthrough curves during steady saturated flow in five subsoils (Hatano et al., 1992). In this case, increased preferential flow was associated with tubular channels and interpedal planar voids having smooth boundaries.

Examinations of tillage influences on macropore continuity using dye staining indicate that long-term no-tillage promotes continuity of cylindrical channels relative to moldboard plowing or chisel tillage (Douglas et al., 1980; Heard et al., 1988; Logsdon et al., 1990). Stained channels >1.0 mm at a depth of 0.1 m in a silty clay loam ranged from 275 to 464 $m^{-2}$ and in a silt loam from 558 to 1124 $m^{-2}$ with no influence of tillage or crop rotation (Heard et al., 1988). The percentage of stained pores at 0.3 m depth relative to those at 0.1 m depth ranged from 23 to 56% and 81 to 109% in the silty clay loam and silt loam soils, with the higher percentages corresponding to the no-tillage treatments. Douglas et al. (1980) also measured greater hydraulic conductivity at the interface of topsoil and subsoil under no-tillage, due to the higher number of continuous earthworm channels at this depth.

An indirect indication of pore continuity results from the difference between measurements of $K_{sat}$ made in situ and then again after removal of the same soil column from the soil profile. A key assumption in this determination of pore continuity is that the continuous macropores in the soil horizon in question subsequently terminate at some lower depth. Measurements of attached vs. detached $K_{sat}$ ranged from no difference, indicating poor macropore continuity through the column, to one order of magnitude difference, indicating large continuity through the horizon in question (Bouma, 1982; Smettem, 1987; Lauren et al., 1988). There have, however, been an insufficient number of studies using this procedure to form quantitative guidelines for interpreting these differences in terms of pore continuity.

An alternative indirect assessment of macropore continuity results from employing air permeability ($K_a$) measurements of soils containing macropores (Ball, 1981a,b; Hamblin and Tennant, 1981; Groenevelt et al., 1984; Ball and O'Sullivan, 1987; Ball et al., 1988; Blackwell et al., 1990, Roseberg and McCoy, 1992). Groenevelt et al. (1984) proposed testing the differences in macropore geometry between samples exhibiting different $K_a$ by comparing the values of air permeability divided by either air-filled porosity ($E_a$) or air-filled porosity squared (i.e. $K_a/E_a$ or $K_a/E_a^2$). Samples having different $K_a$ but similar $K_a/E_a$ would indicate similar air-filled pore size distributions and continuity, with each additional amount of geometrically similar pore space contributing proportionally to $K_a$. According to Poiseuille's law, an increase in pore size would result in a squared increase in air permeability. Thus, samples having different $K_a$ but similar $K_a/E_a^2$ would indicate differences in pore size distribution assuming similar pore continuity. Differences in pore continuity, however, would result in differences in both $K_a/E_a$ and $K_a/E_a^2$ (Ball et al., 1988). These manipulations, therefore, generate indices of pore continuity.

This method was used in a study of cropping history effects, in which a silt loam soil was either cropped to corn continuously for five years or was cropped to forages for three years and to corn for the remaining two years (Groenevelt et al., 1984). During the final two years of the study, $K_a$ and $E_a$ were consistently higher in the forage history treatment, indicating that the effect of cropping history lasted at least two years on this site. Frequent differences in the index

$K_a/E_a$ suggested that the larger $K_a$ values for the forage history did not result from increased porosity at a constant pore size distribution and pore continuity. The absence of differences in the index $K_a/E_a^2$ was explained by uniformly larger pores existing in the forage history treatment, assuming equal pore continuity. Finally, since $K_a/E_a$ was generally different between treatments while $K_a/E_a^2$ was not, there was apparently no difference in pore continuity due to cropping history.

In a field experiment comparing tillage and traffic treatments on a silt loam soil, there were no $K_a$ differences due to tillage in the non-wheel tracked corn interrows over a matric potential range of 0 to -3.0 kPa (Roseberg and McCoy, 1992). Significant reductions in $K_a$ due to wheel traffic in the tilled treatment were associated with consistent differences in $K_a/E_a$ and $K_a/E_a^2$. This analysis indicated that wheel traffic on tilled soil resulted in differences in both air conducting porosity and pore continuity. A problem associated with this technique in detecting differences between field collected data sets is the progressive accumulation of errors in forming $K_a/E_a$ and $K_a/E_a^2$. Thus a breakdown in detecting differences may be due to real differences in pore geometry or from this accumulation of experimental error.

Another approach to analyzing changes in pore geometry uses a generalized form of the Kozeny-Carman equation relating $K_a$ to $E_a$ given by

$$K_a = M \cdot (E_a)^N$$

where M and N are empirical constants (Ball et al., 1988). In this form, the exponent N is considered an index of macropore continuity since it reflects the relative increase in $K_a$ with increasing portions of $E_a$. A high relative increase in $K_a$ would reflect opening of more continuous pathways while a low relative increase in $K_a$ would reflect opening of less continuous pathways. The values of N are found from linear regression of the data pairs log $K_a$ vs. log $E_a$. Thus for a given soil sample, the air permeability is determined for various values of air-filled porosity.

Perhaps the most extensive use of this approach was reported in a tillage and traffic experiment comparing corn cropping in a silt loam soil employing no-tillage or moldboard plowing and sampling in wheel tracked or non-wheel tracked interrows (Roseberg and McCoy, 1992). Results from this study suggest that while tillage increased non-tracked interrow macroporosity (assessed one month after planting) this did not translate into higher $K_a$ values since the macropore continuity index, N, was lower than comparable values for no-tillage. Wheel traffic reduced $K_a$ in the tilled soil but not in the no-tillage soil. This reduction was associated with a loss of macroporosity but with a consequent increase in the continuity index to comparable values for no-tillage.

In summary, quantification of macropore continuity is an important component of measurements of macropore size, number and area, when trying to relate these pore characteristics to water flow in the field. There has been some progress in estimating $K_{sat}$ or macropore flow from various quantitative measurements of these pore characteristics, but as Bouma (1992) stated, the morphological measurements are more laborious than the physical measurements themselves and will not likely replace the needed flow measurements. They are helpful, however, in further elucidating the underlying processes governing macropore flow.

### C. Water Retention and Conductivity

Water retention measurements are often used to calculate pore size distributions, assuming the capillary model describes soil pores. However, Bouma (1991) recently identified a difficulty in using water retention techniques for assessing soil macroporosity, arising from the relatively large influence of the gravitational gradient across a soil sample equilibrated to low (near zero) matric potentials. Consider the example of a 60 mm long soil core equilibrated to -0.6 kPa at the center of the core. At the upper surface, the potential would be -0.9 kPa and at the bottom the potential would be -0.3 kPa, yielding a 100% error in control across the length of the core. Correspondingly, equivalent diameters of air filled pores would range from 0.3 to 1.0 mm across the length of the core. Thus it may not be realistic to attempt measurement of equivalent pore sizes larger than approximately 0.15 mm (corresponding to -2.0 kPa matric potential) using standard water retention methods.

Recently, two methods have been developed to eliminate the influence of the gravitational gradient and extend the range of water retention measurements into the macropore range. In one application, macropore water retention measurements are performed in conjunction with air permeability determinations (Roseberg and McCoy, 1990). Using this device, a positive air pressure is applied to the lower, exposed surface of an undisturbed soil core. The positive air pressure, expressed as the height of a water column, is regulated to equal the vertical height of the core, with the upper surface open to the atmosphere. In this fashion, the downward, gravitational gradient is counteracted by the upward air pressure gradient and, at soil water equilibrium, the matric potential is uniform throughout the soil core. In a second method, matric potential control is established around the circumference of a cylindrical soil core, the core is oriented horizontally and rotated around it's cylindrical axis at approximately 2 rpm (Logsdon et al., 1993). This rotation is thought to even out the gravitational gradient such that during the period of equilibration with the applied matric potential, no one location in the core is higher in the gravitational field than any other location.

Results employing these techniques revealed that undisturbed cores containing macropores generally fail to exhibit negative air entry pressures, and that there is generally a linear decline in water content over a range of 0 to -3.0 kPa (Roseberg and McCoy, 1992; Logsdon et al., 1993). These results confirm an earlier demonstration of the rotated core device over matric pressures -0.3 to +0.4 kPa where observed water contents continued to increase from 0 to +0.4

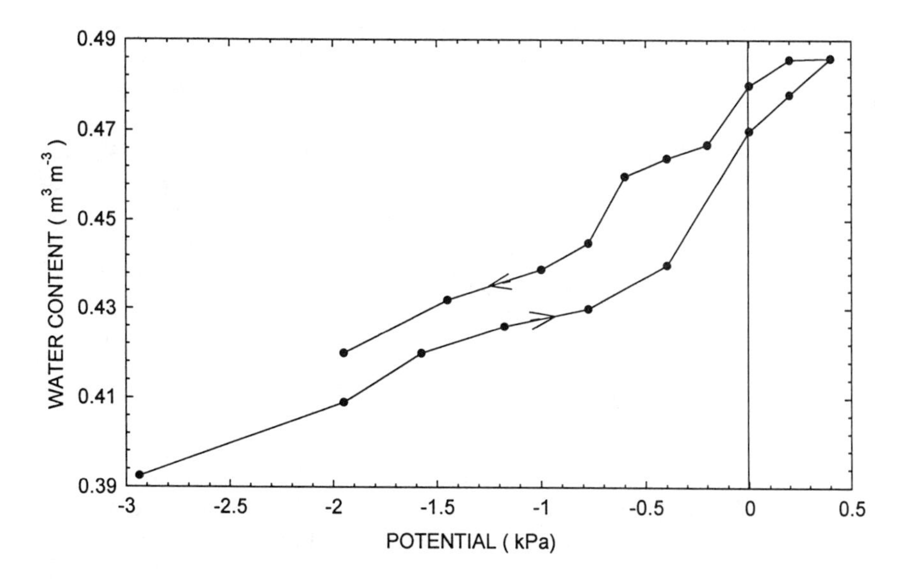


**Figure 4.** The water retention relation for a macroporous soil from a long-term no-tillage experiment. (From McCoy, 1989.)

kPa (Figure 4) (McCoy, 1989). Thus, these studies raise doubts about use of the capillary equation for describing macropores. In addition, soils containing macropores displayed the expected pattern of hysteresis (McCoy, 1989; Logsdon et al., 1993). The absence of negative air entry pressures and the linearity of the water retention curves containing macropores expose a deficiency in applying typical water release curves, such as the van Genuchten equation (van Genuchten, 1980) or the more recent model of Ross and Smettem (1993), to soils containing macropores.

Additional evidence that the capillary equation does not apply to the largest macropores results from the dichotomy between ponded and zero tension hydraulic conductivity measurements for soils containing macropores (Bouma, 1982; Booltink et al., 1991). Studies measuring ponded $K_{sat}$ and using the crust method for measurement of zero tension hydraulic conductivity, $K_{sat}$, exhibit up to 2 orders of magnitude difference for the same, undisturbed soil core (Bouma, 1982; Booltink et al., 1991). The ponded measurements included the participation of large continuous macropores while the zero tension measurements (suggesting soil saturation by the capillary rise equation) apparently excluded full macropore participation. It is impossible to know whether there was flow in incompletely water filled macropores during the $K_{sat}$ measurements. However, the absence of complete water saturation of these pores at zero tension is clearly indicated. Further, an equivalent phenomenon is obtained where measurable air permeabilities were observed at zero tension for soils containing macropores

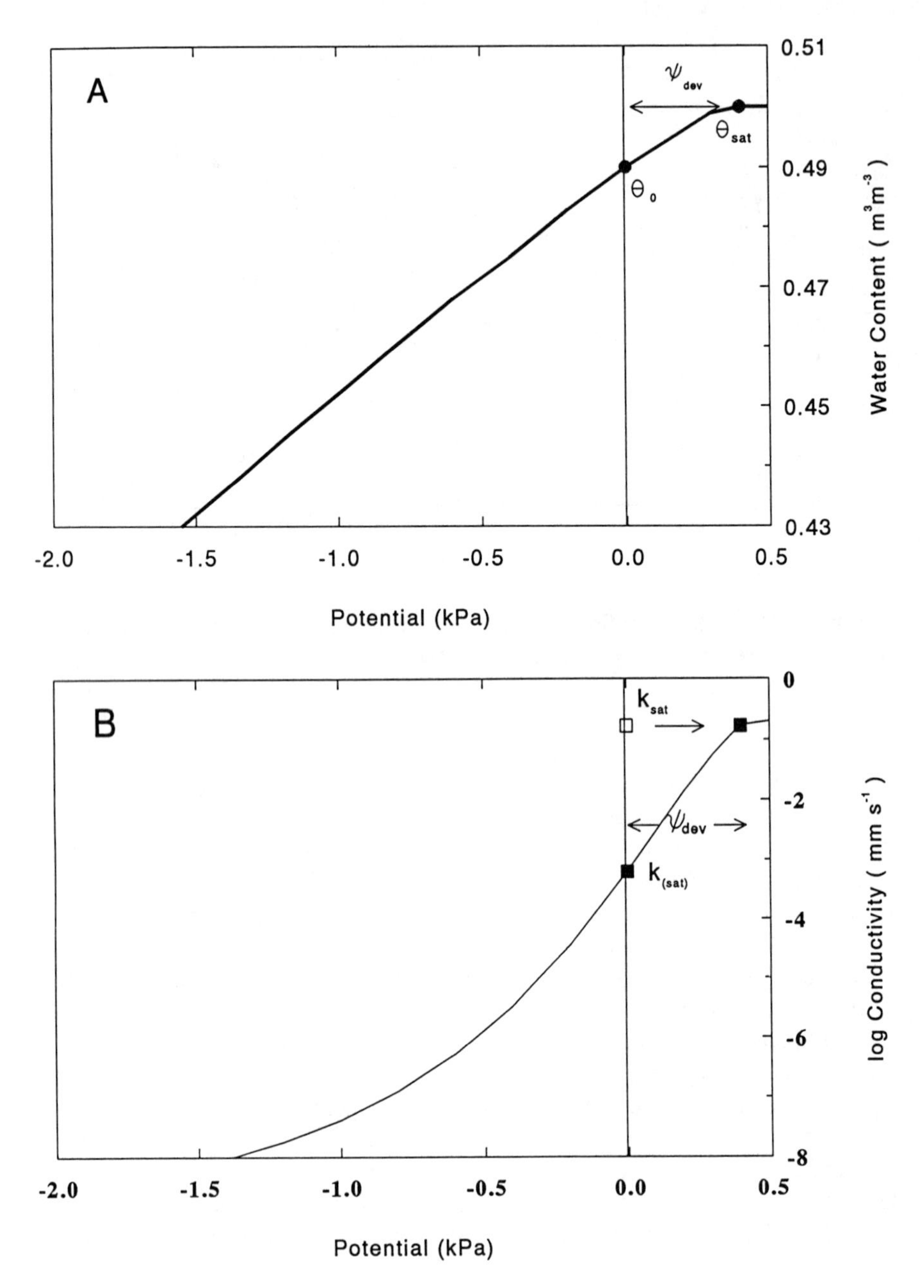


**Figure 5.** A conceptual view of water retention (A) and hydraulic conductivity (B) for soils containing macropores.

(Roseberg and McCoy, 1990; Roseberg and McCoy, 1992). Apparently, some continuous macropores remained air filled at zero tension during these measurements.

A conceptual approach that may resolve the problem of having two different values of hydraulic conductivity at zero tension comes from integrating our view of macropore water retention with a hydraulic conductivity function. This concept is illustrated in Figure 5. In conjunction with the measurements of Figure 4, water content for soils containing macropores exhibit a linear increase as the soil approaches zero-tension from small negative water potentials (Figure 5A). Further, there is a continuing increase in water content for these macroporous soils as positive water pressures are applied. Saturation, indicated by $\theta_{sat}$, occurs, therefore, at some slight positive water pressure. The potential difference between zero-tension and the point of actual saturation, $\Psi_{dev}$, represents the soil water potential deviation from the capillary hypothesis. By applying the concept that there exists a $\Psi_{dev}$ for soils containing macropores, the ponded $K_{sat}$ value is translated into the positive water potential region by the magnitude of $\Psi_{dev}$ (Figure 5B). This manipulation contains the assumption that pores continuing to fill with positive water pressures are in fact water conducting macropores. A continuous function is then drawn between the newly located $K_{sat}$ value and the zero-tension conductivity value, $K_{sat}$.

Following these manipulations that are dependent upon identifying a value for $\Psi_{dev}$, the functional relations for water retention and hydraulic conductivity as a function of potential that rely on an air entry value, can be applied to soils containing macropores by setting the air entry value to positive $\Psi_{dev}$. While $\Psi_{dev}$ may be a useful parameter in providing a functional classification of macropores, we do not anticipate observing a uniform and consistent numerical value for this deviation in all macroporous soils. The natural complexity of the soil pore system and varying proportions of nearly cylindrical channels or highly planar voids in soils suggest that finding a uniform value for $\Psi_{dev}$ is unlikely.

This deviation of the capillary hypothesis for macropores offers supporting evidence for the horizontal void hypothesis of macropore flow initiation. If these horizontal voids are of sufficient dimension to invalidate the capillary theory, then positive water pressures would need to be created in the profile for filling of these pores and continued downward migration of soil water via matrix pores. Thus, saturated loci would be developed in a structured soil during water infiltration. As discussed previously, the development of saturated loci in the neighborhood of a vertically continuous macropore would lead to short-circuit bypass, macropore flow.

Finally, there are questions as to where in the negative range of matric potentials the capillary hypothesis fails. Or more specifically, at what maximum negative matric potential or correspondingly minimum ECD do pores no longer follow the capillary hypothesis. Calculations for a completely wetted capillary and pure water at 20°C yields an ECD value of 13.4 mm equivalent to a matric potential of approximately -0.02 kPa. Alternatively, Hamblin (1985) offers an ECD value of 3.0 mm equivalent to a potential of -0.1 kPa for a capillary under standard temperature and pressure conditions. Actual soil pores will likely fail at a smaller ECD and more negative potential as those offered for ideal capillaries due to the non-ideal nature of soil pores. Luxmoore (1981) has proposed that pores having ECD's greater than 1.0 mm equivalent to a potential of -0.3 kPa function in bypass (channel) flow and, due to the seemingly non-Darcian nature of this flow, the implication is that these pores are not capillaries. Knowing the precise dimension of pores no longer behaving as capillaries is, however, not as important as understanding how non-capillarity influences water flow in soil macropores.

## V. Models of Macropore Flow

Most analyses of water flow and solute transport in porous media are based on a continuum approach, specifically, an analysis where each variable of interest is assumed to be expressible as a function of space and time, with the functions continuous except possibly at isolated points, lines, or surfaces in space. The status of the medium at any given time and position is traditionally analyzed in terms of total water and solute contents ("total", as opposed to contents in macropores, contents in soil matrix, etc.) and overall water and solute flux densities. The law of conservation of mass is expressed in terms of these total content and overall flux density variables.

In actuality, when examined at a sufficiently small scale (called the pore scale), no porous medium is a continuum; it consists of distinct regions: solids and pore spaces. Furthermore, either liquid or gas may occupy a given point in a pore. In principle one could determine the complete geometrical configuration of soil at the pore scale and use a continuum theory to describe, in full detail, the flow of water and transport of solutes. In practice, it is not possible to follow this "detailed" analysis with current measurement technology and computational power. Thus, in order for a continuum theory to be applied to the analysis of processes which occur in soil, the soil is regarded as a continuum at some scale which is much larger than the pore scale. One approach to formalizing this "averaging" viewpoint is to consider a representative elementary volume (REV) of soil (Bear, 1972), a volume large enough (relative to the pore sizes and spacings) so that inclusion or exclusion of a few pores at the edge of the REV does not unduly influence properties which are defined in terms of the REV, and small enough so that any variable which is measured at this scale can be regarded as a continuous function of space to which the tools of differential and integral calculus can be applied. Although other definitions of the REV (Peck, 1983; Wagenet, 1985) and other averaging approaches (Baveye and Sposito, 1985; de Marsily, 1986) have been proposed, the concepts in the following are best visualized in terms of this "continuum" REV.

For a soil with macropores, two physically-based continuum approaches are currently practical, the first of which is a hybrid of the "detailed" and "averaging" methods.

**disjoint-volumes approach**:

Treat the macropores as not being a part of the porous medium, that is, describe movement of water and solutes in each macropore, or in each of a set of macropore categories, individually (i.e., "in detail"), and treat the remainder of the soil as a separate "averaged" continuum. In addition to the equations used to describe flow and transport in these two volumes, equations which describe the exchange of water and solutes at the boundary which separates the two volumes are needed.

**multi-continuum approach:**

Treat the entire soil mass (both macropores and matrix) as more than one continuum. Flow and transport are viewed as occurring in multiple superimposed continua (each occupying the entire region of interest with, for example, one continuum representing large pores, one representing small pores, etc.). This is called a multi-continuum, multi-porosity, or multi-region approach. (In particular, if exactly two continua are considered, the approach is called bi-continuum, dual-porosity, two-region, etc.) In the REV formalization of this approach, the representative volume must be large enough that such REV-defined variables as the density of macropores and the portions of the water and solute flux densities associated with the macropores are not unduly influenced by the vicissitudes of inclusion and non-inclusion at the edge of the REV.

In a disjoint-volumes approach the total volume of space being modeled is partitioned into disjoint volumes, whereas in a multi-continuum approach each continuum occupies the entire volume being modeled. This is not merely an academic distinction: in general the nature of the interactions between the various domains can be more accurately modeled under a disjoint-volumes approach than under a multi-continuum approach. The price to be paid for this accuracy can be an enormous increase in the complexity of the model and its computational demands.

In a disjoint-volumes model, a separate boundary value problem is posed for each sub-volume, and the connections between the sub-volumes are expressed in the form of shared boundary conditions. For example, under this approach, the flow of water from a cylindrical macropore into the surrounding soil matrix could be modeled as a radial flow outward from the macropore boundary into the soil. The interaction between the two volumes need not be expressed in terms of any "average" or "lumped" value of the hydraulic head or solute concentration in the matrix volume; the interaction is expressed in terms of the values of these variables <u>at</u> the boundary between the two volumes. In a bi-continuum model, the hydraulic head or solute concentration is not known at any kind of "boundary" between the two continua. (No such boundary is defined.) Only lumped values, typical of each of the two continua, are known. For example, the flow of water or transport of solute between a continuum which represents macropores and a continuum which represents the soil matrix must be modeled using a "lumped" value of water energy status or solute concentration in each of the continua.

Models can be hybrids of the two approaches and for some models the distinction between the two approaches can be subtle, but the fundamental difference is illustrated by the following examples.

A numerical model which uses the disjoint-volumes approach is described by Edwards et al. (1979). A vertical "wormhole" macropore and a surrounding cylindrical body of soil are modeled, where the cross-sectional area of the cylinder is chosen to represent the average area per macropore in the soil being represented. Retention and flow in the donut-shaped volume of soil matrix are modeled with a finite difference approximation to the two-dimensional Richards equation in cylindrical coordinates. That is, soil water contents and fluxes are known not only as a function of depth (as they would be in a bi-continuum model) but also as a function of distance from the macropore. For purposes of modeling the flow of water between the two sub-volumes, there is no need to represent the water content or flux in the soil sub-volume by a value which is "lumped" over horizontal distance. Instead the appropriate water variables can be evaluated _at_ the boundary between the two sub-volumes In other words, the relationship between variables in the two sub-volumes can be expressed as a boundary condition. In the Edwards, et al. (1979) model the soil is taken to be homogeneous and initially at a horizontally-uniform water content, so the initiation of runoff occurs simultaneously over the entire soil surface of the wormhole's "watershed". This runoff is instantaneously routed to the wormhole, where it infiltrates into the soil via the wormhole wall. The surface over which lateral infiltration occurs moves progressively deeper into the wormhole with time, and when the bottom of the wormhole is reached the excess of runoff at the soil surface over the rate of infiltration into the entire wormhole wall is assumed to progressively fill the wormhole with free water. When the entire wormhole is filled, excess runoff is taken as runoff leaving the wormhole watershed. The key feature which makes this a "disjoint-volumes" model is its two-dimensionality, in particular its modeling of water content and water flow as a function of (radial) distance away from the macropore, as well as vertically.

Hoogmoed and Bouma (1980) present a model which differs subtly from this disjoint-volumes approach. They consider simultaneous vertical flow in both tall (prismatic) peds and in interpedal cracks. Flow from the interpedal cracks into the peds is modeled using a one-dimensional horizontal form of the Richards equation. Thus the interaction between vertical and horizontal flow is not implicit in the numerical flow modeling. The progress of horizontal water infiltration from the interpedal cracks into the peds is reflected as a lumped effect rather than in a way which would facilitate treatment of the ped surface as a "boundary condition". As with Edwards, et al. (1979) all rainfall is initially assumed to enter the upper surface of the peds. After formation of a 2 mm pond, all excess water enters the interpedal cracks, and, using horizontal infiltration theory, is partitioned between horizontal infiltration and drainage out of the 0.2 m deep zone. Simulations of two infiltration rates (0.20 and 0.77 m/day) and two initial volumetric water contents (0.38 and 0.49) indicated very little infiltration through ped walls (at most 4-5% of the rainfall rate) and very high movement through the 0.2 m depth.

In both the Edwards, et al. (1979) and the Hoogmoed and Bouma (1980) models it is assumed that water moves down the macropore wall at a rate equal to the difference between the rate that water is made available for such flow at the soil surface and the rate that water infiltrates horizontally from the macropore into the soil matrix. Beven and Germann (1981), on the other hand, bring in a capacity term by considering a water film of non-zero thickness on the macropore wall, assuming that the water in a partially full macropore is at atmospheric pressure, and modeling the rate of water movement down macropores using laminar flow theory. Water flow from a macropore into the soil matrix is taken to be proportional to the difference between the water pressure in the macropore and the (lumped) water pressure in the matrix continuum. Because of this, the model of Beven and Germann (1981), like the model of Hoogmoed and Bouma (1980), is a bi-continuum model, rather than a disjoint-volumes model. Earlier Philip (1968) and Kutílek and Novák (1976) described bi-continuum models of aggregated soil with a transient entry of water. They adopted the simplifying assumption that the downward movement of water in the larger pores occurs as a sharp front, i.e. the inter-aggregate pore space fills instantaneously with water upon passage of the front. Other multi-continuum models will be considered in section D after a number of specific bypass mechanisms and models are described.

## A. Mechanisms

As we see it, two primary porous medium flow phenomena involving macropores have been part of the motivation for the development of water flow and solute transport models:

**solution bypass**: the process of rapidly-moving water (at some solute concentration) "bypassing" more slowly-moving water (whose solute concentration may be different)

**short-circuit bypass**: the process of water in a large pore not only bypassing slower-moving water, but bypassing dry soil (i.e. water in a large pore "short-circuiting" soil in which smaller pores are empty)

A third phenomenon, not specifically associated with the presence of macropores, is:

**non-equilibrium chemical sorption**: the process of chemical sorption occurring as a kinetic, non-equilibrium, process in a fraction of (or in all of) the soil, and as an equilibrium, instantaneous, process in the remainder.

## B. The Water Retention Hypothesis

The "water retention" hypothesis is so deeply and subtly embedded in most thinking about the behavior of water in soil that it is often not regarded as being a hypothesis. In essence, it is the assumption that locally (i.e. within an REV) soil water is at equilibrium (Philip, 1968; Germann, 1990). If, by whatever mechanism, water were to be found in a large pore while smaller pores located in the same REV were empty, the water retention hypothesis assumes that the water in the large pore is instantaneously transferred to the smaller pores.

(This is a rather loose statement of the hypothesis. Because "pores" are geometrically complicated and intricately interconnected, a single number (pore "size") does not fully describe their complexity. The hysteresis phenomenon is often explained in terms of pores having a body size and a neck size, with the definition of what is meant by a "large pore" and a "small pore" being dependent on the history of wetting and drying. But with this and even more complicated models of hysteresis, the basic "water retention" assumption remains: within an REV, any water found in a "large" pore – however defined – is assumed to be instantaneously transferred to "smaller" pores.)

The dilemma for a soil with macropores is that (a) the REV must be very large in order to accommodate the relatively large spacing between macropores, and (b) it is not reasonable to assume, for a large REV, that water in any large pore contained in the REV is instantaneously transmitted to the smallest empty pores in the REV. In theory, we can imagine the possibility of replacing the water retention hypothesis with a theory which incorporates dynamics of the movement of water between large pores and small pores at the local scale. Lacking such a theory, the phenomenon of short-circuit bypass cannot be represented by a single-volume, single-continuum model. In a bi-continuum model each continuum separately can satisfy the water retention hypothesis, and yet the two continua taken as a whole not necessarily satisfy this hypothesis. The other primary method for modeling a breakdown of the water retention hypothesis is to treat some of the pores as physically separated from the remainder of the soil (disjoint-volumes approach).

Thus, one of the motivations for the development of either disjoint-volumes models or multi-continuum models is a recognition of the occurrence of the short-circuit bypass phenomenon. However, multi-continuum models have also been motivated by the occurrence of non-equilibrium chemical sorption and solution bypass phenomena. Although these two phenomena can, at least in principle, be represented in single-continuum models, it may well be that they are more efficiently represented by multi-continuum models.

## C. Solution Bypass Models

The idea of piston flow, conceptually the simplest solute transfer model, is that incoming water displaces all water ahead of it (e.g., Thomas, et al, 1978). For a one-dimensional downward flow, the piston flow concept is the idea that after a certain amount of water flow through the soil surface, the newly-infiltrated water, and only this water, occupies a sharply demarcated layer in the uppermost portion of the soil. The thickness of the zone occupied by this "new" water at a given instant is determined by integrating the post-inflow volumetric water content at this instant from the soil surface to whatever depth yields an integral equaling the volume of water applied per unit soil surface area. Under the piston flow assumption, the solute flux density $J_s$ [M $L^{-2}$ $T^{-1}$] equals the product of the solute concentration $C$ [M $L^{-3}$] and the liquid flux density $J_w$ [$L^3$ $L^{-2}$ $T^{-1}$].

$$J_s = C\, J_w$$

Early reviews of solute transport (Gardner, 1965; Biggar and Nielsen, 1967; Boast, 1973) document the refinement of this piston flow model into what is called the convection-dispersion equation (CDE) or advection-dispersion equation (ADE). This equation is based on an assumption that whatever mixing (or "hydrodynamic dispersion") occurs (in particular, movement of newly-applied water below the piston-flow depth and retention above this depth of water originally present there) can be represented by adding to the piston flow model a second term wherein the amount of mixing is proportional to the spatial gradient $\nabla C$ of the solute concentration:

$$J_s = C\, J_w \;-\; D_h \nabla C$$

where $D_h$ is the hydrodynamic dispersion coefficient and $C$ can be regarded as the resident concentration of solute (Parker and van Genuchten, 1984). It is striking that the added term, based as it is simply on the solute concentration gradient, serves as well as it does to represent phenomena which are due to a variety of causes, including differences of water flow rates. Models based on this equation tend to predict rather symmetrical S-shaped curves of solute concentration versus depth for situations where a solute front enters soil. Symmetrical curves match experimental data for unstructured or repacked soils fairly well, but structured soils in their natural condition tend to exhibit highly asymmetric concentration profiles.

The first responses to this kind of discrepancy were the development of two types of models: (*i*) "chemical non-equilibrium" models, specifically, models with steady water flow and non-equilibrium chemical sorption (or a "multi-site" combination of equilibrium and non-equilibrium sorption) (e.g., Selim, et al., 1976; Cameron and Klute, 1977) and (*ii*) "physical non-equilibrium" models, in particular, bi-continuum models with no water flow in one ("immobile") continuum and a steady water flow in a mobile continuum (e.g., Coats and Smith, 1964; Passioura, 1971; Skopp and Warrick, 1974; van Genuchten and Wierenga, 1976; Scotter, 1978). For one-dimensional flow in uniform soil, some instances of these two approaches are mathematically equivalent (Nkedi-Kizza, et al., 1984), and a compendium of analytical solutions for the two approaches is available (Toride, et al., 1993).

Numerous additional refinements of the multi-site and multi-continuum approaches have been made. For example, van Genuchten and Wagenet (1989) add microbial degradation of solute as a first-order decay process to both a multi-site model and a multi-continuum model, while retaining the ability to identify analytical solutions of the model equations. It is possible to extend either the multi-site or the multi-continuum approach to conditions of transient water flow, but generally this entails sacrificing the possibility of obtaining an analytical solution to the model equations. A large body of work on solute transport and solution bypass exists and has been reviewed recently (e.g., Nielsen, et al., 1986; Brusseau and Rao, 1990; Selim, 1992). For the most part, only work with possible relevance to the short-circuit bypass phenomenon is considered in the following.

### D. Short-Circuit Bypass Models

A wide array of options are available for the hydrologic component of numerical bi-continuum models:

Consider transient water flow, rather than just steady flow.

Consider water flow in more than one continuum, rather than just considering one mobile and one immobile continuum.

Consider exchange of water between continua, rather than just exchange of solute.

With so many options, it might be useful to group multi-continuum models into two categories: ones which are capable of representing short-circuit bypass and ones which represent at most solution bypass. All steady water flow models fall into the latter category, as do all models where solute is the only entity which moves between the two continua. The representation of the short-circuit bypass phenomenon requires that the model represent interactive continua with respect to water, not just with respect to solute.

Skopp et al, 1981 considered steady water flow in two continua, that is, they dropped the assumption that water in one of the continua be immobile. Earlier, Lindstrom and Boersma (1971) presented a "capillary bundle" model, with multiple flowing domains, but with no transfer of solute from one domain to another. More recently, Steenhuis, et al. (1990) describe a multi-continuum model in which one of the continua contains immobile water, but there is flow in each of the other continua. The unsaturated hydraulic conductivity function is used to estimate rates of flow in the various continua. The water retention hypothesis is obeyed, and there is transfer of solute between continua but no transfer of water between the continua. Russo, et al. (1989) analyzed transient water flow and solute transport using a bi-continuum model consisting of an immobile water fraction of constant water content and a mobile water fraction which is time-varying both in water content and in water flux density. There is transfer of solute between the two continua, but no transfer of water. These are examples of models which go beyond the simple mobile-immobile bi-continuum approach, but which cannot represent short-circuit bypass because water is not exchanged between continua.

The models featured in the above discussion of disjoint-volumes models *versus* multi-continuum models, in particular the disjoint-volumes model of Edwards, et al. (1979) and the bi-continuum models of Hoogmoed and Bouma (1980) and of Beven and Germann (1982), are capable of representing short-circuit bypass. In all three of these models the representation of water retention and flow in the "matrix" volume or continuum differs drastically from the representation of water in the "macropore" volume or continuum.

By contrast, in two more recent models (Jarvis, et al., 1991a and 1991b; and Gerke and van Genuchten, 1993) transient flow and transport in each of two continua are modeled using more similar representations of water retention and flow properties in the two continua. A model in which the representations are essentially identical could be called a "symmetrical" bi-continuum model. In these models there is transfer of both solute and water between the continua. In the model of Gerke and van Genuchten (1993) water transfer between the two continua is taken to be proportional to the difference in matric potential between the two domains. In the simulations performed with this model all the surface input water was placed into the macroporosity. Gerke and van Genuchten (1993) found a strong tendency for the two pore spaces to be constantly in equilibrium unless either the inter-domain water transfer coefficient was taken to be quite small or the effective size of the matrix domains was taken to be quite large. Although this model, like other models mentioned here, is designed to represent only macropores which extend all the way to the soil surface, such models could probably be modified to investigate questions raised earlier in this review regarding the initiation of macropore flow at points below the soil surface.

Another model which can represent short-circuit bypass is that of Chen and Wagenet (1992a). Flow in the macropore domain is considered in some detail, with either the Hagen-Poiseuille or the Chezy-Manning approach taken, depending on flow velocity. The transfer of water from the macropore to the micropore domain is modeled as horizontal infiltration, $i = S/(2t^{1/2})$. Chen and Wagenet (1992b) apply linear filter theory to implement the model.

In summary, there is a wide variety of models which are capable of portraying short-circuit bypass. Such models are only as valid as the assumptions they embody. Crucial among these assumptions are the equations and parameter values for the transfer of solutes and water between the continua. It is very difficult to treat these equations and parameter values as "measurable" soil properties, that is to base the choice of how to model these processes and to determine the associated parameters on measurements and experiments which differ from the measurements and experiments used to verify the complete model. For the non-equilibrium chemical sorption phenomenon there has been an effort to develop independent measurements, or at least to use one set of experiments to determine the parameters in the model and an independent set of experiments to validate the complete model. Without such efforts, it is impossible to tell whether the inter-domain part of a complicated model with a number of fitted parameters is merely a clever exercise in curve-fitting or is a realistic representation of the actual inter-continuum flow or transport phenomena. A key element, both for multi-site, multi-continuum and for disjoint-volumes models, is to discover valid representations of the "hidden" properties (i.e., the inter-continuum flow and transport properties).

Before considering models which are not based on the water retention hypothesis, Darcy's Law, and the Richards equation, two more types of models which are based on these foundations should be mentioned: parallel soil column Monte Carlo simulations (e.g., Dagan and Bresler, 1979; Bresler and Dagan, 1981; Amoozegar-Fard, et al., 1982) and multidimensional heterogeneous models. With the availability of greater and greater computing power it has become feasible to implement two- and three-dimensional finite difference and finite element models of a porous medium, with spatial variability of one or more soil properties. Such models, even if based on an equilibrium sorption assumption and a single-continuum assumption at each node or element, can reflect short-circuit bypass, solution bypass, and/or non-equilibrium chemical sorption phenomena if the spatial variability of soil properties in the model results in separated zones where one or another "type" of chemical sorption occurs or if the spatial variability results in there being a variety of pathways through the soil, with water flow differing from one pathway to the next. Such a model can be thought of as a sort of disjoint-volumes model or a multi-continuum model, with the various flow pathways serving as the volumes or the continua. These pathways are tortuous, randomly located, and randomly oriented. While these complications can be regarded as "good" (i.e., soil-like) features of a model, once again such models will be little more than clever curve-fitting exercises if the choice of how to represent the spatial variability cannot be determined independently from the experiments which are used to validate the model. Generally, the models visualized here are ones where soil properties vary randomly from one position to the next, although spatial correlation can be built into the randomness in such a way that zones of similar property values and systematic pathways result. The model of Ju and Kung (1993) is an extreme example of such spatial correlation. In this model the heterogeneity and the consequent flow pathways are due to inclined coarse layers embedded in a relatively uniform sand, where the randomness comes from generating the positions and orientations of the inclined layers randomly.

### E. "Non-Richards" Models

The remainder of the models considered here are not based on the traditional soil water flow analysis assumptions of the Richards equation, Darcy's Law, and the water retention hypothesis and they share with other models which avoid making the water retention hypothesis the possibility of representing the short-circuit bypass phenomenon.

Application of kinematic wave theory to soil water flow (for example, Sisson, et al., 1980; Smith, 1983; Charbeneau, 1984) requires the assumption that the soil water flux density, $J_w$, at any point in the soil be a function only of the soil water content at that point. According to Darcy's Law, $J_w$ at a given point equals the product of the hydraulic conductivity and the hydraulic gradient at the point. Since there is no reason to suspect that the hydraulic gradient varies simply with soil water content, application of kinematic theory to water flow in soil leads to a drastic simplification of Darcy's Law, namely to the assumption that the hydraulic gradient is constant (generally taken to be a unit vector in the vertical direction). Germann and Beven (1985) make the more easily justified assumption that the hydraulic gradient in a macropore is constant, and, in particular, that the downward water flow rate in a macropore, $q$, is related to the macropore water content, $w$, via a power law, $q = bw^{\alpha}$, where $b$ and $\alpha$ are constants.

In the transfer function approach to modeling flow and transport (Jury, 1982; Jury and Roth, 1990) the arrival of solute at a depth $L$ after an amount $I$ of water is applied at the soil surface is characterized by a probability density function, $f_L(I)$, where the probability that solute applied at the soil surface when $I = 0$ will arrive at depth $L$ between $I$ and $I + dI$ is $f_L(I)\,dI$. If it is assumed that the soil transport properties are uniform over depth, then the probability density function at depth $Z$ is simply related to that at the reference depth, $L$, by:

$$f_Z(I) = (L/Z)\, f_L(IL/Z)$$

or, to quote the example given by Jury (1982), "the probability of reaching a depth of 60 cm after 10 cm of water is applied is equal to the probability of reaching 120 cm after 20 cm of water has been applied." The probability density function can be taken to be a log-normal distribution (for example, Jury, et al., 1982; White, et al., 1984), a gamma function (Jury and Roth, 1990), etc. White, et al. (1984) combined the transfer function characterization of solute transport in a soil, taken as one entity, with an analysis of film flow down vertical cracks and associated diffusion into the soil between the cracks to quantify the effective amount of cracking for the soil they studied.

An advantage of transfer function models over more traditional models, in particular over models based on the water retention hypothesis is that, in principle, short-circuit bypass can be displayed by transfer function models. If short-circuit bypass occurs to a significant extent in an experiment which is used to characterize the probability density function, $f_L(I)$, then short-circuit bypass will be reflected in simulations based on the model. Utermann, et al. (1990) and Grochulska and Kladivko (1993) carry this one step further by using two probability density functions, one for the "fast" water flow region and one for the "slow", thereby explicitly incorporating a two-region approach within the transfer function context. In the work of Grochulska and Kladivko (1993) the largest differences in model parameters between replicates were found for the parameters which represented transfer of chemical between the two domains, another indication that the greatest needs for improvement of multi-domain models lie in their representation of transfer between the domains.

Grant, et al. (1991) characterize potential and actual infiltration rate variability using probability density functions which are based on an assumed log-normal distribution of saturated hydraulic conductivity and on assumed relationships between hydraulic conductivity, sorptivity, and potential infiltration rate. They arbitrarily identify macropore flow as "all soil water fluxes greater than the 0.95th quantile" of the infiltration rate probability density function.

Although network modeling and percolation theory (e.g., Ferrand and Celia, 1992; Berkowitz and Balberg, 1993) are not yet developed to the point that they have contributed to an understanding of macropore flow phenomena, they seem to hold promise. Similarly, the application by Knighton and Wagenet (1987) and by Hornberger, et al. (1990) of time-series models to the analysis of solute transport under transient conditions might lead to models which could represent short-circuit bypass and other macropore phenomena.

A wide variety of models of flow and transport in macropores have been developed. A scheme for classifying such models, and for organizing thinking about them is proposed here. This review of "disjoint-volumes" models, "solution bypass" models, "short-circuit bypass" models, and "non-Richards" models is by no means exhaustive; more generalized reviews are available (e.g., Addiscott and Wagenet, 1985; White, 1985). Rather, the emphasis here is on models which exhibit, or which potentially could exhibit the short-circuit bypass phenomenon. No attempt to cover models of flow in fractured aquifers (e.g., Wang, 1991; Nitao and Buscheck, 1991) was made. Specific analyses of shrinkage cracks and other fracture networks in soils are also ignored.

The single most important question running through the widely divergent literature on macropore flow and transport models is how to model the "internal" processes: flow and transport between the macropore and micropore domains. Resolution of this question, and the associated question of how to measure and characterize these internal processes, independent of the experiments which are used to validate the overall output of a flow and transport model, are keys to the advancement of our knowledge of a number of complicated processes in soil.

## VI. Summary

Macropores are a natural component of many soils in the field and can have a significant impact on the way water and chemicals move through those soils. Recent evidence suggests that flow in macropores may occur under even more conditions than previously thought, including through subsurface macropores not connected to the soil surface or to an extensive saturated zone. Experimental evidence and several hypotheses regarding macropore flow behavior are discussed in this review. Several methods for characterizing the physical soil system in which macropore flow occurs are also evaluated. While much has been learned about macropore flow, it is still not possible to quantitatively predict water and chemical movement in macroporous soils with an adequate degree of confidence. Both experimental and modeling studies are needed to better understand and describe these important processes in field soils.

The age-old question: "What is a macropore?" is not answered here. Instead we would propose that more important questions surround the phenomena which we have called bypass flow, solution bypass, and non-equilibrium sorption. It may be more useful to distinguish pores in terms of the processes which can occur in them, rather than in terms of static properties.

## Acknowledgments

This review is a joint contribution of the Ohio Agricultural Research and Development Center, the University of Illinois Agricultural Experiment Station and the Purdue University Agricultural Experiment Station. Cooperation among these institutions was facilitated by membership of three of us (CWB, EJK and ELM) on North Central Regional Committee no. 60. Partial research support for two of us (CWB and ELM) was provided through cooperative agreement no. 90-34214-5117 from the USDA-CSRS Special Water Quality Grants Program. The findings, opinions and recommendations contained herein are those of the authors and not necessarily those of the cooperating institutions or USDA.

## References

Addiscott, T.M. and R.J. Wagenet. 1985. Concepts of solute leaching in soils: A review of modelling approaches. *J. Soil Sci.* 36:411-424.

Adreini, M.S. and T.S. Steenhuis. 1990. Preferential flow paths under conventional and conservation tillage. *Geoderma* 46:85-102.

Ahuja, L.R., D.G. DeCoursey, B.B. Barnes, and K.W. Rojas. 1991. Characteristics and importance of preferential macropore transport studied with the ARS Root Zone Water Quality Model. p. 32-49. *In* T.J. Gish and A. Shirmohammadi (eds.) Preferential flow (Proceedings of National Symposium). Amer. Soc. Agric. Engin., St. Joseph, MI.

Amoozegar-Fard, A., D.R. Nielsen, and A.W. Warrick. 1982. Soil solute concentration distributions for spatially varying pore water velocities and apparent diffusion coefficients. *Soil Sci. Soc. Am. J.* 46:3-9.

Anderson, C.J. and J. Bouma. 1973. Relationships between saturated hydraulic conductivity and morphometric data of an argillic horizon. *Soil Sci. Soc. Am. Proc.* 37:408-413.

Anderson, C.J. and J. Bouma. 1977a. Water movement through pedal soils. 1. Saturated flow. *Soil Sci. Soc. Am. J.* 41:413-418.

Anderson, C.J. and J. Bouma. 1977b. Water movement through pedal soil. 2. Unsaturated flow. *Soil Sci. Soc. Am. J.* 41:419-423.

Ball, B.C. 1981a. Modelling of soil pores as tubes using gas permeabilities, gas diffusivities and water release. *J. Soil Sci.* 32:465-481.

Ball, B.C. 1981b. Pore characteristics of soils from two cultivation experiments as shown by gas diffusivities and permeabilities and air-filled porosities. *J. Soil Sci.* 32:483-498.

Ball, B.C. and M.F. O'Sullivan. 1987. Cultivation and nitrogen requirements for drilled and broadcast winter barley on a surface water gley (Gleysol). *Soil Tillage Res.* 9:103-122.

Ball, B.C., M.F. O'Sullivan, and R. Hunter. 1988. Gas diffusion, fluid flow and derived pore continuity indices in relation to vehicle traffic and tillage. *J. Soil Sci.* 39:327-339.

Bathke, G.R. and D.K. Cassel. 1991. Anisotropic variation of profile characteristics and saturated hydraulic conductivity in an ultisol landscape. *Soil Sci. Soc. Am. J.* 55:333-339.

Baveye, P. and G. Sposito. 1985. Macroscopic balance equations in soils and aquifers: The case of space- and time-dependent instrumental response. *Water Resour. Res.* 21:1116-1120.

Bear, J. 1972. *Dynamics of Fluids in Porous Media.* American Elsevier, or 1988 republication by Dover Publications, Inc., New York.

Berkowitz, B. and I. Balberg. 1993. Percolation theory and its application to groundwater hydrology. Water Resour. Res. 29:775-795.

Beven, K. 1981. Micro-, meso-, macroporosity and channeling flow phenomena in soils. *Soil Sci. Soc. Am. J.* 45:1245.

Beven, K. and P. Germann. 1981. Water flow in macropores. 2. A combined flow model. *J. Soil Sci.* 32:15-29.

Beven, K. and P. Germann. 1982. Macropores and water flow in soils. *Water Resour. Res.* 18:1311-1325.

Biggar, J.W. and D.R. Nielsen. 1967. Miscible displacement and leaching phenomenon. p. 254-274. *In* R. M. Hagen et al. (eds.), Irrigation of agricultural lands. Am. Soc. Agron. Monogr. No 11. American Society of Agronomy, Madison, WI.

Blackwell, P.S., A.J. Ringrose-Voase, N.S. Jayawardane, K.A. Olsson, D.C. McKenzie, and W.K. Mason. 1990. The use of air-filled porosity and intrinsic permeability to air to characterize structure of macropore space and saturated hydraulic conductivity of clay soils. *J. Soil Sci.* 42:215-228.

Boast, C.W. 1973. Modeling the movement of chemicals in soils by water. *Soil Sci.* 115:224-230.

Boddy, P.L. and J.L. Baker. 1990. Conservation tillage effects on nitrate and atrazine leaching. Amer. Soc. Agric. Engin. Paper No. 90-2505.

Booltink, H.W.G., and J. Bouma. 1991. Physical and morphological characterization of bypass flow a in well-structured clay soil. *Soil Sci. Soc. Am. J.* 55:1249-1254.

Booltink, H.W.G., J. Bouma, and D. Gimenez. 1991. Suction crust infiltrometer for measuring hydraulic conductivities of unsaturated soils near saturation. *Soil Sci. Soc. Am. J.* 55:566-568.

Bottcher, A.B., E.J. Monke, and L.F. Huggins. 1981. Nutrient and sediment loadings from a subsurface drainage system. *Trans. ASAE* 24:1221-1226.

Bouma, J. 1981. Soil morphology and preferential flow along macropores. *Agric. Water Manage.* 3:235-250.

Bouma, J. 1982. Measuring the hydraulic conductivity of soil horizons with continuous macropores. *Soil Sci. Soc. Am. J.* 46:438-441.

Bouma, J. 1991. Influence of soil macroporosity on environmental quality. *Adv. Agron.* 46:1-37.

Bouma, J. 1992. Effect of soil structure, tillage, and aggregation upon soil hydraulic properties. p. 1-36. *In* R.J. Wagenet et al. (eds.) Advances in Soil Science: Interacting Processes in Soil Science. Lewis Pub., Boca Raton, FL.

Bouma J. and L.W. Dekker. 1978. A case study on infiltration into a dry clay soil. I. Morphological observations. *Geoderma* 20:27-40.

Bouma, J. and J.G. Denning. 1974. A comparison of hydraulic conductivities calculated with morphometric and physical methods. *Soil Sci. Soc. Am. Proc.* 38:124-127.

Bouma, J., A. Jongerius, O. Boersma, A. Jager, and D. Schoonderbeek. 1977. The function of different types of macropores during saturated flow through four swelling soil horizons. *Soil Sci. Soc. Am. J.* 41:945-950.

Bouma, J., A. Jongerius, and D. Schoonderbeek. 1979. Calculation of saturated hydraulic conductivity of some pedal clay soils using morphometric data. *Soil Sci. Soc. Am. J.* 43:261-264.

Bouma, J. and J.H.L. Wösten. 1979. Flow patterns during extended saturated flow in two undisturbed swelling clay soils with different macrostructures. *Soil Sci. Soc. Am. J.* 43:16-22.

Brakensiek, D.L., W.J. Rawls, S.D. Logsdon, and W.M. Edwards. 1992. Fractal description of macroporosity. *Soil Sci. Soc. Am. J.* 56:1721-1723.

Bresler, E. and G. Dagan. 1981. Convective and pore scale dispersive solute transport in unsaturated heterogeneous fields. *Water Resour. Res.* 17:1683-1693.

Brusseau, M.L. and P.S.C. Rao. 1990. Modeling solute transport in structured soils: A review. *Geoderma* 46:169-192.

Cameron, D.R. and A. Klute. 1977. Convective-dispersive solute transport with a combined equilibrium and kinetic adsorption model. *Water Resour. Res.* 13:183-188.

Charbeneau, R.J. 1984. Kinematic models for soil moisture and solute transport. *Water Resour. Res.* 20:699-706.

Chen, C. and R.J. Wagenet. 1992a. Simulation of water and chemicals in macropore soils. Part I. Representation of the equivalent macropore influence and its effect on soil water flow. *J. Hydrol.* 130:105-126.

Chen, C. and R.J. Wagenet. 1992b. Simulation of water and chemicals in macropore soils. Part 2. Application of linear filter theory. *J. Hydrol.* 130:127-149.

Childs, E.C., N. Collis-George, and J.W. Holmes. 1957. Permeability measurements in the field as an assessment of anisotropy and structure development. *J. Soil Sci.* 8:27-41.

Clothier, B.E. and K.R.J. Smettem. 1990. Combining laboratory and field measurements to define the hydraulic properties of soil. *Soil Sci. Soc. Am. J.* 54:299-304.

Coats, K.H. and B.D. Smith. 1964. Dead-end pore volume and dispersion in porous media. *Soc. Petrol. Engr. J.* 4:73-84.

Collis-George, N. and R. Lal. 1970. Infiltration into columns of a swelling soil as studied by high speed photography. *Aust. J. Soil Res.* 8:195-207.

Dagan, G. and E. Bresler. 1979. Solute dispersion in unsaturated heterogeneous soil at field scale: I. Theory. *Soil Sci. Soc. Am. J.* 43:461-467.

de Marsily, G. 1986. Quantitative hydrology: groundwater hydrology for engineers. Academic Press. New York.

Dick, W.A., R.J. Roseberg, E.L. McCoy, W.M. Edwards, and F. Haghiri. 1989. Surface hydrologic response of soils to no-tillage. *Soil. Sci. Soc. Am. J.* 53:1520-1526.

Dixon, D.M. and A.E. Peterson. 1971. Water infiltration control: A channel system concept. *Soil Sci. Soc. Am. Proc.* 35:968-973.

Douglas, J.T., M.J. Goss, and D. Hill. 1980. Measurement of pore characteristics in a clay soil under ploughing and direct drilling, including use of a radioactive tracer ($^{14}C$) technique. *Soil Tillage Res.* 1:11-18.

Dunn, G.H. and R.E. Phillips. 1991a. Macroporosity of a well-drained soil under no-till and conventional tillage. *Soil Sci. Soc. Am. J.* 55:817-823.

Dunn, G.H. and R.E. Phillips. 1991b. Equivalent diameter of simulated macropore systems during saturated flow. *Soil Sci. Soc. Am. J.* 55:1244-1248.

Edwards, W.M., L.D. Norton, and C.E. Redmond. 1988. Characterizing macropores that affect infiltration into non-tilled soil. *Soil Sci. Soc. Am. J.* 52:483-487.

Edwards, W.M., M.J. Shipitalo, W.A. Dick, and L.B. Owens. 1992. Rainfall intensity affects transport of water and chemicals through macropores in no-till soil. *Soil Sci. Soc. Am. J.* 56:52-58.

Edwards, W.M., M.J. Shipitalo, L.B. Owens, and L.D. Norton. 1989. Water and nitrate movement in earthworm burrows within long-term no-till cornfields. *J. Soil Water Conser.* 44:240-243.

Edwards, W.M., M.J. Shipitalo, L.B. Owens, and L.D. Norton. 1990. Effect of *Lumbricus terrestris* L. burrows on hydrology of continuous no-till corn fields. *Geoderma.* 46:73-84.

Edwards, W.M., R.R. van der Ploeg, and W. Ehlers. 1979. A numerical study of the effects of noncapillary-sized pores upon infiltration. *Soil Sci. Soc. Am. J.* 43:851-856.

Ehlers, W. 1975. Observations on earthworm channels and infiltration on tilled and untilled loess soil. *Soil Sci.* 119:242-249.

Ela, S.D., S.C. Gupta, and W.J. Rawls. 1992. Macropore and surface seal interactions affecting water infiltration into soil. *Soil Sci. Soc. Am. J.* 56:714-721.

Ferrand, L.A. and M.A. Celia. 1992. The effect of heterogeneity on the drainage capillary pressure-saturation relation. *Water Resour. Res.* 28:859-870.

Gantzer, G.J. and G.R. Blake. 1978. Physical characteristics of Le Sueur clay loam soil following no-till and conventional tillage. *Agron. J.* 70:853-857.

Gardner, W.R. 1965. Movement of nitrogen in soil. p. 555-572. *In* W.V. Bartholomew and F.E. Clark (eds.), Soil nitrogen. Am. Soc. Agron. Monogr. No 10. American Society of Agronomy, Madison, WI.

Gerke, H.H. and M.Th. van Genuchten. 1993. A dual-porosity model for simulating the preferential movement of water and solutes in structured porous media. *Water Resources*. 29:305-319.

Germann, P.F. (ed.). 1988. Rapid and far-reaching hydrologic processes in the vadose zone. Special Issue, *J. Contam. Hydrol.* 3:115-375.

Germann, P. F. 1990. Preferential flow and the generation of runoff. 1. Boundary layer flow theory. *Water Resour. Res.* 26:3055-3063.

Germann, P.F. and K. Beven. 1985. Kinematic wave approximation to infiltration into soils with sorbing macropores. *Water Resour. Res.* 21:990-996.

Germann, P.F., W.M. Edwards, and L.B. Owens. 1984. Profiles of bromide and increased soil moisture after infiltration into soils with macropores. *Soil Sci. Soc. Am. J.* 48:237-244.

Gillham, R.W. 1984. The capillary fringe and its effect on water table response. *J. Hydrol.* 67:307-324.

Gish, T.J. and A. Shirmohammadi (eds.). 1991. Preferential flow (Proceedings of a National Symposium). Amer. Soc. Agric. Engin., St. Joseph, MI. 408 pp.

Granovsky, A.V., E.L. McCoy, W.A. Dick, M.J. Shipitalo, and W.M. Edwards. 1993. Water and chemical transport through long term no-till and plowed soils. *Soil Sci. Soc. Am. J.* (in press).

Grant, S. A., J. D. Jabro, D. D. Fritton, and D. E. Baker. 1991. A stochastic model of infiltration which simulates "macropore" soil water flow. *Water Resour. Res.* 27:1439-1446.

Grochulska, J. and E.J. Kladivko. 1993. A two-region model of preferential flow of chemicals using a transfer function approach. *J. Environ. Qual.* (in press).

Groenevelt, P.H., B.D. Kay, and C.D. Grant. 1984. Physical assessment of a soil with respect to rooting potential. *Geoderma*. 34:101-114.

Hall, J.K., M.R. Murray, and N.L. Hartwig. 1989. Herbicide leaching and distribution in tilled and untilled soil. *J. Environ. Qual.* 18:439-445.

Hamblin, A.P. 1985. The influence of soil structure on water movement, crop growth and water uptake. *Adv. Agron.* 38:95-158.

Hamblin, A.P. and D. Tennant. 1981. The influence of tillage on soil water behavior. *Soil Sci.* 132:233-239.

Hatano, R., N. Kawamura, J. Ikeda, and T. Sakuma. 1992. Evaluation of the effect of morphological features of flow paths on solute transport by using fractal dimensions of methylene blue staining pattern. *Geoderma.* 53:31-44.

Heard, J.R., E.J. Kladivko, and J.V. Mannering. 1988. Soil macroporosity, hydraulic conductivity and air permeability of silty soils under long-term conservation tillage in Indiana. *Soil Tillage Res.* 11:1-18.

Hoogmoed, W.B. and J. Bouma. 1980. A simulation model for predicting infiltration into cracked clay soil. *Soil Sci. Soc. Am. J.* 44:458-461.

Hornberger, G. M., K. J. Beven, and P. F. Germann. 1990. Inferences about solute transport in macroporous forest soils from time series models. *Geoderma* 46:249-262.

Isensee, A.R., R.G. Nash, and C.S. Helling. 1990. Effect of conventional vs. no-tillage on pesticide leaching to shallow groundwater. *J. Environ Qual.* 19:434-440.

Jardine, P.M., G.V. Wilson, and R.J. Luxmoore. 1990. Unsaturated solute transport through a forest soil during rain storm events. *Geoderma.* 46:103-118.

Jarvis, N.J., P.-E. Jansson, P.E. Dik, and I. Messing. 1991a. Modelling water and solute transport in macroporous soil. I. Model description and sensitivity analysis. *J. Soil Sci.* 42:59-70.

Jarvis, N.J., L. Bergstrom, and P.E. Dik. 1991b. Modelling water and solute transport in macroporous soil. II. Chloride breakthrough under non-steady flow. *J. Soil Sci.* 42:71-81.

Jarvis, N.J., P.B. Leeds-Harrison, and J.M. Dosser. 1987. The use of tension infiltrometers to assess routes and rates of infiltration in a clay soil. *J. Soil Sci.* 38:633-640.

Ju, S.-H. and K.-J.S. Kung. 1993. Simulating funnel-type preferential flow and overall flow properties induced by multiple soil layers. *J. Environ. Qual.* (in press).

Jury, W.A. 1982. Simulation of solute transport using a transfer function model. *Water Resour. Res.* 18:363-368.

Jury, W.A. and K. Roth. 1990. Transfer functions and solute movement through soil. Theory and applications. Birkhäuser Verlag, Basel.

Jury, W.A., L.H. Stolzy, and P. Shouse. 1982. A field test of the transfer function model for predicting solute transport. *Water Resour. Res.* 18:369-375.

Kanwar, R.S., J.C. Baker, and J.M. Laflen. 1985. Nitrate movement through soil profiles in relation to tillage system and fertilizer application method. *Trans. ASAE* 28:1802-1807.

Kissel, D.E., J.T. Ritchie, and E. Burnett. 1973. Chloride movement in undisturbed swelling clay soil. *Soil Sci. Soc. Am. Proc.* 37:21-24.

Kladivko, E.J., A.D. Mackay, and J.M. Bradford. 1986. Earthworms as a factor in the reduction of soil crusting. *Soil Sci. Sci. Am. J.* 50:191-196.

Kladivko, E.J., G.E. Van Scoyoc, E.J. Monke, K.M. Oates, and W. Pask. 1991. Pesticide and nutrient movement into subsurface tile drains on a silt loam soil in Indiana. *J. Environ. Qual.* 20:264-270.

Knight, J.H., J.R. Philip, and R.T. Waechter. 1989. The seepage exclusion problem for spherical cavities. *Water Resour. Res.* 25:29-37.

Knighton, R.E. and R.J. Wagenet. 1987. Simulation of solute transport using a continuous time markov process. 2. Application to transient field conditions. *Water Resour. Res.* 23:1917-1925.

Koistra, M.J., R. Miedema, J.H.M. Wösten, J. Versuis, and J. Bouma. 1987. The effect of subsoil cracking on moisture deficits of Pleistocene and Holocene fluvial clay soils in the Netherlands. *J. Soil Sci.* 38:553-563.

Kosmas, C., N. Moustakas, C. Kallianou, and N. Yassoglou. 1991. Cracking patterns, bypass flow and nitrate leaching in Greek irrigated soils. *Geoderma* 49:139-152.

Kung, K.-J.S. 1990. Preferential flow in a sandy vadose zone: 2. Mechanism and implications. *Geoderma* 46:59-71.

Kutílek, M. and V. Novák. 1976. The influence of soil cracks upon infiltration and ponding time. *In* M. Kutílek and J. Šutor (eds.), Proceedings of the Bratislava Symposium on water in heavy soils I:126-134.

Lauren, J.G., R.J. Wagenet, J. Bouma, and J.H.M. Wösten. 1988. Variability of saturated hydraulic conductivity in a glossaquic hapludalf with macropores. *Soil Sci.* 145:20-28.

Lindstrom, F.T. and L. Boersma. 1971. A theory on the mass transport of previously distributed chemicals in a water saturated sorbing porous medium. *Soil Sci.* 111:192-199.

Logsdon, S.D., R.R. Allmaras, L. Wu, J.B. Swan, and G.W. Randall. 1990. Macroporosity and its relation to saturated hydraulic conductivity under different tillage practices. *Soil Sci. Soc. Am. J.* 54:1096-1101.

Logsdon, S.D., E.L. McCoy, R.R. Allmaras, and D.R. Linden. 1993. Macropore characterization by indirect methods. *Soil Sci.* 155:316-324.

Luxmoore, R.J. 1981. Micro-, meso-, and macroporosity of soil. *Soil Sci. Soc. Am. J.* 45:671-672.

Luxmoore, R.J., P.M. Jardine, G.V. Wilson, J.R. Jones, and L.W. Zelazny. 1990. Physical and chemical controls of preferred path flow through a forested hillslope. *Geoderma* 46:139-154.

McCoy, E.L. 1989. Water characteristics of soils containing macropores. p. 246. *In* W. E. Larson et al. (eds.) Mechanics and related processes in structured agriculture soils. Kluwer Academic Pub., Dordrecht, Netherlands.

Moran, C.J. and A.B. McBratney. 1992. Acquisition and analysis of three-component digital images of soil pore structure. I. Method. *J. Soil Sci.* 43:541-549.

Nielsen, D.R., M.Th. van Genuchten, and J.W. Biggar. 1986. Water flow and solute transport processes in the unsaturated zone. *Water Resour. Res.* 22:89S-107S.

Nitao, J.J. and T.A. Buscheck. 1991. Infiltration of a liquid front in an unsaturated fractured porous medium. *Water Resour. Res.* 27:2099-2112.

Nkedi-Kizza, P., J.W. Biggar, H.M. Selim, M.Th. van Genuchten, P.J. Wierenga, J.M. Davidson, and D.R. Nielsen. 1984. On the equivalence of two conceptual models for describing ion exchange during transport through an aggregated oxisol. *Water Resour. Res.* 20:1123-1130.

Parker, J.C. and M.Th. van Genuchten. 1984. Flux-averaged and volume-averaged concentrations in continuum approaches to solute transport. *Water Resour. Res.* 20:866-872.

Parlange, J.Y., T.S. Steenhuis, R.J. Glass, T.L. Richard, N.B. Pickering, W.J. Waltman, N.O. Bailey, M.S. Andreini, and J.A. Throop. 1988. The flow of pesticides through preferential paths in soils. *NY Food Life Sci. Q.* 18:20-23

Passioura, J. B. 1971. Hydrodynamic dispersion in aggregated media. 1. Theory. *Soil Sci.* 111:339-344.

Peck, A. J. 1983. Field variability of soil physical properties. *Advances in Irrigation* 2:189-221.

Perroux, K.M. and I. White. 1988. Designs for disc permeameters. *Soil Sci. Soc. Am. J.* 52:1205-1215.

Philip, J.R. 1968. The theory of absorption in aggregated media. *Aust. J. of Soil Res.* 6:1-19.

Philip, J. R. 1989. Asymptotic solutions of the seepage exclusion problem for elliptic-cylindrical, spheroidal, and strip- and disc-shaped cavities. *Water Resour. Res.* 25:1531-1540.

Philip, J.R., J.H. Knight, and R.T. Waechter. 1989. Unsaturated seepage and subterranean holes: Conspectus, and exclusion problem for circular cylindrical cavities. *Water Resour. Res.* 25:16-28.

Phillips, R.E. and V.L. Quisenberry. 1992. Macropore flow variability among three columns of undisturbed Maury soil. p. 226. *In* Agronomy Abstracts. ASA, Madison, WI.

Phillips, R.E., V.L. Quisenberry, J.M. Zeleznik, and G.H. Dunn. 1989. Mechanism of water entry into simulated macropores. *Soil Sci. Soc. Am. J.* 53:1629-1635.

Quisenberry, V.L. and R.E. Phillips. 1976. Percolation of simulated rainfall under field conditions. *Soil Sci. Soc. Am. J.* 40:484-489.

Quisenberry, V.L. and R.E. Phillips. 1991. Macropore flow of water in a structured soil as influenced by application rate. p. 229. *In* Agronomy Abstracts. ASA, Madison, WI.

Radulovich, R. and P. Sollins. 1987. Improved performance of zero-tension lysimeters. *Soil Sci. Soc. Am. J.* 51:1386-1388.

Radulovich, R., P. Sollins, P. Baveye, and E. Solorzano. 1992. Bypass water flow through unsaturated microaggregated tropical soils. *Soil Sci. Soc. Am. J.* 56:721-726.

Rao, P.S.C., D.E. Rolston, R.E. Jessup, and J.M. Davidson. 1980. Solute transport in aggregated porous media: Theoretical and experimental evaluation. *Soil Sci. Soc. Am. J.* 44:1139-1146.

Richard, T.L. and T.S. Steenhuis. 1988. Tile drain sampling of preferential flow on a field scale. *J. Contam. Hydrol.* 3:307-325.

Ritchie, J.T., D.E. Kissel, and E. Burnett. 1972. Water movement in undisturbed swelling clay soil. *Soil Sci. Soc. Am. Proc.* 36:874-897.

Roseberg, R.J. and E.L. McCoy. 1990. Measurement of soil macropore air permeability. *Soil Sci. Soc. Am. J.* 54:969-974.

Roseberg, R.J. and E.L. McCoy. 1992. Tillage- and traffic-induced changes in macroporosity and macropore continuity: Air permeability assessment. *Soil Sci. Soc. Am. J.* 56:1261-1267.

Ross, P.J. and K.R.J. Smettem. 1993. Describing soil hydraulic properties with sums of simple functions. *Soil Sci. Soc. Am. J.* 57:26-29.

Russell, A.E. and J.J. Ewel. 1985. Leaching from a tropical andept during big storms: A comparison of three methods. *Soil Sci.* 139:181-189.

Russo, D., W.A. Jury, and G.L. Butters. 1989. Numerical analysis of solute transport during transient irrigation. 2. The effect of immobile water. *Water Resour. Res.* 25:2119-2127.

Sadeghi, A.M. and A.R. Isensee. 1992. Effect of tillage systems and rainfall patterns on atrzine distribution in soil. *J. Environ. Qual.* 21:464-469.

Schiegg, H.O. 1979. Verdrängungs-Simulation dreier nicht mischbarer Fluide in poröser Matrix. Mitteilungen Nr.40 der VAW, Eidg. Technische Hochschule Zürich.

Scotter, D. R. 1978. Preferential solute movement through larger soil voids. I. Some computations using simple theory. *Aust. J. Soil Res.* 16:257-267.

Selim, H. M. 1992. Modeling the transport and retention of inorganics in soils. *Advances in Agronomy* 47:331-384.

Selim, H.M., J.M. Davidson, and R.S. Mansell. 1976. Evaluation of a two-site adsorption-desorption model for describing solute transport in soils. Proc., Summer Computer Simul. Conf. 444-448, Am. Inst. of Chem. Engr., Washington, DC.

Seyfried, M.S. and P.S.C. Rao. 1987. Solute transport in undisturbed columns of an aggregated tropical soil: Preferential flow effects. *Soil Sci. Soc. Am. J.* 51:1434-1444.

Shipitalo, M.J. and R. Protz. 1987. Comparison of morphology and porosity of a soil under conventional and zero tillage. *Can. J. Soil Sci.* 67:445-45

Shipitalo, M.J., W.M. Edwards, W.A. Dick, and L.B. Owens. 1990. Initial storm effects on macropore transport of surface-applied chemicals in no-till soil. *Soil Sci. Soc. Am. J.* 54:1530-1536.

Singh, P., R.S. Kanwar, and M.L. Thompson. 1991a. Measurement and characterization of macropores by using AUTOCAD and automatic image analysis. *J. Environ. Qual.* 20:289-294.

Singh, P., R.S. Kanwar, and M.L. Thompson. 1991b. Macropore characterization for two tillage systems using resin-impregnation technique. *Soil Sci. Soc. Am. J.* 55:1674-1679.

Sisson, J.B., A.H. Ferguson, and M.Th. van Genuchten. 1980. Simple method for predicting drainage from field plots. *Soil Sci. Soc. Am. J.* 44:1147-1152.

Skopp, J., W.R. Gardner, and E.J. Tyler. 1981. Solute movement in structured soils: Two-region model with small interaction. *Soil Sci. Soc. Am. J.* 45:837-842.

Skopp, J. and A.W. Warrick. 1974. A two-phase model for the miscible displacement of reactive solutes in soils. *Soil Sci. Soc. Am. Proc.* 38:545-550.

Smettem, K.R.J. 1987. Characterization of water entry into a soil with a contrasting textural class: Spatial variability of infiltration parameters and influence of macroporosity. *Soil Sci.* 144:167-174.

Smettem, K.R.J. and N. Collis-George. 1985. The influence of cylindrical macropores on steady-state infiltration in a soil under pasture. *J. Hydrol.* 79:107-114.

Smettem, K.R.J. and S.T. Trudgill. 1983. An evaluation of some fluorescent and non-fluorescent dyes in the identification of water transmission routes in soils. *J. Soil Sci.* 34:45-56.

Smith, R. E. 1983. Approximate soil water movement by kinematic characteristics. *Soil Sci. Soc. Am. J.* 47:3-8.

Sollins, P. and R. Radulovich. 1988. Effects of soil physical structure on solute transport in a weathered tropical soil. *Soil Sci. Soc. Am. J.* 52:1168-1173.

Steenhuis, T.S. and R.E. Muck. 1988. Preferred movement of nonadsorbed chemicals on wet, shallow, sloping soils. *J. Environ. Qual.* 17:376-384.

Steenhuis, T.S., J.-Y. Parlange, and M.S. Andreini. 1990. A numerical model for preferential solute movement in structured soils. *Geoderma* 46:193-208.

Stehouwer, R.C., W.A. Dick, and S.J. Traina. 1993. Characteristics of earthworm burrow lining affecting atrazine sorption. *J. Environ. Qual.* 22:181-185.

Stehouwer, R.C., W.A. Dick, and S.J. Traina. 1994. Sorption and retention of herbicides in earthworm (*Lumbricus terrestris* L.) and artificial burrows. *J. Environ. Qual.* (in press).

Thomas, G. W., R. E. Phillips, and V. L. Quisenberry. 1978. Characterization of water displacement in soils using simple chromatographic theory. *J. Soil Sci.* 29:32-37.

Toride, N., F. Leij, and M.Th. van Genuchten. 1993. A comprehensive set of analytical solutions for nonequilibrium solute transport with first-order decay and zero-order production. *Water Resour. Res.* 29:2167-2182.

Towner, G. D. 1989. The application of classical physics transport theory to water movement in soil: development and deficiencies. *J. Soil Sci.* 40:251-260.

Trojan, M.D. and D.R. Linden. 1992. Microrelief and rainfall effects on water and solute movement in earthworm burrows. *Soil Sci. Soc. Am. J.* 56:727-733.

Turner, R.R. and K.F. Steele. 1988. Cadmium and manganese sorption by soil macropore linings and fillings. *Soil Sci.* 145:79-86.

Tyler, S.W. and S.W. Wheatcraft. 1990. The consequences of fractal scaling in heterogeneous soils and porous media. p. 109-122. *In* D. Hillel and D.E. Elrick (eds.) Scaling in soil physics: principles and applications. SSSA Special Publication No. 25. SSSA Madison, WI.

Utermann, J., E.J. Kladivko, and W.A. Jury. 1990. Evaluating pesticide migration in tile-drained soils with a transfer function model. *J. Environ. Qual.* 19:707-714.

Vachaud, G., M. Vachaud, and M. Wakil. 1972. A study of the uniqueness of the soil moisture characteristic during desorption by vertical drainage. *Soil Sci. Soc. Am. Proc.* 36:531-532.

van Genuchten, M. Th. 1980. A closed-form equation for predicting the hydraulic conductivity of unsaturated soils. *Soil Sci. Soc. Am. J.* 44:892-898.

van Genuchten, M.Th., D.E. Ralston, and P.F. Germann (eds.). 1990. Transport of water and solutes in macropores. *Geoderma* 46.

van Genuchten, M.Th. and R.J. Wagenet. 1989. Two-site/two-region models for pesticide transport and degradation: Theoretical development and analytical solutions. *Soil Sci. Soc. Am. J.* 53:1303-1310.

van Genuchten, M.Th. and P.J. Wierenga. 1976. Mass transfer studies in sorbing porous media. I. Analytical solutions. *Soil Sci. Soc. Am. J.* 40:473-480.

van Ommen, H.C., M.Th. van Genuchten, W.H. van der Molen, R. Kijksma, and J. Hulshoj. 1989. Experimental and theoretical analysis of solute transport from a diffuse source of pollution. *J. Hydrol.* 105:225-251.

Vepraskas, M.J., A.G. Jongmans, M.T. Hoover, and J. Bouma. 1991. Hydraulic conductivity of saprolite as determined by channels and porous groundmass. *Soil Sci. Soc. Am. J.* 55:932-938.

Wagenet, R. J. 1985. Measurement and interpretation of spatially variable leaching processes. *In* D. R. Nielsen and J. Bouma (eds.), *Soil Spatial Variability*, Pudoc, Wageningen pp. 209-235.

Wang, J.S.Y. 1991. Flow and transport in fractured rocks. *Reviews of Geophys.* 29(Suppl):254-262.

Watson, K.W. and R.J. Luxmoore. 1986. Estimating macroporosity in a forest watershed by use of a tension infiltrometer. *Soil Sci. Soc. Am. J.* 50:578-582.

White, I. and K.M. Perroux. 1987. Use of sorptivity to determine field soil hydraulic properties. *Soil Sci. Soc. Am. J.* 51:1093-1101.

White, I. and K.M. Perroux. 1989. Estimation of unsaturated hydraulic conductivity from field sorptivity measurements. *Soil Sci. Soc. Am. J.* 53:324-329.

White, I. and M.J. Sully. 1987. Macroscopic and microscopic capillary length and time scales from field infiltration. *Water Resour. Res.* 23:1514-1522.

White, R.E. 1985. The influence of macropores on the transport of dissolved and suspended matter through soil. *Adv. Soil Sci.* 3:95-120.

White, R.E., J.S. Dyson, Z. Gerstl, and B. Yaron. 1986. Leaching of herbicides through undisturbed cores of a structured clay soil. *Soil Sci. Soc. Am. J.* 50:277-283.

White, R.E., G.W. Thomas, and M.S. Smith. 1984. Modelling water flow through undisturbed soil cores using a transfer function model derived from $^{3}$HOH and Cl transport. *J. Soil Sci.* 35:159-168.

Wilson, G.V. and R.J. Luxmoore. 1988. Infiltration, macroporosity, and mesoporosity distributions on two forested watersheds. *Soil Sci. Soc. Am. J.* 52:329-335.

Zachmann, J.E. and D.R. Linden. 1989. Earthworm effects on corn residue breakdown and infiltration. *Soil Sci. Soc. Am. J.* 53:1846-1849.

# Water Quality Models for Developing Soil Management Practices

J.R. Williams, J.G. Arnold, C.A. Jones,
V.W. Benson, and R.H. Griggs

I. Introduction . . . . . . . . . . . . . . . . . . . . . . . . . . . 349
II. Simulation Models . . . . . . . . . . . . . . . . . . . . . . . 350
A. Field Scale . . . . . . . . . . . . . . . . . . . . . . . . . . 351
B. Watershed Scale Event Models . . . . . . . . . . . . . . . . . . 354
C. Watershed Scale, Continuous Time Models . . . . . . . . . . . 355
III. Example Models . . . . . . . . . . . . . . . . . . . . . . . . 357
A. EPIC . . . . . . . . . . . . . . . . . . . . . . . . . . . . 357
B. SWRRB . . . . . . . . . . . . . . . . . . . . . . . . . . . . 367
IV. Water Quality Model Applications . . . . . . . . . . . . . . . . 373
A. EPIC . . . . . . . . . . . . . . . . . . . . . . . . . . . . 373
B. SWRRB . . . . . . . . . . . . . . . . . . . . . . . . . . . . 376
References . . . . . . . . . . . . . . . . . . . . . . . . . . . . 378

## I. Introduction

Several water quality models are available for use in assessing the effects of agricultural management on the environment. Since these models are being used to solve a variety of water quality problems, they vary considerably in structure and complexity. For example, simple screening models may be adequate and appropriate for identifying potential pollutant sources. However, more comprehensive models are needed in comparing agricultural management effects on chemical transport by runoff and sediment. Model requirements may also vary depending upon temporal and spatial scales, cost, and risk associated with proposed projects.

Some of the most widely used water quality models, particularly those most useful in agricultural management are described briefly. Two of the models, the Erosion-Productivity Impact Calculator (EPIC) and the Simulator for Water Resources in Rural Basins (SWRRB), are presented in more detail to serve as examples of field and watershed scale water quality models. These models were selected because they feature convenient and comprehensive agricultural and soil management components. For example, EPIC is useful in solving management problems involving crop varieties and rotations, tillage, furrow diking, irrigation, drainage, fertilization, pest control, weather variation, atmospheric $CO_2$ concentration, erosion (wind and water), water quality (nutrients and pesticides), manure handling, crop residue management, liming, and grazing. The model operates on a daily time step and is capable of simulating hundreds of years if necessary. It is also useful in solving short term (within growing season) management problems operating in a real time mode. The SWRRB model was designed for solving watershed scale problems like water supply and quality (nutrients and pesticides), pond and reservoir design, groundwater flow contributions, irrigation water transfer, and stream channel routing of sediment and agrichemicals. SWRRB also operates on a daily time step and allows watershed subdivision. Subdivisions are made to account for spatial variability of soils, land use, weather, and topography. This gives SWRRB the capability to estimate off-site impacts including channel and reservoir deposition and total water supplies. Example applications of EPIC and SWRRB to water quality problems are also described.

## II. Simulation Models

Mathematical models are among the best tools available for analyzing water quality issues. They can project the consequences of alternative management, planning, or policy-level activities and substantially reduce the cost of managing water resources (Office of Technology Assessment, 1982). Barfield et al. (1991) agree that the reason for developing models is to have a tool for analyzing environmental impacts and designing water quality and hydrologic systems. They also note that models are useful for absolute predictions, relative response, developing designs, organizing concepts, improving the understanding of the system, and dealing with parameter uncertainties.

There is also an increasing concern over the off-site impacts of nonpoint source pollution including lake water quality and instream nutrient and toxic concentrations. To estimate off-site loadings, the ability to simulate large basins with heterogeneous soils, land use, and topography is required. Several basin scale water quality models have recently been developed that simulate spatial variability within a watershed. However, these models have been limited by several factors including: 1) computer speed; 2) computer memory; and 3) availability of inputs. These limitations have produced models falling into one of the following three categories:

- Continuous time models with natural subwatershed boundaries that require considerable lumping of subwatershed inputs;
- Single event models subdivided into grid cells that allow more rapid detail;

- Continuous time, spatial models that are so complex that obtaining required inputs inhibits their general use.

Several models were chosen as examples to illustrate the variety of model configurations. The models are placed into three categories for discussion--field scale, watershed scale event, and watershed scale continuous.

## A. Field Scale

### 1. CREAMS

The Chemicals, Runoff, and Erosion from Agricultural Management Systems (CREAMS) model (Knisel, 1980) is a physically based, daily simulation model that estimates runoff, erosion/sediment transport, plant nutrient, and pesticide yield from field-sized areas. The hydrologic component consists of two options. When only daily rainfall data are available to the user, the SCS curve number model is used to estimate surface runoff. If hourly or breakpoint rainfall data are available, an infiltration-based model is used to simulate runoff. Both methods estimate percolation through the root zone of the soil.

The erosion component maintains elements of the USLE, but includes sediment transport capacity for overland flow. A channel erosion/deposition feature of the model permits consideration of concentrated flow within a field. Impoundments are treated in the erosion component also. The plant nutrient submodel of CREAMS has a nitrogen component that considers mineralization, nitrification, and denitrification processes. Plant uptake is estimated, and nitrate leached by percolation out of the root zone is calculated. Both the nitrogen and phosphorus parts of the nutrient component use enrichment ratios to estimate that portion of the two nutrients transported with sediment. The pesticide component considers foliar interception, degradation, and washoff, as well as adsorption, desorption, and degradation in the soil. This method, like the nutrient model, uses enrichment ratios and partitioning coefficients to calculate the separate sediment and water phases of pesticide loss. The CREAMS model is applicable for a field having (1) a single land use; (2) relatively homogeneous soils; (3) spatially uniform rainfall; and (4) a single management system, such as terraces. Normally, a field is less than 100 ha.

CREAMS can estimate the impact management systems, such as planting dates, cropping systems, irrigation scheduling, and tillage operations, have on sediment and nutrient movement. The model is also useful in long-term simulations for pesticide screening of management systems.

### 2. GLEAMS

The Groundwater Loading Effects of Agricultural Management Systems (GLEAMS) model (Leonard et al., 1987) is a continuous simulation, field scale model which was developed as an extension of the CREAMS model. GLEAMS assumes that a field has homogeneous land use, soils, and precipitation. It consists of three major components: hydrology, erosion/sediment yield, and pesticide transport. GLEAMS was developed to evaluate the impact of management practices on potential pesticide leaching within, through, and below the root zone. It also estimates surface runoff and sediment losses from the field. GLEAMS can be used to assess the effect of farm level management decisions on water quality.

GLEAMS considers soil properties and weather characteristics in estimating management impacts, such as planting dates, cropping systems, irrigation scheduling, and tillage operations, have on pesticide movement. The model is also useful in long-term simulations for pesticide screening of soil/management systems.

The model tracks movement of pesticides with percolated water, runoff, and sediment. Upward movement of pesticides and plant uptake are simulated with evaporation and transpiration. Degradation into metabolites is also simulated for compounds that have potentially toxic products. Erosion in overland flow areas is estimated using a modified Universal Soil Loss Equation. Erosion in channels and deposition in temporary impoundments such as tile outlet terraces are considered in estimating sediment yield at the edge of the field.

## 3. EPIC

The EPIC model (Williams et al., 1984) was developed in the early 1980's to assess the effect of erosion on productivity. Since the 1985 RCA application, the model has been expanded and refined to allow simulation of many processes important in agricultural management (Sharpley and Williams, 1990).

EPIC is a continuous simulation model that can be used to determine the effect of management strategies on water quality. The drainage area considered by EPIC is generally a field-sized area, up to 100 ha, where weather, soils, and management systems are assumed to be homogeneous. The major components in EPIC are weather simulation, hydrology, erosion, sedimentation, nutrient cycling, pesticide fate, plant growth, soil temperature, tillage, economics, and plant environment control.

EPIC can be used to compare management systems and their effects on nitrogen, phosphorus, pesticides and sediment. The management components that can be changed are crop rotations, tillage operations, irrigation scheduling, drainage, furrow diking, liming, grazing, manure handling, and nutrient and pesticide application rates and timing.

## 4. NLEAP

Nitrate Leaching and Economic Analysis Package (NLEAP) (Shaffer, 1991) is a field scale computer model that was developed to provide a rapid and efficient method of determining potential nitrate-N leaching associated with agricultural practices. Basic information concerning farm management practices, soils, and climate are translated into N budgets and nitrate-N leaching indices. The model also estimates potential nitrate-N leaching below the root zone and to groundwater supplies, the potential off-site effects of leaching, and the economic impacts of leaching.

NLEAP uses a three-phase approach to determine leaching potential: an annual screening analysis and the more detailed, monthly and event-by-event analyses. The processes modeled include movement of water and nitrate-N, crop uptake, denitrification, ammonia volatilization, mineralization of soil organic matter, nitrification, and mineralization-immobilization associated with crop residue, manure, and other organic wastes.

The screening procedure uses a simplified annual water and nitrogen budget and is designed to give only a general estimate of potential leaching of nitrate-N. The monthly budget analysis calculates leaching with consideration for the seasonal and monthly effects of precipitation, temperature, evapotranspiration, and farm management. The event-by-event analysis provides the best estimate of nitrate-N leaching. Its water and nitrogen budgets track the impacts of each precipitation, irrigation, fertilization, and tillage event on potential nitrate-N leaching.

## 5. PRZM

Pesticide Root Zone Model (PRZM) (Carsel et al., 1984) is a one-dimensional, dynamic, compartmental model that can be used to simulate chemical movement in unsaturated zone within and immediately below the plant root zone. Hydrology and chemical transport are the major components. The hydrology component calculates runoff and erosion based upon the Soil Conservation Service curve number procedure and the Universal Soil Loss Equation, respectively. Evapotranspiration is estimated directly from pan evaporation or by an empirical formula if pan evaporation data is not available.

Pesticide application on soil or on the plant foliage are considered in the chemical transport simulation. Dissolved, adsorbed, and vapor-phase concentrations in the soil are estimated by simultaneously considering the processes of pesticide uptake by plants, surface runoff, erosion, decay, volatilization, foliar washoff, advection, dispersion, and retardation. PRZM considers pulse loads, predicts peak events, and estimates time-varying mass emission of concentration profiles, thus overcoming limitations of the more commonly used steady-state models.

#### 6. RUSTIC

Risk of Unsaturated/Saturated Transport and Transformation of Chemical Concentrations (RUSTIC) (Dean et al., 1989) links three subordinate models in order to predict pesticide fate and transport through the crop root zone, and saturated zone to drinking water wells through PRZM, VADOFT, and SAFTMOD.

VADOFT is a finite-element model for simulating moisture movement and solute transport in the vadose zone. The model simulates one-dimensional, single-phase moisture and solute transport in unconfined, variably saturated porous media. Transport processes include hydrodynamic dispersion, advection, linear equilibrium sorption, and first-order decay. VADOFT predicts infiltration or recharge rate and solute mass flux entering the saturated zone. Parent/daughter chemical relationships may be simulated.

SAFTMOD performs two-dimensional simulations in an areal plane or a vertical cross section. In addition, the model can also perform axisymmetric simulations. Both single (unconfined and confined) and leaky two-aquifer systems can be handled. Transport of dissolved contaminants may also be simulated within the same domain. Transport processes accounted for include hydrodynamic dispersion, advection, linear equilibrium sorption, and first-order decay. Parent/daughter chemical relationships may be simulated.

### B. Watershed Scale Event Models

#### 1. AGNPS

Agricultural Nonpoint Source Pollution Model (AGNPS) was developed (Young et al., 1987) to analyze nonpoint source pollution in agricultural watersheds. AGNPS uses a distributed parameter approach by dividing a watershed into square grid areas called cells. Cell sizes are selected by the user (normally 0.4 to 16 ha) and the model can be applied on watersheds up to 20,000 ha. Runoff is calculated for each cell using the SCS curve number method (USDA Soil Conservation Service, 1972) and sediment yield for each cell is estimated with the Universal Soil Loss Equation (Wischmeier and Smith, 1978) adjusted for slope shape.

AGNPS can compare the effects of implementing various conservation alternatives within the watershed. Cropping systems, fertilizer application rates and timing, point source loads, contributions from feedlots, and the effect of terraced fields can be modeled.

The model partitions soluble nitrogen and phosphorus between surface runoff and infiltration. Chemical oxygen demand and soluble nutrient contributions from feedlots are transported with runoff. When the soluble pollutants reach concentrated flow, they are conservative and accumulate in the flow. Sediment transported nitrogen and phosphorus are also determined. A modified Universal

Soil Loss Equation, adjusting for slope shape, predicts local sediment yield within the originating cell. An estimate of gully erosion occurring in a cell can be added to the total amount of sediment yield in the cell. Sediment and runoff routing through impoundment terrace systems are also simulated.

### 2. ANSWERS

Area Nonpoint Source Watershed Environment Response Simulation (ANSWERS) (Beasley and Huggins, 1982) uses a distributed parameter approach and is limited to a single storm event. ANSWERS consists of a hydrology model, a sediment transport model, and several routing components. The conceptual basis of the model was developed by Huggins and Monke (1966). Hydrology model components include rainfall interception, infiltration, surface detention, and surface retention. A sediment continuity equation is employed which describes the process of soil detachment, transport, and deposition (Foster and Meyer, 1972). A watershed being modeled is divided into a series of small independent elements. The size of the elements normally ranges from 1 to 4 ha and watershed size is limited to approximately 10,000 ha because of computational and input preparation time. The use of small elements allows considerable spatial detail in representing topography, soils, and land use. However, the building of input files and interpreting output requires significant time and considerable knowledge of the model and its operation (Engel and Arnold, 1991).

## C. Watershed Scale, Continuous Time Models

### 1. HSPF

The Hydrological Simulation Program - FORTRAN (HSPF) (Johansen et al., 1984) simulates watershed hydrology and sediment yield. Data requirements for HSPF are extensive (Donigian and Huber, 1990). HSPF is a continuous simulation model and requires continuous data (generally hourly rainfall is required) to drive the simulations. The watershed is divided into land segments and stream channel segments. HSPF has been applied to the 68,000 square mile watershed draining into Chesapeake Bay. Although the watershed can be subdivided, the land segments are normally large enough to require considerable lumping of inputs.

HSPF uses such information as the time history of rainfall, temperature, solar intensity, and parameters related to land use patterns, soil characteristics, and agricultural practices to simulate the processes that occur in a watershed. The initial result of an HSPF simulation is a time history of the quantity and quality of water transported over the land surface and through various soil zones down to the groundwater aquifers. Runoff flow rate, sediment loads, nutrients, pesticides, toxic chemicals and other quality constituent concentrations can be predicted. The model then takes these results and information about the receiving water channels in the watershed and simulates the processes that occur in these channels. This part of the simulation produces a time history of water quantity and quality at any point in the watershed.

## 2. SWRRB

Simulator for Water Resources in Rural Basins (SWRRB) was developed to predict the effect of alternative management decisions on water and sediment yields with reasonable accuracy for ungaged, rural basins (Arnold et al., 1990; Williams et al., 1985). The model was developed by modifying the CREAMS daily rainfall model (Knisel, 1980) for application to large, complex, rural basins. The major changes involved were (a) the model was expanded to allow simultaneous computations on several subwatersheds, and (b) components were added to simulate weather, return flow, pond and reservoir storage, crop growth, transmission losses, and sediment movement through ponds, reservoirs, streams, and valleys. SWRRB operates on a daily time step and is efficient enough to run for many years (100 or more). Since the model is continuous time, it can determine the impacts of management such as crop rotations, planting and harvest dates, and chemical application dates and amounts. Basins can be subdivided into subareas based on differences in land use, soils, topograhpy, and climate. Soil and associated chemical are then routed to the basin outlet. SWRRB has been validated on basins up to 500 square kilometers (Arnold and Williams, 1987). Since SWRRB allows a limited number of subareas, some lumping of inputs is required.

## 3. ROTO

A model called ROTO (Arnold, 1990) was developed to estimate water and sediment yields on large basins (several thousand $Km^2$). ROTO is a continuous time model operating on a daily time step that accepts inputs from continuous time soil water balance models including SWRRB, EPIC, and GLEAMS. ROTO will also accept point sources and withdrawals, measured data, and output from another ROTO run. ROTO uses a command structure to route and add flows down the watershed through channels and reservoirs and operates on continuous (daily time step) allowing management decisions to be evaluated. ROTO also has several advantages over SWRRB including: 1) greater spatial detail; 2) improved watershed routing structure; and 3) greater flexibility, allowing input from several models, point sources, and measured data.

## III. Example Models

One of the major strengths in the modern comprehensive water quality model is the ability to simulate processes and interactions that affect water quality. For example, runoff has a strong influence on water quality, but runoff is affected by many variables including soil and climatic conditions, plant growth, evapotranspiration, plant residue decay, tillage, etc. Continuous simulation models offer many advantages including water quality probability distributions, seasonal variations, and little dependence on initial condition estimates. However, it is important that the major processes and interactions are simulated realistically. The EPIC and SWRRB models were chosen to illustrate the complex process interactions contained in fairly comprehensive continuous simulation water quality models.

### A. EPIC

The components of the field scale EPIC model can be placed into 10 major divisions for discussion--hydrology, weather, erosion, nutrients, pesticide fate, soil temperature, plant growth, tillage, plant environment control, and economics. A detailed description of the EPIC components was given by Williams et al. (1990). A brief description of each of the 10 components is presented here.

#### 1. Hydrology

**a. Surface Runoff**

Surface runoff from daily rainfall is predicted using a procedure similar to the CREAMS runoff model, option one (Knisel 1980; Williams and Nicks, 1982). Like the CREAMS model, runoff volume is estimated with a modification of the SCS curve number method (USDA Soil Conservation Service, 1972). The curve number varies non-linearly from the 1 (dry) condition at wilting point to the 3 (wet) condition at field capacity and approaches 100 at saturation. The EPIC model also includes a provision for estimating runoff from frozen soil.

Peak runoff rate predictions are based on a modification of the Rational Formula. The runoff coefficient is calculated as the ratio of runoff volume to rainfall. The rainfall intensity during the watershed time of concentration is estimated for each storm as a function of total rainfall using a stochastic technique. The watershed time of concentration is estimated using Manning's Formula considering both overland and channel flow.

### b. Percolation

The percolation component of EPIC uses a storage routing technique to predict flow through each soil layer in the root zone. Downward flow occurs when field capacity of a soil layer is exceeded if the layer below is not saturated. The downward flow rate is governed by the saturated conductivity of the soil layer. Upward flow may occur when a lower layer exceeds field capacity. Movement from a lower layer to an adjoining upper layer is regulated by the soil water to field capacity ratios of the two layers.

Percolation is also affected by soil temperature. If the temperature in a particular layer is 0°C or below, no percolation is allowed from that layer.

### c. Lateral Subsurface Flow

Lateral subsurface flow is calculated simultaneously with percolation. A nonlinear function of lateral flow travel time is used to simulate the horizontal component of subsurface flow. The magnitudes of the vertical and horizontal flow components are determined by a simultaneous solution of the two governing equations.

### d. Evapotranspiration

The model offers four options for estimating potential evaporation--Hargreaves and Samani (1985), Penman (1948), Priestley-Taylor (1972), and Penman-Monteith (Monteith, 1965). The Penman and Penman-Monteith methods require solar radiation, air temperature, wind speed, and relative humidity as input. If wind speed, relative humidity, and solar radiation data are not available, the Hargreaves or Priestley-Taylor methods provide options that give realistic results in most cases. The model computes soil and plant evaporation separately as described by Ritchie (1972).

### e. Snow Melt

The EPIC snow melt component is similar to that of the CREAMS model (Knisel, 1980). If snow is present, it is melted on days when the maximum temperature exceeds 0°C, using a linear function of temperature. Melted snow is treated the same as rainfall for estimating runoff and percolation, but rainfall energy is set to 0.0 and peak runoff rate is estimated assuming uniformly distributed rainfall for a 24 h duration.

## 2. Weather

The weather variables necessary for driving the EPIC model are precipitation and air temperature. If the Penman methods are used to estimate potential evaporation, solar radiation, wind speed and relative humidity are also required. Of course, wind speed and direction are also needed when wind erosion is simulated. If daily precipitation, air temperature, and solar radiation data are available, they can be input directly to EPIC. Otherwise, EPIC provides options for simulating various combinations of the five weather variables.

### a. Precipitation

The EPIC precipitation model developed by Nicks (1974) is a first-order Markov chain model. Thus, input to the model must include monthly probabilities of receiving precipitation if the previous day was dry and if the previous day was wet. Given the wet-dry state, the model determines stochastically if precipitation occurs or not. When a precipitation event occurs, the amount is determined by generating from a skewed normal daily precipitation distribution. The amount of daily precipitation is partitioned between rainfall and snowfall using average daily air temperature.

### b. Air Temperature and Solar Radiation

The temperature-radiation model developed by Richardson (1981) was selected for use in EPIC because it simulates temperature and radiation that exhibit proper correlation between one another and rainfall. The residuals of daily maximum and minimum temperature and solar radiation are generated from a multivariate normal distribution. Details of the multivariate generation model were described by Richardson (1981). The dependence structure of daily maximum temperature, minimum temperature, and solar radiation was described by Richardson (1982a).

### c. Wind

The wind simulation model was developed by Richardson (1982b) for use in simulating wind erosion with EPIC. The two wind variables considered are average daily wind velocity and direction. Average daily wind velocity is generated from a two-parameter Gamma distribution. Wind direction expressed as radians from north in a clockwise direction is generated from an empirical distribution specific for each location.

#### d. Relative Humidity

The relative humidity model simulates daily average relative humidity from the monthly average using a triangular distribution. Triangular coordinates are set to produce higher relative humidities on rainy days, lower values on dry days, and to preserve the long-term monthly average.

### 3. Erosion

#### a. Water

The EPIC water erosion model simulates erosion caused by rainfall and runoff and by irrigation (sprinkler and furrow). To simulate rainfall/runoff erosion, EPIC contains three equations--the USLE (Wischmeier and Smith, 1978), the MUSLE (Williams, 1975a), and the Onstad-Foster modification of the USLE (Onstad and Foster, 1975). Only one of the equations (user specified) interacts with other EPIC components.

The hydrology model supplies estimates of runoff volume and peak runoff rate. To estimate the daily rainfall energy in the absence of time-distributed rainfall, it is assumed that the rainfall rate is exponentially distributed. This allows simple substitution of rainfall rates into the USLE equation for estimating rainfall energy. The fraction of rainfall that occurs during 0.5 h is simulated stochastically. The soil erodibility factor is estimated as a function of soil texture and organic content. The crop management factor is evaluated with a function of above-ground biomass, crop residue on the surface, and the minimum C factor for the crop. Other factors of the erosion equation are evaluated as described by Wischmeier and Smith (1978). A nonlinear function of topsoil coarse fragment content is used to adjust the erosion estimates.

#### b. Wind

The Manhattan, Kansas wind erosion equation (Woodruff and Siddoway, 1965), was modified by Cole et al. (1982) for use in the EPIC model. The original equation computes average annual wind erosion as a function of soil erodibility, a climatic factor, soil ridge roughness, field length along the prevailing wind direction, and vegetative cover. The main modification of the model was converting from annual to daily predictions to interface with EPIC.

Two of the variables, the soil erodibility factor for wind erosion and the climatic factor, remain constant for each day of a year. The other variables, however, are subject to change from day to day. The ridge roughness is a function of a ridge height and ridge interval. Field length along the prevailing wind direction is calculated by considering the field dimensions and orientation and the wind direction. The vegetative cover equivalent factor is simulated daily as a function of standing live biomass, standing dead residue, and flat crop residue. Daily wind energy is estimated as a nonlinear function of daily wind velocity.

## 4. Nutrients

### a. Nitrogen

The amount of $NO_3$-N in runoff is estimated by considering the top soil layer only. The decrease in $NO_3$-N concentration caused by water flowing through a soil layer can be simulated satisfactorily using an exponential function. The average concentration for a day can be obtained by integrating the exponential function to give $NO_3$-N yield and dividing by volume of water leaving the layer (runoff, lateral flow, and percolation). Amounts of $NO_3$-N contained in runoff, lateral flow, and percolation are estimated as the products of the volume of water and the average concentration.

Leaching and lateral subsurface flow in lower layers are treated with the same approach used in the upper layer, except that surface runoff is not considered. When water is evaporated from the soil, $NO_3$-N is moved upward into the top soil layer by mass flow.

A loading function developed by McElroy et al. (1976) and modified by Williams and Hann (1978) for application to individual runoff events is used to estimate organic N loss. The loading function estimates the daily organic N runoff loss based on the concentration of organic N in the top soil layer, the sediment yield, and the enrichment ratio.

Nitrification, the conversion of ammonium N to nitrate N is estimated using a combination of the methods of Reddy et al. (1979a) and Godwin et al. (1984). The approach is based on the first-order kinetic rate equation of Reddy et al. (1979a). The rate constant is adjusted to account for variation in soil temperature, water content, and pH. The equation initiates nitrification when soil temperature exceeds 5°C and the rate constant increases linearly with temperature. Nitrification does not occur until soil water content exceeds wilting points (WP). The nitrification rate constant is a linear function of soil water between WP and 25% of the difference between field capacity (FC) and WP. Soil water contents between WP + 0.25 (FC-WP) and FC do not affect the rate constant. The nitrification rate constant declines linearly above FC and approaches zero at saturation. The nitrification rate constant is not affected for pH values between 7.0 and 7.4. The rate constant decreases linearly below pH 7.0 and approaches zero at pH 4.1. Similarly, the rate constant decreases linearly above pH 7.4 and approaches zero at pH 9.0.

Volatilization, the loss of ammonia N to the atmosphere is estimated simultaneously with nitrification using the method of Reddy et al. (1979b). Volatilization of surface applied ammonia is estimated as a function of temperature and wind speed. Depth of ammonia within the soil, cation exchange capacity (CEC) of the soil, and soil temperature are used in estimating below surface volatilization. The volatilization temperature function is the same as used in estimating nitrification. The surface volatilization rate constant increases nonlinearly with windspeed. The subsurface rate constant decreases nonlinearly with soil depth and linearly with CEC.

Denitrification, one of the microbial processes, is a function of temperature and water content. Denitrification is only allowed to occur when the soil water content is 90% of saturation or greater. The denitrification rate is estimated using an exponential function involving temperature, organic carbon, and $NO_3$-N.

The N mineralization model is a modification of the PAPRAN mineralization model (Seligman and van Keulen, 1981). The model considers two sources of mineralization: fresh organic N associated with crop residue and microbial biomass and the stable organic N associated with the soil humus pool. The mineralization rate for fresh organic N is governed by C:N and C:P ratios, soil water, temperature, and the stage of residue decomposition. The N associated with the soil humus pool is divided into two pools (active and stable). Mineralization occurs only in the active pool, but N is allowed to flow very slowly from the stable to the active pool. Mineralization is estimated as a function of organic N weight, soil water, and temperature.

Like mineralization, immobilization is simulated with a modification of the PAPRAN model. Immobilization is a very important process in EPIC because it determines the residue decomposition rate, and residue decomposition has an important effect on erosion. The daily amount of immobilization is computed by subtracting the amount of N contained in the crop residue from the amount assimilated by the microorganisms.

Crop use of N is estimated using a supply and demand approach. The daily crop N demand is estimated as the product of biomass growth and optimal N concentration in the plant. Optimal crop N concentration is a function of growth stage of the crop. Soil supply of N is limited by mass flow of $NO_3$-N to the roots. Actual N uptake is the minimum of supply and demand.

Fixation of N is an important process for legumes. Daily N fixation is estimated as a fraction of daily plant N uptake. The fraction is a function of soil $NO_3$ and water content and plant growth stage. No fixation occurs if the root zone $NO_3$ content is greater than 300 kg $ha^{-1}$ $m^{-1}$. The fraction is allowed to increase to 1.0 as the root zone $NO_3$ content is lowered to 100 kg $ha^{-1}$ $m^{-1}$. The fraction decreases linearly from 1.0 to 0 as soil water increases from 85% of field capacity to saturation. Below 85% of field capacity, the fraction reduces linearly to zero at wilting point. Also, fixation only occurs during the period between 15 and 75% of crop maturity.

To estimate the N contribution from rainfall, EPIC uses an average rainfall N concentration at a location for all storms. The amount of N in rainfall is estimated as the product of rainfall amount and concentration.

**b. Phosphorus**

The EPIC approach to estimating soluble P loss in surface runoff is based on the concept of partitioning pesticides into the solution and sediment phases as described by Leonard and Wauchope (Knisel, 1980). Because P is mostly associated with the sediment phase, the soluble P runoff is predicted using labile P concentration in the top soil layer, runoff volume, and a partitioning factor. Sediment transport of P is simulated with a loading function as described in organic N transport.

The P mineralization model developed by Jones et al. (1984) is similar in structure to the N mineralization model. Mineralization from the fresh organic P pool is governed by C:N and C:P ratios, soil water, temperature, and the stage of residue decomposition. Mineralization from the stable organic P pool associated with humus is estimated as a function of organic P weight, labile P concentration, soil water, and temperature. The P immobilization model also developed by Jones et al. (1984) is similar in structure to the N immobilization model.

The mineral P model was developed by Jones et al. (1984). Mineral P is transferred among three pools: labile, active mineral, and stable mineral. When P fertilizer is applied, it is labile (available for plant use). However, it may be quickly transferred to the active mineral pool. Simultaneously, P flows from the active mineral pool back to the labile pool (usually at a much slower rate). Flow between the labile and active mineral pools is governed by temperature, soil water, a P sorption coefficient, and the amount of material in each pool. Flow between the active and stable mineral P pools is governed by the concentration of P in each pool and the P sorption coefficient.

Crop use of P is estimated with the supply and demand approach described in the N model. However, the P supply is predicted using an equation based on plant demand, labile P concentration, and root weight.

5. Pesticide Fate

GLEAMS technology for simulating pesticide transport by runoff, percolate, soil evaporation, and sediment was added to EPIC. Pesticides may be applied at any time and rate to plant foliage or below the soil surface at any depth. The EPIC simulated plant leaf-area-index determines what fraction of foliar-applied pesticide reaches the soil surface. Also, a fraction of the application rate (called application efficiency) is lost to the atmosphere.

Each pesticide has a unique set of parameters including solubility, half life in soil and on foliage, wash-off fraction, organic carbon adsorption coefficient, and cost. Pesticides on plant foliage and in the soil degrade exponentially according to the appropriate half lives. When rainfall greater than 2.5 mm occurs, pesticide is washed from the foliage to the soil surface according to the wash-off fraction. If the field capacity of the top soil layer (10 mm thick) is exceeded, percolation occurs. Pesticide losses are estimated as a function of water lost/storage volume and the adsorption isotherm ($K_d$) of the chemical. Leaching, runoff, and lateral flow loss amounts are determined by loss volumes and pesticide concentration. Pesticide leaching is estimated for each soil layer when percolation occurs. Soil layers with low storage volumes have high leaching potentials not only because percolation is greater, but also because storage volume displacement is greater (higher concentration). Pesticides with low $K_d$ values and high solubility are transported rapidly with water. Conversely, high $K_d$ value pesticides are adsorbed to soil particles and travel largely with the sediment. Pesticide concentration adsorbed to sediment is estimated as a function of pesticide and soil weight, Kd, and water storage capacity of the soil layer. Pesticide transport by sediment is calculated as a function of pesticide concentration, sediment yield, and an enrichment ratio.

## 6. Soil Temperature

Daily average soil temperature is simulated at the center of each soil layer for use in nutrient cycling and hydrology. The temperature of the soil surface is estimated using daily maximum and minimum air temperature and snow, plant, and residue cover for the day of interest plus the four days immediately preceding. Soil temperature is simulated for each layer using a function of damping depth, surface temperature, and mean annual air temperature. Damping depth is dependent upon bulk density and soil water.

## 7. Crop Growth Model

A single model is used in EPIC for simulating all the crops considered (corn, grain sorghum, wheat, barley, oats, sunflower, soybean, alfalfa, cotton, peanuts, potatoes, durham wheat, winter peas, faba beans, rapeseed, sugarcane, sorghum hay, range grass, rice, casava, lentils, and pine trees). Of course, each crop has unique values for the model parameters. Energy interception is estimated as a function of solar radiation and the crop's leaf area index. The potential increase in biomass for a day is estimated as the product of intercepted energy and a crop parameter for converting energy to biomass. The leaf area index is simulated with equations dependent upon heat units, the maximum leaf area index for the crop, a crop parameter that initiates leaf area index decline, and five stress factors.

Crop yield is estimated using the harvest index concept. Harvest index increases as a non-linear function of heat units from zero at planting to the optimal value at maturity. The harvest index may be reduced by water stress during critical crop stages (usually between 30 and 90% of maturity).

The fraction of daily biomass growth partitioned to roots is estimated to range linearly between two fractions specified for each crop at emergence and at maturity. Root weight in a soil layer is simulated as a function of plant water use within that layer. Root depth increases as a linear function of heat units and potential root zone depth.

The potential biomass is adjusted daily if one of the plant stress factors is less than 1.0 using the product of the minimum stress factor and the potential biomass. The water-stress factor is the ratio of actual to potential plant evaporation. The temperature stress factor is computed with a function dependent upon the daily average temperature, the optimal temperature, and the base temperature for the crop. The N and P stress factors are based on the ratio of accumulated plant N and P to the optimal values. The aeration stress factor is estimated as a function of soil water relative to porosity in the root zone.

Roots are allowed to compensate for water deficits in certain layers by using more water in layers with adequate supplies. Compensation is governed by the minimum root growth stress factor (soil texture and bulk density, temperature, and aluminum toxicity). The soil texture-bulk density relationship was developed by Jones (1983).

## 8. Tillage

The EPIC tillage component was designed to mix nutrients and crop residue within the plow depth, simulate the change in bulk density, and convert standing residue to flat residue. Other functions of the tillage component include simulating ridge height and surface roughness.

Tillage operations convert standing residue to flat residue using an exponential function of tillage depth and mixing efficiency. When a tillage operation is performed, a fraction of the material (equal the mixing efficiency) is mixed uniformly within the plow depth. Also, bulk density is reduced as a function of mixing efficiency, bulk density before tillage, and undisturbed bulk density. After tillage, bulk density returns to the undisturbed value at a rate dependent upon infiltration, tillage depth, and soil texture.

## 9. Plant Environment Control

### a. Drainage

Underground drainage systems are treated as a modification to the natural lateral subsurface flow of the area. Simulation of a drainage system is accomplished by reducing the travel time in a specified soil layer.

#### b. Irrigation

The EPIC user has the option to simulate dryland or irrigated agricultural areas. Sprinkler or furrow irrigation may be simulated and the applications may be user specified or automatic. With the automatic option, the model decides when and how much water to apply. The user must input a plant water stress level or a soil water tension value to trigger automatic irrigation, the maximum volume applied per growing season, and the minimum time interval between applications.

#### c. Fertilization

The EPIC model provides two options for applying fertilizer. With the first option, the user specifies dates, rates, and depths of application of N and P. The second option is more automated--the model decides when and how much fertilizer to apply. The three required inputs are: (1) a plant stress level to trigger nitrogen fertilizer application, (2) the maximum N application per growing season, and (3) the minimum number of days between applications. At planting time, the model takes a soil sample and applies enough N and P to bring the concentrations in the root zone up to the concentration level at the start of the simulation. Additional N may be applied during the growing season.

#### d. Lime

The EPIC model simulates the use of lime to neutralize toxic levels of aluminum in the plow layer. Two sources, KC1-extractable aluminum in the plow layer and acidity caused by ammonia-based fertilizers, are considered. When the sum of acidity due to extractable aluminum and fertilizer N exceeds 4 t $ha^{-1}$, the required amount of lime is added and incorporated into the plow layer.

#### e. Pests

The effects of insects, weeds, and diseases are expressed in the EPIC pest factor. The pest factor is simulated as a function of minimum daily temperature, 30-day moving average rainfall, above ground growing biomass, and crop residue. At the end of the growing season, the pest factor is normalized to fall in the range between the minimum value for the particular crop and 1.0. Crop yields are estimated as the product of the simulated yield and the simulated pest factor.

**f. Grazing**

Livestock grazing is simulated as a daily harvest operation. Users specify daily grazing rate minimum grazing height, harvest efficiency, and date grazing begins and ends. Any number of grazing periods may occur during a year and the grazing schedule may vary from year-to-year within a rotation. Grazing ceases when forage height is reduced to the minimum user specified height.

10. Economics

The crop budgets are calculated using components from the Enterprise Budget Generator (Kletke, 1979). Inputs are divided into two categories: fixed and variable. Fixed inputs include depreciation, interest or return on investment, insurance, and taxes on equipment, land, and capital improvements (terraces, drainage, irrigation systems, etc.). Variable inputs are defined as machinery repairs, fuel and other energy, machine lubricants, seed, fertilizer, pesticides, labor, and irrigation water.

**B. SWRRB**

The components of SWRRB can be placed into eight major divisions--hydrology, weather, sedimentation, soil temperature, crop growth, nutrients, pesticide fate, and agricultural management. A detailed description of the SWRRB components was given by Arnold et al. (1990). A brief description is presented here. Since SWRRB and EPIC use essentially the same components for simulating several processes, descriptions are not repeated.

1. Hydrology

**a. Surface Runoff**

Same as EPIC.

**b. Percolation**

Same as EPIC.

**c. Lateral Subsurface Flow**

Lateral flow is defined as flow that travels laterally within the soil profile that returns to contribute to streamflow. SWRRB uses a kinematic storage routing model (Sloan and Moore, 1984) based on the mass continuity equation with each soil layer defining a control volume. The model is physically-based and accounts for slope, slope length, and saturated conductivity.

**d. Evapotranspiration**

Potential evapotranspiration is estimated with the Priestly-Taylor (1972) method. Required inputs are daily maximum and minimum air temperature and solar radiation. The model computes soil and plant evaporation separately as described by Ritchie (1972).

**e. Snow Melt**

Same as EPIC.

**f. Transmission Losses**

Many semiarid watersheds have alluvial channels that abstract large volumes of streamflow (Lane, 1982). The abstractions, or transmission losses, reduce runoff volumes as the flood wave travels downstream. SWRRB uses Lanes method described in Chapter 19 of the SCS Hydrology Handbook (USDA Soil Conservation Service, 1983) to estimate transmission losses. Channel losses are a function of channel width and length and flow duration. Both runoff volume and peak rate are adjusted when transmission losses occur.

**g. Ponds and Reservoirs**

Farm pond storage is simulated as a function of pond capacity, daily inflows and outflows, seepage, and evaporation. Ponds are assumed to have only emergency spillways. Required inputs are capacity and surface area. Surface area below capacity is estimated as a non-linear function of storage. Reservoirs are treated similarly except they have emergency and principal spillways. Thus, required inputs include volume and surface area at both spillway elevations and the principal spillway release rate.

## 2. Weather

The weather variables necessary for driving SWRRB are precipitation, air temperature, and solar radiation. If daily precipitation data are available, they can be input directly to SWRRB. If not, the weather generator can simulate daily rainfall and temperature. Solar radiation is always simulated. One set of weather variables may be simulated for the entire basin, or different weather may be simulated for each subbasin.

### a. Precipitation

The SWRRB precipitation model developed by Nicks (1974) is a first-order Markov chain model. Thus, input to the model must include monthly probabilities of receiving precipitation if the previous day was dry and if the previous day was wet. Given the wet-dry state, the model determines stochastically if precipitation occurs or not. When a precipitation event occurs, the amount is determined by generating from a skewed normal daily precipitation distribution. The amount of daily precipitation is partitioned between rainfall and snowfall using average daily air temperature.

If precipitation is to be simulated for each subbasin, the amount generated from the skewed normal distribution is assumed to be the mean for all gages for the day. The storm center (location of maximum rainfall) is located in a rectangle with boundaries set at a distance of 100 Km from the basin's maximum and minimum x and y rain gauge coordinates. Thus, the storm center could be located in the basin or as much as 100 Km in any direction from the basin. Each storm center is defined by drawing two random numbers--one for the x scale and one for the y scale. Rainfall at each gage is computed using an area reduction function (Nicks and Igo, 1980). The function reduces rainfall as distance from the storm center increases and as rainfall duration decreases. Finally, the reduction factors are applied to the subbasin rain gauges and adjusted to insure that the mean rainfall agrees with the original skewed normal simulated mean.

### b. Air Temperature and Solar Radiation

Daily maximum and minimum air temperature and solar radiation are generated from a normal distribution corrected for wet-dry probability state. The correction factor is used to provide more deviation in temperatures and radiation when weather changes and for rainy days. Conversely deviations are smaller on dry days. The correction factors are calculated to insure that long-term standard deviations of daily variables are maintained. Monthly values of daily standard deviations of maximum and minimum temperature are input. Monthly values of daily standard deviations of solar radiation are estimated by assuming the difference between mean and maximum daily radiation is four standard deviations.

## 3. Sedimentation

### a. Sediment Yield

Sediment yield is estimated for each subbasin with the Modified Universal Soil Loss Equation (MUSLE) (Williams, 1975a). The hydrology model supplies estimates of runoff volume and peak runoff rate. The crop management factor is evaluated with a function of above-ground biomass, crop residue on the surface, and the minimum C factor for the crop. Other factors of the erosion equation are evaluated as described by Wischmeier and Smith (1978).

### b. Sediment Routing

Ponds and Reservoirs: Inflow sediment yield to ponds and reservoirs (P/R) is computed with MUSLE. The outflow from P/R is calculated as the product of outflow volume and sediment concentration. Outflow P/R concentration is estimated using a simple continuity equation based on volumes and concentrations of inflow, outflow, and pond storage. Initial pond concentration is input and between storm concentration decreases as a function of time and median particle size of inflow sediment.

Channel and Floodplain: The sediment routing model consists of two components operating simultaneously (deposition and degradation). The deposition component is based on fall velocity and the degradation component is based on Bagnold's stream power concept (Williams, 1980).

The particle size distribution of the detached sediment is estimated from the primary particle size distribution (Foster et al., 1980). The particle size of the sediment reaching a subbasin outlet is estimated using the MUSLE derived sediment deposition model (Williams, 1975b). The delivery ratio of each event is estimated as the ratio of the subbasin peak discharge rate to the peak rainfall excess rate raised to the 0.56 power. Since rainfall input is daily and runoff volume is estimated as a daily amount, rainfall excess rate is not simulated directly. However, the peak rainfall excess rate can be estimated as the difference between the peak rainfall rate and the average infiltration rate of the storm. The storm duration and the peak rainfall rate can be estimated by assuming that rainfall rates are exponentially distributed. Note, this rainfall rate assumption is not related to the time distribution of rainfall amount. The delivery ratio is used to estimate a routing coefficient that governs deposition as a function of particle size. Thus, as deposition occurs the particle size distribution is composed of finer material.

Deposition in the channel and floodplain from the subbasin to the basin outlet is based on sediment particle fall velocity. Fall velocity is calculated as a function of particle diameter squared using Stokes Law. The depth of fall through a routing reach is the product of fall velocity and reach travel time. The delivery ratio is estimated for each particle size as a linear function of fall velocity, travel time, and flow depth.

Stream power is used to predict degradation in the routing reaches. Bagnold (1977) defined stream power as the product of water density, flow rate, and water surface slope. Williams (1980) modified Bagnold's equation to place more weight on high values of stream power--stream power raised to 1.5. Also, since hydrographs are not simulated, one flow rate is calculated to represent the total hydrograph. The representative flow rate is determined from simulated peak discharge rate and runoff volume assuming a triangular hydrograph with two recession limbs.

Available stream power is used to reentrain loose and deposited material until all of the material is removed. Excess stream power causes bed degradation. Bed degradation is adjusted by the USLE soil erodibility and cover factors of the channel and floodplain.

### 4. Soil Temperature

Same as EPIC.

### 5. Crop Growth Model

A single model is used in SWRRB for simulating all crops. Energy interception is estimated as a function of solar radiation and the crop's leaf area index. The potential increase in biomass for a day is estimated as the product of intercepted energy and a crop parameter for converting energy to biomass. The leaf area index is simulated with equations dependent upon heat units.

Crop yield is estimated using the harvest index concept. Harvest index increases as a non-linear function of heat units from zero at planting to the optimal value at maturity. The harvest index may be reduced by water stress during critical crop stages (usually between 30 and 90% of maturity).

The fraction of daily biomass growth partitioned to roots is estimated to range linearly from 0.4 at emergence to 0.2 at maturity. Root weight in a soil layer is simulated as a function of plant water use within that layer. Root depth increases as a linear function of heat units and potential root zone depth.

The potential biomass is adjusted daily if one of the plant stress factors is less than 1.0 using the product of the minimum stress factor and the potential biomass. The water-stress factor is the ratio of actual to potential plant evaporation. The temperature stress factor is computed with a function dependent upon the daily average temperature, the optimal temperature, and the base temperature for the crop.

## 6. Nutrients

### a. Nitrogen

The amount of $NO_3$-N in runoff is estimated by considering the top soil layer only. The decrease in $NO_3$-N concentration caused by water flowing through a soil layer can be simulated satisfactorily using an exponential function. The average concentration for a day can be obtained by integrating the exponential function to give $NO_3$-N yield and dividing by volume of water leaving the layer (runoff, lateral flow, and percolation). Amounts of $NO_3$-N contained in runoff, lateral flow, and percolation are estimated as the products of the volume of water and the average concentration.

Leaching and lateral subsurface flow in lower layers are treated with the same approach used in the upper layer, except that surface runoff is not considered. When water is evaporated from the soil, $NO_3$-N is moved upward into the top soil layer by mass flow.

A loading function developed by McElroy et al. (1976) and modified by Williams and Hann (1978) for application to individual runoff events is used to estimate organic N loss. The loading function estimates the daily organic N runoff loss based on the concentration of organic N in the top soil layer, the sediment yield, and the enrichment ratio.

Crop use of N is estimated using a supply and demand approach. The daily crop N demand is estimated as the product of biomass growth and optimal N concentration in the plant. Optimal crop N concentration is a function of growth stage of the crop. Soil supply of N is limited by mass flow of $NO_3$-N to the roots. Actual N uptake is the minimum of supply and demand.

### b. Phosphorus

The SWRRB approach to estimating soluble P loss in surface runoff is based on the concept of partitioning pesticides into the solution and sediment phases as described by Leonard and Wauchope (Knisel, 1980). Because P is mostly associated with the sediment phase, the soluble P runoff is predicted using labile P concentration in the top soil layer, runoff volume, and a partitioning factor. Sediment transport of P is simulated with a loading function as described in organic N transport.

Crop use of P is estimated with the supply and demand approach described in the N model. However, the P supply is predicted using an equation based on plant demand, labile P concentration, and root weight.

7. Pesticides

Same as EPIC.

8. Agricultural Management

**a. Tillage and Residue**

The SWRRB tillage component was designed to partition the above-ground biomass at harvest. Part of the biomass is removed as yield, part is incorporated into the soil, and the remainder is left on the soil surface as residue. The model has no process interactions with incorporated residue. Also, tillage does not effect soil properties.

The fraction of above-ground biomass left on the surface is determined by the tillage strategy selected. Fall plow leaves 5% of the above-ground crop residue on the surface. Percentages assigned to surface residue for the three other tillage strategies are: 25% for spring plow; 50% for conservation tillage; and 95% for zero tillage. The residue decays at a rate determined by soil water content and soil temperature of the top soil layer.

**b. Irrigation**

The user has the option to simulate dryland or irrigated agriculture. If irrigation is selected, he must also specify the runoff ratio (volume of water leaving the field/volume applied) and a plant water stress level to trigger irrigation. The plant water stress factor ranges from 0 to 1.0 (1 means no stress and 0 means no growth). When the user-specified stress level is reached, enough water is applied to fill the root zone to field capacity.

## IV. Water Quality Model Applications

### A. EPIC

1. Water Quality Project Evaluation and Planning

In 1992 many of the 16 Water Quality Demonstration Projects and the 74 Hydrologic Unit Area Projects began using one or more of the water quality models described here. Models are being used in conjunction with monitoring to help identify current water quality problem areas and to estimate the benefits of water quality improvement practices. Without the models, it is difficult to estimate benefits.

## 2. European Economic Community Project

A three-year European Community project involving French, Italian, English, Spanish, and Portugese scientists is using EPIC to simulate complex cropping systems in ten major European production regions. Drs. Guillerono Flichmann at the Mediterranean Agronomic Institute in Montpellier and Dr. Maurice Cabelguenne of the French National Institute of Agronomic Research in Toulouse, are leading the project. EPIC's field-scale crop production and water quality predictions are being utilized in conjunction with farm-scale economic models to assess the effects of common European agricultural policies on the relative regional competitiveness and environmental impacts.

## 3. EPA Nutrient Movement Study

EPA is using EPIC as a tool to examine the impacts of tillage on nitrogen and phosphorus movement. Four tillage strategies were examined using 100-year EPIC simulations at 100 sites in Illinois. No-till corn/soybean rotations were found to reduce sediment bourne nutrients and nitrogen in runoff relative to conventional till continuous corn (Phillips et al., 1993).

## 4. Animal Waste Recycling Planning

The EPIC model is currently being used to study animal waste disposal systems for the Upper North Bosque Hydrological Unit Project in Erath and Hamilton Counties, Texas. Three poster papers have been presented illustrating how EPIC simulates the effects of manure application rates on crop growth, soil fertility, and water quality.

## 5. Crop and Tillage Impacts on Nutrient and Sediment Movement

EPIC has been used to examine the water quality impacts of alternative crop tillage and conservation systems (Benson et al. 1990a). Nitrogen and phosphorus losses were simulated for two soils and five alternative crop/tillage systems. The base system was continuous conventional tillage cotton. The simulated alternative rotations of cotton/sorghum and cotton/sorghum/wheat reduced phosphorus losses for both conventional and conservation tillage. Also, the ratio of N loss to N fertilizer applied was reduced by the rotations, although total N loss increased slightly.

### 6. Sustainable Agriculture's Impact on Production and Water Quality

The Rodale Research Center in cooperation with the World Resource Institute is using EPIC to study sustainable agricultural systems. An expanded sustainable agriculture analysis for the U.S. is now underway.

### 7. Evaluation of New Cropping Systems Influence on Water Quality

Washington State University used EPIC to simulate a field experiment in integrated cropping for a 100-year period to extend the experimental data over time (Benson et al. 1990b). This study examined the feasibility of a spring pear and Medic/Medic/winter wheat rotation as a replacement for the current winter wheat/spring barley/winter wheat/spring pear rotation. The proposed system was studied using relatively short-term experiments, but the results may not represent long-term system behavior. EPIC was used to simulate over 100 years of yields and nitrogen and phosphorus losses. The results indicated that not only were there some potential economic advantages (Goldstein and Young, 1987), but also some water quality enhancement. Both N and P movement losses were reduced.

### 8. Cover Crop Impacts on Water Quality

Meisinger et al. (1991) used EPIC to address the impacts of cover crops on water quality. This study compared EPIC simulations to historic studies to show the reasonableness of EPIC nutrient loss estimates. The EPIC simulations indicated that winter cover crops had the most potential for reducing nitrogen leaching in the humid southeast and in irrigated areas.

### 9. Erosion Impacts on Productivity and Water Quality

Fribourg et al. examined the impact of erosion on nitrate loss by using EPIC to estimate N runoff and leaching for two eroded phases of two Tennessee soils, Memphis and Grenada. The EPIC mean corn yield estimates for both soils for the eroded and non-eroded phases were within 10 percent of the measured for three of the four soils and within 15 percent for the fourth. Estimated nitrogen losses for the eroded phase for the Grenada soil was greater than the non-eroded phase. This is not surprising considering the root restricting fragipan of the Grenada soil.

These examples of past and current use of EPIC indicate that there is tremendous potential for the use of Water Quality models. These models help identify problems, examine systems for improving water quality, estimate potential benefits, and develop national farm programs which will lead to widespread adoption of the water quality measures.

## B. SWRRB

The SWRRB model is currently being used throughout the world by consulting engineers, government agencies, universities, and chemical companies. Several applications follow to demonstrate the model's potential capabilities.

### 1. Hydrology Unit Model of the United States (HUMUS)

The SCS is using the EPIC, SWRRB, and ROTO models as part of the 1997 Resource Conservation Assessment. The models will be linked to national economic models and used for national planning of water supply and quality on the 18 major river basins in the U.S. This system of models allows water, sediment, and attached pollutants to be tracked from their point of origin to major rivers, reservoirs, and coastal zones. Geographic information systems are utilized to integrate the models with national soils, land use, and digital elevation databases. The GIS automatically extracts model input from the map layers and display model output.

### 2. Coastal Pollutant Discharge Inventory

As a part of the National Coastal Pollutant Discharge Inventory, the National Oceanic and Atmospheric Administration (NOAA) is using SWRRB to estimate non-point source loadings from non-urban lands in all coastal counties of the U.S. (Singer et al., 1988). Site-specific data are obtained from the SCS's National Resources Inventory and Soils-5 data bases, NOAA weather stations, U.S. Geological Survey digital land use data tapes, and other local sources. Simulations have been run for cropland, rangeland, and forest land in approximately 770 subwatersheds comprising the Gulf Coast, Eastern, and Western coastal zones of the U.S. Results are compiled by season and added to a comprehensive data base containing pollutant loadings from all significant discharge sources.

### 3. Pesticide Assessment

SWRRB was modified to include simulation of pesticide concentration in the runoff and sediment. The EPA has adopted this version of SWRRB as its pesticide assessment model. They have prepared their own user's manual, "Pesticide Runoff Simulator" (Holst and Kutney, 1987). Many chemical companies and consulting firms are using this version of the model for environmental assessment.

## 4. Effect of Urbanization on Reservoir Loadings

White Rock Lake in Dallas was built in 1910. The SCS completed sediment survey on the lake in 1935, 1956, 1970, 1977, and 1984. Throughout this period, the percentage of urban area has increased from 0 to 77% in 1984. SWRRB was utilized to estimate the effects of urbanization on water and sediment delivery to White Rock Lake (Arnold et al., 1988).

Initially, simulated model results were compared with measured water yields, peak flow rates, and sediment yields. The comparisons showed that the model could do a reasonable job predicting the effect of urbanization on these variables. Also, the effect of urbanization on delivery ratios showed a positive linear correlation.

Once the model had been validated, future simulations, assuming no urbanization since 1935 and all urban since 1910, were compared with actual conditions. It was shown that surface runoff increased and sediment yield decreased as urban area increased. The weather generator in SWRRB was utilized to examine future scenarios. Prediction of reservoir capacity lost to sedimentation to the year 2050 was shown for three scenarios, thus showing the model's usefulness in planning and designing water resources projects.

## 5. Supplemental Pond Irrigation

A comprehensive simulation model was developed to optimize pond size for supplemental irrigation (Arnold and Stockle, 1991). The SWRRB model was chosen as the basis for the simulation model. It was modified to simulate crop yields, double cropping, and supplemental pond irrigation. A simple economic model was also added. SWRRB was then linked with a golden section search to determine the pond size that optimizes average annual return to management. The model also develops frequency distributions for risk assessment. Individual components such as water yield, ET, and crop yields were validated with measured data to insure proper model operation. Arnold and Stockle (1991) apply the model at two locations with extreme differences in climate and soils to demonstrate potential model capabilities.

## 6. Water Rights

SWRRB is currently being used in water rights disputes in New Mexico and Arizona. This is part of an effort being undertaken by the Bureau of Indian Affairs, and the Hopi and Navajo Indian Tribes, to quantify Indian rights to water in reservation basins. SWRRB was selected because the basins are ungaged (lacking measured rainfall, runoff, and diversion data) and because SWRRB is a continuous simulation model.

## References

Arnold, J.G. 1990. ROTO-A continuous water and sediment routing model. ASCE Proc. of the Watershed Management Symposium. Durango, CO. pp. 480-488.

Arnold, J.G. and C.O. Stockle. 1991. Simulation of supplemental irrigation from on-farm ponds. *ASCE J. Irrig. and Drainage* 117(3):408-424.

Arnold, J.G. and J.R. Williams. 1987. Validation of SWRRB--Simulator for water resources in rural basins. *J. Water Resources Planning and Manage.*, ACSW, 113(2):243-256.

Arnold, J.G., M.D. Bircket, J.R. Williams, W.F. Smith, and H.N. McGill. 1988. Modeling the effects of urbanization on basin water yield and reservoir sedimentation. *Water Resources Bull.* 23(6):1021-1029.

Arnold, J.G., J.R. Williams, A.D. Nicks, and N.B. Sammons. 1990. SWRRB: A basin scale simulation model for soil and water resources management. Texas A&M University Press, College Station, TX. 255 pp.

Bagnold, R.A. 1977. Bedload transport by natural rivers. *Water Resources Res.* 13(2):303-312.

Barfield, B.J., C.T. Haan, and D.E. Storm. 1991. Why model? CREAMS-/GLEAMS Symposium, Athens, GA. pp. 3-8.

Beasley, D.B. and L.F. Huggins. 1982. ANSWERS - Users manual. EPA-905/-9-82-001, USEPA, Region 5, Chicago, IL. 54 pp.

Benson, V.W., H.C. Bogusch, Jr., and J.R. Williams. 1990a. Evaluating alternative soil conservation and crop tillage practices with EPIC. p. 91-93 In: P.W. Unger, T.V. Sneed, W.R. Jordan, and R. Jenson (eds.) Proc. Intl. Conf. on Dryland Farming, Challenges in Dryland Agriculture - A Global Perspective, Aug. 1988, Amarillo/Bushland, TX. Texas Agri. Exp. Stn., 965 pp.

Benson, V.W., W.A. Goldstein, D.L. Young, J.R. Williams, C.A. Jones, and J.R. Kiniry. 1990b. Impacts of integrated cropping practices on nitrogen use and movement. p. 426-428 In: P.W. Unger, T.V. Sneed, W.R. Jordan, and R. Jenson (eds.) Proc. Intl. Conf. on Dryland Farming, Challenges in Dryland Agriculture - A Global Perspective, Aug. 1988, Amarillo/Bushland, TX. Texas Agric. Exp. Stn., 965 pp.

Carsel, R.F., C.N. Smith, L.A. Mulkey, J.D. Dean, and P. Jowise. 1984. User's manual for the pesticide root zone model (PRZM): Release 1. EPA-600/3-84-109. U.S. Environmental Protection Agency. Environmental Research Laboratory, Athens, GA.

Cole, G.W., L. Lyles, and L.G. Hagen. 1982. A simulation model of daily wind erosion soil loss. ASAE Paper #82-2575.

Dean, J.D., P.S. Huyakorn, A.S. Donigian, Jr., K.A. Voos, R.W.Schanz, Y.J. Meeks, and R.F. Carsel. 1989. Risk of unsaturated/saturated transport and transformation of chemical concentrations (RUSTIC). EPA/600/3-89/048a. U.S. Environmental Protection Agency. Environmental Research Laboratory, Athens, GA.

Donigian, A.S. and W.C. Huber. 1990. Modeling of nonpoint source water quality in urban and non-urban areas. U.S. Environmental Protection Agency, Athens, GA. Contract No. 68-03-3513 (Draft).

Engel, B.A. and J.G. Arnold. 1991. Agricultural non-point source pollution control using spatial decision support systems. (Draft).

Foster, G.R. and L.D. Meyer. 1972. A closed-form soil erosion equation for upland areas. In: H. Shen (ed.), Sedimentation, Colorado State University, Fort Collins, CO, Chapter 12.

Foster, G.R., L.J. Lane, J.D. Nowlin, J.M. Laflen, and R.A. Young. 1980. A model to estimate sediment yield from field-sized areas: development of model. p. 36-64. In: W.G. Knisel (ed.) CREAMS, A field scale model for chemicals, runoff, and erosion from agricultural management systems. U.S. Dep. Agric. Conserv. Res. Report.

Fribourg, H.A., D.D. Tyler, V.W. Benson, J.R. Williams, J.V. Graveel, J. Logan, and G.R. Wells. Environmentally sound agricultural soil use--prediction of yield, erosion and other off-site effects from corn production. (Mimeo handout.)

Godwin, D.C., C.A. Jones, J.T. Ritchie, P.L.G. Vlek, and L.G. Youngdahl. 1984. The water and nitrogen components of the CERES models. p. 95-100. In: ICRISAT (International Crops Research Institute for the Semi-Arid Tropics). Proc. Internatl. Symp. on Minimum Data Sets for Agrotechnology Transfer. March 21-26, 1983. Patancheru, India: ICRISAT Center.

Goldstein, W.A. and D.L. Young. 1987. An agronomic and economic comparison of a conventional and low-input system in the Palouse. American Journal of Alternative Agriculture. Spring 51-56.

Hargreaves, G.H. and Z.A. Samani. 1985. Reference crop evapotranspiration from temperature. *Appl. Engr. in Agric.* 1:96-99.

Holst, R.W. and L.L. Kutney. 1987. U.S. EPA simulator for water resources in rural basins (EPA-SWRRB). (Draft)

Huggins, L.F. and E.J. Monke. 1966. The mathematical simulation of the hydrology of small watersheds. Technical Report 1, Water Resources Research Center, Purdue University, West Lafayette, IN. 130 pp.

Johansen, N.B., J.C. Imhoff, J.L. Kittle, and A.S. Donigian. 1984. Hydrological Simulation Program--Fortran (HSPF): User's manual for release 8. EPA-600/3-84-066. U.S. Environmental Protection Agency, Athens, GA.

Jones, C.A. 1983. Effect of soil texture on critical bulk densities for root growth. *Soil Sci. Soc. Amer. J.* 47:1208-1211.

Jones, C.A., C.V. Cole, A.N. Sharpley, and J.R. Williams. 1984. A simplified soil and plant phosphorus model. I. Documentation. *Soil Sci. Soc. Am. J.* 48:800-805.

Kletke, D.D. 1979. Operation of the enterprise budget generator. Oklahoma State Univ., Agric. Exp. Sta. Res. Report P-790. Knisel, W.G. (ed.). 1980. CREAMS: A field scale model for chemicals, runoff, and erosion from agricultural management systems. USDA, Conservation Research Report No. 26. 643 pp.

Knisel, W.G. (ed.). 1980. CREAMS: A field scale model for chemicals, runoff, and erosion from agricultural management systems. USDA, Conservation Research Report No. 26. 643 pp.

Lane, L.J. 1982. Distributed model for small semi-arid watersheds. *J. Hydraulic Eng.*, ASCE, 109(HY10):1114-1131.

Leonard, R.A., W.G. Knisel, and D.A. Still. 1987. GLEAMS: Groundwater loading effects of agricultural management systems. Trans. ASAE 30(5):14-03-1428.

McElroy, A.D., S.Y. Chiu, J.W. Nebgen, A. Aleti, and F.W. Bennett. 1976. Loading functions for assessment of water pollution from nonpoint sources. Environmental Protection Tech. Series, USEPA, EPA 600/2-76-151. 445 pp.

Meisinger, J.J., W.L. Hargrove, R.L. Mikkelsen, J.R. Williams, and V.W. Benson. 1991. Effects of cover crops on groundwater quality. p. 57-68. In: W. L. Hargrove (ed.) Cover Crops for Clean Water. Proc. Intl. Conf., April 9-11, 1991, Jackson, TN. Soil Water Conserv. Soc.

Monteith, J.L. 1965. Evaporation and environment. *Symp. Soc. Exp. Biol.* 19:205-234.

Nicks, A.D. 1974. Stochastic generation of the occurrence, pattern, and location of maximum amount of daily rainfall. p. 154-171. In: Symp. on Statistical Hydrology, Aug.-Sept. 1971, Tucson, AZ. Mis. Publ. No. 1275.

Nicks, A.D. and F.A. Igo. 1980. A depth-area-duration model of storm rainfall in the Southern Plains. *Water Resources Res.* 16(5):939-945.

Office of Technology Assessment. 1982. Use of model for water resources management, planning, and policy. Summary, Six Chapters and Appendix. U.S. Gov't. Printing Office, Washington, DC.

Onstad, C.A. and G.R. Foster. 1975. Erosion modeling on a watershed. Trans. ASAE 18:288-292.

Penman, H.L. 1948. Natural Evaporation from Open, Bare Soil and Grass. Proc. Royal Soc. (London) A193:120-145.

Phillips, D.L., P.D. Hardin, V.W. Benson, and J.V. Baglio. 1993. Using the national resources inventory and the EPIC model to evaluate the impact of alternative agricultural management practices in Illinois. *J. Soil Water Conserv.* 48(5):449-457.

Priestley, C.H.B. and R.J. Taylor. 1972. On the assessment of surface heat flux and evaporation using large scale parameters. *Monthly Weather Review.* 100:81-92.

Reddy, K.R., R. Khaleel, M.R. Overcash, and P.W. Westerman. 1979a. A nonpoint source model for land areas receiving animal wastes: I. Mineralization of organic nitrogen. *Trans. ASAE* 22(4):863-872.

Reddy, K.R., R. Khaleel, M.R. Overcash, and P.W. Westerman. 1979b. A nonpoint source model for land areas receiving animal wastes: II. Ammonia volatilization. *Trans. ASAE* 22(6):1398-1404.

Richardson, C.W. 1981. Stochastic simulation of daily precipitation, temperature, and solar radiation. *Water Resources Res.* 17:182-190.

Richardson, C.W. 1982a. Dependence structure of daily temperature and solar radiation. *Trans. ASAE* 25:735-739.

Richardson, C.W. 1982b. A wind simulation model for wind erosion estimation. ASAE Paper No. 82-2576.

Ritchie, J.T. 1972. A model for predicting evaporation from a row crop with incomplete cover. *Water Resources Res.* 8:1204-1213.

Seligman, N.G. and H. van Keulen. 1981. PAPRAN: A simulation model of annual pasture production limited by rainfall and nitrogen. In: Simulation of Nitrogen Behaviour of Soil-Plant Systems, M.J. Frissel and J.A. van Veen, (eds.) (Wageningen, The Netherlands, Jan. 28-Feb 1, 1980), 192-221.

Shaffer, M.J. 1991. Various chapters In: R.F. Follett (ed.) Managing Nitrogen for Groundwater Quality and Farm Profitability. ASAE Monograph (in press).

Sharpley, A.N. and J.R. Williams (eds.). 1990. EPIC--Erosion/Productivity Impact Calculator: 1. Documentation. USDA Tech. Bull. #1768. 235 pp.

Singer, M.P., F.D. Arnold, R.H. Cole, J.G. Arnold, and J.R. Williams. 1988. Use of SWRRB computer model for the national coastal pollutant discharge inventory. In: Proc. AWRA Symp. on Coastal Water Resources, 119-132.

Sloan, P.G. and I.D. Moore. 1984. Modeling subsurface stormflow on steeply sloping forested watersheds. Water Resources Research 20(12):1815-1822.

U.S. Department of Agriculture, Soil Conservation Service. 1972. Hydrology. Chapters 4-10. In: National Engineering Handbook. US Government Printing Office.

U.S. Department of Agriculture, Soil Conservation Service. 1983. Hydrology. Section 4, Chapter 19. National Engineering Handbook. US Government Printing Office.

Williams, J.R. 1975a. Sediment-yield prediction with universal equation using runoff energy factor. p. 224-252. In: Present and Prospective Technology for Predicting Sediment Yield and Sources, USDA, ARS-S-40.

Williams, J.R. 1975b. Sediment routing for agricultural watersheds. *Water Resources Bulletin, AWRA* 11(5):965-974.

Williams, J.R. 1980. SPNM, a model for predicting sediment, phosphorus, and nitrogen yields from agricultural basins. *Water Resources Bulletin, AWRA,* 16(5):843-848.

Williams, J.R. and R.W. Haan. 1978. Optimal operation of large agricultural watersheds with water quality constraints. Texas Water Resources Inst., Texas A&M Univ., TR-96. 152 pp.

Williams, J.R. and A.D. Nicks. 1982. CREAMS hydrology model–Option one. p. 69086 In: V.P. Singh (ed.) Applied Modeling Catchment Hydrology. Proc. Intl. Symp. Rainfall-Runoff Modeling, May 18-21, 1981, Mississippi State, MS.

Williams, J.R., C.A. Jones, and P.T. Dyke. 1984. A modeling approach to determining the relationship between erosion and soil productivity. Trans. ASAE 27(1):129-144.

Williams, J.R., C.A. Jones, and P.T. Dyke. 1990. The EPIC model. Chapter 2, p. 3-02 In: A.N. Sharpley and J.R. Williams (eds.) EPIC-Erosion/Productivity Impact Calculator: 1 Model Documentation. USDA Tech. Bull. No. 1768. 235 pp.

Williams, J.R., A.D. Nicks, and J.G. Arnold. 1985. SWRRB, a simulator or water resources in rural basins. *ASCE Hydraulics Journal,* 111(6):970-986.

Wischmeier, W.H. and D.D. Smith. 1978. Predicting rainfall erosion losses. Agriculture Handbook 537, USDA, SEA, 58 pp.

Woodruff, N.P. and F.H. Siddoway. 1965. A wind erosion equation. *Soil Sci. Soc. Amer. Proc.* 29:602-608.

Young, R.A., C.A. Onstad, D.D. Bosch, and W.P. Anderson. 1987. AGNPS, agricultural non-point-source pollution model. A Watershed Analysis Tool. 1987. U.S. Department of Agriculture, Conservation Research Report 35.

# Research Priorities for Soil Processes and Water Quality in 21st Century

R. Lal and B.A. Stewart

I. Introduction . . . . . . . . . . . . . . . . . . . . . . . . . . . . . . 383
II. Soil Processes Affecting Water Quality . . . . . . . . . . . . . . . . 385
III. Agricultural Practices that Regulate Water Quality . . . . . . . . 387
IV. Standardization of Pollutant Maximum Concentration Limits . . 389
V. Policy Considerations and Public Awareness . . . . . . . . . . . . 389
VI. Conclusions . . . . . . . . . . . . . . . . . . . . . . . . . . . . . . 390
References . . . . . . . . . . . . . . . . . . . . . . . . . . . . . . . . . 391

## I. Introduction

There is a rapid increase in demand for fresh water use, especially in relation to irrigated agriculture. Consequently, as many as 80 nations of the world are faced with shortage of fresh water (Trudgill, 1991). Furthermore, rapid transformation in agriculture experienced since 1950's have a significant adverse impact on water quality. A report by USEPA in 1986 observed that at the time of the survey non-point source pollution was responsible for adverse impact on water quality of 76% of the affected areas of lake water, 65% of the affected stream miles, and 45% of the affected estuarine square miles. The report noted that as much as 50 to 70% of the water resources were polluted due to contamination from agricultural activities. It is likely that similar, if not higher, risks of water pollution exist in developing countries where use of fertilizers is rapidly increasing, accelerated soil erosion is most serious, and stringent regulatory measures are not in place.

Water quality is an important issue. It affects the mere fabric of human life. It is a complex issue involving a range of physical, biological and socio-economic and cultural issues. It involves research in relation to nature and concentrations of substances present in water, their safety limits for different types of water uses, processes and factors responsible for their transport to natural waters, development of cultural practices to reduce concentrations of these contaminants, and identification and implementation of policies to limit these concentrations (Figure 1).

0-87371-980-8/94/$0.00+$.50

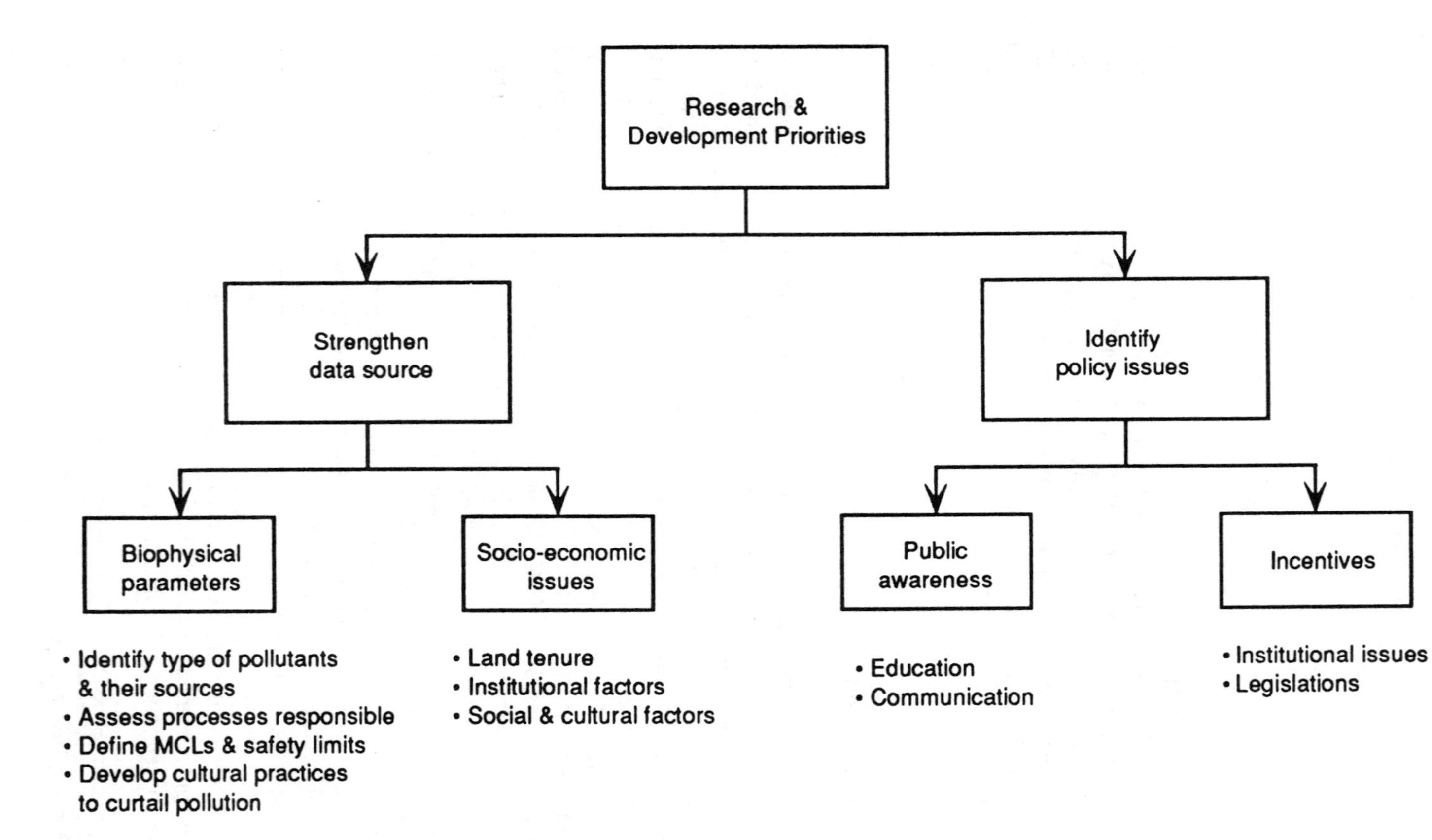


**Figure 1.** Research and development priorities in issues related to water quality.

## II. Soil Processes Affecting Water Quality

Some progress has been made in understanding soil processes that affect water quality in relation to agricultural activities. However, much remains to be done especially with regard to land use and agricultural practices followed in developing countries of the tropics and sub-tropics.

There are several soil processes that affect water quality in relation to non-point source pollution (Figure 2).

1. Accelerated soil erosion

   Accelerated soil erosion, involving transport of suspended and dissolved loads with overland flow from agricultural land, remains to be a major process affecting non-point source pollution. Despite the voluminous literature on soil erosion, the global extent of soil erosion especially with regards to its environmental impacts is least known. Site-specific information on the magnitude of pollutant transported in overland flow, form of pollutant transport whether dissolved or suspended, the timing of pollutant transport with reference to time of application of agricultural chemicals on farmland, and interaction of climate and soil factors for different land uses and management practices are not known.

2. Macropore flow

   The importance of macropore flow or through flow in relation to water quality remains to be a debatable issue. As the argument presented in chapter by McCoy et al. show, macropore flow may adversely affect water quality in some cases and yet have no effect in others. It is thus important to identify those conditions of macropore flow that adversely affect ground water quality.

3. Leaching

   Macropore flow is one process of transport of nutrients and agricultural chemicals from surface into sub-soil and eventually into groundwater and interflow that feeds streams. The magnitude of leaching is generally high only when active vegetation growth is absent. Understanding principal processes of leaching and developing management systems to reduce its magnitude are crucial to decreasing adverse impacts of leaching on water quality.

4. Microbial By-Products and Transformations

   Transport of microbes and microbial by-products from soil into surface and groundwater is another cause of non-point source pollution. Transport of biomass carbon is usually more in overland flow than groundwater because of the nature of the process of soil erosion. Soil erosion is a selective process and leads to a preferential transport of soil organic matter and biomass carbon. Microbial processes that aggravate adverse effects on water quality are mineralization, solubilization, and exudates or microbial by-products.

In addition to their negative impacts on crop growth and agronomic productivity, these four processes have major impact on environment especially

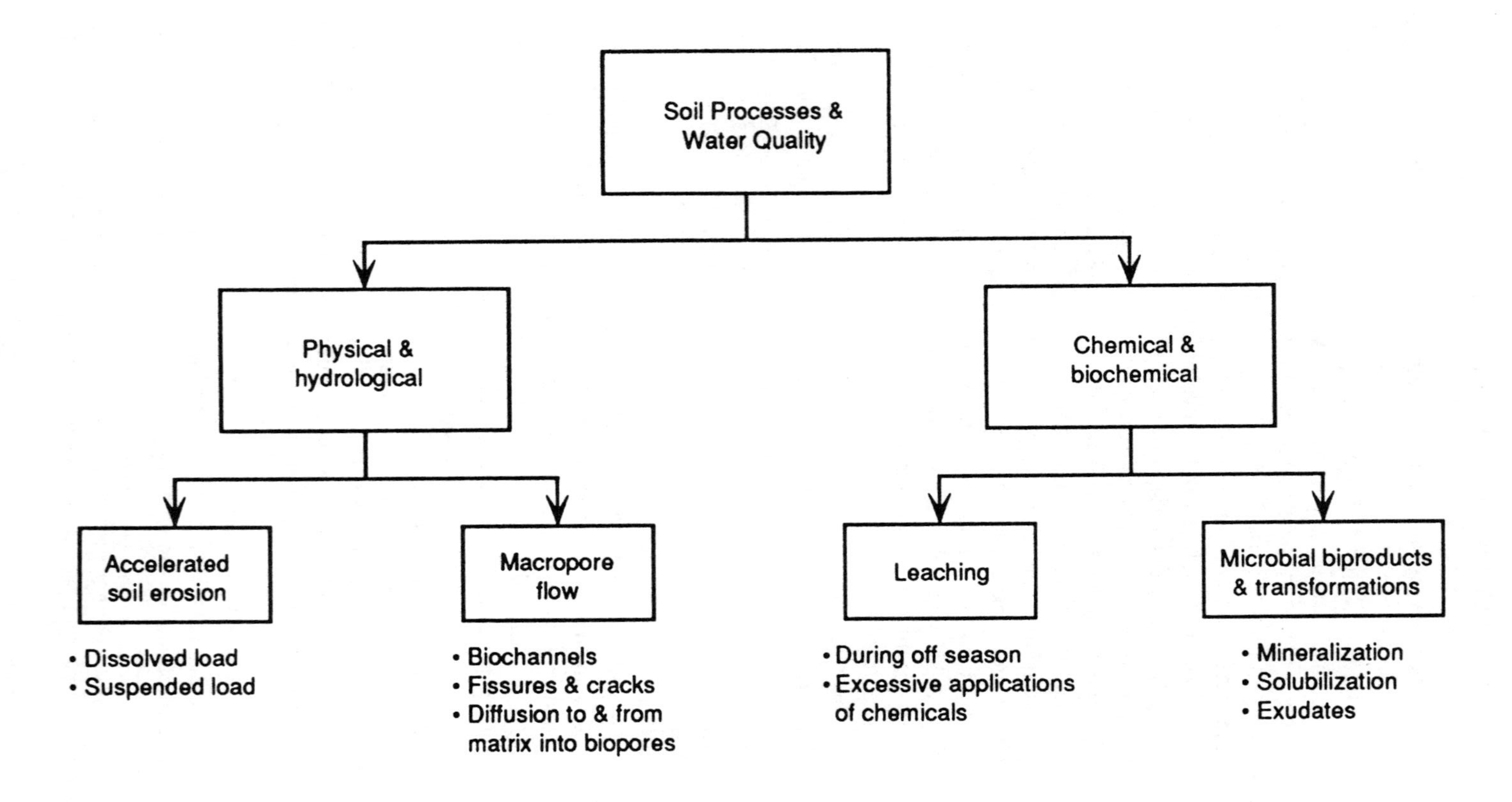


**Figure 2.** Soil processes and water quality.

with regard to the quality of ground and surface waters. It is important to understand factors and causes that affect rate and magnitude of these processes, especially in relation to agronomic practices of soil and crop management.

## III. Agricultural Practices that Regulate Water Quality

Agricultural practices with major impact on water quality include those that involve erosion management, nutrient management, and pest control (Figure 3). Environmentally-friendly practices with positive effects on water quality include conservation-effective management, and techniques of integrated pest management.

1. Erosion Management

   Control of accelerated soil erosion is likely to have a positive effect on water quality especially that of surface water and overland flow. Sediments are a principal source of pollutants transported into overland flow. There are two categories of erosion control measures: (i) agronomic techniques of soil and crop management that decrease runoff water, and reduce rain drop impact and shearing and transport capacity of overland flow, and (ii) engineering techniques designed for safe disposal of excess runoff at reduced velocity. A useful strategy is to combine agronomic and engineering techniques, e.g., establishing vegetative barriers on the contour. Although basic principles of erosion control are known, soil and ecological specific effects of these techniques on water quality are not. Erosion control measures may decrease runoff rate, runoff amount and total sediment and pollutant transport. However, nutrient concentration in runoff may be more for no-till and conservation tillage systems than for plow-based methods of seedbed preparation. Site-specific effects of conservation measures on water quality must be assessed.

2. Nutrient Management

   Developing integrated systems of nutrient management is crucial to enhancing nutrient use efficiency. Systems with low nutrient use efficiency have most drastic adverse effects on water quality. Not only should the nutrient losses by runoff/erosion and leaching be decreased, but nutrient use efficiency should also be improved. Once again, there is a wide range of soil and crop management systems that can be adopted to decrease losses. Rate of fertilizer use can be decreased by strengthening nutrient recycling mechanisms, e.g., biological nitrogen fixation, returning crop residue and other organic wastes to the soil, improving mycorrhizal efficiency, etc. It is important to realize, however, that mismanagement of these techniques can also have severe adverse effects on water quality. Plowing under a green manure, frequent and liberal use of compost and sewage sludge can also increase leaching losses of plant nutrients, contaminate soil and water, and cause high nutrient concentra-

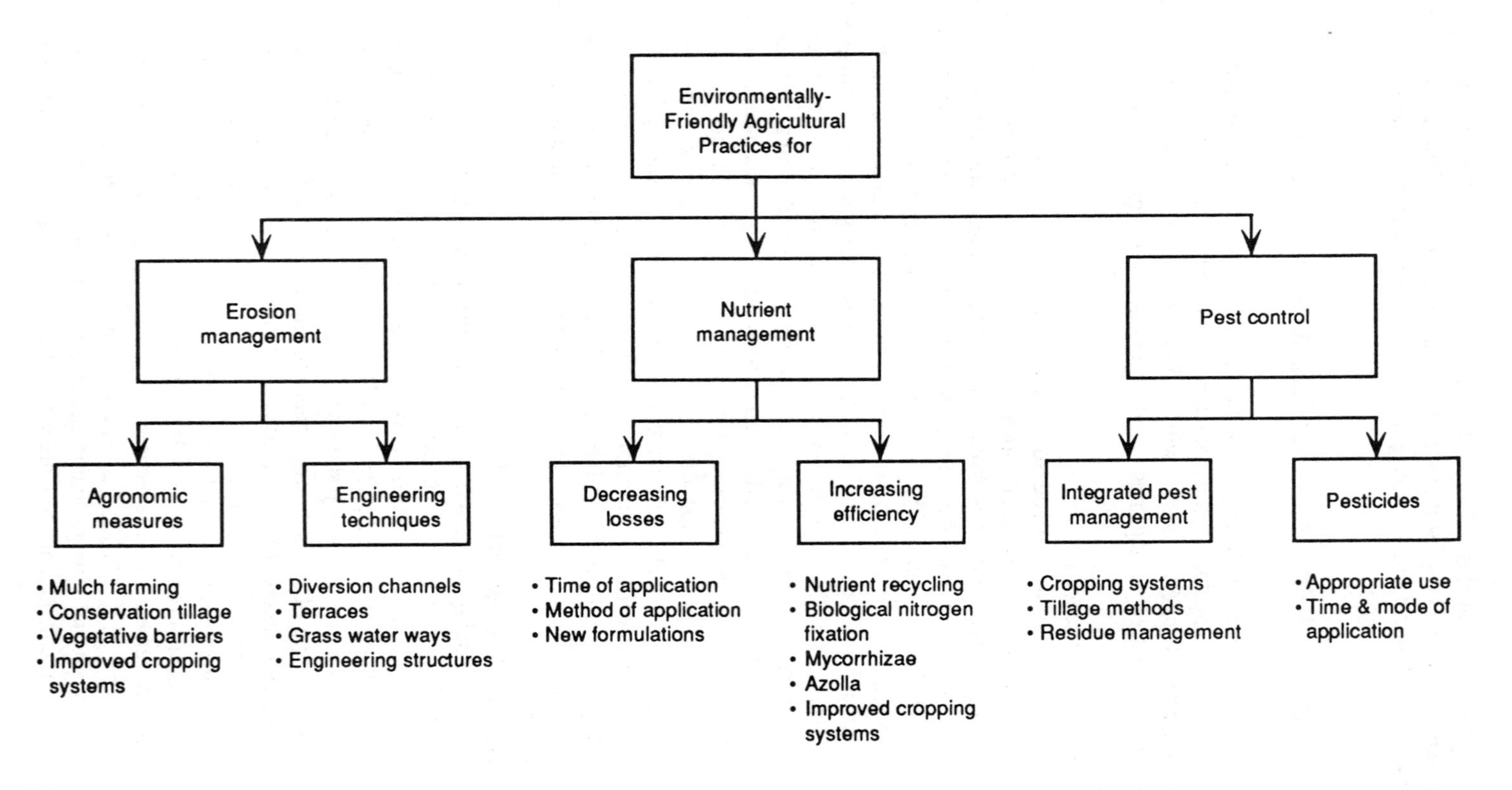


**Figure 3.** Agricultural practices that regulate water quality.

tions in overland flow. It is extremely important to quantify the effects of these practices on water quality for major eco-regions, soils, and farming/cropping systems. Best management practices (BMP's) must be determined for integrated management of nutrients to enhance efficiency and restore losses.

3. Pest Management

   There is a need to develop agricultural practices that decrease use of agricultural chemicals for pest control. In this regard the importance of integrated pest management (IPM) cannot be over-emphasized. Effective weed control plays an important role in the overall objective to enhance productivity and profitability. Similar to the disease and insect control, weed control can also be achieved by cultural practices rather than by herbicides. These practices include plow-based tillage methods, inter-cultivation, crop rotations, etc. However, these practices may also accentuate runoff and accelerate rill and inter-rill erosion. Therefore, the compensatory effects of these practices on water quality must be determined.

## IV. Standardization of Pollutant Maximum Concentration Limits

There is a strong need to develop and standardize criteria and procedures for assessment of water quality for different uses, e.g., human consumption, domestic animal consumption, wildlife habitat, agricultural use and industrial use. The maximum contaminant level (MCL) permitted for each use must be specified with regards to principal pollutants. Principal pollutants with risks to human health are listed in Table 1. The MCL and safe limits must be determined for each of these pollutants for alternate landuses. Because MCL of several of these compounds may be in ppb or less, analytical techniques must be developed and standarized for rapid and accurate determinations of water quality which varies widely over time and space.

## V. Policy Considerations and Public Awareness

In addition to biophysical issues (e.g., chemical use, land use, farming systems, soil and climatic factors), socio-economic and institutional factors also play an important role in water quality. Important socio-economic and institutional factors include educational standards to comprehend the problem, land tenure rights for investments in soil and water conservation, institutional support to develop and advocate environmentally-friendly practices, and incentives to adopt these practices. Institutional support is necessary to identify important policy issues that encourage adoption of research-proven technologies. Incentives can

**Table 1.** Principal non-point source pollutants

| Inorganic | Organic |
|---|---|
| 1. Soil particles, e.g. sand, silt and clay | 1. Dissolved and particulate organic matter including microbial biomass |
| 2. Nutrient cations: Ca, Mg, K, Na, Si, Cu, B, Fe, Zn, Mn | 2. Agricultural chemicals: pesticides, herbicides, e.g., Atrazine, Simazine, Dieldrin, DDT, etc. |
| 3. Nutrient anions: $NO_2$, $NH_4$, $NO_3$, $PO_4$, $SO_4$, Cl | |
| 4. Heavy metals: As, Cd, Cu, Pb, Hg, Se | |

be an important tool in technological adoption, but society has to pay the price for safe environment. Regrettably, there are more pressing problems (e.g., hunger and malnutrition, public health) in several developing countries also where the standards of water quality are extremely poor. The success of developing and implementing policy is also elusive and slow to achieve in industrially developed economies, e.g., USA and western Europe.

Several issues which require policy considerations with regards to water quality are listed in Table 2. Although general issues are similar, there are subtle differences with regards to policy considerations in developing vis-a-vis developed economies. These differences must be carefully considered in identification and implementation of relevant policies.

## VI. Conclusions

There is a need to identify sources, processes, factors and causes of non-point source pollution. The need is especially great in developing countries of the tropics and sub-tropics, where use of fertilizers and agricultural chemicals is rapidly increasing, where techniques for accurate and rapid measurements of pollutant concentrations in natural waters do not exist, and where there is a lack of awareness about the health risks associated with high concentrations of these pollutants. There is a general lack of adequate data required to (i) assess the impact of agricultural practices on water quality, (ii) safe limits of pollutant for different uses, (iii) developing remedial measures for effective control, (iv) educational programs to create public awareness, and (v) identify policy considerations for imposition of safe limits. There is a need to develop systematic programs to strengthen database and understand processes involved in transport of pollutants from agricultural land, especially with regards to soils, cropping/farming systems and land uses in developing countries of the tropics

**Table 2.** Issues that require policy considerations

| Developing economies | Developed economies |
|---|---|
| 1. Methods of deforestation | 1. Application of fertilizers and amendments on the basis of soil test |
| 2. Land use capability assessment and its implementation | 2. Adoption of integrated nutrients and pest management techniques |
| 3. Erosion control measures on steepland | 3. Conservation compliance with regards to farming marginal lands |
| 4. Off farm inputs to enhance productivity and restore degraded lands | 4. Public awareness to replace myths by facts with regard to environmental risks, and health risks associated with poor water quality |
| 5. Environmentally safe formation of agri-chemicals | |
| 6. Waste disposal | |
| 7. Safety procedures for applying chemicals | |

and sub-tropics. It is equally important to identify policy considerations that encourage adoption of environmentally safe but economically profitable agricultural practices. These policies may be different in developing vis-a-vis developed economies especially due to difference in education and economic standards, land tenure systems, resources available, and institutional support available to identify and implement these policies.

## References

Trudgill, S. 1991. Water quality. p. 161-174. In: P. Brackley (ed.), World Guide to Environmental Issues and Organizations. Gale.

# Index

adsorption 22, 26, 32, 38, 39, 175, 181, 200, 233, 234, 252, 254, 257, 258, 314, 351, 363, 364
aerobic fermentation 174
alfalfa 17, 23, 38, 43-45, 55, 66, 117, 149, 150, 364
algae 8, 35, 37, 36, 67, 68, 70
algal growth 70
amines 91, 113, 138, 171, 175
ammonia losses 175, 180, 182, 183, 187, 189, 190, 203-205, 212
ammonium 9, 91-93, 96, 99, 103, 166, 169, 171, 173, 175, 180, 181, 183, 186, 189, 193, 196,197,200, 201, 203, 204, 206, 207, 247, 361
anaerobic fermentation 174, 175, 180
animal manure 9, 20, 26, 58, 60, 61, 102, 173
animal residues 163-165, 169
animal wastes 92, 99, 155, 164-166, 173, 183, 187, 201, 203-205, 207, 209, 211, 212
aquifer 1, 113, 119, 354
atmospheric deposition 233, 239, 240, 248, 249, 257, 259, 261
bacteria 92, 99, 113, 164, 165, 169, 174, 175, 191, 204, 208, 255
bahiagrass 94, 95, 105-107
barley 182, 364, 375
bermudagrass 94, 96, 97, 107, 110, 153
best management practices 61, 91, 114, 389
bioavailable phosphorus 52
biomass 2, 4, 18, 19, 102, 199, 200, 275, 278, 279, 295, 360-366, 370- 373, 385, 390
biosphere 237
bromegrass 100
buffer strips 62
bulk density 51, 248, 364, 365
Canada 31, 43, 45, 114, 153, 244, 246, 260
casava 364
cattle manure 170, 180, 186, 188, 196, 211
chemical availability 7, 19, 32, 33, 66
climate 18, 95, 114, 118, 138, 149, 155, 237, 244, 254, 274, 278, 353, 356, 377, 385
clover 22, 100, 142, 150
coliforms 207, 208
compost 163, 172-174, 180, 387
composting 170, 171, 174, 195
computer model 353
conservation tillage 42, 59, 61, 62, 65-67, 71, 140, 143, 144, 289, 297, 373, 374, 387
contamination 3, 106, 107, 111-113, 117, 119-122, 152-154, 172, 208, 212, 233-235, 238-241, 244, 246, 248, 250, 255, 258-261, 264, 383
conventional tillage 33, 42, 59, 139, 140, 144, 146, 145, 374
corn 17, 21, 22, 38, 42-44, 46, 61, 63, 65, 100, 137, 140-142, 144-149, 153, 154, 208, 320, 321, 364, 374, 375
cotton 3, 12, 38, 364, 374
cover crops 65, 142, 143, 148, 153, 280, 290, 297, 375
crop management 18, 34, 54, 70, 137, 139, 142, 144, 148, 360, 370, 387

crop production 8, 9, 27, 32, 70, 91, 116, 148, 154, 164, 238, 295, 374
crop residue 9, 14, 16, 18, 17, 42, 65, 139, 140, 142, 280, 286, 290, 350, 353, 360-362 , 365, 366, 370, 373, 387
crop residues 20, 32, 65, 97, 148
crop rotations 17, 20, 60, 148, 155, 286, 352, 356, 389
crude protein 95
Darcy's Law 334, 335
denitrification 92, 97, 99, 103,106, 107, 109, 110, 112, 113, 115, 117, 120, 121, 138, 150, 163, 175, 186, 191, 193, 197, 198, 201, 204, 351, 353, 362
desorption 22, 32, 37-39, 51, 71, 233, 252, 254, 257, 351
disease 95, 97, 113, 245, 389
dissolved organic carbon 4, 163, 186, 206, 278
DOC 206, 207, 278
drinking water standards 120
drought 95, 96, 114
earthworms 144, 289, 307, 313, 317
economic sustainability 70
enrichment ratios 39, 51, 52, 254, 351
erosion 3, 4, 8, 22, 25, 28, 39, 47-49, 53-61, 65, 104, 118, 119, 140-143, 148, 274, 278, 279, 283, 285, 286, 290, 292, 296, 297, 349-353, 355, 357, 359, 360, 362, 370, 375, 383, 385, 387, 389, 391
Euglena 67
eutrophication 8, 40, 54, 58, 62, 63, 67, 69, 70, 138, 274
evaporation 16, 65, 148, 189, 352, 353, 358, 359, 363, 365, 368, 371
evapotranspiration 100, 101, 110, 115, 122, 144, 200, 275, 353, 357, 358, 368
extraction 14, 18, 20, 22, 32, 33, 35-38, 49, 51, 66, 67, 95
farmyard manure 240
feces 102, 164-166, 169, 193, 196, 208
feedlot 60, 118
fertilizer availability index 47
fertilizer phosphorus 7, 30
fertilizer placement 20, 22, 32, 46
fertilizer recommendations 21, 30, 32, 33, 115
fertilizers 1, 2, 4, 19, 20, 60, 70, 91-94, 96-99, 107, 109, 111, 112, 115, 116, 121, 122, 152-154, 163, 164, 191, 192, 203-205, 212, 233, 238-240, 261, 280, 284, 289, 296, 297, 299, 310, 366, 383, 390, 391
fescue 101, 107, 109
field capacity 107, 357, 358, 361-363, 373
forages 95, 105, 114, 320
furrow diking 350, 352
Germany 201, 243
golf greens 96
grasslands 95, 99, 100, 102-105, 107, 122, 143, 150, 151, 201, 209
grazing 95, 102-104, 117, 118, 150, 151, 165, 173, 190, 195, 208, 290, 350, 352, 367
groundwater 91, 99, 102, 104-114, 116, 121, 122, 144, 150, 152-155, 163, 200, 205-207, 208, 209, 211, 233, 237, 254, 259, 303, 350, 351, 353, 355, 385
harvest index 364, 371
heavy metals 164, 171, 172, 233-238, 240, 241, 243, 245, 246, 248-250, 252-264, 390

household composts 164
humus 4, 100, 174, 248, 249, 255, 257, 362, 363, 376
hydraulic conductivity 58, 307, 316, 318, 320, 323-325, 332, 335, 336
hydrology 315, 352, 353, 355, 357, 360, 364, 367, 368, 370, 376
immobilization 9, 17, 18, 22, 47, 97, 153, 163, 164, 175, 182, 193-195, 197-200, 206, 252, 353, 362, 363
infiltration 3, 44, 100, 117, 139-141, 144, 148, 186, 189, 205, 286, 304, 307, 311, 317, 318, 319, 325, 328, 329, 333, 336, 351, 354, 355, 365, 370
infiltration rate 186, 205, 317, 336, 370
inhibitors 92, 99, 103, 110, 115, 116, 154
inorganic phosphorus 169
irrigation 56, 57, 94, 99, 107-111, 114-116, 118, 122, 154, 187, 207, 210, 211, 259, 350, 351-353, 360, 366, 367, 373, 377
leaching 3, 4, 27, 38, 42, 64, 65, 91, 92, 94-97, 99-112, 114-119, 122, 139, 142, 148, 149, 150, 152-156, 163, 186, 200-205, 212, 257, 274, 278, 284, 289, 312, 314, 352, 353, 361, 364, 372, 375, 385, 387
leaf area index 364, 371
legumes 20, 60, 100, 101, 142, 148-150, 153, 362
lentils 364
liming 46, 190, 239, 240, 250, 350, 352
litter 2, 155, 166, 169, 170, 177-181, 186, 193, 196, 255
livestock 1, 4, 5, 26, 95, 102, 103, 111, 117, 150, 164, 209, 274, 367
macrophytes 70
macropores 4, 140, 144, 145, 297, 303-305, 307-311, 313-320, 322-327, 329, 330, 333, 336
manure 7-10, 19, 20, 25, 26, 33, 39, 42, 58-66, 69, 70, 99, 102, 117, 118, 137, 153, 155, 164-170, 173-175, 178-183, 186-188, 190, 191, 195, 196, 201-206, 208-212, 240, 316, 350, 352, 353, 374, 387
manure handling 212, 350, 352
manured fields 207
mineralization 4, 7, 13-19, 33, 46, 47, 65, 66, 97, 98, 100, 106, 112, 115, 149, 163, 164, 170, 193-200, 205, 206, 254-256, 274, 275, 278, 283, 284, 351, 353, 362, 363, 385
models 46-49, 52, 53, 71, 118, 119, 155, 163, 199, 258, 303, 305, 315, 326, 328, 329, 330-336, 349-351, 353-357, 373-376
moldboard plowing 139, 140, 144, 320, 321
municipal wastes 164, 172
Netherlands 101, 104, 143, 151
nitrate 65, 66, 91, 93, 99-102, 104-109, 111, 114-116, 118, 120, 138, 146, 148, 149, 151, 154, 155, 166, 171, 173, 186, 193, 197, 200-202, 204-207, 209, 259, 283, 314, 351, 353, 361, 375
nitrate leaching 91, 99, 101, 102, 111, 115, 116, 118, 155, 201, 202, 205, 353
nitrification inhibitors 92, 103, 110, 115, 116, 154

nitrogen 7, 8, 28, 91, 93-98, 101, 105, 108, 111, 112, 115, 119, 121, 137-139, 141, 143, 148, 152, 164-166, 168-171, 173, 175, 176, 180-184, 186, 190, 191, 193, 195, 196-201, 203-207, 209, 212, 248, 256, 284, 351-354, 361, 366, 372, 374, 375, 387
nitrogen budgets 137, 138, 353
no-till 23, 33, 34, 38, 42,43, 51-52, 66, 70, 141, 142, 149, 285, 289, 290, 292-295, 374, 387
nonpoint source pollution 8, 71, 350, 354
Norway 239, 240, 243, 246-249, 260, 261, 263
nutrient flux 46
OM 99, 100, 106, 115, 116
orchardgrass 62, 63, 102, 108, 150
organic carbon 4, 163, 186, 206, 275, 278, 362, 363
organic matter 4, 15, 17, 19, 20, 25, 35, 37, 39, 47, 52, 66, 96-99, 152, 169, 174, 175, 191, 195, 204, 205, 207, 212, 241, 249-255, 261, 274, 278, 289, 353, 385
organic phosphorus 169
particulate phosphorus 51
peat 179, 180, 182, 181-183, 186, 254, 260
pesticide fate 352, 354, 357, 363, 367
pesticides 1-4, 171, 233, 238, 244, 289, 296, 299, 310, 350, 352, 356, 363, 364, 367, 372, 373, 390
phosphate rock 238
phytotoxic 244
pollution 8, 48, 71, 99, 102, 106, 109-112, 114, 117, 119, 120, 122, 153, 155, 163, 164, 209, 212, 234, 247-249, 257, 259-261, 264, 296-298, 303, 350, 354, 383, 385, 390
porosity 286, 316, 317, 320, 321, 327, 365
positional availability 7, 22, 33
poultry litter 155, 169, 170
poultry manure 26, 64, 65, 170, 182, 188, 195, 196, 202, 208, 210
rain fall 37-39, 41, 42, 52, 54, 93, 100, 102, 103, 106, 111, 138, 140, 141, 145, 200, 205, 274, 278, 283, 289, 290, 307, 310-314, 328, 329, 351, 355-363, 366 370, 377
rainfall intensity 37, 52, 205, 311-314, 357
rainfall interception 355
redox 170, 233, 250, 252, 254
Rhodes grass 108
Richards equation 328, 334
riparian zones 41, 62, 63, 121, 143
rock phosphate 238
root zone 9, 22, 34, 94, 100, 106, 107, 109, 111, 114-117, 200, 319, 351-354, 358, 362, 365, 366, 371, 373
rooting zone 116, 119
runoff 3, 4, 7-10, 12, 19, 20, 22, 25, 35, 37, 36-46, 48-56, 58-71, 101-105, 108, 109, 112, 115, 117, 118, 122, 137, 139-144, 146, 163, 164, 205-210, 212, 233, 257, 258, 275, 278-283, 289, 290, 292-297, 307, 328, 349, 351-355, 357, 358, 360, 361, 363, 364, 367, 368, 370-377, 387, 389
ryegrass 96, 97, 100-103, 107, 148, 150, 151, 201
S-coated ureas 96

sediment 3, 20, 35, 37, 36, 39, 51, 52, 55, 62, 67, 68, 71, 103, 108, 109, 113, 140, 141, 142, 257-260, 275, 279, 280, 283-286, 289, 290, 292, 293, 295-297, 349, 350, 351, 352, 354-356, 361, 363, 364, 370-372, 374, 376, 377, 387
sewage sludge 26, 69, 93, 163, 164, 167, 168, 170-172, 181, 187, 188, 197, 198, 207-209, 210, 211, 233, 238, 241, 243, 261, 387
simulation models 118, 349, 350, 357
slow-release N 92, 96, 110, 122
slurry 103, 166, 169, 170, 183-191, 193, 197, 201, 202, 204, 206, 208, 210, 211
small grains 95
soil erodibility 3, 360, 371
soil loss 42, 51, 52, 54, 58, 60, 103, 140-142, 352-355, 370
soil temperature 18, 46, 191, 197, 275, 289, 352, 357, 358, 361, 362, 364, 367, 371, 373
soil testing 7, 8, 22, 30-32, 34
soil texture 26, 54, 99, 107, 110, 141, 311, 318, 360, 365
soil water 26, 47, 51, 65, 98, 99, 106, 107, 110, 116, 117, 122, 149, 254, 275, 304, 322, 325, 328, 330, 334-336, 356, 358, 361-366, 373
solar radiation 358, 359, 364, 368, 369, 371
soluble phosphorus 51
sorghum 3, 23, 38, 42, 62, 105, 364, 374
sorption 37, 39, 40, 44, 54, 55, 66, 98, 252, 253, 315, 329-331, 333, 334, 337, 354, 363
soybean 22, 38, 46, 140, 141, 149, 364, 374
stargrass 106
subsurface runoff 42-44, 104
sudangrass 105
sugarcane 364
sunflower 364
surface runoff 3, 4, 12, 19, 42-44, 47, 59, 63, 66, 67, 101, 103-105, 108, 112, 115, 117, 122, 137, 139-144, 146, 163, 205, 208, 209, 212, 258, 275, 279, 283, 296, 307, 351-354, 357, 361, 363, 367, 372, 377
sustainable agriculture 375
Sweden 99, 209, 240, 259
swine manure 26, 60, 195
Tanzania 247, 253
thatch 97, 109
tile drain 45, 303
tile drainage 43-45, 47, 147
tillage 2, 8, 13, 16, 17, 21, 23, 33, 42, 44, 53, 54, 59, 61, 62, 64-67, 71, 137, 139, 140-144, 146, 145, 149, 153, 273, 286, 288-290, 297, 311, 314-317, 320, 321, 323, 350-353, 357, 365, 373, 374, 387, 389
turf 96, 97, 109, 110, 122, 150, 155, 245
urea 92, 93, 96, 97, 99, 108-110, 153, 154, 165, 166, 168, 169, 182, 189, 190, 195, 197, 203
uric acid 166, 168, 169, 190, 195
urine 102, 150, 164-166, 169, 180, 188-191, 193, 196, 197, 201, 202
vegetative filter strips 62, 63
volatile fatty acid 170, 191
volatilization 96-100, 102, 103, 109, 163, 164, 175, 181, 182, 186-191, 193, 197, 198, 201, 203, 209, 212, 353, 361, 362

water quality 1-5, 7, 8, 23, 29, 38, 41, 46, 53, 58, 59, 63, 70, 71, 91, 101, 103, 104, 107, 111, 118, 119, 122, 137, 139, 144, 154, 155, 163, 207, 208, 258, 263, 264, 273-280, 283, 284, 286-291, 296-298, 319, 349, 350, 352, 357, 373-375, 383-390
water rights 377
weed species 96
wheat 17, 18, 21-24, 26, 28, 27, 29, 28, 29, 31, 34, 42, 43, 48, 50, 62, 95, 140, 146, 148, 154, 364, 374, 375
wind erosion 359, 360